Lecture Notes in Computer Scie

Commenced Publication in 1973
Founding and Former Series Editors:
Gerhard Goos, Juris Hartmanis, and Jan van Leeuwen

Rana Barua   Tanja Lange (Eds.)

Progress in Cryptology – INDOCRYPT 2006

7th International Conference on Cryptology in India
Kolkata, India, December 11-13, 2006
Proceedings

 Springer

Volume Editors

Rana Barua
Indian Statistical Institute
Division of Theoretical Statistics and Mathematics
Kolkata, India
E-mail: rana@isical.ac.in

Tanja Lange
Eindhoven University of Technology
Department of Mathematics and Computer Science
Eindhoven, Netherlands
E-mail: tanja@hyperelliptic.org

Library of Congress Control Number: 2006937160

CR Subject Classification (1998): E.3, G.2.1, D.4.6, K.6.5, K.4, F.2.1-2, C.2

LNCS Sublibrary: SL 4 – Security and Cryptology

ISSN   0302-9743
ISBN-10   3-540-49767-6 Springer Berlin Heidelberg New York
ISBN-13   978-3-540-49767-7 Springer Berlin Heidelberg New York

Springer is a part of Springer Science+Business Media

springer.com

© Springer-Verlag Berlin Heidelberg 2006
Printed in Germany

Typesetting: Camera-ready by author, data conversion by Scientific Publishing Services, Chennai, India
Printed on acid-free paper   SPIN: 11941378   06/3142   5 4 3 2 1 0

Preface

Indocrypt 2006, the 7th International Conference on Cryptology in India, took place December 11-13, 2006 in Kolkata, India. As in previous years, it was organized by the Cryptology Research Society of India, and the General Chair Bimal Roy did an excellent job in keeping all strands together by ensuring an excellent collaboration between the local organizers and the Program Committee and making the conference memorable for the talks and the social program.

Two invited lectures were presented at Indocrypt 2006: James L. Massey spoke about *Whither Cryptography?* and Alfred J. Menezes presented *Another Look at "Provable Security". II*, which is a joint work with Neal Koblitz.

The submission deadline for Indocrypt was on August 18 and we received 186 submissions. To give the authors maximal time to modify their papers and at the same time guarantee the maximal review time for the Program Committee, we had separate submission and revision deadlines. After the submission deadline it was no longer possible to submit a new paper and so the PC members could enter the selection phase during which they only got to see the abstracts of the papers. Out of the 186 papers originally submitted, 20 were withdrawn before the revision deadline on August 21 and 81 were revised at least once. Our experience with this approach of separating the deadlines was entirely positive. We would also like to take this opportunity to thank the developers of iChair, Thomas Baignéres and Matthieu Finiasz at EPFL, for making iChair available and for answering several questions which allowed us to extend the functionality of iChair to handle separate deadlines. The software was very useful for the submission process, the collection of reviews, and the discussion.

The Program Committee did a remarkable job of finishing refereeing almost all papers by September 8, even though 166 papers marked a new record in submissions. In the following discussion phase many more reports were added and the Program Committee worked intensively to gain confidence in its decisions. The 39 Program Committee members produced 422 comments during the 2 weeks of discussion. It is our pleasure to thank all Program Committee members for their very timely and concentrated effort that allowed us to notify the authors on time on September 22 and even send out the comments the same day.

Finally, we would like to thank all authors for submitting interesting new research papers to Indocrypt, providing us with an embarrassment of riches out of which we could only accept 29 contributed papers, even though many more would have been worth publishing. It is a pleasure to see Indocrypt being a well-accepted cryptography conference where fresh results are submitted.

December 2006

Rana Barua and Tanja Lange
Program Chairs, Indocrypt 2006

Organization

Indocrypt 2006 was organized by the Indian Statistical Institute, Kolkata, in collaboration with the Cryptology Research Society of India.

General Chair

Bimal Roy   Indian Statistical Institute, Kolkata, India

Local Organizers

Sumanta Sarkar	Indian Statistical Institute, Kolkata, India
Somitra Sanadhya	Indian Statistical Institute, Kolkata, India
Susmita Ruj	Indian Statistical Institute, Kolkata, India
Siddheswar Rathi	Indian Statistical Institute, Kolkata, India
Snehashis Mukherjee	Indian Statistical Institute, Kolkata, India
Kuntal Ghosh	Indian Statistical Institute, Kolkata, India
Amitava Sinha	Indian Statistical Institute, Kolkata, India
Mahabir Prasad	Indian Statistical Institute, Kolkata, India
Deepak Kumar Dalai	Indian Institute of Science Education and Research, Kolkata, India
Sanjit Chatterjee	Indian Institute of Science Education and Research, Kolkata, India
Subhomay Maitra	Indian Statistical Institute, India
Goutam Paul	Jadavpur University, Kolkata, India
Pinaki Sarkar	Jadavpur University, Kolkata, India

Program Chairs

Rana Barua	Indian Statistical Institute, Kolkata, India
Tanja Lange	Technical University of Denmark and Eindhoven Technical University, Netherlands

Program Committee

Michel Abdalla	Ecole Normale Supérieure and CNRS, France
Roberto M. Avanzi	Ruhr University Bochum, Germany
Paulo Barreto	University of Saõ Paulo, Brazil
Daniel J. Bernstein	University of Illinois at Chicago, USA

Colin Boyd	Queensland University of Technology, Australia
Anne Canteaut	INRIA Rocquencourt, France
Claude Carlet	University of Paris 8, France
Yvo Desmedt	University College London, UK
Orr Dunkelman	Technion - Israel Institute of Technology, Israel
Krishnan Gopalakrishnan	East Carolina University, USA
Kishan C. Gupta	University of Waterloo, Canada
Tom Høholdt	Technical University of Denmark, Denmark
Jin Hong	Seoul National University, Korea
Laurent Imbert	LIRMM, CRNS, France
Tetsu Iwata	Nagoya University, Japan
Antoine Joux	DGA and Université de Versailles Saint-Quentin-en-Yvelines, France
Marc Joye	Thomson R&D France, France
Charanjit Jutla	IBM T.J. Watson Research Center, USA
Chi Sung Laih	National Cheng Kung University, Taiwan
Kerstin Lemke-Rust	Ruhr University Bochum, Germany
C.E. Veni Madhavan	Indian Institute of Science, Bangalore, India
John Malone-Lee	UK
Pradeep Kumar Mishra	University of Calgary, Canada
Gregory Neven	Katholieke Universiteit Leuven, Belgium and Ecole Normale Supérieure, France
Bart Preneel	Katholieke Universiteit Leuven, Belgium
C. Pandu Rangan	Department of Computer Science and Engineering, IIT Madras, India
Bimal Roy	Indian Statistical Institute, Kolkata, India
Ahmad-Reza Sadeghi	Ruhr University Bochum, Germany
Rei Safavi-Naini	University of Wollongong, Australia
Pramod K. Saxena	Scientific Analysis Group, Delhi, India
Jennifer Seberry	University of Wollongong, Australia
Nicolas Sendrier	INRIA Rocquencourt, France
Nigel Smart	University of Bristol, UK
Martijn Stam	EPFL, Switzerland
Bo-Yin Yang	Academia Sinica, Taiwan
Melek D. Yücel	Middle East Technical University, Turkey
Jianying Zhou	Institute for Infocomm Research, Singapore

Referees

Andre Adelsbach	Daniel Augot	Jean-Claude Bajard
Andris Ambainis	Joonsang Baek	Lejla Batina

S.S. Bedi
Peter Birkner
Carlo Blundo
Jean Christian
  Boileau
Alexandra Boldyreva
Xavier Boyen
Emmanuel Bresson
Jan Cappaert
Sanjit Chatterjee
Liqun Chen
Benoit Chevallier-Mames
Jung-Hui Chiu
Yvonne Cliff
Deepak Kumar Dalai
Ivan Damgaard
Tanmoy Kanti Das
Tom St Denis
Alex Dent
Vassil Dimitrov
Christophe Doche
Gwenael Doerr
Sylvain Duquesne
Alain Durand
Thomas Eisenbarth
Xinxin Fan
Pooya Farshim
Serge Fehr
Dacio Luiz Gazzoni
  Filho
Caroline Fontaine
Pierre-Alain Fouque
Philippe Gaborit
Sebastian Gajek
Fabien Galand
Steven Galbraith
Gagan Garg
Rosario Gennaro
Pascal Giorgi
Philippe Golle
Louis Goubin
Robert Granger
Johannes Groszschaedl
Indivar Gupta
Robbert de Haan

Tor Helleseth
Katrin Hoeper
Susan Hohenberger
Nick Howgrave-Graham
Po-Yi Huang
Ulrich Huber
Shaoquan Jiang
Thomas Johansson
Shri Kant
Guruprasad Kar
Stefan Katzenbeisser
Jonathan Katz
John Kelsey
Dalia Khader
Alexander Kholosha
Eike Kiltz
Lars Ramkilde
  Knudsen
Ulrich Kuehn
H.V. Kumar Swamy
Meena Kumari
Sandeep Kumar
Sébastien Kunz-Jacques
Kaoru Kurosawa
Fabien Laguillaumie
Joseph Lano
Cédric Lauradoux
Dong Hoon Lee
Frédéric Lefebvre
Stephane Lemieux
Arjen K. Lenstra
Pierre-Yvan Liardet
Helger Lipmaa
Joseph Liu
Zhijun Li
Pierre Loidreau
Spyros Magliveras
Subhamoy Maitra
Stéphane Manuel
Mark Manulis
Keith Martin
Nicolas Meloni
Sihem Mesnager
J. Mohapatra
Abhradeep Mondal

Michele Mosca
Mridul Nandi
Yassir Nawaz
Christophe Negre
Afonso Araújo Neto
Michael Neve
Rafail Ostrovsky
Daniel Page
Pascal Paillier
Saibal K. Pal
Je Hong Park
Kenny Paterson
Anindya Patthak
Souradyuti Paul
Jan Pelzl
Kun Peng
Pino Persiano
Duong Hieu Phan
N. Rajesh Pillai
Alessandro Piva
David Pointcheval
Axel Poschmann
Emmanuel Prouff
Frederic Raynal
Vincent Rijmen
Matt Robshaw
Jörg Rothe
Pieter Rozenhart
Andy Rupp
Palash Sarkar
Berry Schoenmakers
Jörg Schwenk
Siamak Fayyaz
  Shahandashti
Nicholas Sheppard
Igor Shparlinski
Thomas Shrimpton
Herve Sibert
Francesco Sica
Alice Silverberg
Rainer Steinwandt
Hung-Min Sun
V. Suresh
Michael Szydlo
Adrian Tang

Nicolas Thériault	Nicolas	Kjell Wooding
Soeren S. Thomsen	Veyrat-Charvillon	Hongjun Wu
Dongvu Tonien	Charlotte Vikkelsoe	Brecht Wyseur
Mårten Trolin	Ryan Vogt	Yongjin Yeom
Boaz Tsaban	Melanie Volkamer	Aaram Yun
Wen-Guey Tzeng	Brent Waters	Nam Yul Yu
Damien Vergnaud	Andreas Westfeld	Moti Yung

Sponsoring Institutions

Cranes Software
Microsoft India
Metalogic Systems
Tata Consultancy Services

Table of Contents

Provable Security: Key Agreement

Invited Talk

Provable Security: Public Key Cryptography

Symmetric Cryptography: Design

Modes of Operation and Message Authentication Codes

Fast Implementation of Public Key Cryptography

ID-Based Cryptography

Embedded System and Side Channel Attacks

Whither Cryptography?

James L. Massey

Prof.-em. ETH-Zurich,
Adj. Prof.: Lund Univ., Sweden, and
Tech. Univ. of Denmark
jamesmassey@compuserve.com

Abstract. Diffie and Hellman's famous 1976 paper, "New Directions in Cryptography," lived up to its title in providing the directions that cryptography has followed in the past thirty years. Where will, or should, cryptography go next? This talk will examine this question and consider many possible answers including: more of the same, number-theoretic algorithms, computational-complexity approaches, quantum cryptography, circuit-complexity methods, and new computational models. Opinions will be offered on what is most likely to happen and what could be most fruitful. These opinions rest not on any special competence by the speaker but rather on his experience as a dabbler in, and spectator of, cryptography for more than forty years.

R. Barua and T. Lange (Eds.): INDOCRYPT 2006, LNCS 4329, p. 1, 2006.

Non-randomness in eSTREAM Candidates Salsa20 and TSC-4

Simon Fischer[1], Willi Meier[1], Côme Berbain[2], Jean-François Biasse[2], and M.J.B. Robshaw[2]

[1] FHNW, 5210 Windisch, Switzerland
{simon.fischer, willi.meier}@fhnw.ch
[2] FTRD, 38–40 rue du Général Leclerc, 92794 Issy les Moulineaux, France
{come.berbain, jeanfrancois.biasse, matt.robshaw}@orange-ft.com

Abstract. Stream cipher initialisation should ensure that the initial state or keystream is not detectably related to the key and initialisation vector. In this paper we analyse the key/IV setup of the eSTREAM Phase 2 candidates Salsa20 and TSC-4. In the case of Salsa20 we demonstrate a key recovery attack on six rounds and observe non-randomness after seven. For TSC-4, non-randomness over the full eight-round initialisation phase is detected, but would also persist for more rounds.

Keywords: Stream Cipher, eSTREAM, Salsa20, TSC-4, Chosen IV Attack.

1 Introduction

Many synchronous stream ciphers use two inputs for keystream generation; a secret key K and a non-secret initialisation vector IV. The IV allows different keystreams to be derived from a single secret key and facilitates resynchronization. In the general model of a synchronous stream cipher there are three functions. During initialisation a function F maps the input pair (K, IV) to a secret initial state X. The state of the cipher then evolves at time t under the action of a function f that updates the state X according to $X^{t+1} = f(X^t)$. Keystream is generated using an output function g to give a block of keystream $z^t = g(X^t)$. While TSC-4 follows this model, Salsa20 has no state update function f and g involves reading out the state X. Instead, we view the IV to Salsa20 as being the combination of a 64-bit *nonce* and a 64-bit *counter* and keystream is generated by repeatedly computing $F(K, IV)$ for an incremented counter.

In the analysis of keystream generators (*i.e.* in the analysis of f and g) it is typical to assume that the initial state X is random. Hence for a stream cipher we require that F has suitable randomness properties, and in particular, that it has good diffusion with regards to both IV and K. (Clearly this applies equally to the case when the output of F is the keystream.) Indeed, if diffusion of the IV is not complete then there may well be statistical or algebraic dependences in the keystreams for different IV's, as chosen-IV attacks on numerous stream ciphers demonstrate (*e.g.*, [6,9,8]). Good mixing of the secret key is similarly required and there should not be any identifiable subsets of keys that have a traceable

R. Barua and T. Lange (Eds.): INDOCRYPT 2006, LNCS 4329, pp. 2–16, 2006.

influence on the initial state (or on the generated keystream), see [7]. Since, in many cases, F is constructed from the repeated application of a relatively simple function over r rounds, determining the required number of rounds r can be difficult. For a well-designed scheme, we would expect the security of the mechanism to increase with r, though there is a clear cost in reduced efficiency.

In this paper we investigate the initialisation of Salsa20 and TSC-4. We consider a set of well-chosen inputs (K, IV) and compute the outputs $F(K, IV)$. Under an appropriate measure we aim to detect non-random behaviour in the output. Throughout we assume that the IV's can be chosen and that most, or all, of the key bits are unknown. The paper is organized as follows. The specification of Salsa20 is recalled in Section 2 with an analysis up to seven rounds in Section 3. TSC-4 is described in Section 4 with analysis in Section 5. We draw our conclusions in Section 6. For notation we use $+$ for addition modulo 2^{32}, $\oplus$ for bitwise XOR, $\wedge$ for bitwise AND, $\lll$ for bitwise left-rotation, and $\ll$ for bitwise left-shift. The most (least) significant bit will be denoted msb (lsb).

2   Description of Salsa20

A full description of Salsa20 can be found in [1]. As mentioned in the introduction, we view the initialisation vector as $IV = (v_0, v_1, i_0, i_1)$ where (v_0, v_1) denotes the nonce and (i_0, i_1) the counter. Throughout we consider the 256-bit key version of Salsa20 and we denote the key by $K = (k_0, \ldots, k_7)$ and four constants by $c_0, \ldots, c_3$ (see [1]). We denote the cipher state by $X = (x_0, \ldots, x_{15})$ where each x_i is a 32-bit word.

$$X = \begin{pmatrix} x_0 & x_1 & x_2 & x_3 \\ x_4 & x_5 & x_6 & x_7 \\ x_8 & x_9 & x_{10} & x_{11} \\ x_{12} & x_{13} & x_{14} & x_{15} \end{pmatrix} \text{ where } X^0 = \begin{pmatrix} c_0 & k_0 & k_1 & k_2 \\ k_3 & c_1 & v_0 & v_1 \\ i_0 & i_1 & c_2 & k_4 \\ k_5 & k_6 & k_7 & c_3 \end{pmatrix}.$$

At each application of the initialisation process $F(K, IV)$ 512 bits of keystream are generated by using the entirety of the final state as the keystream. The computation F is built around the quarterround function illustrated in Fig. 1 with quarterround$(y_0, y_1, y_2, y_3) = (z_0, z_1, z_2, z_3)$.

The operation columnround function updates all 16 words of the state X and can be described as follows. Each column i, $0 \le i \le 3$, is rotated upwards by i array positions. Each column is then used independently as input to the quarterround function. The resulting set of four columns, $0 \le i \le 3$, are then rotated down by i array positions. The operation rowround can be viewed as being identical to the columnround operation except that the state array is transposed both before and after using the columnround operation. Salsa20 updates the internal state by using columnround and rowround one after the other. After r rounds, the state is denoted X^r and the keystream given by $z = X^0 + X^r$ using wordwise addition modulo 2^{32}. The original version of Salsa20 has $r = 20$, i.e. 10 rounds of columnround interleaved with 10 rounds of rowround, though shorter versions with $r = 8$ and $r = 12$ have been proposed [2].

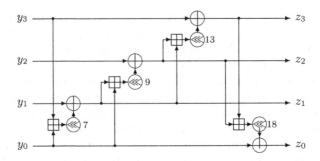

Fig. 1. The quarterround function of Salsa20

3   Analysis of Salsa20

In this section we will demonstrate two related instances of non-random behaviour. These are detectable over six and seven rounds of Salsa20 respectively. Depending on the attack model this also permits key recovery. To start, we illustrate our approach by building on the earlier work of Crowley [5] and we describe a framework that allows a more sophisticated analysis to take place. This is achieved in two steps. First, we identify interesting differential effects in a simplified version of Salsa20. Second, we identify key and IV choices that allow us to ensure that the behaviour of the genuine Salsa20 is reasonably well-approximated by the simplified version. This technique allows us to make a systematic research of possible input differences $\mathcal{ID}$'s and consequently to find $\mathcal{ID}$'s with optimal properties.

As mentioned, our observations are differential in nature. We will work with two copies of the state where X^0 is filled with the input (K, IV) and a second state Y^0 is initialized according to $Y^0 = X^0 \oplus \Delta^0$ where $\Delta^0 = (\Delta^0_0, \ldots, \Delta^0_{15})$ is the $\mathcal{ID}$. Note that the specifications of Salsa20 require that any $\mathcal{ID}$ must be zero in the diagonal words Δ^0_0, Δ^0_5, Δ^0_{10}, and Δ^0_{15}. After r rounds of Salsa20 the output difference $\mathcal{OD}$ is given[1] by $\Delta^r = X^r \oplus Y^r$.

3.1   A Linearised Version of Salsa20

In previous work, Crowley [5] identified a truncated differential over three rounds of Salsa20. Consider setting $\Delta^0_i = 0$ for $i \neq 9$ and $\Delta^0_9 = $ 0x80000000. Then the following truncated differential for the first three rounds holds with a theoretical probability 2^{-12}. In practice a variety of effects conspire to give an average probability of 2^{-9}.

$$
\begin{pmatrix}
0 & 0 & 0 & 0 \\
0 & 0 & 0 & 0 \\
0 & \text{0x80000000} & 0 & 0 \\
0 & 0 & 0 & 0
\end{pmatrix}
\xrightarrow[\substack{\text{row} \\ \text{col}}]{\substack{\text{col} \\ \text{col}}}
\begin{pmatrix}
? & ? & ? & \text{0x02002802} \\
? & ? & ? & ? \\
? & ? & ? & ? \\
? & ? & ? & ?
\end{pmatrix}
$$

---

[1] Note that due to the feedforward in Salsa20 that uses addition modulo 2^{32} this is not necessarily the same as the difference in the corresponding keystream.

Given the behaviour exhibited in $x_3^3 \oplus y_3^3$ it is tempting to look for some impact in the next round. Yet, it is not clear how to proceed in a methodical manner.

To establish an appropriate framework for analysis, we introduce an alternative algorithm LinSalsa20. This is identical to Salsa20 except that all integer additions have been replaced by exclusive-or. The corresponding round functions are denoted LinColumnround and LinRowround. Assume that two initial states X^0 and $Y^0 = X^0 \oplus \Delta^0$ are iterated by LinSalsa20. Then since LinSalsa20 is completely linear in $\mathrm{GF}(2)$, the difference $\Delta^r = X^r \oplus Y^r$ coincides exactly with computing r iterations of Δ^0 with LinSalsa20. This computation does not require knowledge of the key and we refer to a differential path generated by LinSalsa20 as a *linear differential*. It is straightforward to see that there are many (admissible) input differences for which the output of LinSalsa20 is trivially non-random.

Proposition 1. *Consider an input* $\Delta_i^0 \in \{\text{0xFFFFFFFF}, \text{0x00000000}\}$ *for all words* $i = 0, \ldots, 15$. *Then, for any number of updates with* LinSalsa20, *one has* $\Delta_i^r \in \{\text{0xFFFFFFFF}, \text{0x00000000}\}$.

However we need to be more careful. While LinSalsa20 allows some straightforward analysis, the further the behaviour of LinSalsa20 is from the true Salsa20, the less useful it will be. Since a differential of large Hamming weight is likely to induce carries and hence non-linear behaviour to the genuine Salsa20, we will need a linear differential of low Hamming weight. Such a differential is intended to offer a reasonably good approximation to the same differential in genuine Salsa20. We will consider a linear differential to be of low weight if any computation involving active words in the difference only uses words of low Hamming weight ($\ll 16$). Let us consider Crowley's differential within this linear model.

Example 1. Consider an input difference with $\Delta_9^0 = \text{0x80000000}$ as the one non-zero word. The weight of differences for the first four rounds is as follows.

$$
\begin{pmatrix} 0\,0\,0\,0 \\ 0\,0\,0\,0 \\ 0\,1\,0\,0 \\ 0\,0\,0\,0 \end{pmatrix} \xrightarrow{\text{col}}
\begin{pmatrix} 0\,2\,0\,0 \\ 0\,3\,0\,0 \\ 0\,1\,0\,0 \\ 0\,1\,0\,0 \end{pmatrix} \xrightarrow{\text{row}}
\begin{pmatrix} 4\,\ 2\,2\,2 \\ 7\,10\,3\,6 \\ 1\,\ 3\,4\,1 \\ 0\,\ 1\,1\,2 \end{pmatrix} \xrightarrow{\text{col}}
\begin{pmatrix} 9\,19\,\ 6\,\ 5 \\ 3\,13\,\ 5\,\ 5 \\ 4\,11\,11\,\ 7 \\ 1\,16\,\ 2\,10 \end{pmatrix}
$$

$$
\xrightarrow{\text{row}}
\begin{pmatrix} 13\ 13\ 14\ 10 \\ 13\ 13\ 13\ 19 \\ 16\ 18\ 19\ 11 \\ \boxed{11}\ 17\ 20\ 15 \end{pmatrix}
$$

The top line of this differential is as far as Crowley goes, but when using LinSalsa20 it appears we can go one round further. Indeed, one can identify a low-weight linear differential for word x_{12}^4, among others. Note that x_{12} is a right-to-diagonal word (with wrap) and is updated first in round four; the 16 in x_{13}^3 has no effect on x_{12}^4. □

The linear model can also be used to find longer differentials. A well-chosen multi-bit input may cause smaller diffusion than a single-bit input; non-zero bits can be placed in positions where they are annihilated in the update process. To illustrate, we focus again on a single column where the weight of the input (starting with the diagonal element) is $(0, 2, 1, 1)$. With a fixed relative position of the non-zero bits in this input, one can obtain an output after the first round of the form $(0, 1, 0, 0)$. The absolute position of the non-zero bits and the choice of column are free parameters and naturally leads to an identified sub-class of inputs. These all have the same properties in LinSalsa20.

Example 2. Consider an input difference with non-zero words $\Delta_2^0 = \text{0x00000100}$, $\Delta_6^0 = \text{0x00001000}$, and $\Delta_{14}^0 = \text{0x80080000}$.

$$
\begin{pmatrix} 0 & 0 & 1 & 0 \\ 0 & 0 & 1 & 0 \\ 0 & 0 & 0 & 0 \\ 0 & 0 & 2 & 0 \end{pmatrix} \xrightarrow{\text{col}} \begin{pmatrix} 0 & 0 & 0 & 0 \\ 0 & 0 & 0 & 0 \\ 0 & 0 & 0 & 0 \\ 0 & 0 & 1 & 0 \end{pmatrix} \xrightarrow{\text{row}} \begin{pmatrix} 0 & 0 & 0 & 0 \\ 0 & 0 & 0 & 0 \\ 0 & 0 & 0 & 0 \\ 1 & 1 & 3 & 4 \end{pmatrix} \xrightarrow{\text{col}} \begin{pmatrix} 4 & 1 & 3 & 4 \\ 1 & 2 & 4 & 8 \\ 1 & 0 & 7 & 10 \\ 3 & 1 & 3 & 14 \end{pmatrix}
$$

$$
\xrightarrow{\text{row}} \begin{pmatrix} 13 & 1 & 6 & 7 \\ 11 & 14 & 5 & 7 \\ 7 & 4 & 14 & 5 \\ 14 & 21 & 18 & 17 \end{pmatrix} \xrightarrow{\text{col}} \begin{pmatrix} 13 & 16 & 17 & 17 \\ 6 & 16 & 19 & 23 \\ 14 & \boxed{13} & 18 & 15 \\ 18 & 16 & 15 & 15 \end{pmatrix}
$$

One can identify a truncated low-weight linear differential for x_9^5 which is an out-of-diagonal word. Note that some words in the final array may have a lower Hamming weight, but their generation required computations using average-weighted words and so they are unlikely to be relevant to genuine Salsa20.   □

3.2   Differentials in True Salsa20

In Section 3.1 we identified classes of inputs (with a required $\mathcal{ID}$) which gave low-weight truncated linear differentials after four and five rounds of LinSalsa20. For genuine (nonlinear) Salsa20, the same differentials might not behave in the same way and a differential trail will depend on the input. Therefore to find optimal $\mathcal{ID}$'s and inputs we will need to consider which conditions allow the non-linear differential trail to be closely approximated by the linear differential.

   The only non-linear operation in Salsa20 is integer addition in the quarterround function, denoted $x_a + x_b$. Given an $\mathcal{ID}$ (Δ_a, Δ_b), the nonlinear $\mathcal{OD}$ corresponds to the XOR of $x_a + x_b$ and $(x_a \oplus \Delta_a) + (x_b \oplus \Delta_b)$. Thus, the nonlinear $\mathcal{OD}$ is identical to the linear $\mathcal{OD}$, if

$$
(x_a + x_b) \oplus ((x_a \oplus \Delta_a) + (x_b \oplus \Delta_b)) = \Delta_a \oplus \Delta_b . \tag{1}
$$

Each non-zero bit in Δ_a and Δ_b may cause integer addition to create or annihilate a sequence of carry bits. Hence we focus on low-weight trails to keep more control of such events. Note that a difference in the most significant bit is always linear.

   In the following sections, we will be indirectly considering Eq. 1 when placing conditions on the inputs so that a differential in Salsa20 follows a linear differential in LinSalsa20 for some steps before diverging. Such conditions might be on the nonce, on the counter (conditions which can be satisfied by sampling a keystream in the appropriate way), or on the key (thereby establishing classes of weak keys). While many of these issues are complex, for instance, conditions aimed at linearising the $\mathcal{ID}$ must not conflict with the way the counter is likely to be incremented, some results are given in the subsequent sections.

3.3   Non-randomness in Four and Five Rounds of Salsa20

Consider the linear differential of Ex. 1 and set $\mathcal{ID}$ to be identical to that of [5]. By using LinSalsa20 we suspect a statistical imbalance in $x_{12}^4 \oplus y_{12}^4$. Given a set of N different pairs of (K, IV), where each pair takes the same fixed $\mathcal{ID}$, the distribution of the output difference for the N pairs can be analysed. However, we might consider a subset of the bits or even a single bit, and by examining each bit in $x_{12}^4 \oplus y_{12}^4$ one finds that bit 26 is heavily unbalanced[2]. This imbalance can be detected using a χ^2 test (see Appendix A) where a χ^2 score greater than some threshold is good evidence of non-randomness.

The behaviour of the differential is heavily key-dependent. The presence or absence of carries, on which Salsa20 relies, depends on the actual values of the operands. Thus some keys will dampen, and others amplify, the evolution of a differential. The imbalance in bit 26 is greater the closer Salsa20 is to LinSalsa20. A close inspection of the first round of the differential reveals that the first two additions, differentially speaking, act as XOR while the third does not. However, depending on how i_1 is incremented, we can establish conditions on the key to ensure that it does. Thus there are keys for which the imbalance in bit 26 is boosted. We refer to this as partially *linearising* the first round of the differential[3] and key conditions that achieve this are presented in Appendix B.

Example 3. Take N inputs (K, IV) where $IV = (v_0, v_1, i_0, i_1)$. The key K and nonce (v_0, v_1) are chosen at random though in the second experiment some bits of k_0 and k_6 are adapted. The counter (i_0, i_1) starts at zero and we sample the keystream so that the counter i_0 increments from 0 to $N - 1$. For each input, we use values of i_1 to generate an associate input with $\mathcal{ID}\ \Delta_9^0 = $ 0x80000000 (and zero otherwise). Compute the $\mathcal{OD}$ after four rounds of Salsa20 and evaluate the χ^2 statistic on bit 26 of Δ_{12}^4. In a χ^2 test on a single bit with threshold $T = 40$, the probability a uniform random source gives $\chi^2 > 40$ is around 2^{-32}, thus the probability of false alarm is 2^{-32}. For 100 experiments using random keys and nonces, the results are listed in Tab. 1.   □

Next we consider five rounds of Salsa20 and we use the differential of Ex. 2. The non-zero bits are located in column two. Word x_{14} is updated first by $x_{14}^1 = x_{14}^0 \oplus (x_{10}^0 + x_6^0)_{\lll 7}$. A second state $y_i^0 = x_i^0 \oplus \Delta_i^0$ is updated in the same way and, according to Eq. 1, the difference of this first update will follow the linear differential if

$$(x_{10}^0 + x_6^0) \oplus \left((x_{10}^0 \oplus \Delta_{10}^0) + (x_6^0 \oplus \Delta_6^0) \right) = \Delta_{10}^0 \oplus \Delta_6^0 \ . \tag{2}$$

---

[2] In fact there are many unbalanced bits in the state of Salsa20 after four rounds.

[3] A more sophisticated set of conditions can be derived to linearise the entirety of the first round. However for clarity we restrict ourselves to the simpler case.

Table 1. Non-randomness in four rounds of Salsa20

N	All Keys and Nonces		Weak Key Class	
	average χ^2 value	% values > 40	average χ^2 value	% values > 40
2^{12}	33	20	51	34
2^{14}	123	41	192	46
2^{16}	315	46	656	68

Notice that Δ_{10}^0 is zero and that Δ_6^0 has a single non-zero bit in position 12. Further, $x_{10}^0 = c_2$ and $x_6^0 = v_0$. Bits $12\ldots9$ of c_2 are defined as $(\ldots 0110\ldots)_2$. Consequently, if bits $11\ldots9$ of v_0 are chosen as $(\ldots 000\ldots)_2$, then no carry is produced from the right, and Eq. 1 is satisfied. Subsequently x_2 is updated and so provided the previous update followed the linear differential, the only non-zero bit in the difference will be in bit 31 and the linear trial will be followed. Updating x_6 is similar while updating x_{11} only involves zero differences.

Thus we have identified conditions on three bits of v_0, part of the nonce, so that the first round of genuine Salsa20 with the $\mathcal{ID}$ of Ex. 2 follows the linear trail. In fact, the $\mathcal{ID}$ of Ex. 2 turns out to be optimal, *i.e.* it seems to have minimum weight after two rounds of Salsa20; bitwise rotations of $\mathcal{ID}$ reduce the number of msb's while shifting the difference to another column shifts the input-condition to a key word instead of v_0. Without input conditions on v_0, the first round would follow the linear trail with a probability of about $\Pr = 0.175$.

Example 4. Take N inputs (K, IV) where $IV = (v_0, v_1, i_0, i_1)$. The key K and nonce (v_0, v_1) are chosen at random though in the second experiment bits 9–11 of v_0 are zero. The counter (i_0, i_1) starts at zero and we sample the keystream so that the counter i_0 increments from 0 to $N - 1$. For each input, we use values of k_1, v_0, k_7 to generate an associate input with $\mathcal{ID}$ $\Delta_2^0 = $ 0x00000100, $\Delta_6^0 = $ 0x00001000, $\Delta_{14}^0 = $ 0x80080000 (and zero otherwise). Compute the $\mathcal{OD}$ after five rounds of Salsa20 and evaluate the χ^2 statistic on bit 1 of Δ_9^5. In a χ^2 test on a single bit with threshold $T = 40$, the probability a uniform random source gives $\chi^2 > 40$ is around 2^{-32}. For 100 experiments using random keys and nonces, the results are listed in Tab. 2.   □

Table 2. Non-randomness in five rounds of Salsa20

N	All Keys and Nonces		Weak Nonce Class	
	average χ^2 value	% values > 40	average χ^2 value	% values > 40
2^{20}	5	4	27	26
2^{22}	16	11	105	73
2^{24}	78	17	383	89

3.4   Non-randomness in Six and Seven Rounds of Salsa20

The results presented in Section 3.3 give statistical weaknesses, as measured by the χ^2 test on a single bit, over four rounds and five rounds of Salsa20. To create these biased distributions, we used $\mathcal{ID}$'s of slightly different types. For the four round imbalance we use a non-zero difference in the counter values while for the five round imbalance we use non-zero differences in part of the key k_1 and k_7. We will comment on this later since it has an impact on the attacks we can mount.

Both statistical anomalies can be detected two rounds later. In both cases we intercept the required keystream and we guess the necessary key words to unwind the last two rounds of state update. Thus, for a single guess of the relevant words of key, the backwards computation is carried out over two rounds for N pairs of output, where each output was generated using the chosen input difference. The χ^2 statistic of the target bit of the target word is evaluated, and a χ^2 test with some threshold T applied. Our analysis tells us that a correct key guess will yield a significant χ^2 score. We assume that an incorrect key guess results in essentially random candidate values for the bit we test. Thus, a significantly large χ^2 value suggests that the key guess may be correct. The remaining key words can be searched exhaustively and the entire key guess verified against the keystream. If the χ^2 value for a key guess is not significant we move on to a new guess. The target word and bit as well as the key words to guess are given in Tab. 3.

Table 3. Key words to guess to partially unwind the last two rounds

Differential	Word	Bit	# Rounds	Key Words to Guess
Example 3	Δ_{12}^4	26	Salsa20/6	$k_3 \ k_4 \ k_5 \ k_6 \ k_7$
Example 4	Δ_9^5	1	Salsa20/7	$k_0 \ k_2 \ k_3 \ k_4 \ k_5 \ k_6$

Clearly, the scale of the imbalance in the target bit is important to the success of this method. The closer Salsa20 behaves to LinSalsa20 then the greater the imbalance in the target bit, and the greater the χ^2 score we expect to observe. This helps an attacker in two ways:

1. If certain keys and IV's give a high χ^2 score, then a greater proportion of the keys from an identified set should be susceptible to attack.
2. Higher χ^2 scores permit less keystream or greater precision in an attack.

To begin to get a picture of how things might behave in practice, we have implemented a restricted version of this style of attack. In principle we could use the four round differential of Ex. 3 to attack six rounds of Salsa20. To keep the experiments tractable, however, we use the same differential to attack a restricted five-round version as a demonstration (*i.e.* we unwind one round only).

Example 5. We recover nine bits (bits 4 to 12) of k_3 under the assumption that k_5 has been correctly guessed. Over 100 random keys and nonces and N pairs, we give the success rate when assuming the correct key lies among the candidate values giving the three highest χ^2 values. We repeat the experiment for the weak key class identified in Ex. 3. For the weak key class we observe that the same proportion of keys can be recovered when using one quarter of the text, see

Table 4. Demonstration of a key recovery attack on five rounds of Salsa20

	All Keys and Nonces	Weak Key Class
N	% success rate	% success rate
2^{12}	20	28
2^{14}	29	41
2^{16}	44	54

Tab. 4. We recall that the weak keys only improve the differential propagation and that our attack is also working for other keys.   □

As demonstrated in Ex. 5, at least in principle, our observations can be used in the way we intend. In the case of Salsa20/6 we estimate the work effort for a key-recovery attack to be around 2^{177} operations using 2^{16} pairs of keystream blocks sampled appropriately from the same keystream[4]. This is a crude estimate. Since such an attack requires guessing more key bits, more text may well be required. However, since the entirety of the target word can be recovered for any single key guess, using a single bit to test a key will miss much of the information available. Thus, it seems prudent to anticipate a final complexity close to these initial estimates. Under a related-key attack Salsa20/7 might be broken in around 2^{217} operations using 2^{24} pairs of keystream blocks taken from two sets of keystream. However, the practical validity of such an attack is debatable [3], so we merely observe that over seven of the 20 rounds in Salsa20, statistical imbalances can be detected.

4   Description of TSC-4

The stream cipher TSC-4 is specified in [12]. It consists of two states X and Y of 4×32 bits each, denoted $X = (x_0, x_1, x_2, x_3)^T$ and $Y = (y_0, y_1, y_2, y_3)^T$. Let $[x]_i$ denote bit i of a single 32 bit word x, then a bit-slice i of state X is defined as $([x_3]_i, [x_2]_i, [x_1]_i, [x_0]_i)$. We first describe the regular update function f, which updates the two states X and Y independently by single-cycle T-functions. In the case of state X, a 32-bit parameter α_X is computed as a function of X. It is defined by $\alpha_X = p \oplus (p + c_X) \oplus s$ with $p = x_0 \wedge x_1 \wedge x_2 \wedge x_3$ and $s = (x_0 + x_1 + x_2 + x_3) \lll 1$ and constant $c_X = $ 0x51291089. If $[\alpha_X]_i = 1$, then a fixed S-box S is applied to bit-slice i of X, and if $[\alpha_X]_i = 0$, then a fixed S-box S^6 is applied to bit-slice i of x (for all $i = 0, \ldots, 31$). The state Y is similarly updated where parameter α_Y has constant $c_Y = $ 0x12910895. Notice that the least significant bit-slice is always mapped by S. The S-boxes are defined as

$$\mathsf{S} = \{9, 2, 11, 15, 3, 0, 14, 4, 10, 13, 12, 5, 6, 8, 7, 1\}$$
$$\mathsf{S}^6 = \{6, 13, 8, 0, 5, 12, 1, 11, 4, 14, 3, 10, 15, 7, 2, 9\}\,.$$

---

[4] We note in passing that we can recover the key for the 128-bit version of Salsa20/5 in 2^{81} operations using 2^{16} pairs of keystream blocks.

The output function g produces a keystream byte z by combining some bytes of both states (using integer addition, XOR, shift and rotation), see [12] for more details.

Consider the initialisation function F of TSC-4. To start, the internal state of 256 bits is loaded with $K = (k_0, \ldots, k_9)$ and $IV = (i_0, \ldots, i_9)$ each of 10×8 bits (a single 32-bit word is denoted as a concatenation of four 8-bit words).

$$X = \begin{pmatrix} x_0 \\ x_1 \\ x_2 \\ x_3 \end{pmatrix} = \begin{pmatrix} k_3 & k_2 & k_1 & k_0 \\ k_7 & k_6 & k_5 & k_4 \\ i_3 & i_2 & i_1 & i_0 \\ i_7 & i_6 & i_5 & i_4 \end{pmatrix}, \quad Y = \begin{pmatrix} y_0 \\ y_1 \\ y_2 \\ y_3 \end{pmatrix} = \begin{pmatrix} i_1 & i_0 & i_9 & i_8 \\ i_5 & i_4 & i_3 & i_2 \\ k_1 & k_0 & k_9 & k_8 \\ k_5 & k_4 & k_3 & k_2 \end{pmatrix}$$

A single round of the initialisation function (denoted as a *warm-up* update) consists of a regular update and some additional operations: A byte $z = g(X)$ is produced, x_1 and y_0 are rotated to the left by eight positions, and then byte z is XOR-ed to the 8 least significant bits of x_1 and y_0. The specifications of TSC-4 propose $r = 8$ rounds.

5 Analysis of TSC-4

In [10,11], predecessors of TSC-4 have been attacked by exploiting a bit-flip bias for multiple applications of the state update function f. This bias still exists for regular updates of TSC-4, but the strong filter function g prevents from an attack. In this section, we disregard the details of the filter function and investigate the statistical properties of multiple warm-up updates of TSC-4: While the regular updates have some guaranteed properties, the warm-up updates use additional *ad hoc* operations that are designed to accelerate diffusion. Notice that our analysis is embedded in a more general context: we actually consider the initialisation function F of TSC-4 and try to detect some non-random behaviour in a set of outputs (*i.e.* in the TSC-4 initial states) that are produced by a set of well-chosen inputs (*i.e.* in the IV's).

5.1 Statistical Model of Initialisation

We investigate the statistical properties of the initialisation process. In our simple statistical model, we assume that α (with exception of the lsb) and the feedback z are uniformly randomly. Consider a single bit-slice i (not the least significant one) in the state X, then our assumptions imply for each round:

1. Bit-slice i is mapped uniformly randomly by S or by S^6.
2. After application of the S-box, bit 1 of bit-slice i is chosen uniformly randomly.

With a fixed input $w \in \{0, \ldots, 15\}$, these two steps are repeated for r rounds, so we can analyse the distribution of the output $v \in \{0, \ldots, 15\}$. Within this model, the distribution can be computed exactly in 2^{2r} steps. The other cases (*i.e.* the

Table 5. Average bias ε^2 in the statistical model for $r = 6, \ldots, 12$ rounds, and for different bit-slices

	lsb in X	lsb in Y	non-lsb in X	non-lsb in Y
$r = 6$	7.1×10^{-3}	4.2×10^{-3}	7.7×10^{-5}	1.8×10^{-5}
$r = 8$	9.7×10^{-4}	2.1×10^{-4}	4.5×10^{-6}	4.6×10^{-7}
$r = 10$	1.3×10^{-4}	7.6×10^{-6}	2.1×10^{-7}	9.8×10^{-9}
$r = 12$	2.3×10^{-5}	1.0×10^{-6}	5.5×10^{-9}	2.1×10^{-10}

least-significant bit-slice and the state Y) are treated similarly. The bias of the distribution is measured with the *Euclidean Squared Distance* $\varepsilon^2 := \sum \varepsilon_v^2$ with $\varepsilon_v := \Pr(v) - 1/16$, where $\Pr(v)$ denotes the probability for output v (given some fixed parameters). In Tab. 5, the bias ε^2 is shown for different parameters. To simplify the presentation we compute ε^2 for all inputs w and show the *average* values only.

As expected, the average bias is decreasing with the number of rounds r. In the case of the least-significant bit-slice in the state X, it is reduced by a factor of about 2.6 with each additional round. Interestingly, the position of the random bit (step 2) has a notable influence on the distribution and diffusion is better for state Y. And, as expected, diffusion is better for bit-slices which are not on the least-significant position (intuitively a combination of S and S^6 results in larger diffusion than using S only).

5.2 Experimental Results

Now we attempt to detect the bias of the previous subsection in the genuine initialisation function $\mathsf{F}(K, IV)$ of TSC-4. We need N different inputs (K, IV) where the value of a fixed bit-slice i is the same for all inputs. Each bit-slice consists of two key bits and two IV bits. Consequently, bit-slice i is the same for all inputs, if the key is fixed (and unknown), and if the IV bits of bit-slice i are fixed (though the other IV bits can be varied). The N outputs can then be used to evaluate a χ^2 statistic on bit-slice i. Provided that the assumptions on the model of the previous section are valid, bit-slice $i = 0$ of the state X is expected to have maximum bias. Here is an example for $r = 8$ rounds.

Example 6. Take N different inputs (K, IV) where $IV = (i_0, \ldots, i_9)$. The key is fixed, IV bytes $i_0, i_1 \ldots, i_7$ are zero, and i_8, i_9 increments from 0 to $N - 1$. Compute all N outputs after $r = 8$ rounds of $\mathsf{F}(K, IV)$ and evaluate the χ^2 statistic on the least-significant bit-slice in the state X of the output. In a χ^2 test on 4 bits with threshold $T = 80$, the probability a uniform random source gives $\chi^2 > 80$ is around 2^{-34}. For 100 experiments using random keys, the results are listed in Tab. 6. □

For all three choices of N, the assigned bias $\varepsilon^2 = (\chi^2 - 15)/16N$ becomes about $2^{-9.2}$ (see Appendix A). This is in good agreement with the model of Section 5.1,

Table 6. Average χ^2 statistic in the experiment for $r = 8$ rounds and a varying number of samples

N	All Keys	
	average χ^2 value	% values > 80
2^{10}	40	3
2^{12}	119	67
2^{14}	421	100

which predicts an average bias of $\varepsilon^2 = 2^{-9.8}$ in this setup[5]. Of course, the initial state cannot be accessed by an attacker, so the χ^2 test has perhaps a certificational character. However, the setup of Ex. 6 does not require any key bit to be known, and the number of samples N is very small. Consequently, this non-randomness may be a basis for future attacks that includes analysis of the filter function g.

The non-randomness is not limited to the least significant bit-slice. A notable example is $i = 8$ (and with other parameters as in Ex. 6), which results in an average value of $\chi^2 = 45$ for $N = 2^{10}$. This is a consequence of the specific setup in Ex. 6 where bit-slices $i = 8, 9 \ldots$ of X after the first round are the same for all N states and so the effective number of rounds is only $r - 1$ (in addition, the biased bit 1 of bit-slice 0 is rotated into bit-slice 8).

The above experiment with $i = 8$ was carried out for a varying number of rounds, see Fig. 2. In order to measure χ^2 values for a larger number of rounds, we increased the number of samples to $N = 2^{18}$. This was done by choosing zero IV bytes $i_0, i_1, \ldots, i_5$ and counting up i_6, i_7, i_8, i_9 from 0 to $N-1$. Supplementary

r	χ^2	ε^2
6	47593	1.1×10^{-2}
8	6067	1.4×10^{-3}
10	1437	3.4×10^{-4}
12	260	5.8×10^{-5}
14	44	6.8×10^{-6}

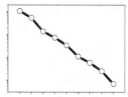

Fig. 2. Average χ^2 and the assigned bias ε^2 for $r = 6, \ldots, 14$ rounds (where the bias ε^2 is plotted on the right)

experiments revealed that ε^2 is approximately constant for different values of N, hence ε^2 is a good measure for the diffusion of F. The bias ε^2 in terms of r can be approximated by an exponential decay and in one round ε^2 is reduced by a factor of about 2.5. By extrapolation, we expect that about $r = 32$ rounds would be necessary to obtain a bias of $\varepsilon^2 = 2^{-40}$. In an extended experiment one could also measure the effectiveness of the combined initialisation function F and update function f^t. For example, with $r = 8$, $t = 50$ and $N = 2^{18}$, we

---

[5] Notice that two input bits of bit-slice $i = 0$ are always zero in the setup of Ex. 6. This has a small influence on the modeled bias in Tab. 5.

observed an average value of $\chi^2 = 32$ when using the same setup as previously. However we did not observe a bias in the keystream.

6   Conclusions

In this paper we considered the way the key and the initialisation information is used in two Phase 2 candidates in eSTREAM. In the case of TSC-4, the initial cipher state is derived using eight applications of a warm-up function. Non-randomness over all eight iterations can be detected in the initial state with about 1000 inputs. Each additional round increases the data requirements by a factor of about 2.5 and this non-randomness requires the attacker to choose IV bits only. However no bias in the keystream of TSC-4 resulting from this non-randomness has yet been detected, so it remains to be seen if our observations can form the basis for an attack in the future. As the rating of *Focus* candidate in eSTREAM Phase 2 testifies, Salsa20 is widely viewed as a very promising proposal. Nothing in this paper affects the security of the full version of the cipher. However we expect that the key can be recovered from five rounds of 128-bit Salsa20 with around 2^{81} operations and six rounds of 256-bit Salsa20 with around 2^{177} operations. Both attacks would require very moderate amounts of text. If we allow related-key attacks then the security of seven rounds of 256-bit Salsa20 might be in question with around 2^{217} operations. However, given divided opinions on such an attack model, we prefer to observe that a statistical weakness has been observed over seven rounds. While we anticipate some progress, we are doubtful that many more rounds can be attacked using the methods of this paper. Thus Salsa20 still appears to be a conservative design. Given our results, however, we are doubtful that Salsa20/8 will offer adequate security in the future, though Salsa20/12 could turn out to be a well-balanced proposal.

Acknowledgments

The first and second author are supported in part by grant 5005-67322 of NCCR-MICS (a center of the Swiss National Science Foundation). The second author also receives partial funding through GEBERT RÜF STIFTUNG. Other authors are supported by the French Ministry of Research RNRT X-CRYPT project and by the European Commission through the IST Program under Contract IST-2002-507932 ECRYPT.

References

1. D.J. Bernstein. Salsa20. In *eSTREAM, ECRYPT Stream Cipher Project*, Report 2005/025.
2. D.J. Bernstein. Salsa20/8 and Salsa20/12. In *eSTREAM, ECRYPT Stream Cipher Project*, Report 2006/007.
3. D.J. Bernstein. Related-key attacks: who cares? In *eSTREAM discussion forum*, June 22, 2005. http://www.ecrypt.eu.org/stream/phorum/read.php?1,23.

4. A. Biryukov. A New 128 Bit Key Stream Cipher: LEX. In *eSTREAM, ECRYPT Stream Cipher Project*, Report 2005/013.
5. P. Crowley. Truncated Differential Cryptanalysis of Five Rounds of Salsa20. In *eSTREAM, ECRYPT Stream Cipher Project*, Report 2005/073.
6. J. Daemen, R. Goverts, and J. Vandewalle. Resynchronization Weaknesses in Synchronous Stream Ciphers. In *Advances in Cryptology - EUROCRYPT 1993*, LNCS 765, pages 159–167. Springer-Verlag, 1993.
7. M. Dichtl and M. Schafheutle. Linearity Properties of the SOBER-t32 Key Loading. In *Fast Software Encryption 2002*, LNCS 765, pages 159–167. Springer-Verlag, 1993.
8. P. Ekdahl and T. Johansson. Another Attack on A5/1. In *IEEE Transactions on Information Theory*, volume 49/1, pages 284–289, 2003.
9. S.R. Fluhrer, I. Mantin, and A. Shamir. Weaknesses in the Key Scheduling Algorithm of RC4. In *Selected Areas in Cryptography 2001*, LNCS 2259, pages 1–24. Springer-Verlag, 2001.
10. S. Künzli, P. Junod, and W. Meier. Distinguishing Attacks on T-functions. In *Progress in Cryptology - Mycrypt 2005*, LNCS 3715, pages 2–15. Springer-Verlag, 2005.
11. F. Muller and T. Peyrin. Linear Cryptanalysis of the TSC Family of Stream Ciphers. In *Advances in Cryptology - ASIACRYPT 2005*, LNCS 3788, pages 373–394. Springer-Verlag, 2005.
12. D. Moon, D. Kwon, D. Han, J. Lee, G.H. Ryu, D.W. Lee, Y. Yeom, and S. Chee. T-function Based Streamcipher TSC-4. In *eSTREAM, ECRYPT Stream Cipher Project*, Report 2006/024.

A The χ^2 Test

Let $X := X_1, X_2, \ldots, X_N$ denote N *i.i.d.* random variables where each $X_i \in \{x_0, \ldots, x_m\}$ and with unknown distribution. A χ^2 test is applied on the observation X, in order to decide if the observation is consistent with the hypothesis that X_i have distribution D_U. Let N_i be the number of observations x_i in X, and E_i the expectation for x_i under distribution D_U. Then, the χ^2 statistic is a random variable defined by

$$\chi^2 := \sum_{i=1}^{m} \frac{(N_i - E_i)^2}{E_i} \ . \tag{3}$$

In the case of a uniform distribution D_U, one has $E_i = N/m$. The χ^2 statistic (for large N) is then compared with the threshold of the $\chi^2_{\alpha,m-1}$ distribution having $m-1$ degrees of freedom and significance level α. Consequently, a χ^2 test can be defined by a threshold T, where the hypothesis is accepted if $\chi^2(X) < T$.

If X has uniform distribution D_U, the expectation of χ^2 becomes $E_U(\chi^2) = m - 1$. If X has another distribution D_X (which is assumed to be close to D_U), then the expectation of χ^2 becomes about $E_X(\chi^2) = (c + 1)m - 1$, where $c := N\varepsilon^2$ and $\varepsilon^2 := \sum \varepsilon_i^2$ with probability bias $\varepsilon_i := \Pr_X(x_i) - \Pr_U(x_i)$. Notice that $E_X(\chi^2)$ differs from $E_U(\chi^2)$ significantly, if $c = \mathcal{O}(1)$. Consequently, about $N = \mathcal{O}(1/\varepsilon^2)$ samples are required to distinguish a source with distribution D_X from a source with distribution D_U.

B   Weak Key Conditions for Example 3

The key conditions for the weak key class of Ex. 3 are on k_0 and k_6. First set the following bits of k_0 to the values shown:

$$
\begin{array}{lrrrrrr}
\textit{bit number:} & 0 & 1 & 20 & 21 & 22 & 23 \\
\textit{value:} & 0 & 1 & 0 & 0 & 1 & 1
\end{array}
$$

Next set bit 7 of k_6 equal to bit 7 of T where $c_1 = \texttt{0x3320646E}$ and

$$T = (((k_0 + c_1) \lll 7) + c_1) \lll 9.$$

Note that all these conditions are randomly satisfied with a probability of 2^{-7}.

Differential and Rectangle Attacks on Reduced-Round SHACAL-1

Jiqiang Lu[1,*], Jongsung Kim[2,3,**], Nathan Keller[4,***], and Orr Dunkelman[5,†]

[1] Information Security Group, Royal Holloway, University of London
Egham, Surrey TW20 0EX, UK
Jiqiang.Lu@rhul.ac.uk
[2] ESAT/SCD-COSIC, Katholieke Universiteit Leuven
Kasteelpark Arenberg 10, B-3001 Leuven-Heverlee, Belgium
Kim.Jongsung@esat.kuleuven.be
[3] Center for Information Security Technologies(CIST), Korea University
Anam Dong, Sungbuk Gu, Seoul, Korea
joshep@cist.korea.ac.kr
[4] Einstein Institute of Mathematics, Hebrew University
Jerusalem 91904, Israel
nkeller@math.huji.ac.il
[5] Computer Science Department, Technion
Haifa 32000, Israel
orrd@cs.technion.ac.il

Abstract. SHACAL-1 is an 80-round block cipher with a 160-bit block size and a key of up to 512 bits. In this paper, we mount rectangle attacks on the first 51 rounds and a series of inner 52 rounds of SHACAL-1, and also mount differential attacks on the first 49 rounds and a series of inner 55 rounds of SHACAL-1. These are the best currently known cryptanalytic results on SHACAL-1 in an one key attack scenario.

Keywords: Block cipher, SHACAL-1, Differential cryptanalysis, Amplified boomerang attack, Rectangle attack.

1 Introduction

The 160-bit block cipher SHACAL-1 was proposed by Handschuh and Naccache [9,10] based on the compression function of the standardized hash function

---

* This author as well as his work was supported by a Royal Holloway Scholarship and the European Commission under contract IST-2002-507932 (ECRYPT).
** This author was financed by a Ph.D grant of the Katholieke Universiteit Leuven and by the Korea Research Foundation Grant funded by the Korean Government(MOEHRD) (KRF-2005-213-D00077) and supported by the Concerted Research Action (GOA) Ambiorics 2005/11 of the Flemish Government and by the European Commission through the IST Programme under Contract IST2002507932 ECRYPT.
*** This author was supported by the Adams fellowship.
† This author was partially supported by the Israel MOD Research and Technology Unit.

R. Barua and T. Lange (Eds.): INDOCRYPT 2006, LNCS 4329, pp. 17–31, 2006.
© Springer-Verlag Berlin Heidelberg 2006

SHA-1 [20]. It was selected for the second phase of the NESSIE (New European Schemes for Signatures, Integrity, and Encryption) project [18], but was not recommended for the NESSIE portfolio in 2003 because of concerns about its key schedule. Since SHACAL-1 is the compression function of SHA-1 used in encryption mode, there is much significance to investigate its security against different cryptanalytic attacks.

The security of SHACAL-1 against differential cryptanalysis [2] and linear cryptanalysis [17] was first analyzed by the proposers. Subsequently, Nakahara Jr. [19] conducted a statistical evaluation of the cipher. In 2002, Kim *et al.* [15] presented a differential attack on the first 41 rounds of SHACAL-1 with 512 key bits and an amplified boomerang attack on the first 47 rounds of SHACAL-1 with 512 key bits, where the former attack is due to a 30-round differential characteristic with probability 2^{-138}, while the latter attack is based on a 36-round amplified boomerang distinguisher (see Ref. [15] for the two differentials) that was conjectured by the authors to be the longest distinguisher (*i.e.*, the distinguisher with the greatest number of rounds). However, in 2003, Biham *et al.* [5] pointed out that the step for judging whether a final candidate subkey is the right one in the amplified boomerang attacks presented in [15] is incorrect due to a flaw in the analysis on the number of wrong quartets that satisfy the conditions of a right quartet. They then corrected it with the fact that all the subkeys of SHACAL-1 are linearly dependent on the user key. Finally, by converting the Kim *et al.*'s 36-round boomerang distinguisher to a 36-round rectangle distinguisher, Biham *et al.* presented rectangle attacks on the first 47 rounds and two series of inner 49 rounds of SHACAL-1 with 512 key bits. These are the best cryptanalytic results on SHACAL-1 in an one key attack scenario, prior to the work described in this paper. Other cryptanalytic results on SHACAL-1 include the related-key rectangle attacks [7,11,14]; however, these related-key attacks [1] are very difficult or even infeasible to be conducted in most cryptographic applications, though certain current applications may allow for them, say key-exchange protocols [13].

In this paper, we exploit some better differential characteristics than those previously known in SHACAL-1. More specifically, we exploit a 24-round differential characteristic with probability 2^{-50} for rounds 0 to 23 such that we construct a 38-round rectangle distinguisher with probability $2^{-302.3}$. Based on this distinguisher, we mount rectangle attacks on the first 51 rounds and a series of inner 52 rounds of SHACAL-1 with 512 key bits. We also exploit a 34-round differential characteristic with probability 2^{-148} for rounds 0 to 33 and a 40-round differential characteristic with probability 2^{-154} for rounds 30 to 69, which can be used to mount differential attacks on the first 49 rounds and a series of inner 55 rounds of SHACAL-1 with 512 key bits, respectively.

The rest of this paper is organised as follows. In the next section, we briefly describe the SHACAL-1 cipher, the amplified boomerang attack and the rectangle attack. In Sections 3 and 4, we present rectangle and differential attacks on the aforementioned reduced-round versions of SHACAL-1, respectively. Section 5 concludes this paper.

2   Preliminaries

2.1   The SHACAL-1 Cipher

The encryption procedure of SHACAL-1 can be described as follows,

1. The 160-bit plaintext P is divided into five 32-bit words $A_0||B_0||C_0||D_0||E_0$.
2. For $i = 0$ to 79:
$$A_{i+1} = K_i \boxplus ROT_5(A_i) \boxplus f_i(B_i, C_i, D_i) \boxplus E_i \boxplus W_i,$$
$$B_{i+1} = A_i,$$
$$C_{i+1} = ROT_{30}(B_i),$$
$$D_{i+1} = C_i,$$
$$E_{i+1} = D_i.$$
3. The ciphertext is $(A_{80}||B_{80}||C_{80}||D_{80}||E_{80})$,

where $\boxplus$ denotes addition modulo 2^{32}, $ROT_i(X)$ represents left rotation of X by i bits, $||$ denotes string concatenation, K_i is the i-th round key, W_i is the i-th round constant,[1] and the function f_i is defined as,

$$f_i(B, C, D) = \begin{cases} f_{\mathrm{if}} = (B\&C)|(\neg B\&D) & 0 \le i \le 19 \\ f_{\mathrm{xor}} = B \oplus C \oplus D & 20 \le i \le 39, 60 \le i \le 79 \\ f_{\mathrm{maj}} = (B\&C)|(B\&D)|(C\&D) & 40 \le i \le 59 \end{cases}$$

where $\&$ denotes the bitwise logical AND, $\oplus$ denotes the bitwise logical exclusive OR (XOR), $\neg$ denotes the complement, and $|$ represents the bitwise OR operations.

The key schedule of SHACAL-1 takes as input a variable length key of up to 512 bits; Shorter keys can be used by padding them with zeros to produce a 512-bit key string, however, the proposers recommend that the key should not be shorter than 128 bits. The 512-bit user key K is divided into sixteen 32-bit words $K_0, K_1, \cdots, K_{15}$, which are the round keys for the first 16 rounds. Each of the remaining round keys is generated as $K_i = ROT_1(K_{i-3} \oplus K_{i-8} \oplus K_{i-14} \oplus K_{i-16})$.

2.2   Amplified Boomerang and Rectangle Attacks

Amplified boomerang attack [12] and rectangle attack [3] are both variants of the boomerang attack [21]. As a result, they share the same basic idea of using two short differentials with larger probabilities instead of a long differential with a smaller probability.

Amplified boomerang attack treats a block cipher $E : \{0,1\}^n \times \{0,1\}^k \to \{0,1\}^n$ as a cascade of two sub-ciphers $E = E^1 \circ E^0$. It assumes that there exist two differentials: one differential $\alpha \to \beta$ through E^0 with probability p (i.e., $Pr[E^0(X) \oplus E^0(X^*) = \beta | X \oplus X^* = \alpha] = p$), and the other differential $\gamma \to \delta$ through E^1 with probability q (i.e., $Pr[E^1(X) \oplus E^1(X^*) = \delta | X \oplus X^* = \gamma] = q$), with p and q satisfying $p \cdot q \gg 2^{-n/2}$. Two pairs of plaintexts $(P_1, P_2 = P_1 \oplus \alpha)$ and $(P_3, P_4 = P_3 \oplus \alpha)$ is called a right quartet if the following three conditions hold:

---

[1] We note that this is the opposite to Refs. [9,10,20]; however, we decide to stick to the common notation K_i as a round subkey.

C1: $E^0(P_1) \oplus E^0(P_2) = E^0(P_3) \oplus E^0(P_4) = \beta$;
C2: $E^0(P_1) \oplus E^0(P_3) = E^0(P_2) \oplus E^0(P_4) = \gamma$;
C3: $E^1(E^0(P_1)) \oplus E^1(E^0(P_3)) = E^1(E^0(P_2)) \oplus E^1(E^0(P_4)) = \delta$.

If we take N pairs of plaintexts with the difference α, then we have approximately $N \cdot p$ pairs with the output difference β after E^0, which generate about $\frac{(N \cdot p)^2}{2}$ candidate quartets. Assuming that the intermediate values after E^0 distribute uniformly over all possible values, we get $E^0(P_1) \oplus E^0(P_3) = \gamma$ with probability 2^{-n}. Once this occurs, $E^0(P_2) \oplus E^0(P_4) = \gamma$ holds as well, as $E^0(P_2) \oplus E^0(P_4) = E^0(P_1) \oplus E^0(P_2) \oplus E^0(P_3) \oplus E^0(P_4) \oplus E^0(P_1) \oplus E^0(P_3) = \gamma$. As a result, the expected number of right quartets is about $\frac{(N \cdot p)^2}{2} \cdot 2^{-n} \cdot q^2 = N^2 \cdot 2^{-n-1} \cdot (p \cdot q)^2$. On the other hand, for a random cipher, the expected number of right quartets is approximately $N^2 \cdot 2^{-2n}$. Therefore, if $p \cdot q > 2^{-n/2}$ and N is sufficiently large, the amplified boomerang distinguisher can effectively distinguish between E and a random cipher with an enough bias.

Rectangle attack achieves advantage over an amplified boomerang attack by allowing β to take any possible value β' in E^0 and γ to take any possible value γ' in E^1, as long as $\beta' \neq \gamma'$. Starting with N pairs of plaintexts with the difference α, the expected number of right quartets is about $N^2 \cdot (\widehat{p} \cdot \widehat{q})^2 \cdot 2^{-n}$, where $\widehat{p} = (\sum_{\beta'} Pr^2(\alpha \to \beta'))^{\frac{1}{2}}$, $\widehat{q} = (\sum_{\gamma'} Pr^2(\gamma' \to \delta))^{\frac{1}{2}}$.

3   Rectangle Attacks on Reduced-Round SHACAL-1

We exploit a 24-round differential characteristic with probability 2^{-50} for rounds 0–23: $(e_{29}, 0, 0, 0, e_{2,7}) \to (e_{14,29}, e_{9,31}, e_2, e_{29}, 0)$. Table 1 describes the full differential.

- By combining the 24-round differential with a differential composed of rounds 24–35 of the second differential of [15] (which has probability 2^{-20} in these rounds), a 36-round distinguisher with probability $2^{-300}(= (2^{-50} \cdot 2^{-20})^2 \cdot 2^{-160})$ is obtained, gaining a factor of 2^{12} over the probability of the most powerful currently known 36-round one due to Kim et al..
- By combining the 24-round differential with a differential composed of rounds 23–35 of the second differential of [15] (which has probability 2^{-24} in these rounds), a 37-round distinguisher with probability $2^{-308}(= (2^{-50} \cdot 2^{-24})^2 \cdot 2^{-160})$ is obtained.
- By combining the 24-round differential with a differential composed of rounds 21–34 of the second differential of [15] (which has probability 2^{-27} in these rounds), a 38-round distinguisher with probability $2^{-314}(= (2^{-50} \cdot 2^{-27})^2 \cdot 2^{-160})$ is obtained.

These amplified boomerang distinguishers can be used to mount amplified boomerang attacks on certain reduced-round versions of SHACAL-1 with different lengths of user keys. Nevertheless, due to the nature that all the possible β and γ (as long as they are different) can be used in a rectangle distinguisher,

Table 1. A 24-round differential with probability 2^{-50} for Rounds 0 to 23

Round(i)	ΔA_i	ΔB_i	ΔC_i	ΔD_i	ΔE_i	Prob.	Round(i)	ΔA_i	ΔB_i	ΔC_i	ΔD_i	ΔE_i	Prob.
input	e_{29}	0	0	0	$e_{2,7}$	2^{-2}	13	0	e_8	e_1	0	0	2^{-2}
1	e_7	e_{29}	0	0	0	2^{-2}	14	0	0	e_6	e_1	0	2^{-2}
2	e_{12}	e_7	e_{27}	0	0	2^{-3}	15	0	0	0	e_6	e_1	2^{-2}
3	e_{17}	e_{12}	e_5	e_{27}	0	2^{-4}	16	e_1	0	0	0	e_6	2^{-1}
4	e_{22}	e_{17}	e_{10}	e_5	e_{27}	2^{-4}	17	0	e_1	0	0	0	2^{-1}
5	0	e_{22}	e_{15}	e_{10}	e_5	2^{-4}	18	0	0	e_{31}	0	0	2^{-1}
6	e_5	0	e_{20}	e_{15}	e_{10}	2^{-3}	19	0	0	0	e_{31}	0	2^{-1}
7	0	e_5	0	e_{20}	e_{15}	2^{-3}	20	0	0	0	0	e_{31}	1
8	e_{15}	0	e_3	0	e_{20}	2^{-2}	21	e_{31}	0	0	0	0	2^{-1}
9	0	e_{15}	0	e_3	0	2^{-2}	22	e_4	e_{31}	0	0	0	2^{-1}
10	0	0	e_{13}	0	e_3	2^{-2}	23	$e_{9,31}$	e_4	e_{29}	0	0	2^{-3}
11	e_3	0	0	e_{13}	0	2^{-2}	output	$e_{14,29}$	$e_{9,31}$	e_2	e_{29}	0	/
12	e_8	e_3	0	0	e_{13}	2^{-2}							

these amplified boomerang distinguishers can be converted into rectangle distinguishers so that the resultant rectangle attacks can work more efficiently. Here, we will just present rectangle attacks on SHACAL-1 with 512 key bits based on the 38-round distinguisher.

3.1   Attacking Rounds 0 to 50

Let $E_f \circ E^1 \circ E^0$ be the 51-round SHACAL-1 with 512 key bits, where E^0 denotes rounds 0 to 23, E^1 denotes rounds 24 to 37, and E_f denotes rounds 38 to 50.

To compute $\widehat{p}$ (resp., $\widehat{q}$) (defined in Section 2.2) in such an attack, we need to summarize all the possible output differences β' for the input difference α through E^0 (resp., all the possible input differences γ' having an output difference δ through E^1), which is computationally infeasible. As a countermeasure, we can count as many such possible differentials as we can.

For simplicity, we compute $\widehat{p}$ by just counting the 24-round differentials that only have variable output differences $(\Delta A_{24}, e_{9,31}, e_2, e_{29}, 0)$ compared with the

24-round differential, where ΔA_{24} is an element from the set $\{(\overbrace{0, \cdots, 0, \overbrace{1, \cdots, 1,}^{2}}^{m}$

$\overbrace{1, 0, \cdots, 0, \underbrace{1, \cdots, 1,}_{j}}^{14} \overbrace{1, 0, \cdots, 0, \underbrace{1, \cdots, 1,}_{k}}^{9} 0, 0, 0, 0, 0) | 0 \leq m \leq 2, 0 \leq j \leq 14, 0 \leq$

$k \leq 9\}$, for such an output difference with the form is possible for the input difference $(e_{9,31}, e_4, e_{29}, 0, 0)$ to round 23. It was shown in [16] that the following Theorem 1 holds for the addition difference,

Theorem 1. [16] Given three 32-bit differences ΔX, ΔY and ΔZ. If the probability $Prob[(\Delta X, \Delta Y) \overset{\boxplus}{\to} \Delta Z] > 0$, then

$$Prob[(\Delta X, \Delta Y) \overset{\boxplus}{\to} \Delta Z] = 2^s,$$

where the integer s is given by $s = \#\{i|0 \leq i \leq 30, not((\Delta X)_i = (\Delta Y)_i = (\Delta Z)_i)\}$.

Thus, we can compute a loose lower bound $\widehat{p} = 2^{-49.39}$ by only counting the 46 differentials with $k + j + m \leq 5$; when $k + j + m > 5$ the contribution is negligible. We note that the more the counted possible differentials, the better the resultant $\widehat{p}$, but according to our results the improvement is negligible.

Biham et al. [5] got a lower bound $\widehat{q}$ in their attack as $\widehat{q} = 2^{-30.28}$ by only changing the first one or two rounds in the Kim et al.'s second differential. Since our 38-round distinguisher just uses the first 14 rounds from round 21 to 34 in the Kim et al.'s second differential, throwing round 35 away, therefore, $2^{-26.28}(= 2^{-30.28} \cdot 2^4)$ is the right value for the $\widehat{q}$ in our attack.

Now, we conclude that the distinguisher holds a lower bound probability $2^{-311.34}(\approx (2^{-49.39} \cdot 2^{-26.28})^2 \cdot 2^{-160})$. However, we can adopt the following two techniques to further reduce the complexity of the attack:

T1) Fix the four fixed bits $a_9 = a_9^* = 0$, $b_9 = b_9^* = 0$, $b_{31} = b_{31}^* = 0$ and $c_{29} = c_{29}^* = 0$ in any pair of plaintexts $P = (A, B, C, D, E)$ and $P^* = (A^*, B^*, C^*, D^*, E^*)$, where x_i is the i-th bit of X. This increases the probability of the characteristic in the first round by a factor of 4. Thus, a lower bound probability $2^{-47.39}(= 2^2 \cdot 2^{-49.39})$ is obtained for the above 46 possible 24-round differentials with such four bits fixed in any pair.

T2) Count many possible 14-round differentials $\gamma' \rightarrow \delta'$ for each input difference γ' to round 24 in our distinguisher. For expediency, we count those 14-round differentials that only have variable output differences $(\Delta A_{38}, e_{9,31}, e_2, e_{29}, 0)$ compared with the 14-round differential from round 21 to 34 in the Kim et al.'s second differential. In our observation on this 1-round difference, there are at least two possible ΔA_{38} (i.e., $e_{29}, e_{14,29}$) with probability 2^{-3}, four possible ΔA_{38} (i.e., $e_{5,14,29}, e_{14,15,29}, e_{14,29,30}, e_{14,29,30,31}$) with probability 2^{-4}, and seven possible ΔA_{38} (i.e., $e_{5,14,29,30,31}, e_{14,15,29,30,31}, e_{5,6,14,29}, e_{5,14,15,29}, e_{14,15,16,29}, e_{5,14,29,30}, e_{14,15,29,30}$) with probability 2^{-5}. We denote the set of these 13 differences by $\mathcal{S}$. Thus, these 13 possible 14-round differentials hold a lower bound probability of $2^{-23.76}(\approx 2 \cdot 2^{-26.28} + 4 \cdot 2^{-27.28} + 7 \cdot 2^{-28.28})$.

Finally, this rectangle distinguisher holds a lower bound probability $2^{-302.3}(\approx (2^{-47.39} \cdot 2^{-23.76})^2 \cdot 2^{-160})$ for the right key, while it now holds with a probability of $2^{-312.6}(\approx (2^{-160} \cdot (2 + 4 + 7))^2)$ for a wrong key. The number of available plaintext pairs decreases to 2^{155} due to the four fixed bits.

Consequently, we can apply this rectangle distinguisher to break the first 51 rounds of SHACAL-1.

Attack Procedure

1. Choose $2^{152.65}$ pairs of plaintexts with difference $\alpha = (e_{29}, 0, 0, 0, e_{2,7})$ and four fixed bits as described above: (P_i, P_i'), for $i = 1, 2, \cdots, 2^{152.65}$. Ask for their encryption under 51-round SHACAL-1 to obtain their corresponding ciphertext pairs (C_i, C_i'). The $2^{152.65}$ pairs generate about $2^{305.3}$ candidate quartets $((P_{i_1}, P_{i_1}'), (P_{i_2}, P_{i_2}'))$, where $1 \leq i_1, i_2 \leq 2^{152.65}$.

2. Guess a 352-bit key K_f for rounds 40 to 50 in E_f, do follows,

   2.1 Partially decrypt all the ciphertext pairs (C_i, C_i') with K_f to get their intermediate values just before round 40: $(E_{K_f}^{-1}(C_i), E_{K_f}^{-1}(C_i'))$. Then, for each quartet $((C_{i_1}, C_{i_1}'), (C_{i_2}, C_{i_2}'))$, check if both the two 96-bit differences in words C, D and E positions of $E_{K_f}^{-1}(C_{i_1}) \oplus E_{K_f}^{-1}(C_{i_2})$ and $E_{K_f}^{-1}(C_{i_1}') \oplus E_{K_f}^{-1}(C_{i_2}')$ belong to the set $\{(u, e_{7,29}, e_2) | ROT_{30}(u) \in \mathcal{S}\}$. If the number of the quartets passing this test is greater than or equal to 6, then go to Step 2.2; Otherwise, repeat Step 2 with another guess for K_f.

   2.2 Guess a 32-bit subkey K_{39} for round 39, and then decrypt each remaining quartet $((E_{K_f}^{-1}(C_{i_1}), E_{K_f}^{-1}(C_{i_1}')), (E_{K_f}^{-1}(C_{i_2}), E_{K_f}^{-1}(C_{i_2}')))$ with K_{39} to get their intermediate values just before round 39: $((E_{K_{39}}^{-1}(E_{K_f}^{-1}(C_{i_1})), E_{K_{39}}^{-1}(E_{K_f}^{-1}(C_{i_1}'))), (E_{K_{39}}^{-1}(E_{K_f}^{-1}(C_{i_2})), E_{K_{39}}^{-1}(E_{K_f}^{-1}(C_{i_2}'))))$. We denote them by $((X_{i_1}, X_{i_1}'), (X_{i_2}, X_{i_2}'))$. Finally, check if both the two 128-bit differences in words B, C, D and E positions of $X_{i_1} \oplus X_{i_2}$ and $X_{i_1}' \oplus X_{i_2}'$ belong to the set $\{(u, e_{7,29}, e_2, e_{29})\}$. If the number of the quartets passing this test is greater than or equal to 6, then go to Step 2.3; Otherwise, repeat this step with another guess for K_{39} (If all the values of K_{39} fail, then go to Step 2).

   2.3 Guess a 32-bit subkey K_{38} for round 38, and then decrypt each remaining quartet $((X_{i_1}, X_{i_1}'), (X_{i_2}, X_{i_2}'))$ with K_{38} to get their intermediate values just before round 38: $((E_{K_{38}}^{-1}(X_{i_1}), E_{K_{38}}^{-1}(X_{i_1}')), (E_{K_{38}}^{-1}(X_{i_2}), E_{K_{38}}^{-1}(X_{i_2}')))$. We denote them by $((\overline{X}_{i_1}, \overline{X'}_{i_1}), (\overline{X}_{i_2}, \overline{X'}_{i_2}))$. Finally, check if both the two 160-bit differences $\overline{X}_{i_1} \oplus \overline{X}_{i_2}$ and $\overline{X'}_{i_1} \oplus \overline{X'}_{i_2}$ belong to the set $\{(u, e_{7,29}, e_2, e_{29}, 0)\}$. If the number of the quartets passing this test is greater than or equal to 6, then record (K_f, K_{38}, K_{39}) and go to Step 3; Otherwise, repeat this step with another guess for K_{38} (If all the values of K_{38} fail, then go to Step 2.2; If all the values of K_{39} fail, then go to Step 2).

3. For a suggested (K_{38}, K_{39}, K_f), exhaustively search the remaining 96 key bits using trial encryption. Three known pairs of plaintexts and ciphertexts are enough for this trial process. If a 512-bit key is suggested, output it as the master key of the 51-round SHACAL-1. Otherwise, go to Step 2.

This attack requires $2^{153.65}$ chosen plaintexts. The required memory for this attack is dominated by the ciphertext pairs, which is about $2^{153.65} \cdot 20 \approx 2^{157.97}$ memory bytes.

The time complexity of Step 1 is $2^{153.65}$ 51-round SHACAL-1 encryptions; The time complexity of Step 2.1 is dominated by the partial decryptions, which is about $2^{352} \cdot 2^{153.65} \cdot \frac{11}{51} \approx 2^{503.44}$. In Step 2.1, since the probability that a quartet meets the filtering condition in this step is $(\frac{13}{96})^2 \approx 2^{-184.6}$, the expected number of the quartets passing the test for each subkey candidate is $2^{305.3} \cdot 2^{-184.6} \approx 2^{120.7}$, and it is evident that the probability that the number of quartets passing the test for a wrong subkey is no less than 6 is about 1. Thus, almost all the 2^{352} subkeys pass through Step 2.1. In Step 2.2, the time complexity is about $2^{352} \cdot 2^{32} \cdot 2^{120.7} \cdot 4 \cdot \frac{1}{51} \approx 2^{501.03}$. In this step, since the probability that a remaining quartet meets the filtering condition in this step is $2^{-32} \cdot 2^{-32} \approx 2^{-64}$, the expected number of the quartets passing the test for each subkey candidate is $2^{120.7} \cdot 2^{-64} \approx 2^{56.7}$. Again, almost all the 2^{384} subkeys pass through Step 2.2. In Step 2.3, the time complexity is about $2^{384} \cdot 2^{32} \cdot 2^{56.7} \cdot 4 \cdot \frac{1}{51} \approx 2^{469.03}$. In this step, since the probability that a remaining quartet meets the filtering condition in this step is also 2^{-64}, the expected number of the quartets passing the test for each subkey candidate is $2^{56.7} \cdot 2^{-64} \approx 2^{-7.3}$, and the probability that the number of quartets passing the test for a wrong subkey is no less than 6 is about $\sum_{i=6}^{2^{56.7}} \left(\binom{2^{56.7}}{i} \cdot (2^{-64})^i \cdot (1 - 2^{-64})^{2^{56.7}-i} \right) \approx 2^{-53.29}$. Thus, on average, about $2^{416} \cdot 2^{-53.29} = 2^{362.71}$ subkeys pass through Step 2.3, which result in $2^{362.71} \cdot 2^{96} \approx 2^{458.71}$ 51-round encryptions in Step 3. Therefore, this attack totally requires about $2^{153.65} + 2^{503.44} + 2^{501.03} + 2^{469.03} + 2^{458.71} \approx 2^{503.7}$ encryptions.

Since the probability that a wrong 512-bit key is suggested in Step 3 is about $2^{-480}(= 2^{-160.3})$, the expected number of suggested wrong 512-bit keys is about $2^{-480} \cdot 2^{458.71} \approx 2^{-21.29}$, which is quite low. While the expected number of quartets passing the difference test in Step 2.5 for the right key is 8 $(= 2^{305.3} \cdot 2^{-302.3})$, and the probability that the number of quartets passing the difference test in Step 2.5 for the right subkey is no less than 6 is about $\sum_{i=6}^{2^{305.3}} \left(\binom{2^{305.3}}{i} \cdot (2^{-302.3})^i \cdot (1 - 2^{-302.3})^{2^{305.3}-i} \right) \approx 0.81$. Therefore, with a probability of 0.81, we can break the 51-round SHACAL-1 with 512 key bits by using the amplified boomerang attack, faster than an exhaustive search.

3.2   Attacking Rounds 28–79

A generic key recovery algorithm based on a rectangle distinguisher was presented by Biham *et al.* in [4] and then updated in [6] recently, which treats a block cipher $E : \{0,1\}^n \times \{0,1\}^k \rightarrow \{0,1\}^n$ as $E = E_f \circ E^1 \circ E^0 \circ E_b$, where E^0 and E^1 constitute the rectangle distinguisher, while E_b and E_f are some rounds before and after the rectangle distinguisher, respectively. In this subsection, we will use their results to break the 52 rounds from round 28 to 79 of SHACAL-1.

To apply the generic attack procedure [4], we need to determine the following six parameters:

- m_b: the number of subkey bits in E_b to be attacked.
- m_f: the number of subkey bits in E_f to be attacked.
- r_b: the number of bits that are active or can be active before the attacked round, given that a pair has the difference α at the entrance of the rectangle distinguisher.
- r_f: the number of bits that are active or can be active after the attacked round, given that a pair has the difference δ at the output of the rectangle distinguisher.
- 2^{t_b}: the number of possible differences before the attacked round, given that a pair has the difference α at the entrance of the rectangle distinguisher.
- 2^{t_f}: the number of possible differences after the attacked round, given that a pair has the difference δ at the output of the rectangle distinguisher.

Our attack is applied in the backward direction, that is to say, it is a chosen ciphertext attack. Anyway, as the data requirement of the attack is the entire code book, it can be easily used as a known plaintext attack.

Let E_b denote round 79, E^0 denote rounds 64 to 78, E^1 denote rounds 41 to 63, and E_f denote rounds 38 to 40. We first describe the two differentials to be used in this rectangle distinguisher. By cyclically rotating the last 23-round differential in the 24-round differential to the right by 9 bit positions, we can get a 23-round differential with probability $q = 2^{-49}$: $(e_{30}, e_{20}, 0, 0, 0) \rightarrow (e_{5,20}, e_{0,22}, e_{25}, e_{20}, 0)$. This 23-round differential is used in E^1, while the Kim et al.'s second differential with probability $p = 2^{-31}$ in [15] is used in E^0. Similarly, we can compute a lower bound probability $\widehat{q} = 2^{-47.77}$ for the 23-round differentials that only have variable output differences compared with the 23-round differential described above. As mentioned before, a lower bound $\widehat{p} = 2^{-30.28}$ has been got by only changing the first one or two rounds in the Kim et al.'s second differential. Therefore, this 38-round rectangle distinguisher holds at least a probability of $2^{-316.1} (\approx (2^{-47.77} \cdot 2^{-30.28})^2 \cdot 2^{-160})$ for the right key, while it holds probability 2^{-320} for a wrong key.

As we attack one round (i.e., round 79) before the distinguisher, we can compute m_b, r_b, and t_b as follows: There is only one 32-bit subkey K_{79} in E_b, therefore, $m_b = 32$. A pair with a difference $(e_{9,19,29,31}, e_{14,29}, e_{7,29}, e_2, e_{29})$ before round 79 has a difference with the form $(R, e_{9,19,29,31}, e_{12,27}, e_{7,29}, e_2)$ after round 79. Obviously, the bit differences in the three least significant bits of R will definitely be 0, while the bit differences in the other 29 bit positions will be variable. As a result, $r_b = 29 + 4 + 2 + 2 + 1 = 38$. In our analysis, R has exactly 15648 possible values. So, $t_b = log_2^{15648} \approx 13.9$.

There are three rounds (i.e., rounds 38 to 40) after the distinguisher, thus $m_f = 96$. A pair that has a difference $(e_{30}, e_{20}, 0, 0, 0)$ before round 41 has a difference with the form $(e_{20}, 0, 0, 0, S)$ before round 40, where S has the following 12 possible values: $e_{25,30}$, $e_{25,30,31}$, $e_{25,26,30}$, $e_{25,26,30,31}$, $e_{25,26,27,30}$, $e_{25,26,27,30,31}$, $e_{25,26,27,28,30}$, $e_{25,26,27,28,30,31}$, $e_{25,26,27,28,29,30}$, $e_{25,26,27,28,29,30,31}$, $e_{25,26,27,28,29,31}$, $e_{25,26,27,28,29}$. These differences can be reached from a difference with the form $(0, 0, 0, S, T)$ before round 39, where T has bits 20 to 31 active, of which bits 21 to 24 must take one of the five possible values $1_x, 3_x, 7_x, F_x$, and $1F_x$ according to the carry, while bits 25 to 31 cannot be predicted as they all depend on the exact value of S. This set of differences can be caused by differences with the form $(0, 0, S, T, U)$ before round 38, where U has bits 20 to 31 active. Thus, $r_f = 7 + 12 + 12 = 31$, and there are at most $12 \cdot (5 \cdot 2^7) \cdot 2^{12} = 31457280$ possible differences with the form $(0, 0, S, T, U)$ before round 38, so $t_f = log_2(31457280) \approx 24.9$.

Assigning these parameters to the Biham et al.'s generic attack procedure leads to a rectangle attack on rounds 38 to 79. Then, with an exhaustive key search for the remaining 10 rounds, we can attack 52-round SHACAL-1. The attack procedure is summarized as follows.

Attack Procedure

(a) Based on the above 38-round rectangle distinguisher, apply the Biham et al.'s generic attack procedure [6] on the 42 rounds from round 38 to 79 of

SHACAL-1. Output the four 32-bit subkey candidates for rounds 38, 39, 40 and 79 with the maximal counter number.

(b) Find the ten 32-bit subkeys for rounds 28 to 37 using an exhaustive search.

According to [6], the time complexity of Step (a) in our attack is about $2^{m_b+m_f+1} + N + N^2 \cdot (2^{r_f-n-1} + 2^{t_f-n-1} + 2^{2t_f+2r_b-2n-3} + 2^{m_b+t_b+2t_f-2n-2} + 2^{m_f+t_f+2t_b-2n-2}) = 2^{129} + 2^{160} + 2^{320} \cdot (2^{31-161} + 2^{24.9-161} + 2^{2\cdot24.9+2\cdot38-323} + 2^{32+13.9+2\cdot24.9-322} + 2^{96+24.9+2\cdot13.9-322}) \approx 2^{190.02}$ memory accesses. In Step (b), by guessing the subkeys of rounds 28 to 37, it is possible to partially encrypt all the plaintexts and then apply the previous Step (a). Each subkey guess requires 2^{160} partial encryptions and $2^{190.02}$ memory accesses, therefore, the total time complexity is $2^{320} \cdot 2^{160} \cdot \frac{10}{52} \approx 2^{477.6}$ 52-round SHACAL-1 encryptions and $2^{320} \cdot 2^{190.02} = 2^{510.02}$ memory accesses.

Note: There exists another attack on the 52 rounds from round 28 to 79, which is composed of a similar rectangle attack on rounds 35 to 77, followed by an exhaustive search on the 288-bit subkeys of rounds 28 to 34, 78 and 79. Let E_b denote round 77, E^0 denote rounds 64 to 76, E^1 denote rounds 38 to 63, and E_f denote rounds 35 to 37. For E^0 we use the 13-round differential composed of rounds 23 to 35 in the second differential of [15], which holds probability $p = 2^{-24}$. The 26-round differential $(0, 0, e_{19,24}, e_{14,19,24}, e_{14}) \rightarrow (e_{14,31}, e_{16,26}, e_{19}, e_{14}, 0)$ with probability $q = 2^{-55}$ is used in E^1, which is obtained by cyclically rotating the 24-round differential to the left by 17 bit positions and appending two more rounds before the input. We computed a lower bound on the related probabilities $\widehat{p} = 2^{-23.48}$ and $\widehat{q} = 2^{-53.77}$. Therefore, the distinguisher holds at least a probability of $2^{-314.5} (\approx (2^{-53.77} \cdot 2^{-23.48})^2 \cdot 2^{-160})$ for the right key, while it holds probability 2^{-320} for a wrong key. As before we computed that $m_b = 32$, $r_b = 38$, $t_b = 13.9$, $m_f = 96$, $r_f = 12+17+18 = 47$, and $t_f = log_2^{\prime}9\cdot64\cdot12\cdot2^{13}\cdot2^{18}) \approx 43.8$. Finally, we can break 52-round SHACAL-1. According to [6], the data complexity is $N = 2^{\frac{n}{2}+2}/(\widehat{p} \cdot \widehat{q}) = 2^{80+53.77+23.48+2} = 2^{159.25}$ chosen plaintexts/ciphertexts with difference $(e_{9,19,29,31}, e_{14,29}, e_{7,29}, e_2, e_{29})$ before round 76, however, this cannot be guaranteed if we start with chosen ciphertexts. Alternatively, we apply the attack as a known plaintext attack. With $2^{159.625}$ known plaintexts, we can get $2^{318.25}$ pairs, of which about $2^{158.25} (= 2^{318.25} \cdot 2^{-160})$ would have the desired difference. This attack requires $2^{288} \cdot 2^{159.625} \cdot \frac{9}{52} \approx 2^{445.1}$ encryptions and the time complexity is about $2^{288} \cdot [2^{m_b+m_f+1} + N + N^2 \cdot (2^{r_f-n-1} + 2^{t_f-n-1} + 2^{2t_f+2r_b-2n-3} + 2^{m_b+t_b+2t_f-2n-2} + 2^{m_f+t_f+2t_b-2n-2})] = 2^{288} \cdot [2^{129} + 2^{159.25} + 2^{318.5} \cdot (2^{-114} + 2^{-117.2} + 2^{-157.4} + 2^{-185} + 2^{-147.4}) \approx 2^{204.65}] = 2^{492.65}$ memory accesses.

4   Differential Attacks on Reduced-Round SHACAL-1

The 24-round differential in Table 1 can be extended to a 30-round differential $(e_{29}, 0, 0, 0, e_{2,7}) \rightarrow (e_{0,4,12,17,24,25,27,29}, e_{7,17,19,31}, e_{0,5,15,27,30}, e_{5,17,25,27,29}, e_{2,5,22,27})$ with probability 2^{-93}, which has a significantly higher probability than the longest currently known (30-round) differential with probability 2^{-138} due to Kim *et al.*. More importantly, it can be extended to as long as a 34-round differential $(e_{29}, 0, 0, 0, e_{2,7}) \rightarrow (e_{0,5,7,12,13,15,17,20,28,29}, e_{5,7,9,23,25,29}, e_{3,12,15,18,20,25,27,30}, e_{5,7,13,15,17,23,25,29}, e_{2,10,15,22,23,25,27,30})$ with probability 2^{-148}.

These differentials with different rounds can be used to attack different reduced round variants of SHACAL-1. Here, we just present the differential attack on SHACAL-1 with 512 key bits based on the 34-round differential.

4.1 Attacking Rounds 15–69

The 34-round differential can be applied to the 34 rounds from round 40 to 73, due to the differential distribution of the two functions f_{if} and f_{maj}. Then, by appending 10 more rounds before round 40 and removing the last 4 rounds in the above 34-round differential, we exploit a 40-round differential characteristic with probability 2^{-154} for rounds 30 to 69: $(e_{4,8,11,13,16}, e_{3,8,11,13,31}, e_{1,6,11,16,21,29,31}, e_{1,4,8,11,13,16,21}, e_{3,9,11,13,16,18,21,29,31}) \rightarrow (e_{0,4,12,17,24,25,27,29}, e_{7,17,19,31}, e_{0,5,15,27,30}, e_{5,17,25,27,29}, e_{2,5,22,27})$.

This 40-round differential can be used to mount a chosen ciphertext attack on the 55 rounds from round 15 to 69. By counting the 30 possible 40-round differentials that only have variable input differences $(e_{4,8,11,13,16}, e_{3,8,11,13,31}, e_{1,6,11,16,21,29,31}, e_{1,4,8,11,13,16,21}, \Delta E_{30})$ compared with the 40-round differential described above (where ΔE_{30} are shown in Table 2), we can conclude these 40-round differentials hold a lower bound probability $2^{-150} (= 2 \cdot 2^{-154} + 28 \cdot 2^{-155})$ for a right key, while they hold a probability of $2^{-155.09} (\approx 30 \cdot 2^{-160})$ for a wrong key. Consequently, we can break the 55-round SHACAL-1 as follows.

Table 2. Possible input differences ΔE_{30} in Round 30 with their respective probabilities

Prob.	ΔE_{30}
2^{-154}	$e_{3,9,11,13,16,18,21,29,31}$, $e_{3,4,9,11,13,16,18,21,29,31}$
2^{-155}	$e_{3,4,5,9,11,13,16,18,21,29,31}$, $e_{3,5,9,11,13,16,18,21,29,31}$, $e_{3,5,6,9,11,13,16,18,21,29,31}$,
	$e_{3,4,5,6,9,11,13,16,18,21,29,31}$, $e_{3,7,9,11,13,16,18,21,29,31}$, $e_{3,4,7,9,11,13,16,18,21,29,31}$,
	$e_{3,9,10,11,13,16,18,21,29,31}$, $e_{3,4,9,10,11,13,16,18,21,29,31}$, $e_{3,9,10,13,16,18,21,29,31}$,
	$e_{3,4,9,10,13,16,18,21,29,31}$, $e_{3,9,11,12,13,16,18,21,29,31}$, $e_{3,4,9,11,12,13,16,18,21,29,31}$,
	$e_{3,9,11,12,16,18,21,29,31}$, $e_{3,4,9,11,12,16,18,21,29,31}$, $e_{3,9,11,13,14,16,18,21,29,31}$,
	$e_{3,4,9,11,13,14,16,18,21,29,31}$, $e_{3,9,11,13,16,17,18,21,29,31}$, $e_{3,4,9,11,13,16,17,18,21,29,31}$,
	$e_{3,9,11,13,16,17,21,29,31}$, $e_{3,4,9,11,13,16,17,21,29,31}$, $e_{3,9,11,13,16,18,19,21,29,31}$,
	$e_{3,4,9,11,13,16,18,19,21,29,31}$, $e_{3,9,11,13,16,18,21,22,29,31}$, $e_{3,4,9,11,13,16,18,21,22,29,31}$,
	$e_{3,9,11,13,16,18,21,29,30,31}$, $e_{3,4,9,11,13,16,18,21,29,30,31}$, $e_{3,9,11,13,16,18,21,29,31}$,
	$e_{3,4,9,11,13,16,18,21,29,30}$

Attack Procedure

1. Choose 2^{153} pairs of ciphertexts with difference $(e_{0,4,12,17,24,25,27,29}, e_{7,17,19,31}, e_{0,5,15,27,30}, e_{5,17,25,27,29}, e_{2,5,22,27})$: (C_i, C_i'), for $i = 1, \cdots, 2^{153}$. Decrypt them to get their corresponding plaintext pairs (P_i, P_i').

2. Guess a 352-bit key K_f for rounds 15 to 25, do follows,

    2.1 Partially encrypt each pair (P_i, P_i') using K_f to get their intermediate values just after round 25: $(E_{K_f}(P_i), E_{K_f}(P_i'))$. Then, check if the 32-bit difference ΔA_{26} in $E_{K_f}(P_i) \oplus E_{K_f}(P_i')$ belongs to $\{ROT_2(\Delta E_{30}) | \Delta E_{30}$ are those in Table 2$\}$. If the number of the pairs (P_i, P_i') passing this test is greater than or equal to 6, then record K_f and all the qualified pairs (P_i, P_i') and go to Step 2.2; Otherwise, repeat this step with another K_f.

    2.2 Guess a 32-bit subkey K_{26} for round 26, then partially encrypt each pair $(E_{K_f}(P_i), E_{K_f}(P_i'))$ with K_{26} to get their intermediate values just after round 26. We denote these values by (X_i, X_i'). Finally, check if the 64-bit difference $(\Delta A_{27}, \Delta B_{27})$ in $X_i \oplus X_i'$ belongs to $\{(e_{3,6,10,13,15,18,23}, ROT_2 (\Delta E_{30}))\}$. If the number of the pairs $(E_{K_f}(P_i), E_{K_f}(P_i'))$ passing this test is greater than or equal to 6, then record (K_f, K_{26}) and all the qualified pairs (X_i, X_i') and go to Step 2.3; Otherwise, repeat this step with another K_{26}.

    2.3 Guess a 32-bit subkey K_{27} for round 27, then partially encrypt each remaining pair (X_i, X_i') with K_{27} to get their intermediate values just after round 27. We denote them by $(\overline{X}_i, \overline{X}_i')$. Finally, check if the 96-bit difference $(\Delta A_{28}, \Delta B_{28}, \Delta C_{28})$ in $\overline{X}_i \oplus \overline{X}_i'$ belongs to the set $\{(e_{1,3,8,13,18,23,31}, e_{3,6,10,13,15,18,23}, ROT_2(\Delta E_{30}))\}$. If the number of the pairs (X_i, X_i') passing this test is greater than or equal to 6, then record (K_f, K_{26}, K_{27}) and all the qualified pairs $(\overline{X}_i, \overline{X}_i')$ and go to Step 2.4; Otherwise, repeat this step with another K_{27}.

    2.4 Guess a 32-bit subkey K_{28} for round 28, then partially encrypt each remaining pair $(\overline{X}_i, \overline{X}_i')$ with K_{28} to get their intermediate values just after round 28. We denote them by $(\widehat{X}_i, \widehat{X}_i')$. Finally, check if the 128-bit difference $(\Delta A_{29}, \Delta B_{29}, \Delta C_{29}, \Delta D_{29})$ in $\widehat{X}_i \oplus \widehat{X}_i'$ belongs to $\{(e_{3,8,11,13,31}, e_{1,3,8,13,18,23,31}, e_{3,6,10,13,15,18,23}, ROT_2(\Delta E_{30}))\}$. If the number of the pairs $(\overline{X}_i, \overline{X}_i')$ passing this test is greater than or equal to 6, then record $(K_f, K_{26}, K_{27}, K_{28})$ and all the qualified pairs $(\widehat{X}_i, \widehat{X}_i')$ and go to Step 2.5; Otherwise, repeat this step with another K_{28}.

    2.5 Guess a 32-bit subkey K_{29} for round 29, then partially encrypt each remaining pair $(\widehat{X}_i, \widehat{X}_i')$ with K_{29}, and finally check if the 160-bit difference $E_{K_{29}}(\widehat{X}_i) \oplus E_{K_{29}}(\widehat{X}_i')$ belongs to $\{(e_{4,8,11,13,16}, e_{3,8,11,13,31}, e_{1,3,8,13,18,23,31}, e_{3,6,10,13,15,18,23}, ROT_2(\Delta E_{30}))\}$. If the number of the pairs $(\widehat{X}_i, \widehat{X}_i')$ passing this test is greater than or equal to 6, then record $(K_f, K_{26}, K_{27}, K_{28}, K_{29})$; Otherwise, repeat Step 2 with another 352-bit key.

3. For a suggested $(K_f, K_{26}, K_{27}, K_{28}, K_{29})$, do an exhaustive search for the remaining 32 key bits using trial encryption. Four known pairs of plaintexts and ciphertexts are enough for this trial process. If a 512-bit key is suggested, output it as the master key of the 55-round SHACAL-1; Otherwise, repeat Step 2 with another 352-bit key.

This attack requires 2^{154} chosen plaintexts. The memory for this attack is also dominated by the ciphertext pairs, so it requires about $2^{154} \cdot 20 \approx 2^{158.32}$ memory bytes.

The time complexity of Step 1 is 2^{154} 55-round SHACAL-1 encryptions; The time complexity of Step 2.1 is dominated by the partial decryptions, which is about $2^{352} \cdot 2^{154} \cdot \frac{11}{55} \approx 2^{503.68}$. In Step 2.1, since the probability that a pair meets the filtering condition in this step is $\frac{30}{2^{32}} \approx 2^{-27.09}$, the expected number of the pairs passing the test for each subkey candidate is $2^{153} \cdot 2^{-27.09} \approx 2^{125.91}$, and the probability that the number of pairs passing this test for a wrong subkey is no less than 6 is about $\sum_{i=6}^{2^{153}} \left(\binom{2^{153}}{i} \cdot (2^{-27.09})^i \cdot (1 - 2^{-27.09})^{2^{153}-i} \right) \approx 1$. Thus, almost all the 2^{352} subkeys pass through Step 2.1. In Step 2.2, the time complexity is about $2^{352} \cdot 2^{32} \cdot 2^{125.91} \cdot 2 \cdot \frac{1}{55} \approx 2^{505.13}$. In this step, since the probability that a remaining pair meets the filtering condition in this step is 2^{-32}, the expected number of the pairs passing the test for each subkey candidate is $2^{125.91} \cdot 2^{-32} \approx 2^{93.91}$, and the probability that the number of pairs passing the test for a wrong subkey is no less than 6 is about $\sum_{i=6}^{2^{125.91}} \left(\binom{2^{125.91}}{i} \cdot (2^{-32})^i \cdot (1 - 2^{-32})^{2^{125.91}-i} \right) \approx 1$. Thus, almost all the 2^{384} subkeys pass through Step 2.2. Similarly, we can get that the time complexity in either of Step 2.3, 2.4 and 2.5 is also $2^{505.13}$; Besides, almost all the 2^{448} subkeys pass through Step 2.4, and the expected number of the pairs passing the test in Step 2.4 for each subkey candidate is $2^{93.91} \cdot 2^{-32 \times 2} \approx 2^{29.91}$. In Step 2.5, since the probability that a remaining pair meets the filtering condition in this step is also 2^{-32}, the expected number of the pairs passing the test for each subkey candidate is $2^{29.91} \cdot 2^{-32} \approx 2^{-2.09}$, and the probability that the number of pairs passing the test for a wrong subkey is no less than 6 is about $\sum_{i=6}^{2^{29.91}} \left(\binom{2^{29.91}}{i} \cdot (2^{-32})^i \cdot (1 - 2^{-32})^{2^{29.91}-i} \right) \approx 2^{-22.03}$. Thus, on average, about $2^{448} \cdot 2^{32} \cdot 2^{-22.03} \approx 2^{457.97}$ subkeys pass through Step 2.5, which result in $2^{457.97} \cdot 2^{32} \approx 2^{489.97}$ encryptions in Step 3. Therefore, this attack totally requires about $2^{154} + 2^{503.68} + 4 \cdot 2^{505.13} + 2^{489.97} \approx 2^{507.26}$ encryptions.

Since the probability that a wrong 512-bit key is suggested in Step 3 is about $2^{-640} (= 2^{-160.4})$, the expected number of suggested wrong 512-bit keys is about $2^{-640} \cdot 2^{489.97} \approx 2^{-150.03}$, which is extremely low. The expected number of the pairs passing the test in Step 2.5 for the right key is $8 \ (= 2^{153} \cdot 2^{-150})$ and the probability that the number of the pairs passing the test in Step 2.5 for the right subkey is no less than 6 is about $\sum_{i=6}^{2^{153}} \left(\binom{2^{153}}{i} \cdot (2^{-150})^i \cdot (1 - 2^{-150})^{2^{153}-i} \right) \approx 0.8$. Therefore, with a probability of 0.8, we can break the 55-round SHACAL-1 with 512 key bits by using the differential attack.

4.2   Attacking Rounds 0–48

We can learn that the 64 possible 34-round differentials that have only variable output differences $(e_{0,5,7,12???,17,20,28???}, e_{5,7,9,23,25,29}, e_{3,12,15,18,20,25,27,30},$ $e_{5,7,13,15,17,23,25,29}, e_{2,10,15,22,23,25,27,30})$ compared with the one described earlier hold a probability of $2^{-138} (= 64 \cdot 2^2 \cdot 2^{-148})$ for the right key, and hold a probability of $2^{-154} (= 64 \cdot 2^{-160})$ for a wrong key, where "$i???$" $(i = 12, 28)$ means that the bit in i position takes 1 and each of the three bits in $i+1$, $i+2$ and

$i + 3$ positions takes an arbitrary value from $\{0, 1\}$. Similarly, using 2^{141} pairs of plaintexts with difference $(e_{29}, 0, 0, 0, e_{2,7})$ and such four fixed bits as described in Section 3.1, the attack requires about $2^{146.32}(\approx 2^{142} \cdot 20)$ memory bytes and $2^{496.45}(\approx 2^{352} \cdot 2^{142} \cdot \frac{11}{49} + 4 \cdot 2^{384} \cdot 2^{141} \cdot \frac{64}{2^{32}} \cdot 2 \cdot \frac{1}{49})$ encryptions.

5 Conclusions

In this paper, we exploit some better rectangle distinguishers and differential characteristics than those previously known in SHACAL-1. Based on them, we finally mount rectangle attacks on the first 51 rounds and a series of inner 52 rounds of SHACAL-1, and mount differential attacks on the first 49 rounds and a series of inner 55 rounds of SHACAL-1. These are the best currently known cryptanalytic results on SHACAL-1 in an one key attack scenario.

Acknowledgments

The authors are very grateful to Jiqiang Lu's supervisor Prof. Chris Mitchell for his valuable editorial comments and to the anonymous referees for their comments. Jiqiang Lu would like to thank Prof. Eli Biham for his help.

References

1. E. Biham, New types of cryptanalytic attacks using related keys, Advances in Cryptology — EUROCRYPT'93, T. Helleseth (ed.), Volume 765 of Lecture Notes in Computer Science, pp. 398–409, Springer-Verlag, 1993.
2. E. Biham and A. Shamir, Differential cryptanalysis of the Data Encryption Standard, Springer-Verlag, 1993.
3. E. Biham, O. Dunkelman, and N. Keller, The rectangle attack — rectangling the Serpent, Advances in Cryptology — EUROCRYPT'01, B. Pfitzmann (ed.), Volume 2045 of Lecture Notes in Computer Science, pp. 340–357, Springer-Verlag, 2001.
4. E. Biham, O. Dunkelman, and N. Keller, New results on boomerang and rectangle attacks, Proceedings of FSE'02, J. Daemen and V. Rijmen (eds.), Volume 2365 of Lecture Notes in Computer Science, pp. 1–16, Springer-Verlag, 2002.
5. E. Biham, O. Dunkelman, and N. Keller, Rectangle attacks on 49-round SHACAL-1, Proceedings of FSE'03, T. Johansson (ed.), Volume 2887 of Lecture Notes in Computer Science, pp. 22–35, Springer-Verlag, 2003.
6. O. Dunkelman, Techniques for cryptanalysis of block ciphers, Ph.D dissertation of Technion, 2006. Available at *http://www.cs.technion.ac.il/users/wwwb/cgi-bin/tr-info.cgi?2006/PHD/PHD-2006-02*
7. O. Dunkelman, N. Keller, and J. Kim, Related-key rectangle attack on the full SHACAL-1, Proceedings of SAC'06, to appear in Lecture Notes in Computer Science, Springer-Verlag, 2006.
8. H. Handschuh, L. R. Knudsen, and M. J. Robshaw, Analysis of SHA-1 in encryption mode, Proceedings of CT-RSA'01, D. Naccache (ed.), Volume 2020 of Lecture Notes in Computer Science, pp. 70–83, Springer-Verlag, 2001.

9. H. Handschuh and D. Naccache, SHACAL, Proceedings of The First Open NESSIE Workshop, 2000. Available at *https://www.cosic.esat.kuleuven.be/nessie/workshop/submissions.html*

10. H. Handschuh and D. Naccache, SHACAL, NESSIE, 2001. Available at *https://www.cosic.esat.kuleuven.be/nessie/tweaks.html*

11. S. Hong, J. Kim, S. Lee, and B. Preneel, Related-key rectangle attacks on reduced versions of SHACAL-1 and AES-192, Proceedings of FSE'05, H. Gilbert and H. Handschuh (eds.), Volume 3557 of Lecture Notes in Computer Science, pp. 368–383, Springer-Verlag, 2005.

12. J. Kelsey, T. Kohno, and B. Schneier, Amplified boomerang attacks against reduced-round MARS and Serpent, Proceedings of FSE'00, B. Schneier (ed.), Volume 1978 of Lecture Notes in Computer Science, pp. 75–93, Springer-Verlag, 2001

13. J. Kelsey, B. Schneier, and D. Wagner, Key-schedule cryptanalysis of IDEA, G-DES, GOST, SAFER, and Triple-DES, Advances in Cryptology — CRYPTO'96, N. Koblitz (ed.), Volume 1109 of Lecture Notes in Computer Science, pp. 237–251, Springer–Verlag, 1996.

14. J. Kim, G. Kim, S. Hong, S. Lee, and D. Hong, The related-key rectangle attack — application to SHACAL-1, Proceedings of ACISP'04, H. Wang, J. Pieprzyk, and V. Varadharajan (eds.), Volume 3108 of Lecture Notes in Computer Science, pp. 123–136, Springer-Verlag, 2004.

15. J. Kim, D. Moon, W. Lee, S. Hong, S. Lee, and S. Jung, Amplified boomerang attack against reduced-round SHACAL, Advances in Cryptology — ASIACRYPT'02, Y. Zheng (ed.), Volume 2501 of Lecture Notes in Computer Science, pp. 243–253, Springer-Verlag, 2002.

16. H. Lipmaa and S. Moriai, Efficient algorithms for computing differential properties of addition, Proceedings of FSE'01, M. Matsui (ed.), Volume 2355 of Lecture Notes in Computer Science, pp. 336–350, Springer-Verlag, 2001.

17. M. Matsui, Linear cryptanalysis method for DES cipher, Advances in Cryptology — EUROCRYPT'93, T. Helleseth (ed.), Volume 765 of Lecture Notes in Computer Science, pp. 386–397, Springer-Verlag, 1994.

18. NESSIE, *https://www.cosic.esat.kuleuven.be/nessie/*

19. J. Nakahara Jr, The statistical evaluation of the NESSIE submission, 2001.

20. U.S. Department of Commerce, Secure Hash Standard FIPS 180-1, N.I.S.T., 1995.

21. D. Wagner, The boomerang attack, Proceedings of FSE'99, L. Knudsen (ed.), Volume 1636 of Lecture Notes in Computer Science, pp. 156–170, Springer-Verlag, 1999.

Algebraic Attacks on Clock-Controlled Cascade Ciphers

Kenneth Koon-Ho Wong[1], Bernard Colbert[2], Lynn Batten[2],
and Sultan Al-Hinai[1]

[1] Information Security Institute (ISI), Queensland University of Technology (QUT),
Brisbane, Australia
[2] Deakin University, Melbourne, Australia

Abstract. In this paper, we mount the first algebraic attacks against
clock controlled cascade stream ciphers. We first show how to obtain
relations between the internal state bits and the output bits of the Gollmann clock controlled cascade stream ciphers. We demonstrate that the
initial states of the last two shift registers can be determined by the
initial states of the others. An alternative attack on the Gollmann cascade is also described, which requires solving quadratic equations. We
then present an algebraic analysis of Pomaranch, one of the phase two
proposals to eSTREAM. A system of equations of maximum degree four
that describes the full cipher is derived. We also present weaknesses in
the filter functions of Pomaranch by successfully computing annihilators
and low degree multiples of the functions.

1 Introduction

Algebraic attacks, in which the initial states are solved for as a system of multivariate polynomial equations derived from the target cipher, were introduced by
Courtois and Meier in [10,12] as a new method of analyzing cipher output. This
method of attack was first applied to block ciphers and public key cryptosystems
by Courtois and Pieprzyk [9,15]. Many *regularly* clocked linear feedback shift register (LFSR) based stream ciphers have since then fallen under algebraic attacks,
as demonstrated in [3,4,7,15,11], whereas *irregularly* clocked stream ciphers have
been more resistant. There are, to our knowledge, only two papers in the literature dealing with algebraic attacks on irregularly clocked stream ciphers, first
in [12] and then in [1], dealing with separate classes of clock controlled stream
ciphers. Our interest in this paper is in extending algebraic attacks to a third
class of clock controlled stream ciphers that has not yet been examined under
algebraic attacks — the clock controlled LFSR-based *cascade* stream ciphers.

In an LFSR based clock control cascade cipher, the output of the first LFSR
controls the clocking of a second LFSR, both outputs together control the clocking of a third LFSR, and so on. In this paper, we present algebraic analyses
of two such ciphers, the Gollmann cascade generator [17] and Pomaranch, an
eSTREAM project candidate [19,21].

The idea of cascading a set of LFSRs was due to Gollmann [17] and was further studied by Chambers and Gollmann in 1988 [18]. In the latter study, they

R. Barua and T. Lange (Eds.): INDOCRYPT 2006, LNCS 4329, pp. 32–47, 2006.

conclude that better security is achieved with a large number of short LFSRs instead of a small number of long ones. Park, Lee and Goh [24], having extended the attack of Menicocci on a 2-register cascade using statistical techniques [23], successfully broke 9-register cascades where each register has fixed length 100. They suggested also that 10-register cascades might be insecure. In 1994, Chambers [6] proposed a clock-controlled cascade cipher in which each 32-bit portion of the output sequence of each LFSR passes through an invertible s-box with the result being used to clock the next register. Several years later, the idea of cascade ciphers resurfaced in a proposal by Jansen, Helleseth and Kholosha [19] to the 2005 SKEW workshop, which became the eSTREAM candidate Pomaranch. Pomaranch can be viewed as a variant of the Gollmann cascade in which a number of bits from each register are filtered using a nonlinear function, and the result is used to control the clocking of the next register.

In the case of the Gollmann cascade, the key is the combined initial states of the registers, and the keystream output is the output of the final register. Pomaranch uses an initialization vector for key loading and its keystream output is the sum of certain bits taken from each of the registers. In subsequent sections of this paper, we present our algebraic attacks on the Gollmann cascade generator. This leads us into our algebraic analysis of Pomaranch, where the cipher construction is more complicated than the Gollmann cascade. Unless otherwise specified, additions and multiplications presented in this paper are defined over $GF(2)$.

2 Clock-Controlled Gollmann Cascade Generator

The Gollmann cascade generator, introduced in [17], employs k LFSRs arranged serially such that each register except for the first one is clock-controlled by an input bit, which is the sum of the output bits of its predecessors. This structure is shown in Figure 1. Initially, all registers are filled independently with key bits. Let the input bit to the i-th register at time t be a_i^t, for $i \geq 2$. The i-th register is clocked if and only if $a_i^t = 1$. The output bit of the i-th register is then added to a_i^t, and the result becomes a_{i+1}^t. The keystream output of the generator is the output of the k-th LFSR.

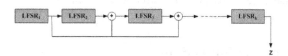

Fig. 1. The Gollmann Cascade Generator

Gollmann first proposed cascading k cyclic registers of the same prime length p with feedback polynomial $f(x) = x^p + 1$. This is known as the p-cycle. A variation of the Gollmann cascade, called an m-sequence cascade, has the cyclic registers replaced by maximum length LFSRs of the same length l. We will algebraically analyse this type of Gollmann cascade generator with registers of variable length, and present an attack showing how we can recover the initial states of the registers.

2.1   Algebraic Attack on the Gollmann Cascade Generator

In this section, we present a basic algebraic analysis of the clock-controlled Gollmann cascade generator. We first consider a cascade generator of $k = 4$ LFSRs of any lengths. However, the analysis presented here can be adapted to a cascade with any number of registers. For this particular case, we have LFSRs A, B, C, D of lengths l, m, n, r respectively. Let, for example, A_i^t be the i-th bit of register A at time t. $A_l^t, B_m^t, C_n^t, D_r^t$ are then the outputs of the respective registers. A is clocked regularly as a traditional LFSR. The states of the other LFSRs at time t can be expressed as follows.

$$B_i^t = B_i^{t-1}(A_l^{t-1} + 1) + B_{i-1}^{t-1}A_l^{t-1} \tag{1}$$
$$C_i^t = C_i^{t-1}(A_l^{t-1} + B_m^{t-1} + 1) + C_{i-1}^{t-1}(A_l^{t-1} + B_m^{t-1}) \tag{2}$$
$$D_i^t = D_i^{t-1}(A_l^{t-1} + B_m^{t-1} + C_n^{t-1} + 1) + D_{i-1}^{t-1}(A_l^{t-1} + B_m^{t-1} + C_n^{t-1}), \tag{3}$$

where B_0^t, C_0^t, D_0^t are the feedback sum of the corresponding registers. Since the keystream of the generator is given by the output of the final register, we have

$$z^t = D_r^t. \tag{4}$$

Using (3) with $i = r$ and (4) with $t + 1$ instead of t we get

$$z^{t+1} = z^t(A_l^t + B_m^t + C_n^t + 1) + D_{r-1}^t(A_l^t + B_m^t + C_n^t). \tag{5}$$

which can be expressed as

$$z^t + z^{t+1} = (z^t + D_{r-1}^t)(A_l^t + B_m^t + C_n^t) \tag{6}$$

From (6), we see that if $z^t + z^{t+1} = 1$, then

$$1 = A_l^t + B_m^t + C_n^t. \tag{7}$$

We will make use of this observation to recover the initial state bits in our attack presented in the next section.

2.2   Recovering the Initial State

Based on the observations in the previous section, we propose recovering the internal state of all registers as follows. Consider again the case where $k = 4$. We know that whenever $z^t + z^{t+1} = 1$, then $A_l^t + B_m^t + C_n^t = 1$. Therefore, we can start by guessing the initial states of A, B, which will enable us generate linear equations (2) for C with the substitution $C_n^t = A_l^t + B_m^t + 1$ when $z^t + z^{t+1} = 1$. Once we have enough linearly independent equations we can then solve for the initial states of C. Using these values of C, we can then recover the internal state of register D by solving the linear equations (3) by substituting the values for A, B, C and the keystream. If we obtain a solution for both C and D, then we have found the initial states of all registers. The complexity of this approach is given by the complexity of guessing A and B multiplied by the complexity of solving linear equations in the other registers. Gaussian elimination can be in general performed in $O(n^\omega)$ operations, with $2 \leq \omega \leq 3$. For simplicity, throughout this paper we assume the worst case with $\omega = 3$. The time complexity of this attack is then $2^{l+m}(n^3 + r^3)$. The minimum keystream requirement for recovering C is $2n$ on average, since we only get an equation if $z^t + z^{t+1} = 1$, and that for recovering D is r. The same keystream can be used to construct equations for C and D, so the attack needs $\max(2n, r)$ bits of keystream. In practice we might need a small percentage more than this requirement due to linear dependencies among the linear equations generated. For a cascade cipher with k LFSRs, the complexity of the presented approach will be $2^u(l_1^3 + l_2^3)$, where l_1, l_2 are the length of the last two registers and u is the total length of the remaining registers. Clearly, the proposed attack is better than exhaustive key search, which has complexity $2^{u+l_1+l_2}$.

2.3   Comparison of Attacks on the Gollmann Cascade

As far as we are aware, there are four published attacks on the Gollmann cascade generator. These are the lock-in effect attack by Gollmann and Chambers [5], the attack by Menococci [23], the clock control guessing attack by Zenner [26], and the attack by Park et. al. [24]. The complexities of the lock-in effect attack and the Menicocci attack are far higher than ours, so here we will only make comparison of our attack with the ones that are more effective. The clock control guessing attack has a relatively closer complexity and has a similar approach to ours. Zenner applies linear consistency tests to the Gollmann cascade using a technique of guessing the clock control bits resulting in a search tree. This attack is similar to ours in that it forms a set of equations after guessing a number of bits. It then solves the equations discarding those that are inconsistent. The complexity of the attack is of order $2^{(l+m+n+r)/2}(l + m + n + r)^3$.

The only attack that outperforms ours is the attack by Park et. al. [24]. This attack is based on guessing the initial states of each register successively in the cascade with some desired probability. Their analysis starts by building a matrix of conditional probabilities between the inputs and outputs of a register. This matrix is used to determine the probabilities of particular outputs of the cipher given particular inputs to the registers. These probabilities are biased when a run of zeros or ones occurs, yielding an efficient algorithm for finding the initial states of the registers by scanning the given keystream for runs of at least u consecutive zeros or ones, where u is determined by the desired error rate of the algorithm.

The algorithm builds linear equations in the unknowns of the initial state bits of the first register of the cascade. Random equations from these are solved until a solution with high probability is found. The algorithm then uses this solution to build equations for the next register. This process is repeated until the initial states of every register are recovered except for the last. Finally, the intial states of the last register are recovered by using previously known techniques. The theoretical complexity of this attack is not given in the paper, so we have estimated it from the algorithm and the experimental data. Gaussian elimination is present, so $O(l^3)$ is used for the asymptotic time complexity. From the experimental data we deduce that $O(l^2)$ bits of keystream are required to find the desired run of zeros or ones.

Although Park's attack is very efficient, it does have some limitations. Park's attack requires consecutive keystream bits with runs of zeros or ones, but our attack does not even require consecutive keystream bits. We can use any subset of the keystream as long as the equations describing them are dependent on the values of all of the initial state bits. Our keystream requirement is also much less than that of Park's attack. Hence, our attack would be the only one that is feasible for implementations where rekeying occurs frequently such as the frame based communication systems for mobile telephones and wireless networks. Table 1 summarizes the abovementioned attacks on the Gollmann cascade generator for $k = 4$.

Table 1. Best known attacks on Gollmann cascade generator

Attack	Minimum keystream required (MK)	Time complexity (TC)	MK $l = m = n$ $= r = 64$	TC $l = m = n$ $= r = 64$
Park [24]	$6l^2$	$36l^3$	2^{15}	2^{24}
Lock-in-effect [5]	$4l^2$	$\frac{16}{l^2}(2^l - 2)^4$	2^{14}	2^{248}
Clock control guessing [26]	$l + m + n + r$	$2^{(l+m+n+r)/2}(l + m + n + r)^3$	2^8	2^{152}
Ours	$\max(2n, r)$	$2^{l+m}(n^3 + r^3)$	2^7	2^{147}

2.4 An Alternative Algebraic Attack

In this section we describe another algebraic attack on the Gollmann cascade. The resistance of clock-controlled stream ciphers to traditional algebraic attacks is due to the fact that clock controls cause increases in the degrees of the equations generated. In the case of the Gollmann cascade, we obtain (1), (2), (3). The clock control bits are multiplied with the bit states of the registers causing an increase in the degree of the equations. Since the clock control of the Gollmann cascade is linear in the register states, the degrees of the equations increase by one at every clock. For example, let the initial states bits of A, B be a_i, b_i respectively. The outputs of A at the first two clocks are then

$$A_l^0 = a_l$$
$$A_l^1 = a_{l-1},$$

and the states of B at the first three clocks are expressed as

$$B_i^0 = b_i$$
$$B_i^1 = b_i(a_l + 1) + b_{i-1}a_l$$
$$B_i^2 = b_i(a_l + 1)(a_{l-1} + 1) + b_{i-1}\{(a_l + 1)a_{l-1} + a_l(a_{l-1} + 1)\} + b_{i-2}a_la_{l-1}.$$

The equations formed in the variables a_i, b_i will increase in degree as the cipher is clocked. However, if we instead use new variables for the register states at every clock, this degree accumulation can be prevented and we obtain quadratic equations in terms of the register states at each time t, as shown in (1), (2), (3). Initially, we have $l + m + n + r$ variables representing the initial states of the registers. Since A is regularly clocked, we do not need to introduce new variables for A. Also, since $D_r^t = z^t$, we can just replace that state with the keystream output. This means that a total of $m + n + r - 1$ new variables are introduced at each time t. Then $m + n + r$ equations are introduced as relations between old and new variables. Since the key size is $l + m + n + r$, we would need at least $l + m + n + r$ clocks to form a system of equations with a unique solution. From the analysis above, we obtain $(m + n + r)(l + m + n + r)$ quadratic equations in $(m + n + r)(l + m + n + r)$ unknowns for the Gollmann cascade. These equations can be solved by techniques such as Gröbner basis methods, but not by linearisation. This is because as we use more keystream, the number of variables and therefore monomials in the system increases at a much higher rate, and it is not possible to obtain enough equations in the linearised variables.

Not so much is known about the practical complexities of solving polynomial equations by Gröbner basis methods, such as Faugère's F_5 and the XL and related algorithms. It is known that the worst case complexity of these algorithms on random systems of equations in $\mathrm{GF}(2)$ are $O(2^{2v})$, where v is the number of equations. However, for specific systems and as the number of equations exceeds the number of variables, the complexity of XL and its variants can drop significantly, to even polynomial time. See, for example, [2,13,14,25] for descriptions and analyses of these algorithms. In our case, since the variables are in $\mathrm{GF}(2)$, we do not need to consider solutions in the algebraic closure of $\mathrm{GF}(2)$ and can solve the equations subject to the field equations $x^2 + x = 0$ for each variable x in the system. Therefore, we obtain another $(m+n+r)(l+m+n+r)$ equations, giving us twice as many equations as variables. Furthermore, the equations are very sparse, since each variable is used only for two clocks. These properties might prove useful at reducing the complexity of finding the solution to the system. The actual efficiency of this attack is yet to be gauged and is the subject of further research in this area. We will use this concept of attack in a later section on Pomaranch.

3   An Algebraic Analysis of Pomaranch

In this section, we provide an algebraic analysis of version 3 of Pomaranch [19,21], an eSTREAM stream cipher candidate. We begin by describing the cipher. Here we mainly discuss in the context of the 128-bit version of Pomaranch, but the analysis also holds for the 80-bit version. Although only the 80-bit version of Pomaranch is officially in phase 2 of the eSTREAM project, the 128-bit version will also be important since there are doubts about the security of 80-bit ciphers, as discussed in the eSTREAM project. Earlier versions of Pomaranch have been cryptanalysed in [8,20,22], but as of the beginning of phase 2 of the project, version 3 of Pomaranch has been published to address weaknesses found in the above papers. As far as we are aware, this paper is the first to analyse version 3 of the cipher.

3.1   Pomaranch Description

The overall structure of Pomaranch is shown in Figure 2. Pomaranch is a clock-controlled cascade generator consisting of 9 jump registers $R_1, R_2, \ldots, R_9$. The jump registers are implemented as autonomous Linear Finite State Machines (LFSM), each containing 18 memory cells, and each cell contains a bit value. Each cell in a jump register behaves as a shift cell or a feedback cell. At each clock, a shift cell simply shifts its state to the next cell, whereas a feedback cell feeds its state back to itself and adds it to the state from the previous cell, as well as performing a normal shift like a shift cell. The behaviour of each cell depends on the value of a Jump Control (JC) signal to the jump register where the cell belongs. Algebraically, a transition matrix A governs the behaviour of an LFSM. The transition matrices of the jump registers in Pomaranch take the form

$$
A = \begin{pmatrix}
d_{18} & 0 & 0 & \ldots & 0 & 1 \\
1 & d_{17} & 0 & \ldots & 0 & t_{17} \\
0 & 1 & d_{16} & \ddots & \vdots & \vdots \\
0 & 0 & \ddots & \ddots & 0 & \vdots \\
\vdots & \vdots & \ddots & 1 & d_2 & t_2 \\
0 & 0 & \ldots & 0 & 1 & d_1 + t_1
\end{pmatrix},
$$

where t_i determines the positions of the feedback taps, and d_i determines whether the cells are shift cells or feedback cells. At any moment, half of the cells in the registers behave as shift cells, and the other half as feedback cells, so half the d_i are 0 and the other half are 1. If $JC = 0$ for a certain register, the register is clocked according to its transition matrix A. If $JC = 1$, all cells are switched to the opposite behaviour. This is equivalent to switching the transition matrix to $A+I$, where I is the identity matrix. Two different transition matrices A_1, A_2 are used for odd (type 1) and even (type 2) numbered registers respectively, with different values of t_i, d_i. Each jump register is then connected to a key map, which consists of an s-box and a nonlinear filter. Key bits are diffused into the key map and a one bit output is drawn. The jump control of a register is then taken as the sum of all outputs from the key maps from all previous registers. The keystream output is given by the sum of the contents of the 17th cells of all registers. The 128-bit key k of Pomaranch is divided into eight 16-bit subkeys

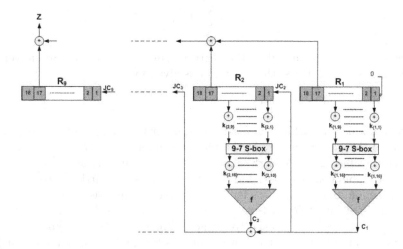

Fig. 2. Pomaranch Version 3

K_1 to K_8, where K_i represents the key bits $k_{i,1}, k_{i,2}, \ldots, k_{i,16}$. Each section of Pomaranch except the last contains a jump register of length 18 and a nonlinear function composed of an s-box and a degree 4 boolean filter function f. The last section of Pomaranch only has a jump register. The key map at section i first takes a 9-bit vector v from cells numbered $1, 2, 4, 5, 6, 7, 9, 10, 11$ from jump registers of type 1 or $1, 2, 3, 4, 5, 7, 9, 10, 11$ from jump registers of type 2. Then the 9 least significant bits of K_i are added to v. The sum is passed through a 9-to-7-bit inversion s-box over $\mathrm{GF}(2^9)$ defined by the primitive polynomial $x^9 + x + 1$. The resulting 7 bits are then added to the 7 most significant bits of K_i. This is fed into the boolean function f of degree 4, and the output of f is called the jump control out bit of the section and is denoted as c_i. The c_i from each section is used to produce the jump control bits JC_2 to JC_9 controlling the registers R_2 to R_9 at time t respectively, as follows.

$$JC_i^t = \sum_{j=2}^{i-1} c_j^t, \quad 2 \le i \le 9.$$

The jump control bit JC_1 of register R_1 is permanently set to zero.

3.2   Algebraic Analysis of Pomaranch

In this section we provide an algebraic analysis of Pomaranch. From the description of the way that the cipher is clocked, each register can be represented as

$$R^t = (A + JC^t \cdot I)R^{t-1}, \tag{8}$$

where R^t is the state of a register at time t, A is its transition matrix, JC^t is its jump control, and I is the identity matrix. The keystream is given by

$$z^t = \sum_{i=1}^{9} r^t_{i,17}, \tag{9}$$

where $r^t_{i,17}$ is the 17-th cell of the register i at time t. In the 80-bit version employing only 6 sections, the keystream is given by

$$z^t = g(r^t_{1,17}, r^t_{2,17}, \ldots, r^t_{5,17}) + r^t_{6,17}, \tag{10}$$

where g is a nonlinear filter of degree 3. In order to understand how the key is related to the output and what the relations among different subkey bits are, we analyse what happens in the first section containing R_1 as the key bits are mixed in a nonlinear manner, and how the key bits are carried across into the second section. Similar analysis follows for the remaining sections.

Let the i-th bit of R_1 at time t be r_i for $1 \leq i \leq 18$. Nine selected bits r_j are added to $k_1, k_2, \ldots, k_9$ respectively, according to the register type. These are then fed into the key map that consists of an s-box and a boolean function f of degree 4. Firstly, the bits pass through an inversion s-box over $GF(2^9)$. Let the input to the s-box be $a_1, a_2, \ldots, a_9$, then we have $a_i = r_j + k_i$ for $1 \leq i \leq 9$. Let the output from the s-box be $b_1, b_2, \ldots, b_9$. Each b_i can be represented with equations in a_i. An explicit function of each b_i in terms of a_i will be of high degree, in our case it is of degree 9, i.e.

$$b_i = \sum_{e \in GF(2)^9} \left(s_e \prod_{i=1}^{9} a_i^{e_i} \right), \quad 1 \leq i \leq 7,$$

where s_e are coefficients in $GF(2)$. We can also form implicit relations between the inputs and outputs of the s-boxes of degree 2, as was shown in [16]. The relations for this s-box are presented in Appendix A. From the truth table of the s-box shown in [21], we can see that the seven output bits $b_2, b_3, \ldots, b_8$ are used for the next component, discarding b_1, b_9. Those output bits from the s-box are then added to the next 7 key bits and the result $b_i + k_{8+i}$ for $2 \leq i \leq 8$ is filtered through f, a degree 4 function. We have computed the algebraic normal form (ANF) of f as shown in Appendix A. Assuming we take the explicit functions for the s-box and the ANF of the filter, the output c_1 from the filter function will be an expression of degree 13 in 16 key bits. This expression is then fed into the next register R_2 as JC_2, and also added to the output c_2 from R_2 to be fed into R_3 as JC_3, and so forth.

From the analysis above, a degree 13 equation in 16 key bits is generated from the first section, which becomes an input in the second section. As JC_2 is fed forward into R_2, the high degree expression and variables from the first section are carried across as well. The key map of the second section will mix the next

16-bit subkey K_2 into the expression, raising the degree by 13 to 26 and the number of variables by 16 to 32. This accumulation continues to carry across to R_3 and beyond. The number of possible monomials each JC_i possesses at the first clock is then

$$M_{JC_i} = \max\left\{\sum_{j=0}^{13i} \binom{16i}{j}, 2^{128}\right\}, \quad 1 \le i \le 9. \qquad (11)$$

By the end of the first clock, we will have equations of degree 104 in 128 variables. This degree accumulation does not carry over to the next clock, since JC_1 is set to be constantly zero. Therefore, in order to perform an algebraic attack in this manner, we need at least 128 bits of keystream, giving us 128 equations of degree 104. Although the keystream requirement is low, generating such equations is time and memory consuming and the effort needed in solving them is likely to be much more than that of exhaustive key search.

3.3 Overcoming the Problem of the Degree Accumulation

There are three main components in each section of Pomaranch, namely a jump register, an s-box, and a nonlinear filter. The outputs of each component are nonlinear in its inputs, and the expressions describing the outputs will have a higher degree than those of the inputs. Since each output is fed into the next component, the degree of the expressions accumulates. To prevent this degree accumulation, we introduce new variables so that the nonlinearities are not carried across the components. Let $r_{i,m}^t$ be the m-th bit of register R_i at time t, $b_{i,m}^t$ be the output bits of the s-box at the i-th section at time t, JC_i^t be the jump control input to R_i at time t, and $k_{i,m}$ be the key bits used in the i-th section. We sucessively introduce the above variables as we step through the keystream generation. At the start, we have 128 variables $k_{i,m}$ whose values are to be determined. We go through each component and try to discover equations that relate to $k_{i,m}$. At time t, we proceed as follows. The relations between the jump controls and the registers are

$$R_i^t = \begin{cases} A_1 R_{i-1}^t & i = 1 \\ (A_2 + JC_{i-1}^t I)R_{i-1}^t & i \in \{2,4,6,8\} \\ (A_1 + JC_{i-1}^t I)R_{i-1}^t & i \in \{3,5,7,9\} \end{cases}.$$

The first register will always contain known bits, since it is not under the effect of jump controls. Hence, we have 8 registers with 18 variables each to introduce in each clock. Each new R_i^t is a function of R_{i-1}^t. This yields 144 quadratic equations in 144 variables. The s-box is defined by inversion over $GF(2^9)$, which means that the relations between its input and output bits can be expressed as a system of 9 quadratic equations. The inputs are linear sums of certain register bits and key bits, so we have the equations

$$s_0(r_{i,1}^t + k_{i,1}, \ldots, r_{i,11}^t + k_{i,9}, b_{i,1}^t, \ldots, b_{i,9}^t) = 1$$
$$s_j(r_{i,1}^t + k_{i,1}, \ldots, r_{i,11}^t + k_{i,9}, b_{i,1}^t, \ldots, b_{i,9}^t) = 0, \quad 1 \le j \le 8.$$

In each clock we have 8 s-boxes, which yield 72 quadratic equations and 72 new variables $b_{i,m}^t$. Let c_i^t be the jump control output from the nonlinear filter f of the i-th section at time t. Then

$$c_i^t = f(b_{i,2}^t + k_{i,10}, b_{i,2}^t + k_{i,11}, \ldots, b_{i,8}^t + k_{i,16}) \qquad (12)$$

giving a degree 4 equation. The jump control input to the next register is then

$$JC_{i+1}^t = JC_i^t + c_i^t.$$

We do not assign new variables to c_i^t. Therefore, we obtain a degree 4 equation with the new variable JC_{i+1}^t. Since there are 8 filters in one clock, we get 8 equations and 8 variables. Finally, at the end of each clock we get the keystream bit

$$z^t = \sum_{i=0}^{9} r_{i,17}^t.$$

We can use the above equation and rewrite it as, for example,

$$r_{9,17}^t = z^t + \sum_{i=0}^{8} r_{i,17}^t,$$

and we can replace $r_{9,17}$ with the above expression, thereby eliminating a variable at each clock. As a whole, we obtain 224 equations and 223 new variables at each clock, plus the original 128 variables which represent the key bits. In order to obtain a unique solution, we would require at least 128 bits of keystream, giving 28672 equations in 28672 variables. Of these equations, 1024 are of degree four and 27648 are quadratic. In the case of the 80-bit version, we have

$$z^t = g(r_{1,17}^t, r_{2,17}^t, \ldots, r_{5,17}^t) + r_{6,17}^t,$$

and we can rewrite it as

$$r_{6,17}^t = z^t + g(r_{1,17}^t, r_{2,17}^t, \ldots, r_{5,17}^t),$$

where g is the cubic nonlinear filter. This contributes a cubic equation to the system. We then obtain 140 equations and 139 new variables at each clock, plus the original 80 variables which represent the key bits. In order to obtain a unique solution, we would require at least 80 bits of keystream, giving 11200 equations in 11200 variables. Of these equations, 400 are of degree four, 80 are cubic and 10800 are quadratic.

   In order to solve the set of equations, we would require techniques such as Gröbner basis methods, since we cannot get enough equations for linearisation. With the addition of field equations into the system, we obtain twice as many equations as variables. The sparse structure of the equations may also give rise to complexity reductions in solving the system by the XL and related algorithms. This would reduce the complexity of finding the solution significantly from $O(2^{2v})$ for random systems, where v is the number of equations. See, for example, [2,13,14,25] for analyses and details of implementing these algorithms. All key bits can be recovered when the solution to this set of equations is found. While there is yet no evidence for or against whether this type of attack would be better than exhaustive key search, the size and the form of the equations generated can be used intuitively to judge the cipher's strength. Algebraic attacks are still a widely discussed and controversial topic in the cryptographic community, so its consequences should not be overlooked.

We note here that the designers of Pomaranch have increased the size of the registers from 14 to 18. This increase has no effect on the size or the degree of the final equations. In fact, the degrees of the equations are independent of the size of the registers. They are affected by the clock control mechanism and nonlinear components.

3.4 Algebraic Analysis of the Filter Function

We have found a cubic annihilator y_f of the degree four filter f such that $fy_f = 0$ and cubic polynomials $e_{f,1}, e_{f,2}, e_{f,3}$ such that $h_{f,i} = fe_{f,i}$ are cubic for all e_i. These are shown in Appendix A. This shows a potential weakness in the filter function f. However, in terms of our algebraic attack these multipliers are not useful, because of the newly introduced variables. If we apply a cubic mutiplier e to (12), we obtain

$$c_i^t e = fe,$$

which is still of degree 4, due to the presence of c_i^t. To successfully reduce the degree of our equations, we need to find annihilators or low degree multiples of $f + c_i^t$, which we have not been able to do. For the 80-bit version of Pomaranch, we have also found a quadratic annihilator y_g of the cubic filter g such that $gy_g = 0$ and quadratic polynomials $e_{g,1}, e_{g,2}$ such that $h_{g,i} = ge_{g,i}$ are cubic for all $e_{g,i}$. These are shown in Appendix A. Again, there could be weaknesses in the filter function g. However, our algebraic attack cannot make use of this since we need annihilators or low degree multiples of $g + r_{6,17}^t$ in order to reduce the equations describing the cipher.

We leave as an open question other possible uses of these results. However, we believe that the filter function f, g should be changed so that no annihilators or low degree multiples exist for f, g. This would resist possible future attacks based on algebraic techniques.

4   Conclusion

In this paper, we have described the first algebraic attacks against clock controlled cascade stream ciphers, in particular the Gollmann cascade and the eSTREAM candidate Pomaranch. We have established relations between the initial state bits and the output bits of the Gollmann clock controlled cascade stream ciphers, and demonstrated that the initial state bits of the last two registers can be determined from those of the previous registers, which yields an attack better than exhaustive key search.

For the Gollmann cascade we also showed that the effect of a clock control on algebraic attacks is that the degree of the equations generated increases with each clock. An alternative attack on the cascade was developed to eliminate the effect of this degree accumulation, resulting in a large sparse low degree polynomial system, which can be generated and solved more efficiently using algorithms such as Faugère's F_5 and XL and its variants.

Our algebraic analysis on Pomaranch further showed that a cipher with non-linear components can be expressed as a system of equations with maximum degree no higher than the maximum degree of the components. In the analysis, we showed how to generate degree four equations in the key bits and other component bits of the system. The bottleneck to reducing this degree further is due to the filter function of degree four. The input and output sizes of the components determine the number of new variables to be introduced and therefore the size of the system of equations needed to describe the cipher. We also make the observation that increasing the size of the jump registers has no effect on increasing the degree of the equations.

Finally, we have found annihilators and low degree multiples of both the filter function in the keymap and the filter function for the keystream contributions in the 80-bit version of Pomaranch, which indicates a possible weakness in the cipher.

References

1. S. Al-Hinai, L. Batten, B. Colbert and K. Wong. Algebraic attacks on clock controlled stream ciphers. In L. M. Batten and R. Safavi-Naini, editors, *Proceedings of Information Security and Privacy - 11th Australasian Conference, ACISP 2006*, volume 4058 of *Lecture Notes in Computer Science*, pages 1–16. Springer-Verlag, 2006.
2. G. Ars, J.-C. Faugère, H. Imai, M. Kawazoe, M. Sugita. Comparison between XL and Gröbner basis algorithms. In P. J. Lee, editor, *Advances in Cryptology - ASIACRYPT 2004*, volume 3329 of Lecture Notes in Computer Science, pages 338–353. Springer-Verlag, 2004.
3. F. Armknecht. Improving fast algebraic attacks. In B. Roy and W. Meier, editors, *Fast Software Encryption - FSE 2004*, volume 3017 of *Lecture Notes in Computer Science*, pages 65–82, Springer-Verlag, 2004.
4. F. Armknecht and M. Krause. Algebraic attacks on combiners with memory. In D. Boneh, editor, *Advances in Cryptology - CRYPTO 2003*, volume 2729 of *Lecture Notes in Computer Science*, pages 162–175, Springer-Verlag, 2003.
5. W. G. Chambers and D. Gollmann. Lock-in Effect in Cascades of Clock-Controlled Shift-Registers. In Christoph G. Günther, editor, *Advances in Cryptology - Proceedings of EUROCRYPT 88*, volume 330 of *Lecture Notes in Computer Science*, pages 331–344. Springer-Verlag, 1988.
6. W. G. Chambers. Two stream ciphers. In R. Anderson, editor, *Fast Software Encryption - FSE 1994*, volume 809 of *Lecture Notes in Computer Science*, pages 51-55, Springer-Verlag, 1994.
7. J. Y. Cho and J. Pieprzyk. Algebraic attacks on SOBER-t32 and SOBER-t16 without stuttering. In B. Roy and W. Meier, editor *Fast Software Encryption - FSE 2004*, volume 3017 of *Lecture Notes in Computer Science*, pages 49–64, Springer-Verlag, 2004.

8. C. Cid, H. Gilbert and T. Johansson. Cryptanalysis of Pomaranch. *eSTREAM, ECRYPT Stream Cipher Project, Report 2005/060*, 2005.
9. N. Courtois. The security of hidden field equations (HFE). In D. Naccache, editor, *Progress in Cryptology - CT-RSA 2001*, volume 2020 of *Lecture Notes in Computer Science*, pages 266–281, Springer-Verlag, 2001.
10. N. Courtois. Algebraic attacks on combiners with memory and several outputs. In C. Park, S. Chee, editors, *Information Security and Cryptology ICISC 2004*, volume 3506 of *Lecture Notes in Computer Science*, pages 3–20, Springer-Verlag, 2005.
11. N. Courtois. Fast algebraic attacks on stream ciphers with linear feedback. In D. Boneh, editor, *Advances in Cryptology - CRYPTO 2003*, volume 2729 of *Lecture Notes in Computer Science*, pages 176–194 , Springer-Verlag, 2003.
12. N. Courtois and W. Meier. Algebraic attacks on stream ciphers with linear feedback. In E. Biham, editor, *Advances in Cryptology - Eurocrypt 2003* , volume 2656 of *Lecture Notes in Computer Science*, pages 346–359, Springer-Verlag, 2003.
13. N. Courtois, A. Klimov, J. Patarin, and A. Shamir. Efficient algorithms for solving overdefined systems of multivariate polynomial equations. In B. Preneel, editor, *Eurocrypt 2000*. volume 1807 of *Lecture Notes in Computer Science*, pages 392-407, Springer-Verlag 2000.
14. N. Courtois, J. Patarin. About the *XL* algorithm over GF(2). In M. Joye, editor, *Topics in Cryptology - CT-RSA 2003*. volume 2612 of *Lecture Notes in Computer Science*, pages 141-157, Springer-Verlag 2003.
15. N. Courtois and J. Pieprzyk. Cryptanalysis of block ciphers with overdefined systems of equations. In Y. Zheng, editor, *Advances in Cryptology - ASIACRYPT 2002*, volume 2501 of *Lecture Notes in Computer Science*, pages 267–287, Springer-Verlag, 2002.
16. N. Courtois, B. Debraize, and E. Garrido. On Exact Algebraic [Non-]Immunity of S-Boxes Based on Power Functions. In L. M. Batten and R. Safavi-Naini, editors, *Proceedings of Information Security and Privacy - 11th Australasian Conference, ACISP 2006*, volume 4058 of *Lecture Notes in Computer Science*, pages 76–86, Springer-Verlag, 2006.
17. D. Gollmann. Pseudo Random Properties of Cascade Connections of Clock Controlled Shift Registers. In T. Beth, N. Cot, and I. Ingemarsson, editors, *Advances in Cryptology - Proceedings of EUROCRYPT 84*, volume 209 of *Lecture Notes in Computer Science*, pages 93–98, Springer-Verlag, 1985.
18. D. Gollmann and W. G. Chambers. Clock-controlled shift registers: a review. *IEEE Journal on Selected Areas in Communications*, 7, pages 525–533, 1989.
19. T. Helleseth, C. Jansen and A. Kholosha. Pomaranch - Design and Analysis of a Family of Stream Ciphers. *eSTREAM, ECRYPT Stream Cipher Project, Report 2006/008*, 2005.
20. M. Hasanzadeh, S. Khazaei and A. Kholosha. On IV Setup of Pomaranch. *eSTREAM, ECRYPT Stream Cipher Project, Report 2005/082*, 2005.
21. C. Jansen, T. Helleseth and A. Kholosha. Cascade Jump Controlled Sequence Generator and Pomaranch Stream Cipher (Version 3). *eSTREAM, ECRYPT Stream Cipher Project, Report 2006/006*, 2006.
22. S. Khazaei. Cryptanalysis of Pomaranch (CJCSG). *eSTREAM, ECRYPT Stream Cipher Project, Report 2005/065*. 2005.
23. R. Menicocci. Cryptanalysis of a two stage Gollmann cascade generator. In W. Wolfowicz, editor, *Proceedings of the 3rd Symposium on State and Progress of Research in Cryptography*, 6269, 1993.

24. S.J. Park, S.J. Lee, and S.C. Goh. On the security of the Gollmann cascades. In D. Coppersmith, aditor, *Advances in Cryptology CRYPTO 95*, volume 963 of *Lecture Notes in Computer Science*, pages 148-156, Springer-Verlag, 1995.
25. B. Yang and J. Chen. All in the XL Family: Theory and Practice. In C. Park, S. Chee, editors, *Information Security and Cryptology ICISC 2004*. volume 3506 of *Lecture Notes in Computer Science*, pages 67-86, Springer-Verlag 2005.
26. E. Zenner. On the efficiency of the clock control guessing attack. In P.J. Lee, C.H. Lim, editors, *Information Security and Cryptology - ICISC 2002*, volume 2587 of *Lecture Notes in Computer Science*, pages 200–212, Springer-Verlag, 2002.

A   Algebraic Relations in Pomaranch

The algebraic normal form (ANF) of the Pomaranch filter function is as follows.

$$f(x_1,x_2,x_3,x_4,x_5,x_6,x_7)$$

$$= x_1x_2x_3x_4 + x_1x_2x_3x_5 + x_1x_2x_3x_6 + x_1x_2x_3x_7 + x_1x_2x_4x_5 + x_1x_2x_4x_7$$

$$+ x_1x_2x_6x_7 + x_1x_3x_4x_6 + x_1x_3x_4x_7 + x_1x_3x_5x_6 + x_1x_3x_5x_7 + x_1x_3x_6x_7$$

$$+ x_1x_4x_5x_7 + x_2x_3x_4x_5 + x_2x_3x_4x_6 + x_2x_3x_4x_7 + x_2x_3x_5x_6 + x_2x_3x_5x_7$$

$$+ x_2x_3x_6x_7 + x_2x_4x_5x_6 + x_2x_4x_6x_7 + x_3x_4x_5x_6 + x_3x_4x_5x_7 + x_3x_4x_6x_7 + x_4x_5x_6x_7$$

$$+ x_1x_2x_3 + x_1x_2x_4 + x_1x_2x_6 + x_1x_2x_7 + x_1x_3x_5 + x_1x_3x_6$$

$$+ x_1x_3x_7 + x_1x_4x_6 + x_1x_4x_7 + x_1x_6x_7 + x_2x_3x_5 + x_2x_3x_6$$

$$+ x_2x_3x_7 + x_2x_4x_7 + x_2x_5x_7 + x_2x_6x_7 + x_3x_4x_6 + x_3x_5x_7 + x_5x_6x_7$$

$$+ x_1x_2 + x_1x_6 + x_2x_7 + x_3x_7 + x_4x_5 + x_4x_7$$

$$+ x_1 + x_2 + x_3 + x_5 + x_6 + x_7$$

The relations between the inputs a_i and the outputs b_i of the Pomaranch s-box is as follows.

$$a_1b_1 + a_2b_9 + a_3b_8 + a_4b_7 + a_5b_6 + a_6b_5 + a_7b_4 + a_8b_3 + a_9b_2 = 1$$

$$a_1b_2 + a_2b_1 + a_2b_9 + a_3b_8 + a_3b_9 + a_4b_7 + a_4b_8 + a_5b_6 + a_5b_7$$
$$+ a_6b_5 + a_6b_6 + a_7b_4 + a_7b_5 + a_8b_3 + a_8b_4 + a_9b_2 + a_9b_3 = 0$$

$$a_1b_3 + a_2b_2 + a_3b_1 + a_3b_9 + a_4b_8 + a_4b_9 + a_5b_7 + a_5b_8 + a_6b_6$$
$$+ a_6b_7 + a_7b_5 + a_7b_6 + a_8b_4 + a_8b_5 + a_9b_3 + a_9b_4 = 0$$

$$a_1b_4 + a_2b_3 + a_3b_2 + a_4b_1 + a_4b_9 + a_5b_8 + a_5b_9 + a_6b_7 + a_6b_8$$
$$+ a_7b_6 + a_7b_7 + a_8b_5 + a_8b_6 + a_9b_4 + a_9b_5 = 0$$

$$a_1b_5 + a_2b_4 + a_3b_3 + a_4b_2 + a_5b_1 + a_5b_9 + a_6b_8 + a_6b_9 + a_7b_7$$
$$+ a_7b_8 + a_8b_6 + a_8b_7 + a_9b_5 + a_9b_6 = 0$$

$$a_1b_6 + a_2b_5 + a_3b_4 + a_4b_3 + a_5b_2 + a_6b_1 + a_6b_9 + a_7b_8 + a_7b_9 + a_8b_7 + a_8b_8 + a_9b_6 + a_9b_7 = 0$$

$$a_1b_7 + a_2b_6 + a_3b_5 + a_4b_4 + a_5b_3 + a_6b_2 + a_7b_1 + a_7b_9 + a_8b_8 + a_8b_9 + a_9b_7 + a_9b_8 = 0$$

$$a_1b_8 + a_2b_7 + a_3b_6 + a_4b_5 + a_5b_4 + a_6b_3 + a_7b_2 + a_8b_1 + a_8b_9 + a_9b_8 + a_9b_9 = 0$$

$$a_1b_9 + a_2b_8 + a_3b_7 + a_4b_6 + a_5b_5 + a_6b_4 + a_7b_3 + a_8b_2 + a_9b_1 + a_9b_9 = 0$$

The annihilator y_f and low degree multiples $e_{f,i}$ of the filter function f in Pomaranch are as follows.

$y_f(x_1,x_2,x_3,x_4,x_5,x_6,x_7)$

$= x_1x_3x_5+x_2x_4x_5+x_3x_4x_5+x_2x_3x_6+x_1x_4x_6+x_1x_5x_6+x_3x_5x_6+x_4x_5x_6$

$\quad +x_1x_3x_7+x_1x_4x_7+x_2x_5x_7+x_3x_5x_7+x_1x_6x_7+x_2x_6x_7+x_4x_6x_7+x_5x_6x_7$

$\quad +x_1x_3+x_2x_3+x_1x_4+x_2x_4+x_3x_4+x_1x_5+x_4x_5+x_3x_6$

$\quad +x_4x_6+x_5x_7+x_6x_7+x_4$

$e_{f,1}(x_1,x_2,x_3,x_4,x_5,x_6,x_7)$

$= x_1x_2x_4+x_1x_3x_5+x_2x_4x_5+x_3x_4x_5+x_2x_3x_6+x_2x_4x_6+x_3x_4x_6+x_4x_5x_6$

$\quad +x_1x_2x_7+x_1x_3x_7+x_3x_4x_7+x_4x_5x_7+x_1x_6x_7+x_1x_3+x_1x_4+x_2x_4$

$\quad +x_3x_4+x_2x_5+x_3x_5+x_5x_6+x_1x_7+x_3x_7+x_4x_7+x_5x_7$

$\quad +x_2+x_3+x_6+x_7$

$e_{f,2}(x_1,x_2,x_3,x_4,x_5,x_6,x_7)$

$= x_1x_2x_4+x_1x_4x_6+x_2x_4x_6+x_3x_4x_6+x_1x_5x_6+x_3x_5x_6+x_1x_2x_7+x_1x_4x_7$

$\quad +x_3x_4x_7+x_2x_5x_7+x_3x_5x_7+x_4x_5x_7+x_2x_6x_7+x_4x_6x_7+x_5x_6x_7+x_2x_3$

$\quad +x_1x_5+x_2x_5+x_3x_5+x_4x_5+x_3x_6+x_4x_6+x_5x_6+x_1x_7$

$\quad +x_3x_7+x_4x_7+x_6x_7+x_2+x_3+x_4+x_6+x_7$

$e_{f,3}(x_1,x_2,x_3,x_4,x_5,x_6,x_7)$

$= x_1x_2x_3+x_1x_4x_5+x_1x_2x_6+x_1x_3x_6+x_2x_3x_6+x_1x_5x_6+x_4x_5x_6+x_2x_3x_7$

$\quad +x_1x_4x_7+x_1x_5x_7+x_4x_5x_7+x_1x_6x_7+x_2x_6x_7+x_3x_6x_7+x_4x_6x_7+x_1x_6$

$\quad +x_2x_6+x_3x_6+x_5x_6+x_1x_7+x_4x_7+x_5x_7+x_6+x_7$

In particular we have $y_f = g_1 + g_2, (f+1)g_1 = 0, (f+1)g_3 = 0$. The annihilator y_g and low degree multiples $e_{g,i}$ of the filter function f in Pomaranch are as follows.

$$y_g(x_1,x_2,x_3,x_4,x_5) = x_1x_2+x_1x_3+x_2x_3+x_3x_4+x_2x_5+x_2+x_3$$

$$e_{g,1}(x_1,x_2,x_3,x_4,x_5) = x_1x_2+x_1x_3+x_3x_4+x_2x_5$$

$$e_{g,2}(x_1,x_2,x_3,x_4,x_5) = x_2x_3+x_2+x_3$$

In particular we have $y = e_1 + e_2, (f+1)g_1 = 0$

An Algorithm for Solving the LPN Problem and Its Application to Security Evaluation of the HB Protocols for RFID Authentication

Marc P.C. Fossorier[1], Miodrag J. Mihaljević[2,4], Hideki Imai[3,4],
Yang Cui[5], and Kanta Matsuura[5]

[1] University of Hawaii, Department of Electrical Engineering, Honolulu, USA
[2] Mathematical Institute, Serbian Academy of Sciences and Arts, Belgrade, Serbia
[3] Chuo University, Faculty of Science and Engineering, Tokyo, Japan
[4] Research Center for Information Security (RCIS), National Institute of Advanced
Industrial Science and Technology (AIST), Tokyo, Japan
[5] University of Tokyo, Institute of Industrial Science (IIS), Tokyo, Japan

Abstract. An algorithm for solving the "learning parity with noise" (LPN) problem is proposed and analyzed. The algorithm originates from the recently proposed advanced fast correlation attacks, and it employs the concepts of decimation, linear combining, hypothesizing and minimum distance decoding. However, as opposed to fast correlation attacks, no preprocessing phase is allowed for the LPN problem. The proposed algorithm appears as more powerful than the best one previously reported known as the BKW algorithm proposed by Blum, Kalai and Wasserman. In fact the BKW algorithm is shown to be a special instance of the proposed algorithm, but without optimized parameters. An improved security evaluation, assuming the passive attacks, of Hopper and Blum HB and HB$^+$ protocols for radio-frequency identification (RFID) authentication is then developed. Employing the proposed algorithm, the security of the HB protocols is reevaluated, implying that the previously reported security margins appear as overestimated.

Keywords: cryptanalysis, LPN problem, fast correlation attacks, HB protocols, RFID authentication.

1 Introduction

In [12] (following the prior work of Hopper and Blum [10]), two shared-key authentication protocols have been proposed and analyzed. Their extremely low computational cost makes them attractive for low-cost devices such as radio-frequency identification (RFID) tags. The first protocol (called the HB protocol) is proven secure against a passive (eavesdropping) adversary, while the second (called HB$^+$) is proven secure against the stronger class of active adversaries. Security of these protocols is based on the conjectured hardness of the "learning parity with noise" (LPN) problem (see [3], for example). In [13], the security of the HB and HB$^+$ protocols under parallel/concurrent executions has been proven.

R. Barua and T. Lange (Eds.): INDOCRYPT 2006, LNCS 4329, pp. 48–62, 2006.

The underlying paradigm of the HB protocol is the following. Suppose Alice and a computing device C share an k-bit secret $\mathbf{x}$, and Alice would like to authenticate herself to C. Then C selects a random challenge $\mathbf{a} \in \{0,1\}^k$ and sends it to Alice. Alice computes the binary inner-product $\mathbf{a} \cdot \mathbf{x}$, then sends the result back to C. Finally, C computes $\mathbf{a} \cdot \mathbf{x}$, and accepts the single round authentication if Alice's parity bit is correct. In a single round, someone imitating Alice who does not know the secret $\mathbf{x}$ will guess the correct value $\mathbf{a} \cdot \mathbf{x}$ half the time. By repeating for r rounds, Alice can lower the probability of naively guessing the correct parity bits for all r rounds to 2^{-r}. However, an eavesdropper capturing $O(k)$ valid challenge-response pairs between Alice and C can quickly calculate the value of $\mathbf{x}$ through Gaussian elimination. To prevent revealing $\mathbf{x}$ to passive eavesdroppers, Alice can inject noise into her response. Alice intentionally sends the wrong response with constant probability $p \in (0, 1/2)$. Then C authenticates Alice's identity if fewer than pr of her responses are incorrect.

Suppose that an eavesdropper, i.e., a passive adversary, captures q rounds of the HB protocol over several authentications and wishes to make the impersonation. The goal of the adversary in this case is equivalent to the core problem investigated in this paper. This problem is known as the learning parity in the presence of noise, or LPN problem. It is shown in [12] that the security of the both HB and HB$^+$ protocols against the passive attack depends on the hardness of LPN problem.

On the other hand, the results reported in [9] have recently shown a man-in-the-middle attack on the HB$^+$ protocol. However, the arguments given in [12] and [13] limit the impact of such attack.

Accordingly, this paper is focused only to the LPN problem and the passive attacking of HB and HB$^+$ protocols.

Motivation for the Work. Despite certain differences, both the LPN problem and the underlying problem of fast correlation attack can be viewed as the problem of solving an overdefined system of noisy linear equations. However, it appears that the currently reported approaches for solving the LPN problem do not take into account the approaches developed for fast correlation attacks. Accordingly, a goal of this work is to consider employment of fast correlation attack approaches for solving the LPN problem. Another motivation of this work is the security re-evaluation of the HB protocol for RFID authentication as its security level appears as a direct consequence of the LPN problem hardness.

Summary of the Contributions. This paper proposes a generic algorithm for solving the LPN problem. The proposed algorithm originates from the recently proposed advanced fast correlation attacks and it employs the following concepts: decimation, linear combining, hypothesizing and decoding. However, as opposed to fast correlation attacks, no preprocessing can be performed, which introduces an additional constraint. The following main characteristics of the proposed algorithm have been analytically established: (i) average time complexity; and (ii) average space complexity. The proposed algorithm has been compared with the best previously reported one, namely the BKW algorithm, and its advantages for solving the LPN problem have been pointed out. The proposed algorithm has been applied for security reevaluation of the HB protocol for RFID authentication implying that the previously reported security margins obtained based on the BKW algorithm are overestimated, and more realistic security margins have been derived.

Organization of the Paper. Section 2 provides a brief review of the LPN problem and specifies it in a form relevant for further analysis. The BKW algorithm is presented in Section 3. An algorithm for solving the LPN problem is proposed and analyzed in Section 4. Comparisons between this algorithm and the BKW algorithms are made in Section 5. Security reevaluation of the HB protocol for RFID authentication via employment of the proposed algorithm is given in Section 6.

2   The LPN Problem

Let k be a security parameter. If $\mathbf{x}, \mathbf{g}_1, ..., \mathbf{g}_n$ are binary vectors of length k, let $y_i = <\mathbf{x} \cdot \mathbf{g}_i>$ denote the dot product of $\mathbf{x}$ and $\mathbf{g}_i$ (modulo 2). Given the values $\mathbf{g}_1, y_1; \mathbf{g}_2, y_2; ..., \mathbf{g}_n, y_n$, for randomly-chosen $\{\mathbf{g}_i\}_i^n$ and $n = O(k)$, it is possible to efficiently solve for $\mathbf{x}$ using standard linear-algebraic techniques. However, in the presence of noise where each y_i is flipped (independently) with probability p, finding $\mathbf{x}$ becomes much more difficult. We refer to the problem of learning $\mathbf{x}$ in this latter case as the LPN problem.

Note that p is usually taken to be a fixed constant independent of k, as will be the case in this work. The value of p to use depends on a number of tradeoffs and design decisions. Indeed the LPN problem becomes *harder* as p increases. However in certain authentication protocols where the security appears as a consequence of the LPN problem hardness, the larger the value of p is, the more often the honest prover becomes rejected. The hardness of the LPN_p problem (for constant $p \in (0, \frac{1}{2})$) has been studied in many previous works. Particularly note that the LPN problem can be formulated also as the problem of decoding a random linear block code [1,18] and has been shown to be NP-complete [1]. Beside the worst-case hardness results, there are numerous studies about the average-case hardness of the problem (see for example [2,3,4,10,18]).

In this work, we further investigate the formulation of the LPN_p problem as that of decoding a random linear block code by noticing that the rate k/n of this code is quite low and that the noise level p of the underlying binary symmetric channel is quite high. Both observations also hold for the fast correlation attack for which a similar parallelism with decoding a random linear code has been exploited [19,15]. For this work, we reformulate the LPN problem after introducing the following notations:

- $\mathbf{x} = [x_i]_{i=1}^k$ is a k-dimensional binary vector;
- $\mathbf{G} = [\mathbf{g}_i]_{i=1}^n$ is a $k \times n$ binary matrix and each $\mathbf{g}_i = [g_i(j)]_{j=1}^k$ is a k-dimensional binary column vector;
- $\mathbf{y} = [y_i]_{i=1}^n$ and $\mathbf{z} = [z_i]_{i=1}^n$ are n-dimensional binary vectors;
- For each $i = 1, 2, ..., n$, e_i is a realization of a binary random variable E_i such that $Pr(E_i = 1) = p$ and $Pr(E_i = 0) = 1 - p$, and all the random variables E_i are mutually independent.

For given $\mathbf{G}$ and $\mathbf{x}$, the vectors $\mathbf{y}$ and $\mathbf{z}$ are specified as follows: $\mathbf{y} = \mathbf{x} \cdot \mathbf{G}$, $z_i = y_i \oplus e_i$, $i = 1, 2, ..., n$. Accordingly we have the following formulation of the LPN problem relevant for this paper.

LPN Problem. For given $\mathbf{G}$, $\mathbf{z}$ and p, recover $\mathbf{x}$.

3   The BKW Algorithm

3.1   Preliminaries

In [3], the BKW algorithm has been reported for solving the LPN problem. It is based on the following paradigm:

a) draw s strings (columns of the matrix $\mathbf{G}$);
b) partition the s strings into groups based on their last b bits;
c) choose a string from each partition, add it to every other string within that partition, and then discard it;
d) repeat on the procedure to the second-to-last b bits, third-to-last ... until all that remain are partitions in which only the first b bits are non-zero.

In every step of the algorithm $\mathrm{poly}(k, 2^b, s)$ operations are performed, since for each of the partitions (at most 2^b partitions), at most s k-bit strings are added together.

Based on the explanations given in [3], the BKW algorithm is re-written in an explicit algorithmic form in the next section. The core underlying idea of the BKW algorithm can be considered as an instance of the generalized birthday paradox approach [20].

3.2   Algorithm

- *Input*
  - matrix $\mathbf{G}$, vector $\mathbf{z}$ and the probability of error p.
- *Initialization*
  - Set the algorithm parameters: integers a and b such that[1] $ab \geq k$.
  - Select the parameter q according to $q = f((1 - 2p)^{-2^a}, b)$, where $f(\cdot)$ is a polynomial function.
  - Consider each $\mathbf{g_i}$ from $\mathbf{G}$ as consisting of a concatenated segments, labeled as $1, 2, ..., a$, each composed of b bits.
  - Set the algorithm parameter $\alpha = 1$.
- *Processing*
  1. Repeat the following steps (a) - (g) q times :
     (a) Randomly select a subset Ω of $a2^b$ previously not considered columns of $\mathbf{G}$.

---

[1] The method for selecting the values a and b is not relevant for the algorithm description itself.

(b) Classify the elements of Ω into at most 2^b categories Ω_j such that all the vectors $\mathbf{g_i} \in \Omega_j$ have the identical segment of b bits labeled as a.

(c) In each category Ω_j, randomly select a vector and do the following:
- modify the vectors within the category by performing bit-by-bit XOR-ing of the selected vector with all other vectors in the category, yielding that all the modified vectors have all zeros in the last, a-th segment;
- remove the selected vector from Ω_j;
- form the updated / modified Ω as the union of all Ω_i; the expected number of elements in the updated Ω is $(a-1)2^b$.

(d) Classify the elements of the current set Ω into at most 2^b categories Ω_j such that all the vectors $\mathbf{g_i} \in \Omega_j$ have the identical segment labeled as $a-1$, recalling that all the vectors contain the all zero b-tuple in the segment labeled as a.

(e) In each category Ω_j, randomly select a vector and do the following:
- modify the vectors within the category by performing bit-by-bit XOR-ing of the selected vector with all other vectors in the category, yielding that all the modified vectors have all zeros in the segment $a-1$ (as well as in the segment with label a) ;
- remove the selected vector from Ω_j;
- form the updated / modified Ω as the union of all Ω_i; the expected number of elements in the updated Ω is $(a-2)2^b$.

(f) Repeat $a-3$ times the procedure performed in the previous two steps, so that the last modification of Ω contains on average 2^b vectors with only zeros in all the segments with labels from 2 to a.

(g) For each $\ell = 1, 2, ..., b$, do the following:
  i. Based on the vector, if it exists, from the current Ω with all zeros at the positions $1, 2, ..., \ell-1$ and $\ell+1, \ell+2, ..., k$, generate an estimate about x_ℓ which is correct with probability equal to $0.5 + 0.5(1-2p)^{2^{a-1}}$;
  ii. If the targeting vector does not exist in the currently considered collection repeat the steps (a) - (f).

2. For each $\ell = 1, 2, ..., b$, do the following:
Employing majority logic based decoding on q individual estimates of x_ℓ, generate the final estimate on x_ℓ which is assumed correct with a probability close to 1 for q large enough.

3. Set $\alpha \rightarrow \alpha + 1$ and do the following:
- if $\alpha > a$ go to Output step;
- if $\alpha \leq a$ perform the following re-labeling and go to Processing Step 1:
  - $1 \rightarrow a$;
  - for each $i = 2, 3, ..., a$, re-label $i \rightarrow i - 1$.

− *Output*
Estimation of **x**.

3.3   Complexity Analysis

According to the structure of the BKW algorithm, and the results reported in [3,12] we have the following statements.

Proposition 1 [3]. The required sample size and the time complexity of the BKW algorithm can be estimated as $f((1-2p)^{-2^a}, 2^b)$ where $f(\cdot)$ is a polynomial function.

Proposition 2 [12]. The BKW algorithm requires a sample of dimension proportional to $a^3 m 2^b$ where $m = \max\{(1-2p)^{-2^a}, b\}$.

Proposition 3 [12]. The time complexity of the BKW algorithm is proportional to $C a^3 m 2^b$ where $m = \max\{(1-2p)^{-2^a}, b\}$, and C is a constant.

These estimates can be further refined. In particular, Proposition 3 can be put into a more precise form, and the space complexity of the BKW algorithm can be estimated via analysis of the re-written BKW algorithm as follows.

Required Sample
The required sample, i.e. the dimension n of the matrix $\mathbf{G}$ for the BKW algorithm execution depends on the following:
• Each execution of Step 1.(a) requires random drawing of $a 2^b$ previously not considered columns of the matrix $\mathbf{G}$;
• The structure of the BKW algorithm implies that the expected number of executions of Step 1.(a) executions is proportional to $a^2 q = a^2 (1-2p)^{-2^a}$.
   These considerations imply that the required sample of the BKW algorithm is proportional to $a^3 (1-2p)^{-2^a} 2^b$ which is in accordance with Proposition 2.

Time Complexity
The time complexity of the BKW algorithm depends on the following:
• Each repetition of Steps 1.(a) to 1.(f) has time complexity proportional to $a 2^b$;
• The expected number of repetitions of Steps 1.(a) to 1.(f) implied by Step 1.(g).(ii) is a (according to [3,12]);
• Steps 1.(a) to 1.(g) should be repeated q times, with $q = (1-2p)^{-2^a}$;
• Step 3 requires that Steps 1 and 2 should be repeated a times;
• Each bit-by-bit mod2 addition of two k-dimensional vectors with all zeros in the last αb positions has cost proportional to $k - \alpha b$;
• The decoding of a bit of the vector $\mathbf{x}$ involves $(1-2p)^{-2^a}$ parity-checks, implying a complexity proportional to $(1-2p)^{-2^a}$ with a direct approach.
   Based on these remarks, Proposition 2 can be reformulated in the following more precise form.

Proposition 4. The average time complexity of the BKW algorithm is proportional to $a^3 (k/2)(1-2p)^{-2^a} 2^b + k(1-2p)^{-2^a}$.

According to Proposition 4, the decoding time complexity per bit of the BKW algorithm is proportional to $a^3 (1-2p)^{-2^a} 2^{b-1} + (1-2p)^{-2^a}$.

Space Complexity

The space complexity of the BKW algorithm is dominated by the dimension of the matrix $\mathbf{G}$ noting that its parameter n should be greater than the required sample.

Proposition 5. The space complexity of the BKW algorithm is proportional to $ka^3(1-2p)^{-2^a}2^b$.

4   An Algorithm for Solving the LPN Problem

The proposed algorithm for solving the LPN problem originates from the algorithms developed for the fast correlation attack against certain stream ciphers. However, there are also a few differences. The most substantial difference is that the developed algorithm does not contain any pre-processing phase. As a result, the pre-processing phase employed in fast correlation attacks has to be performed so that its computational cost becomes close to that of the processing phase itself.

4.1   Advanced Fast Correlation Attack Approach for the LPN Problem

The LPN problem is equivalent to the model of the underlying problem regarding cryptanalysis of certain stream ciphers. For this model of cryptanalysis, a number of powerful fast correlation attacks have been developed including those reported in [17,5,8,16].

Both, the LPN problem and the fast correlation attack require solving a heavily overdefined but consistent system of linear equations whose variables are corrupted by noise.

In its general presentation, the advanced fast correlation attack (see [16]) is a certain decoding technique based on a pre-processing phase and a processing phase which employs: (i) sample decimation; (ii) mapping based on linear combining; (iii) hypothesis testing; (iv) final decoding.

The pre-processing phase of a fast correlation attack is done only once, and during that phase suitable linear equations (parity-checks) are determined. The pre-processing phase is independent of the sample which is the input for the processing phase. In general, the complexity of preprocessing can be much higher than that of processing as done only once.

The processing phase can be considered as the minimum distance decoding based on the parity checks determined during the pre-processing.

In solving LPN problem no preprocessing is allowed as the parity-checks are sample dependent. This constitutes the main difference between the two approaches.

In the following section, an algorithm for solving the LPN problem is presented based on these steps [17,5,8,16].

4.2   Algorithm

The following algorithm uses three parameters b_H, b_0 and w which need to be optimized. The optimization of these values follows the presentation of the algorithm.

- *Input*
  - the matrix $\mathbf{G}$, the vector $\mathbf{z}$ and the probability of error p .
- *Initialization*
  - Select the algorithm parameters: b_H, b_0 and w.
  - Select the integer parameter q such that [8]:

$$q \geq (1 - 2p)^{-2w} .$$

  - Form the $(k + 1) \times n$ matrix $\mathbf{G_e}$ obtained by adding the vector $\mathbf{z}$ to the top row of $\mathbf{G}$ (z_i becomes the 0-th element of the i-th column of $\mathbf{G_e}$).
- *Processing*
  1. **Phase I: Decimation of the matrix $\mathbf{G_e}$**
     (a) Search over the columns of the matrix $\mathbf{G_e}$ and select all the columns $\mathbf{g}_i$ such that $g_i(j) = 0$ for $j = k - b_0 + 2, k - b_0 + 3, ..., k + 1$.
     (b) Form the decimated version $\mathbf{G^*}$ of the matrix $\mathbf{G_e}$ based on the columns selected in the previous step; $\mathbf{G^*}$ is a $(k + 1) \times n^*$ matrix with on average, $n^* = 2^{-b_0} n$.
  2. **Phase II: Linear Combining of the Decimated Columns**
     (a) Search for mod2 sums of up to w columns of $\mathbf{G^*}$ such that the resultant vector has any weight in positions $j = 0$ to $j = b_H$ and weight 1 in positions $j = b_H + 1$ to $k - b_0$.
     (b) For each $j = b_H + 1$ to $k - b_0$, record at least q such columns to form the matrix Ω_j.
  3. **Phase III: Hypothesizing and Partial Decoding**
     Consider all 2^{b_H} possible hypotheses for bits $x_0, \cdots, x_{b_H - 1}$ of $\mathbf{x}$ which correspond to rows $j = 1$ to $j = b_H$ in $\mathbf{G^*}$ and perform the following:
     (a) Select a previously not considered hypothesis on the first b_H bits of the vector $\mathbf{x}$; if no one new hypothesis is possible go to the Phase IV.
     (b) For the given hypothesis on the first b_H bits of the vector $\mathbf{x}$, employing the matrix Ω_j, estimate x_j based on

$$(1 \ \mathbf{x}) \cdot \Omega_j = \mathbf{0} \tag{1}$$

     for each $j = b_H + 1, \cdots k - b_0$. To this end, a majority rule on all the decisions x_j needed to satisfy (1), is used.
     (c) Compute the Hamming weight of $(1 \ \mathbf{x}) \cdot \mathbf{G^*}$ to check the validity of the given hypothesis (see [19,17] for details).
     (d) - Memorize the first $k - b_0$ positions of the L most likely vectors $\mathbf{x}$ found so far (list decoding of size L, with $L \ll 2^{b_H}$);
         - Go to the step (a) of Phase III.

4. **Phase IV: Final Decoding**
   (a) For the L vectors $\mathbf{x}$ recorded in Phase-III, repeat Phase-II based on $\mathbf{G_e}$ (or a punctured version $\mathbf{G_p}$) and Phase-III to estimate the decimated bits $x_{k-b_0}, \cdots, x_{k-1}$.
   (b) Select the most likely vector $\mathbf{x}$ based on the Hamming weight of $(1 \ \mathbf{x}) \cdot \mathbf{G_p}$.

- *Output*
  Estimation of $\mathbf{x}$.

4.3 Complexity Analysis of the Proposed Algorithm

For given parameters b_0, b_H and w, we obtain the following results.

Theorem 1. The average time complexity C of the proposed algorithm is dominated by:

$$C \sim (k - b_0) \left(\binom{n^*/2}{\lceil w/2 \rceil} + 2^{b_H} w \log_2(1 - 2p)^{-2} \right) \tag{2}$$

Proof. Denote by C_I, C_{II}, C_{III} and C_{IV} the average time complexities of the algorithm phases I to IV, respectively. According to the algorithm structure, the overall time complexity C is given as: $C = C_I + C_{II} + C_{III} + C_{IV}$.

The identification of the columns of the matrix $\mathbf{G}$ to construct $\mathbf{G}^*$ can be done during the sample collection phase and therefore has high level of parallelism. It follows that $C_I = O(n)$ but this complexity (which may become dominant compared to C) is discarded as it can be assumed that $\mathbf{G}^*$ is given along with $\mathbf{G}$ at the beginning of the algorithm. Note that neglecting this scanning task is a common practices employed and reported in the literature: see e.g. [11, Table I], [6, p. 354], [7, Table 2], [14, Remark 5]). (An additional justification for neglecting the screening complexity originates from the fact that in certain scenarios where the same secret key is assigned to a group of entities, the sample collection-screening can be performed in parallel.)

Phase-IV can be viewed as repeating the LPN problem to retrieve b_0 bits instead of $k - b_0$. As a result, for $b_0 < k/2$, C_{IV} can also be discarded as corresponding to solving the same problem, but for a smaller size. It follows $C \sim C_{II} + C_{III}$.

Phase-II can be viewed as constructing parity check equations of weight[2] $w+1$. Using the square-root algorithm proposed in [5] in conjunction with the hashing approach described in [20], the time complexity of this part is $O\left(\binom{n^*/2}{\lceil w/2 \rceil}\right)$ for each position $j = b_H + 1, \cdots, k - b_0$.

Phase-III can be viewed as evaluating the parity check equations found in Phase-II for each position $j = b_H + 1, \cdots, k - b_0$. Again using the results of [5] based on Walsh transform, the resulting complexity for q parity check equations is $O\left(2^{b_H} \log_2 q\right)$, where it is implicitly assumed $b_H \geq \log_2 q$ (which corresponds

---

[2] Recall that columns of $\mathbf{G}$ summing to zero form codewords of the dual code, or equivalently parity check equations.

to most cases of interest). Based on [8], we need $q \geq (1 - 2p)^{-2w}$, so that for each j, the complexity of Phase-III becomes $O\left(2^{b_H} w \log_2(1 - 2p)^{-2}\right)$.

Joining together the established partial dominating complexities, we obtain the theorem statement.

Theorem 2. The average space complexity M of the proposed algorithm is dominated by:

$$M \sim (k - b_0)\left(n^* + \binom{n^*/2}{\lfloor w/2 \rfloor}\right) \tag{3}$$

Proof. The space complexity to store the matrix $\mathbf{G}^*$ is simply $(k - b_0)n2^{-b_0}$. The memory requirement needed to construct each matrix Ω_j based on the method of [5] is $O\left(\binom{n^*/2}{\lfloor w/2 \rfloor}\right)$. Finally assuming again that $b_0 < k/2$, the memory requirement for Phase-IV can be discarded as in the worst case, it is of the same order as that considered in Phases I to III. The previous discussion directly imply the theorem statement.

Based on [8], we readily verify that the size of the sample given by Proposition 2 is larger than that required for the proposed algorithm with $w \leq 2^{a-1}$.

4.4   Optimization of the Parameters b_0 and b_H

Based on Theorem 1, we have $C \sim C_{II} + C_{III}$. With respect to the fast correlation attack, it can be observed that C_{II} corresponds to the pre-processing cost (searching for a sufficiently large number of parity check equations of a given form) and C_{III} corresponds to the processing cost (solving the system of parity check equations for a given sample). As a result, for the LPN problem, we have the additional constraint $C_{II} \sim C_{III}$.

For given values w and b_0, the value b_H which minimizes C_{III} is given by [8]

$$b_{H,opt} = 2w \log_2(1 - 2p)^{-1} + k - b_0 - \log_2\binom{n^*}{w}. \tag{4}$$

For this value, we have

$$\log_2 C_{III,opt} \sim$$

$$2w \log_2(1 - 2p)^{-1} + k - b_0 - \log_2\binom{n^*}{w} + \log_2(k - b_0) + \log_2 w + \log_2 \log_2(1 - 2p)^{-2}. \tag{5}$$

Equating $C_{III,opt}$ with C_{II} given by

$$\log_2 C_{II} \sim \log_2(k - b_0) + \log_2\binom{n^*/2}{\lceil w/2 \rceil} \tag{6}$$

and solving for b_0, we obtain

$$b_{0,opt} \approx \left(\frac{3w}{2} - 1\right)^{-1}\left(\frac{3w}{2}(\log_2 n^* - \log_2 w) - 2w \log_2(1 - 2p)^{-1}\right.$$
$$\left. -k + \frac{w}{2} - \log_2 w - \log_2 \log_2(1 - 2p)^{-2}\right) \tag{7}$$

Table 1. Time complexities C_{II} and C_{III} for the LPN problem with $p = 0.25$

n	k	w	$b_{0,opt}$	$b_{H,opt}$	$\log_2 C_{II}$	$\log_2 C_{III}$
2^{24}	32	2	16	5 ($> \log_2 q = 4$)	11	11
		4	18	3 ($< \log_2 q = 8$)		
2^{80}	224	4	47	58	70.5	67.9
		6	58	56	67.8	66.6
		8	63	57	66.7	67.7

Based on (4) and (7), Table 1 summarizes the complexities C_{II} and C_{III} for $p = 0.25$ and values of n and k relevant to the LPN approach to attack the HB protocol [12]. For the $(2^{24}, 32)$ code, we observe that while for $w = 2$, C_{II} and C_{III} are efficiently balanced, the optimum value $b_{H,opt}$ for $w = 4$ no longer verifies $b_{H,opt} > w \log_2(1 - 2p)^{-2}$ so that the approach of [5] has reached its minimum cost. For the $(2^{80}, 224)$ code, we observe that different selections of w provide complexities C_{II} and C_{III} very close after proper optimization of both b_0 and b_H. In that case, selecting the smallest value w minimizes memory requirements.

5   Comparison of the Proposed Algorithm and the BKW Algorithm

It can be observed that the $(a - 1)$ repetitions of the procedure to obtain vectors of weight 1 in the BKW algorithm is equivalent to constructing parity-check equations (or codewords in the dual codes) of weight $w = 2^{a-1} + 1$. Interestingly, both approaches require exactly the same number $(1 - 2p)^{2w}$ of check sums per bit. The results of Table 1 indicate that the choice of w is not critical in minimizing the time complexity of the proposed algorithm. However the techniques used to generate these check sums are very different as well as that to estimate the bits. Table 2 compares the time and space complexities per information bit obtained from the results of Sections 3 and 4.

In order to establish a clearer comparison, we consider the special case $w = 2^{a-1}$ so that the same number of check sums is required by both approaches and $b = b_H = b_0$ (again this case corresponds to a meaningful comparison when applying these algorithms to the HB protocol). The corresponding complexities are given in Table 3. We observe that even in this non optimized case, the proposed algorithm is more efficient in most cases. In particular for $a = 4$, $(k - 2b)/2 = b$ so that a factor $(\log_2 w + 1)^3 (1 - 2p)^{-w}$ is gained both in time and space complexity. The gain $(1 - 2p)^{-w}$ is mostly due to the application of the square-root algorithm to determine the check sums based on [5,20]. However, gains even larger than $(\log_2 w + 1)^3 (1 - 2p)^{-w}$ are available by proper selection of the parameters b_0 and b_H due to both decimation and hypothesis testing.

Note that in a particular case considered previously when both algorithms employ the same number of parity checks of the same weight, the same probability of success is expected, although the complexities of the algorithms are

Table 2. Time and space complexities per information bit of the BKW algorithm and the proposed algorithm

	time complexity	space complexity
BKW algorithm	$\sim (k/b)^3 (1-2p)^{-2^{k/b}} 2^b$	$\sim (k/b)^3 (1-2p)^{-2^{k/b}} 2^b$
Proposed algorithm	$\sim \binom{n^*/2}{\lceil w/2 \rceil} + 2^{b_H} \log_2(1-2p)^{-2w}$	$\sim n^* + \binom{n^*/2}{\lfloor w/2 \rfloor}$

Table 3. Time and space complexities per information bit of the BKW algorithm and the proposed algorithm with $w = 2^{a-1}$ and $b = b_H = b_0$

	time complexity	space complexity
BKW algorithm	$\sim (\log_2 w + 1)^3 (1-2p)^{-2w} 2^b$	$\sim (\log_2 w + 1)^3 (1-2p)^{-2w} 2^b$
Proposed algorithm	$\sim (1-2p)^{-w} 2^{(k-2b)/2} + 2^b \log_2(1-2p)^{-2w}$	$\sim (1-2p)^{-w} 2^{(k-2b)/2}$

different. A general discussion of the success probability is out of the scope due to the paper length limitation.

6   Security Re-evaluation of the HB Protocol for RFID Authentication

6.1   Security of the HB Protocol and LPN Problem

It is well known that the security of the HB protocol, as well as HB$^+$ protocol, depends on the complexity of solving the LPN problem (see for example [12,13]). The two main issues regarding the security evaluation of the HB protocol can be summarized as follows:

• Collecting the sample for cryptanalysis via recording the challenges and responses exchanged during the protocol executions and forming the matrix $\mathbf{G}$ and the vector $\mathbf{z}$ which define the underlying LPN problem; each challenge is recorded as a certain column $\mathbf{g_i}$ of the matrix $\mathbf{G}$, and its corresponding response is recorded as the element z_i of the vector $\mathbf{z}$;
• Solving the obtained LPN problem.

Regarding the sample collection for cryptanalysis note the following issues: (i) Each authentication session involves the same secret key, and consequently all the available challenge-response pairs can be jointly employed for recovering the secret key; (ii) Each authentication session involves r mutually independent challenges and responses, providing r columns of the matrix $\mathbf{G}$ and the corresponding r elements of the vector $\mathbf{z}$, so that via collecting the pairs from

Table 4. Security margins of the HB protocol against a passive attack, when the employed key consists of k bits and the employed noise corresponds to $p = 0.25$, based on the BKW and proposed algorithms

number of secret key bits: k	security margin for the BKW algorithm [12]	security margin for the proposed algorithm
$k = 32$	$\sim 2^{23}$	$\sim 2^8$
$k = 224$	$\sim 2^{80}$	$\sim 2^{61}$

s authentication sessions, we obtain the matrix $\mathbf{G}$ and the vector $\mathbf{z}$ of dimensions $k \times n$ and n, respectively, where $n = s \cdot r$.

Importantly, note that for this particular application, the decimation can be realized directly by recording only the challenges with zeros in their last b_0 positions if n is known a-priori. This further justifies that C_I can be discarded in Theorem 1 as well as the corresponding memory requirement in Theorem 2.

6.2   Security Evaluation of the HB Protocol Based on the BKW and Proposed Algorithms

The security evaluation considered in this section assumes the passive attacking scenario only. Following the approach for security evaluation employed in [12], we consider the security margin which measures the complexity of recovering a bit of the secret key. As a simple illustrative example, we first consider the security evaluation of the HB protocol with $k = 32$, $p = 0.25$, assuming we can collect a sample of size $n = 2^{24}$. Accordingly, we consider the security margin of the protocol employing the BKW algorithm with parameters $a = 2$ and $b = 16$, as suggested in [12]. Based on Propositions 3 and 4, the expected time complexity of of the BKW algorithm is proportional to $ka^3(1 - 2p)^{-2^a}2^b = 32 \cdot 2^3 \cdot 2^4 \cdot 2^{16}$, so that the security margin (expected complexity per recovered secret key bit) is 2^{23}. For the proposed algorithm, we select $w = 2$, $b_0 = 16$ and $b_H = 5$ based on Table 1, so that the expected complexity per recovered secret key bit becomes $2 \cdot 2^{11}/16 = 2^8$. As a result, the proposed algorithm reduces the time complexity of the BKW algorithm almost to its cubic root for this simple example.

For security evaluation of the HB protocol with $k = 224$, $n = 2^{80}$ and $p = 0.25$, we have considered the BKW algorithm with $a = 4$ and $b = 56$, (see [12]) , while based on Table 1, we selected $w = 6$, $b_0 = 58$ and $b_H = 56$ for the proposed algorithm. The corresponding security margins are 2^{80} and 2^{61}, respectively. Again a significant gain has been achieved by the proposed algorithm. These results of security evaluation are summarized in Table 4

With respect to the special case considered Table 3, we observe that by proper optimization of all parameters, gains beyond the factor $(\log_2 w + 1)^3(1 - 2p)^{-w}$ have been achieved by the proposed algorithm over the BKW algorithm as this factor takes values 2^5 and 2^{14} for $k = 32$ and $k = 224$, respectively, while the corresponding gains are 2^{15} and 2^{19}.

7 Concluding Discussion

In this paper an advanced algorithm for solving the LPN problem has been proposed. This algorithm originates from the fast correlation attacks. However, as opposed to fast correlation attacks, no preprocessing phase is allowed for solving the LPN problem. As a result, step equivalent to the pre-processing phase has been included as initialization step of the proposed algorithm. The complexity of this step has to be balanced with that of the step corresponding to processing in the fast correlation attack.

Assuming the same sample as input for the BKW algorithm and the algorithm proposed in this paper, the origins for the gains obtained by the proposed algorithm appear from its more sophisticated structure with additional parameters to be optimized. Particularly, the origins of the gain could be summarized as follows: (i) Employed selection (decimation) of the parity checks during the sample collection so that only suitable parity-checks are involved in processing, yielding a reduction in the complexity; (ii) Appropriate balancing of the dedicated hypotheses testing and the complexity of decoding under these given hypotheses.

The proposed algorithm for solving the LPN problem can be reduced to the BKW algorithm, i.e. the BKW algorithm becomes a special case of the proposed algorithm. Consequently, the proposed algorithm always works at least as well as the BKW algorithm, and in many scenarios it yields an additional computational savings (as a consequence of possibility for optimization its flexible structure according to the given inputs).

The developed algorithm has been employed for security evaluation of the HB protocols because the security of both HB protocols, HB and HB$^+$, depends on the hardness of the underlying LPN problem. It has been shown that, assuming the same scenario for cryptanalysis, the developed algorithm for the LPN problem implies lower complexity of recovering the secret key than the BKW algorithm. As result the security margins of the HB protocols reported in [12, Appendix D] become overestimated.

References

1. E.R. Berlekamp, R.J. McEliece, and H.C.A. van Tilborg, "On the Inherent Intractability of Certain Coding Problems", *IEEE Trans. Info. Theory*, vol. 24, pp. 384-386, 1978.
2. A. Blum, M. Furst, M. Kearns, and R. Lipton, "Cryptographic Primitives Based on Hard Learning Problems", CRYPTO '93, *Lecture Notes in Computer Science*, vol. 773, pp. 278-291, 1994.
3. A. Blum, A. Kalai and H. Wasserman, "Noise-Tolerant Learning, the Parity Problem, and the Statistical Query Model", *Journal of the ACM*, vol. 50, no. 4, pp. 506-519, July 2003.
4. F. Chabaud, "On the Security of Some Cryptosystems Based on Error-Correcting Codes. EUROCRYT '94, *Lecture Notes in Computer Science*, vol. 950, pp. 113-139, 1995.

5. P. Chose, A. Joux and M. Mitton, "Fast Correlation Attacks: An Algorithmic Point of View", EUROCRYPT2002, *Lecture Notes in Computer Science*, vol. 2332, pp. 209-221, 2002.
6. N.T. Courtois and W. Meier, "Algebraic attacks on stream ciphers with linear feedback", EUROCRYPT'2003, *Lecture Notes in Computer Science*, vol. 2656, pp. 345-359, 2003.
7. N.T. Courtois, "Fast algebraic attacks on stream ciphers with linear feedback", CRYPTO'2003, *Lecture Notes in Computer Science*, vol. 2729, pp. 176-194, 2003.
8. M.P.C. Fossorier, M.J. Mihaljević and H. Imai, "A Unified Analysis for the Fast Correlation Attack", *2005 IEEE Int. Symp. Inform. Theory - ISIT'2005*, Adelaide, Australia, Sept. 2005, Proceedings, pp. 2012-2015 (ISBN 0-7803-9151-9).
9. H. Gilbert, M. Robshaw and H. Sibert, "An Active Attack against HB+ a Provably Secure Lightweight Authentication Protocol", IACR, Cryptology ePrint Archive, Report 2005/237, July 2005. Available at http://eprint.iacr.org/2005/237 .
10. N. Hopper and M. Blum, "Secure Human Identification Protocols", ASIACRYPT 2001, *Lecture Notes in Computer Science*, vol. 2248, pp. 52-66, 2001.
11. P. Hawkes and G. Rose, "Rewriting variables: the complexity of fast algebraic attacks on stream ciphers", CRYPTO 2004, *Lecture Notes in Computer Science*, vol. 3159, pp. 390-406, Aug. 2004.
12. A. Juels and S. Weis, "Authenticating Pervasive Devices with Human Protocols", CRYPTO2005, *Lecture Notes in Computer Science*, vol. 3621, pp. 293-308, 2005. Updated version available at: http://www.rsasecurity.com/rsalabs/staff/bios/ajuels/publications/pdfs/lpn.pdf
13. J. Katz and J.S. Shin, "Parallel and Concurrent Security of the HB and HB+ Protocols", EUROCRYPT2006, *Lecture Notes in Computer Science*, vol. 4004, pp. 73-87, 2006.
14. K. Khoo, G. Gong and H.-K. Lee, "The Rainbow Attack on Stream Ciphers Based on Maiorana-McFarland Functions", Applied Cryptography and Network Security - ACNS 2006, *Lecture Notes in Computer Science*, vol. 3989, pp. 194-209, 2006.
15. W. Meier and O. Staffelbach, "Fast Correlation Attacks on Certain Stream Ciphers," *Journal of Cryptology*, vol. 1, pp. 159-176, 1989.
16. M.J. Mihaljević, M.P.C. Fossorier and H. Imai, "A General Formulation of Algebraic and Fast Correlation Attacks Based on Dedicated Sample Decimation", AAECC2006, *Lecture Notes in Computer Science*, vol. 3857, pp. 203-214, 2006.
17. M.J. Mihaljević, M.P.C. Fossorier and H. Imai, "Fast Correlation Attack Algorithm with List Decoding and an Application", FSE2001, *Lecture Notes in Computer Science*, vol. 2355, pp. 196-210, 2002.
18. O. Regev, "On Lattices, Learning with Errors, Random Linear Codes, and Cryptography", *37th ACM Symposium on Theory of Computing*, Proceedings, pp. 84-93, 2005.
19. T. Siegenthaler, "Decrypting a Class of Stream Ciphers Using Ciphertext Only," *IEEE Trans. Comput.*, vol. C-34, pp. 81-85, 1985.
20. D. Wagner, "A Generalized Birthday Problem," CRYPTO '02, *Lecture Notes in Computer Science*, vol. 2442, pp. 288-304, 2002.

Update on Tiger[*]

Florian Mendel[1], Bart Preneel[2], Vincent Rijmen[1],
Hirotaka Yoshida[3], and Dai Watanabe[3]

[1] Graz University of Technology
Institute for Applied Information Processing and Communications
Inffeldgasse 16a, A–8010 Graz, Austria
{Florian.Mendel, Vincent.Rijmen}@iaik.tugraz.at
[2] Katholieke Universiteit Leuven, Dept. ESAT/SCD-COSIC,
Kasteelpark Arenberg 10, B–3001 Heverlee, Belgium
Bart.Preneel@esat.kuleuven.be
[3] Systems Development Laboratory, Hitachi, Ltd.,
1099 Ohzenji, Asao-ku, Kawasaki-shi, Kanagawa-ken, 215-0013 Japan
{hirotaka.yoshida.qv, dai.watanabe.td}@hitachi.com

Abstract. Tiger is a cryptographic hash function with a 192-bit hash value which was proposed by Anderson and Biham in 1996. At FSE 2006, Kelsey and Lucks presented a collision attack on Tiger reduced to 16 (out of 24) rounds with complexity of about 2^{44}. Furthermore, they showed that a pseudo-near-collision can be found for a variant of Tiger with 20 rounds with complexity of about 2^{48}.

In this article, we show how their attack method can be extended to construct a collision in the Tiger hash function reduced to 19 rounds. We present two different attack strategies for constructing collisions in Tiger-19 with complexity of about 2^{62} and 2^{69}. Furthermore, we present a pseudo-near-collision for a variant of Tiger with 22 rounds with complexity of about 2^{44}.

Keywords: cryptanalysis, hash functions, differential attack, collision, near-collision, pseudo-collision, pseudo-near-collision.

1 Introduction

Recent results in cryptanalysis of hash function show weaknesses in many commonly used hash functions, such as SHA-1 and MD5 [4,5]. Therefore, the cryptanalysis of alternative hash functions, such as Tiger, is of great interest.

In [2], Kelsey and Lucks presented a collision attack on Tiger-16, a round reduced variant of Tiger (only 16 out of 24 rounds), with complexity of about 2^{44}. In the attack they used a kind of message modification technique developed

---

[*] This work was supported in part by the Austrian Science Fund (FWF), project P18138. This work was supported in part by a consignment research from the National Institute on Information and Communications Technology (NiCT), Japan. This work was supported in part by the Concerted Research Action (GOA) Ambiorics 2005/11 of the Flemish Government.

R. Barua and T. Lange (Eds.): INDOCRYPT 2006, LNCS 4329, pp. 63–79, 2006.

for Tiger to force a differential pattern in the chaining variables after round 7, which can then be canceled by the differences in the expanded message words in the following rounds. This led to a collision in the Tiger hash function after 16 rounds. Furthermore, they showed that a pseudo-near-collision can be found in a variant of Tiger with 20 rounds in about 2^{48} applications of the compression function.

In this article, we extend the attack to construct a collision in Tiger-19. We present two different collision attacks on Tiger-19 with complexity of 2^{62} and 2^{69}. Furthermore, we present a pseudo-near-collision attack for a variant of Tiger with 22 rounds with complexity of about 2^{44} and a pseudo-collision attack for Tiger-23/128, a version of Tiger reduced to 23 rounds with truncated output, with complexity 2^{44}. A summary of our results is given in Table 1.

Table 1. Overview of attacks on the Tiger hash function

number of rounds	type	complexity	
Tiger-16	collision	2^{44}	in [2]
Tiger-19	collision	2^{62} and 2^{69}	in this article
Tiger-19	pseudo-collision	2^{44}	in this article
Tiger-21	pseudo-collision	2^{66}	in this article
Tiger-23/128	pseudo-collision	2^{44}	in this article
Tiger-20 [1]	pseudo-near-collision	2^{48}	in [2]
Tiger-21	pseudo-near-collision	2^{44}	in this article
Tiger-22	pseudo-near-collision	2^{44}	in this article

The remainder of this article is structured as follows. A description of the Tiger hash function is given in Section 2. The attack of Kelsey and Lucks on Tiger-16 is described in Section 3. In Section 4, we describe a method to construct collisions in Tiger-19. Another method for construction collisions in Tiger-19 is described in Section 5. Furthermore, we present a pseudo-near-collision for Tiger-22 in Section 6 and a pseudo-collision for Tiger-23/128 in Section 7. Finally, we present conclusions in Section 8.

2   Description of the Hash Function Tiger

Tiger is a cryptographic hash function that was designed by Ross Anderson and Eli Biham in 1996 [1]. It is an iterative hash function that processes 512-bit input message blocks and produces a 192-bit hash value. In the following, we briefly describe the hash function. It basically consists of two parts: the key-schedule and the state update transformation. A detailed description of the hash function is given in [1]. For the remainder of this article we use the same notation as is used in [2]. The notation is given in Table 2.

---

[1] Kelsey and Lucks show a pseudo-near-collision for the last 20 rounds of Tiger.

Table 2. Notation

Notation	Meaning
$A + B$	addition of A and B modulo 2^{64}
$A - B$	subtraction of A and B modulo 2^{64}
$A * B$	multiplication of A and B modulo 2^{64}
$A \oplus B$	bit-wise XOR-operation of A and B
$\neg A$	bit-wise NOT-operation of A
$A \ll n$	bit-shift of A by n positions to the left
$A \gg n$	bit-shift of A by n positions to the right
X_i	message word i (64-bits)
$X_i[\text{even}]$	the even bytes of message word X_i (32-bits)
$X_i[\text{odd}]$	the odd bytes of message word X_i (32-bits)

2.1 State Update Transformation

The state update transformation starts from a (fixed) initial value IV of three 64-bit registers and updates them in three passes of eight rounds each. In each round one 64-bit word X is used to update the three chaining variables A, B and C as follows.

$$C = C \oplus X$$
$$A = A - \mathbf{even}(C)$$
$$B = B + \mathbf{odd}(C)$$
$$B = B \times \mathtt{mult}$$

The results are then shifted such that A, B, C become B, C, A. Fig. 1 shows one round of the state update transformation of Tiger.

The non-linear functions **even** and **odd** used in each round are defined as follows.

$$\mathbf{even}(C) = T_1[c_0] \oplus T_2[c_2] \oplus T_3[c_4] \oplus T_4[c_6]$$
$$\mathbf{odd}(C) = T_4[c_1] \oplus T_3[c_3] \oplus T_2[c_5] \oplus T_1[c_7]$$

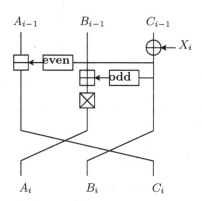

Fig. 1. The round function of Tiger

where C is split into eight bytes $c_0, \ldots, c_7$ where c_0 is the most significant byte. The four S-boxes $T_1, \ldots, T_4 : \{0,1\}^8 \rightarrow \{0,1\}^{64}$ are used to compute the output of the non-linear functions **even** and **odd**. For the definition of the four S-boxes we refer to [1]. Note that chaining variable B is multiplied with the constant mult $\in \{5, 7, 9\}$ at the end of each round. The value of the constant is different in each pass of the Tiger hash function.

After the last round of the state update transformation, the chaining variables A_{-1}, B_{-1}, C_{-1} and the output values of the last pass A_{23}, B_{23}, C_{23} are combined, resulting in the final value of one iteration (feed forward). The result is the final hash value or the initial value for the next message block.

$$A'_{23} = A_{-1} \oplus A_{23}$$
$$B'_{23} = B_{-1} - B_{23}$$
$$C'_{23} = C_{-1} + C_{23}$$

2.2 Key Schedule

Between two passes of Tiger, there is one key schedule. The key schedule is an invertible function which ensures that changing a small number of bits in the message will affect a lot of bits in the next pass. While the message words $X_0, \ldots, X_7$ are used in the first pass to update the chaining variables, the remaining 16 message words, 8 for the second pass and 8 for the third pass, are generated by applying the key schedule as shown below.

$$(X_8, \ldots, X_{15}) = \text{KeySchedule}(X_0, \ldots, X_7)$$
$$(X_{16}, \ldots, X_{23}) = \text{KeySchedule}(X_8, \ldots, X_{15})$$

The key schedule modifies the inputs $(Y_0, \ldots, Y_7)$ in two steps, as shown below.

first step

$$Y_0 = Y_0 - (Y_7 \oplus \text{A5A5A5A5A5A5A5A5})$$
$$Y_1 = Y_1 \oplus Y_0$$
$$Y_2 = Y_2 + Y_1$$
$$Y_3 = Y_3 - (Y_2 \oplus ((\neg Y_1) \ll 19))$$
$$Y_4 = Y_4 \oplus Y_3$$
$$Y_5 = Y_5 + Y_4$$
$$Y_6 = Y_6 - (Y_5 \oplus ((\neg Y_4) \gg 23))$$
$$Y_7 = Y_7 \oplus Y_6$$

second step

$$Y_0 = Y_0 + Y_7$$
$$Y_1 = Y_1 - (Y_0 \oplus ((\neg Y_7) \ll 19))$$
$$Y_2 = Y_2 \oplus Y_1$$
$$Y_3 = Y_3 + Y_2$$
$$Y_4 = Y_4 - (Y_3 \oplus ((\neg Y_2) \gg 23))$$
$$Y_5 = Y_5 \oplus Y_4$$
$$Y_6 = Y_6 + Y_5$$
$$Y_7 = Y_7 - (Y_6 \oplus \text{0123456789ABCDEF})$$

The final values $(Y_0, \ldots, Y_7)$ are the output of the key schedule and the message words for the next pass.

3 Previous Attack on Tiger

In this section, we will briefly describe the attack of Kelsey and Lucks on Tiger-16. A detailed description of the attack is given in [2]. For a good understanding of our results, it is recommended to study it very carefully. Space restrictions do not permit us to copy all the important details of the original attack. The attack on Tiger-16 can be summarized as follows.

1. Choose a characteristic for the key schedule of Tiger that holds with high probability (ideally with probability 1).
2. Use a kind of message modification technique [5] developed for Tiger to construct certain differences in the chaining variables for round 7, which can then be canceled by the differences in the message words in the following rounds. This leads to a collision in the Tiger hash function after 16 rounds.

In the following we will describe both parts of the attack in detail.

3.1 High Probability Characteristic for the Key Schedule of Tiger

For the attack Kelsey and Lucks used the key schedule difference given in (1). It has probability 1 to hold in the key schedule of Tiger. This facilitates the attack.

$$(I, I, I, I, 0, 0, 0, 0) \rightarrow (I, I, 0, 0, 0, 0, 0, 0) \tag{1}$$

Note that I denotes a difference in the MSB of the message word. Hence, the XOR difference (denoted by $\Delta^{\oplus}$) and the additive difference (denoted by Δ^{+}) is the same in this particular case.

To have a collision after 16 rounds, there has to be a collision after round 9 as well. Hence, the following differences are needed in the chaining variables for round 7 of Tiger.

$$\Delta^{\oplus}(A_6) = I, \quad \Delta^{\oplus}(B_6) = I, \quad \Delta^{\oplus}(C_6) = 0 \tag{2}$$

Constructing these differences in the chaining variables after round 6 is the most difficult part of the attack. Therefore, Kelsey and Lucks adapted the idea of message modification from the MD-family to Tiger. The main idea of message modification is to use the degrees of freedom we have in the choice of the message words to control the differences in the chaining variables. In the case of Tiger, the differential pattern given in (2) has to be met in order to have a collision after 16 rounds of Tiger.

3.2 Message Modification by Meeting in the Middle

In this section, we explain the idea of message modification in Tiger according to Fig. 2. Assume that the values of $(A_{i-1}, B_{i-1}, C_{i-1})$ and the additive differences $\Delta^{+}(A_{i-1})$, $\Delta^{+}(B_{i-1})$, $\Delta^{+}(C_{i-1})$ are known as well as the additive differences in the message words X_i and X_{i+1}. Then the additive difference $\Delta^{+}(C_{i+1})$ can be forced to be any difference δ^* with probability $1/2$ by applying the birthday attack. As depicted in Fig. 2, the additive difference $\Delta^{+}(C_{i+1})$ depends on the additive differences $\Delta^{+}(B_{i-1})$, $\Delta^{+}(\mathbf{odd}(B_i))$, and $\Delta^{+}(\mathbf{even}(B_{i+1}))$.

For any nonzero XOR difference $\Delta^{\oplus}(B_{i+1}[\mathbf{even}])$, one expect about 2^{32} *different* corresponding additive output differences $\Delta^{+}(\mathbf{even}(B_{i+1}))$. Similarly, for

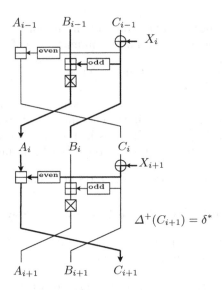

Fig. 2. Outline of the message modification step in Tiger

any nonzero XOR difference $\Delta^{\oplus}(\mathbf{odd}(B_i))$, one expect close to 2^{32} corresponding *different* additive output differences $\Delta^+(\mathbf{odd}(B_i))$.

Thus, if the XOR differences $\Delta^{\oplus}(B_{i+1}[\mathbf{even}])$ and $\Delta^{\oplus}(B_i[\mathbf{odd}])$ both are nonzero, a meet-in-the-middle (MITM) approach can be applied to solve the following equation:

$$\mathbf{mult} \times (\Delta^+(B_{i-1}) + \Delta^+(\mathbf{odd}(B_i))) - \Delta^+(\mathbf{even}(B_{i+1})) = \delta^* \ .$$

This is done by performing the following two steps:

1. Store the 2^{32} candidates for $\Delta^+(\mathbf{odd}(B_i))$ in a table.
2. For all 2^{32} candidates for $\Delta^+(\mathbf{even}(B_{i+1}))$, test if some $\Delta^+(\mathbf{odd}(B_i))$ exists with $\Delta^+(\mathbf{odd}(B_i)) = (\Delta^+(\mathbf{even}(B_{i+1})) + \delta^*)/(\mathbf{mult}) - \Delta^+(B_{i-1})$.

This technique takes 2^{33} evaluations of each of the functions **odd** and **even**, which is equivalent to about 2^{29} evaluations of the compression function of Tiger reduced to 16 rounds and some 2^{33} 64-bit word units of storage space.

Note that if the choice of the values of the message words X_i and X_{i+1} is constrained by k-bits then the success probability of the message modification step is reduced by a factor of 2^k. This is referred to as a constrained message modification step.

3.3 The Collision Attack on Tiger-16

With the key schedule difference given in Section 3.1 and the new developed message modification technique for Tiger described in Section 3.2, Kelsey and Lucks show a collision attack on Tiger reduced to 16 rounds. The method can be summarized as follows (see [2]).

0. Precomputation: Find an additive difference L^+ with a low Hamming weight XOR difference $L^\oplus$ which can be canceled out by a suitable choice for $X_6[\textbf{even}]$. In the analysis Kelsey and Lucks assume, that an additive difference L can be found which is consistent to an 8-bit XOR difference L^{xor}. This step of the attack has a complexity of about 2^{27}.

1. Choose suitable values for $X_0, X_1, X_2[\textbf{even}]$ such that $\Delta^\oplus(A_2)$, $\Delta^\oplus(B_2)$, $\Delta^\oplus(C_2)$ are *useful*. A difference is called *useful* if there are differences in the even and odd bytes of the word. This step adds negligible cost to the attack complexity.

2. Do a message modification step to get a suitable XOR-difference L^{xor} in C_3 which is consistent with the additive difference L of the precomputation step. This step has complexity of about 2^{36} and determines the message words $X_2[\textbf{odd}]$ and $X_3[\textbf{even}]$.

3. Do a constrained message modification step to get $\Delta^\oplus(C_4) = I$. This determines $X_3[\textbf{odd}]$ and $X_4[\textbf{even}]$. Completing this step has complexity of about 2^{40}. This is due to the fact that 8 bits of X_4 (4 bits in $X_4[\textbf{even}]$ and 4 bits in $X_4[\textbf{odd}]$) are constrained by the transition of the XOR difference L^{xor} in C_3 to the additive difference L in B_4.

4. Do a constrained message modification step to get $\Delta^\oplus(C_5) = I$. This determines $X_4[\textbf{odd}]$ and $X_5[\textbf{even}]$. Completing this step has complexity of about 2^{44}.

5. Determine $X_6[\textbf{even}]$ by using C_5 and the results of the precomputation step. This adds no additional cost to the attack complexity.

Hence, a collision in Tiger-16 can be found with a complexity of about 2^{44} applications of the compression function. In the attack a characteristic for the key schedule differences is used which has probability 1 as well as a message modification technique developed for Tiger to force certain differences in the chaining variables after round 6 which can then be canceled by the differences in the expanded message words X_8 and X_9. For a detailed description of the attack we refer to [2].

4   A Collision Attack on Tiger-19 – Method 1

In this section we present a collision attack on Tiger-19 with complexity of about 2^{62} hash computations. First, we show how the attack of Kelsey and Lucks can be extended to construct a pseudo-collision in Tiger-19 with complexity of about 2^{44} hash computations. Second, we show how this pseudo-collision can be turned into a collision for Tiger-19 by using a kind of neutral bit technique. The collision attack on Tiger-19 has a complexity of 2^{62} hash computations.

4.1   A Pseudo-collision for Tiger-19

In this section we will show how to construct a pseudo-collision for Tiger-19 with a complexity of about 2^{44}. The attack is an extension of the attack of Kelsey and Lucks on Tiger-16.

To construct a pseudo-collision in Tiger-19 we use the key schedule difference given in (3). It has probability 1 to hold in the key schedule of Tiger which facilitates the attack.

$$(0, 0, 0, I, I, I, I, 0) \rightarrow (0, 0, 0, I, I, 0, 0, 0) \rightarrow (0, 0, 0, I, I, I, I, I) \qquad (3)$$

Note that the key schedule difference from round 3 to 18 is the 16-round difference used by Kelsey and Lucks in the attack on Tiger-16. Hence, we can use the same attack strategy which was used to break Tiger-16 in the attack on Tiger-19 as well. The attack work as follows:

1. Choose arbitrary values for the chaining variables A_2, B_2, C_2 for round 3.
2. Employ the attack on 16 rounds, to find message words $X_3, \ldots, X_7$ and $X_8[\text{even}], X_9[\text{even}]$ such that the output after round 18 collides.
3. To compute the real message words $X_0, \ldots, X_7$, we have to choose suitable values for $X_8[\text{odd}], X_9[\text{odd}]$ and $X_{10}, \ldots, X_{15}$ such that X_4, X_5, X_6 and X_7 are correct after computing the key schedule backward. Note that X_3 can be chosen freely, because we can modify C_2 such that $C_2 \oplus X_3$ stay constant. In detail, we choose arbitrary values for $X_8[\text{odd}], X_9[\text{odd}], X_{10}, X_{11}$ and calculate $X_{12}, \ldots, X_{15}$ as follows.

$$X_{12} = (X_4 \oplus (X_{11} - X_{10})) - (X_{11} \oplus (\neg X_{10} \gg 23))$$
$$X_{13} = (X_5 + (X_{12} + (X_{11} \oplus (\neg X_{10} \gg 23)))) \oplus X_{12}$$
$$X_{14} = (X_6 - (X_{13} \oplus X_{12} \oplus (\neg(X_{12} + (X_{11} \oplus (\neg X_{10} \gg 23))) \gg 23))) + X_{13}$$
$$X_{15} = (X_7 \oplus (X_{14} - X_{13})) - (X_{14} \oplus \texttt{0123456789ABCDEF})$$

This adds negligible cost to the attack complexity and guarantees that X_4, X_5, X_6 and X_7 are always correct after computing the key schedule backward.
4. To compute the initial chaining values A_{-1}, B_{-1} and C_{-1} run the rounds 2, 1 and 0 backwards.

Hence, we can construct a pseudo-collision for Tiger-19 with a complexity of about 2^{44} applications of the compression function. We can turn this pseudo-collision into a collision for Tiger-19. This is described in detail in the next section.

4.2   From a Pseudo-collision to a Collision in Tiger-19

Constructing a collision in Tiger-19 works quite similar as constructing the pseudo-collision. Again we use the key schedule difference given in (3) and employ the attack on 16 rounds of Tiger. The attack can be summarized as follows.

1. Choose arbitrary values for X_0, X_1 and X_2 and compute the chaining variables A_2, B_2, C_2 for round 3.
2. Employ the attack on 16 rounds, to find the message words $X_3, \ldots, X_7$ and $X_8[\text{even}], X_9[\text{even}]$ such that the output after round 18 collides.

3. To guarantee the $X_8[\mathbf{even}], X_9[\mathbf{even}]$ are correct after applying the key schedule, we use the degrees of freedom we have in the choice of X_0, X_1, X_2, X_3. Note that for any difference we introduce into X_0, you can introduce canceling differences into X_1, X_2, X_3 such that A_2, B_2 and $B_3 = C_2 \oplus X_3$ stay constant. This is a kind of local collision for the first 4 rounds of Tiger.

$$X_0^{\mathbf{new}} = \text{arbitrary}$$
$$X_1^{\mathbf{new}} = C_0^{\mathbf{new}} \oplus C_0 \oplus X_1$$
$$X_2^{\mathbf{new}} = C_1^{\mathbf{new}} \oplus C_1 \oplus X_2$$
$$X_3^{\mathbf{new}} = C_2^{\mathbf{new}} \oplus C_2 \oplus X_3$$

After testing all 2^{64} possible choices for X_0 and changing X_1, X_2, and X_3 accordingly such that A_2, B_2 and B_3 stay constant, we expect to get the correct values for $X_8[\mathbf{even}], X_9[\mathbf{even}]$ after applying the key schedule of Tiger.

Hence, this step of the attack has a complexity of at about 2^{64} key schedule computations and 3×2^{64} round computations. This is equivalent to about 2^{62} applications of the compression function of Tiger-19.

Thus, we can construct a collision in Tiger-19 with complexity of about $2^{62} + 2^{44} \approx 2^{62}$ applications of the compression function. We are not aware of any other collision attack on Tiger which works for so many rounds. The best collision attack on Tiger so far was for 16 rounds by Kelsey and Lucks described in [2].

5 Collision Attack on Tiger-19 – Method 2

We now present another method to find collisions for the 19-round Tiger. The attack complexity of this attack method is slightly higher than the one in the previous attack method. One difference from the previous method is that the first method uses larger space of message than the second one. This can been seen where X_0 is used in each attack. The first method uses whole 64 bits of X_0 and the second one uses less bits of X_0.

The attack described here is also an extension of the attack by Kelsey and Lucks. However, our attack is in a different situation from their attack. Their attack precomputes the additive difference L and then use X_6 to cancel it out in the main phase. Similarly, our attack precomputes the additive difference α and then use X_9 to cancel it out in the main phase. The key difference is that their attack controls X_6 in a deterministic way but our attack has to do in a probabilistic way due to the key schedule. This causes the main difficulty we have to solve here.

The outline of the attack is as follows:

1. Search for a good differential characteristic of the message words for 19 rounds.
2. Construct a good differential characteristic for 19 rounds by considering the message word differences expected from the characteristic in Step 1.

Table 3. A collision-producing differential characteristic

i	$\Delta(A_i)$	$\Delta(B_i)$	$\Delta(C_i)$	$\Delta(X_i)$
3	0	I	*	I
4	*	*	*	I
5	*	*	*	I
6	*	*	γ	I
7	*	γ	I	0
8	α	*	I	0
9	I	I	0	0
10	I	0	I	0
11	0	0	I	I
12	0	0	0	I

3. Divide this characteristic for round 3-9 into two consecutive characteristics (characteristic for round 3-7 and characteristic for round 8-9) so that we work on them *independently*.
4. Do the MITM step for the characteristic for round 3-7. Determine the chaining values A_3, B_3, B_3 and the message words X_4, X_5, X_6, X_7[**even**].
5. Do the MITM step for the characteristic for round 8-9 by varying the message words X_0, X_1, X_2, X_3, X_7[**odd**] while keeping the previously determined values unchanged. Determine all of the values.

In the attack, we use the same characteristic for the key schedule as in Section 4 and then construct a differential characteristic as shown in Table 3, where α and γ are some useful values in our attack. We will explain how these value are chosen in the next section.

5.1   The Precomputation Phase of the Attack

Before performing our attack, we need an algorithm to find a good differential characteristic starting with $\Delta^+(C_6)$ and ending with $\Delta^+(C_9)$ as shown in Table 3. We need the additive difference $\Delta^+(\text{even}(B_9))$ to be equal to $\Delta^+(A_8)$. The question we have here is what difference we want in C_6 for obtaining a high probability. A solution to this is to compute the differences backward starting from the additive difference $\Delta^+(B_9) = I$. By performing experiments, we searched for α and γ such that the corresponding differential probabilities p_1, p_2 are high[2]. As a result, we found a high probability differential characteristic which is shown in the following:

$$\Delta^+(B_9) = I \overset{\text{even}}{\to} \Delta^+(\text{even}(B_9)) = \Delta^+(A_8) = \alpha \text{ with probability } p_2 \ ,$$

$$\Delta^+(A_8) = \alpha \overset{\div,+}{\to} \Delta(B_7)^+ = \gamma \text{ with probability } 1 \ ,$$

$$\Delta^+(B_7) = \gamma \overset{\oplus}{\to} \Delta(C_6)^+ = \Delta^+(B_7 \oplus X_7) = \gamma \text{ with probability } p_1 \ .$$

---

[2] We have searched some sub space for the values α and γ so far. Searching the whole space could give us the better values for both of two.

Here the additive differences are

$$\alpha = \texttt{0x80c02103d43214d6} \, ,$$
$$\gamma = \alpha/7 \mod 2^{64} = \texttt{0xedd24ddbf9be02fa} \, ,$$

and probabilities are $p_1 = 2^{-26}$ and $p_2 = 2^{-28}$. We here study the above characteristic with probability p_1 in detail.

In general, for a pair of data (J, J') and some constant value Q, if we assume the Hamming weight of $\Delta^{\oplus}(J, J')$ to be k, then the probability that $\Delta^+(J, J') = \Delta^+(J \oplus Q, J' \oplus Q)$ is 2^{-k}. This means that k bits of Q are constrained[3] to hold the above equation. Therefore, in the case of the characteristic with probability $p_1 = 2^{-26}$, we expect α to have 26 active bits as a XOR difference, which imposes a 26-bit condition on X_7.

Because of the large number of active bits, it seems plausible to assume that there is a 13-bit condition on $X_7[\textbf{even}]$ and a 13-bit condition on $X_7[\textbf{odd}]$. We denote the probabilities that these two conditions hold by $p_{1,even} = p_{1,odd} = 2^{-13}$ respectively.

5.2   The Main Phase of the Attack

We here describe how the main attack phase is performed. For a preparation we present the following lemma explaining the generic birthday attack which will be used for the MITM technique to work.

Lemma 1. *Consider two functions f and g having the same output space of n bit length. If we assume that f and g are random and we have r_1 inputs for f and r_2 inputs for g, the probability of having a pair of inputs (x, y) producing a collision $f(x) = g(y)$ is given by $p = 1 - \exp(-r_1 r_2/2^n)$ [3].*

This tells us that the MITM step works with some probability even if the number of output differences of the **odd** or **even** is less than 2^{32}. The main attack phase is performed as follows.

1. Arbitrarily choose the chaining values A_3, B_3, C_3 for round 4.
2. Choose $X_4[\textbf{even}]$ and ensure that the difference $\Delta^{\oplus} C_4$ is useful. By useful we mean that the corresponding XOR difference has at least 1 active bit in each odd byte for having the 2^{32} values for the additive difference $\Delta^+ \textbf{odd}(B_5)$. The work here is negligible.
3. Choose $X_4[\textbf{odd}]$ and $X_5[\textbf{even}]$ to ensure that the difference is $\Delta^{\oplus} C_5$ useful.
4. Perform a MITM step by choosing $X_5[\textbf{odd}]$ and $X_6[\textbf{even}]$ to get an additive difference γ in C_6. The expected work here is approximately 2^{33} evaluations of both of the **odd** function and the **even** function, and we determine $X_5[\textbf{odd}]$ and $X_6[\textbf{even}]$. Each failure requires that we go back to Step 3.

---

[3] For example, an XOR difference of 1 is consistent with an additive difference of either -1 or $+1$. If the low bit in J is 0, the low bit in J' will be 1, and reaching an additive difference of -1 will require fixing the low bit of Q to 1.

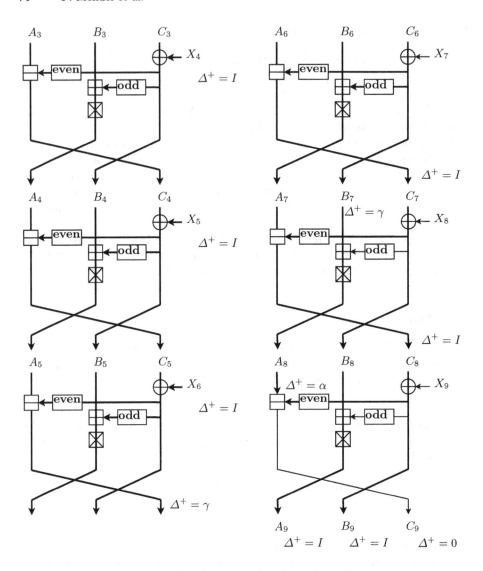

Fig. 3. The information flow from C_6 to C_9

5. Set 13 bits of $X_7[\textbf{even}]$ to hold the 13-bit condition on $X_7[\textbf{even}]$ derived in the precomputation phase in Sect. 5.1 and then perform a MITM step using the generic birthday attack of Lemma 1. This is performed by choosing $X_6[\textbf{odd}]$ and the rest bits of $X_7[\textbf{even}]$ to get additive difference I in C_7.

Each failure requires that we go back to Step 3. The expected work here is about 2^{13} computations, each of which consists of two kinds of evaluations: 2^{33} evaluations of the **odd** function and 2^{19} evaluations of the **even** function. We determine $X_6[\textbf{odd}]$ and $X_7[\textbf{even}]$ at the end of this step.

6. Set 13 bits of $X_7[\textbf{odd}]$ to hold the 13-bit condition on $X_7[\textbf{odd}]$ derived in the precomputation phase in Sect. 5.1 and then perform a MITM step to get the additive difference I in C_8. This is done by randomly choosing the rest bits of $X_7[\textbf{odd}]$ and randomly generating $X_8[\textbf{even}]$.

   The message word $X_8[\textbf{even}]$ is generated in the following way: Randomly choose the message word X_0 and determine X_1, X_2, and X_3 so that the resulting A_3, B_3, C_3 are consistent with A_3, B_3, C_3 chosen in Step 1. We then determine $X_8[\textbf{even}]$ from the key schedule.

   The above MITM step is performed with 2^{19} values for $\Delta^+(A_7)$ and 2^{28} values for $\Delta^+(\textbf{even}(B_8))$ [4]. According to Lemma 1, the success probability of this attack is 2^{-17}. Therefore the expected work here is about 2^{18} computations, each of which consists of two kinds of evaluations: 2^{19} evaluations of the **odd** function and 2^{28} evaluations of the **even** function.

7. Compute $X_9[\textbf{even}]$ by processing the key schedule and check if $\Delta^+\textbf{even}(B_9) = \Delta^+(A_8)$, which means $\Delta^+(C_9) = 0$. Each failure requires that we go back to Step 6.

5.3   Complexity Analysis

We discuss the attack complexity in the attack in Sect. 5.2. The important thing to consider when we estimate the complexity is that the task of Steps 1-5 can be performed *independently* of the task of Steps 6-7. We first perform Steps 1-5 and then perform Steps 6-7 without changing the values which have been determined in Steps 1-5.

In order to determine $X_4, X_5, X_6, X_7[\textbf{even}]$ by performing from Step 1 to Step 5, the required time complexity is equivalent to $p_{1,even}^{-1} \cdot 2^{33}$ evaluations of the **odd** function. In order to determine $X_0, X_1, X_2, X_3, X_7[\textbf{odd}]$ by performing from Step 6 to Step 7, the required time complexity is equivalent to $p_2^{-1} \cdot 2^{18} \cdot 2^{28} = 2^{28} \cdot 2^{46} = 2^{74}$ evaluations of the **odd** function.

The time complexity required by this attack is dominated by the latter part, which is equivalent to 2^{69} computations of the compression function of Tiger reduced to 19 rounds.

6   A Pseudo-Near-Collision for Tiger-22

In this section we present a pseudo-near-collision for Tiger-22 with complexity of about 2^{44}. Similar as we construct a pseudo-collision in Tiger-19, we can construct a pseudo-near-collision in Tiger-22. Again we use a key schedule difference that holds with probability 1 in the key schedule of Tiger and employ the attack on 16 rounds of Tiger. The key-schedule difference used in the attack is given in (4).

$$(0,0,I,0,0,0,I,I) \rightarrow (I,0,0,0,0,0,I,I) \rightarrow (0,0,0,0,0,0,I,I) \qquad (4)$$

---

[4] Because of the XOR difference $\Delta^{\oplus}(B_8) = I$, there is only one active S-box at the input of the **even**(B_8). This makes the number of the additive difference smaller than 2^{32}.

The attack work as follows:

1. Choose arbitrary values for the chaining variables A_5, B_5, C_5 for round 6.
2. Employ the attack on 16 rounds, to find message words $X_6, \ldots, X_{10}$ and $X_{11}[\mathbf{even}], X_{12}[\mathbf{even}]$ such that the output after round 21 collides.
3. To compute the real message words $X_0, \ldots, X_7$, we have to choose suitable values for $X_{11}[\mathbf{odd}], X_{12}[\mathbf{odd}]$ and $X_{13}, \ldots, X_{15}$ to guarantee that X_7 is correct after computing the key schedule backward. Therefore, we choose arbitrary values for $X_{11}[\mathbf{odd}], X_{12}[\mathbf{odd}], X_{13}, X_{14}$ and calculate X_{15} as follows:

$$X_{15} = (X_7 \oplus (X_{14} - X_{13})) - (X_{14} \oplus \mathtt{0123456789ABCDEF})$$

   This adds negligible cost to the attack complexity and guarantees that X_7 is correct after computing the key schedule backward. Note that X_6 can be chosen freely, because we can modify C_5 such that $C_5 \oplus X_6$ stays constant.
4. Run the rounds 5, 4, 3, 2, 1 and 0 backwards to compute the initial values A_{-1}, B_{-1} and C_{-1}. Since there is a difference in the message word X_2 in the MSB, we have to introduce the same difference in the initial value to cancel it out, namely

$$\Delta^{\oplus}(B_{-1}) = I \ .$$

   Since the difference is in the MSB this happens with probability 1.
5. Of course, the feed forward destroys the pseudo-collision. After the feed forward we get the same output differences as in the initial values. Since the difference is in the MSB this has probability 1.

$$\Delta^{\oplus}(B'_{21}) = \Delta^{\oplus}(B_{-1} - B_{21}) = I$$

Hence, we can construct a pseudo-near-collision for Tiger-22 with complexity of about 2^{44}. For an ideal hash function with a hash value of 192-bit we would expect a complexity of about 2^{90} to construct a pseudo-near-collision with a one bit difference. Note that a pseudo-near-collision for Tiger-21 with a one bit difference can be found with the same complexity. A detailed description of the attack is given in the appendix.

7   A Pseudo-collision for Tiger-23/128

Tiger/128 is a variant of Tiger, where the final hash value is truncated to 128 bit. This variant was specified in [1] to make Tiger compatible to MD5. In this section, we present a pseudo-collision for 23 rounds of Tiger/128. In detail, we can turn the pseudo-near-collision for Tiger-22 into a pseudo-collision for Tiger-23/128 by adding one additional round. If we add one round then the output after 23 rounds has the following differences in the chaining variables:

$$\Delta^{\oplus}(A_{22}) = 0, \quad \Delta^{\oplus}(B_{22}) = I, \quad \Delta^{\oplus}(C_{22}) \neq 0 \text{ (arbitrary)} \ .$$

Due to the feed-forward the difference in B_{22} cancels out with probability 1. Hence, we have a pseudo-collision in Tiger-23/128, since only register A and B are used for the final hash value of Tiger-128. The attack has a complexity of about 2^{44} applications of the compression function.

8   Conclusion

In [2], Kelsey and Lucks discussed the possibility of extending their attack to more rounds of Tiger and the applicability of their attack techniques to the full hash function.

In this article, we presented two strategies for constructing collision in the Tiger-19 hash function. The first has a complexity of about 2^{62} hash computations and the second has a slightly higher complexity of about 2^{69} hash computations.

The best attack on a reduced variant of Tiger so far was proposed by Kelsey and Lucks in [2]. They showed a collision attack on Tiger-16 with a complexity of about 2^{44} and a pseudo-near-collision for a variant of Tiger with 20 rounds with a complexity of about 2^{48}. We have extended their approach to show collision attacks on Tiger-19 and presented a pseudo-near-collision for Tiger-22 and a pseudo-collision for Tiger-23/128. Based on this we conclude that the security margin of Tiger is not as large as one could hope for. It remains a topic of further research to determine whether the attacks can be extended to Tiger variants with more than 23 rounds.

Acknowledgement

The authors wish to thank Antoine Joux, Elisabeth Oswald, and the anonymous referees for useful comments and discussions.

References

1. Ross J. Anderson and Eli Biham. TIGER: A Fast New Hash Function. In Dieter Gollmann, editor, *Fast Software Encryption, Third International Workshop, Cambridge, UK, February 21-23, 1996, Proceedings*, volume 1039 of *Lecture Notes in Computer Science*, pages 89–97. Springer, 1996.
2. John Kelsey and Stefan Lucks. Collisions and Near-Collisions for Reduced-Round Tiger. In Matt Robshaw, editor, *Fast Software Encryption, 13th International Workshop, FSE 2006, Graz, Austria, March 15-17, 2006*, volume 4047 of *LNCS*, pages 111–125, 2006.
3. Alfred J. Menezes, Paul C. van Oorschot, and Scott A. Vanstone. *Handbook of Applied Cryptography*. CRC Press, 1997. Available online at http://www.cacr.math.uwaterloo.ca/hac/.
4. Xiaoyun Wang, Yiqun Lisa Yin, and Hongbo Yu. Finding Collisions in the Full SHA-1. In Victor Shoup, editor, *Advances in Cryptology - CRYPTO 2005, 25th Annual International Cryptology Conference, Santa Barbara, California, USA, August 14-18, 2005, Proceedings*, volume 3621 of *LNCS*, pages 17–36. Springer, 2005.
5. Xiaoyun Wang and Hongbo Yu. How to Break MD5 and Other Hash Functions. In Ronald Cramer, editor, *Advances in Cryptology - EUROCRYPT 2005: 24th Annual International Conference on the Theory and Applications of Cryptographic Techniques, Aarhus, Denmark, May 22-26, 2005. Proceedings*, volume 3494 of *LNCS*, pages 19–35. Springer, 2005.

A   A Pseudo-Near-Collision for Tiger-21

In a similar way as we construct a pseudo-near-collision in Tiger-22, we can construct a pseudo-near-collision for Tiger-21. For the attack we use the key-schedule difference given in (5). It has probability 1 to hold in the key-schedule of Tiger.

$$(0, I, 0, 0, 0, I, I, I) \rightarrow (0, 0, 0, 0, 0, 0, I, I, 0) \rightarrow (0, 0, 0, 0, 0, 0, I, I, I) \tag{5}$$

Again we use the attack on 16 rounds of Tiger (described in Section 3) to construct a pseudo-near-collision in Tiger-21. The attack work as follows:

1. Choose arbitrary values for the chaining variables A_4, B_4, C_4 for round 5.
2. Employ the attack on 16 rounds, to find message words $X_5, \ldots, X_9$ and $X_{10}[\mathbf{even}], X_{11}[\mathbf{even}]$ such that the output after round 20 collides.
3. To compute the real message words $X_0, \ldots, X_7$, we have to choose suitable values for $X_{10}[\mathbf{odd}], X_{11}[\mathbf{odd}]$ and $X_{12}, \ldots, X_{15}$ such that X_6 and X_7 is correct after computing the key schedule backward. Therefore, we choose arbitrary values for $X_{10}[\mathbf{odd}], X_{11}[\mathbf{odd}]$ and X_{12}, X_{13} and calculate X_{14}, X_{15} as follows:

$$X_{14} = (X_6 - (X_{13} \oplus X_{12} \oplus (\neg (X_{12} + (X_{11} \oplus (\neg X_{10} \gg 23))) \gg 23))) + X_{13}$$
$$X_{15} = (X_7 \oplus (X_{14} - X_{13})) - (X_{14} \oplus \mathtt{0123456789ABCDEF})$$

This adds negligible cost to the attack complexity and X_6, X_7 are always correct after computing the key schedule backward. Note that X_5 can be chosen freely, because we can modify C_4 such that $C_4 \oplus X_5$ stay constant.
4. Run the rounds 4, 3, 2, 1 and 0 backwards to compute the initial values A_{-1}, B_{-1} and C_{-1}. Since there is a difference in the message word X_1 in the MSB, we introduce the same difference in the initial value to cancel it out. Since the difference is in the MSB, this happens with probability 1.

$$\Delta^{\oplus}(A_{-1}) = I$$

5. Of course, the feed forward destroys the pseudo-collision. After the feed forward we get the same output differences as in the initial values:

$$\Delta^{\oplus}(A'_{20}) = \Delta^{\oplus}(A_{-1} \oplus A_{20}) = I \ .$$

Hence, we can construct a pseudo-near-collision for Tiger-21 with complexity of about 2^{44} applications of the compression function. For an ideal hash function with a hash value of 192-bit we would expect a complexity of about 2^{90} applications of the compression function instead of 2^{44}.

B   A Pseudo-collision for Tiger-21

In a similar way as we construct a pseudo-near-collision in Tiger-21, we can construct a pseudo-collision in Tiger-21. For the attack we use again the key-schedule difference given in (5). The attack can be summarized as follows:

1. Choose arbitrary values for the chaining variables A_0, B_0, C_0 for round 1.
2. Choose random values for X_1, X_2, X_3, X_4 and calculate A_4, B_4, C_4.
3. Employ the attack on 16 rounds of Tiger, to find message words $X_5, \ldots, X_9$ and $X_{10}[\text{even}], X_{11}[\text{even}]$ such that the output after round 20 collides.
4. To compute the real message words $X_0, \ldots, X_7$, we have to choose suitable values for X_0, X_1 and X_2 such that X_8, X_9 and $X_{10}[\text{even}], X_{11}[\text{even}]$ are correct after computing the key schedule. Note that X_0 and X_1 can be chosen freely, because we can modify C_0 and C_1 such that $C_{-1} \oplus X_0$ and $C_0 \oplus X_1$ stay constant. Since a difference is introduced by X_1, we have after round 1 that $\Delta^{\oplus}(C_1) \neq 0$. Hence, X_2 can not be chosen freely.

However, since we can choose the value of $C_0 \oplus X_1$ in the beginning of the attack, we can guarantee that the Hamming weight of $\Delta^{\oplus}(C_1)$ is small. Computer experiments show that the smallest weight we can get is 22. Consequential there are $2^{64-22} = 2^{42}$ possible choices for C_1 and X_2 such that $\Delta^{\oplus}(C_1 \oplus X_2)$ and $C_1 \oplus X_2$ stay constant. Hence, we have $2^{64+64+42} = 2^{170}$ degrees of freedom in the key schedule of Tiger. Therefore, we have to repeat the attack at most 2^{22} times to guarantee that X_8, X_9 and $X_{10}[\text{even}], X_{11}[\text{even}]$ are correct after applying the key schedule.

Hence, we can find a pseudo-collision in Tiger-21 with a complexity of about $2^{44+22} = 2^{66}$ applications of the compression function. Note that we assume in the analysis that it is computational easy to find suitable values for X_0, X_1, X_2.

RC4-Hash: A New Hash Function Based on RC4

(Extended Abstract)

Donghoon Chang[1], Kishan Chand Gupta[2], and Mridul Nandi[3]

[1] Center for Information Security Technologies(CIST), Korea University, Korea
dhchang@cist.korea.ac.kr
[2] Department of Combinatorics and Optimization, University of Waterloo, Canada
kgupta@math.uwaterloo.ca
[3] David R. Cheriton School of Computer Science, University of Waterloo, Canada
m2nandi@cs.uwaterloo.ca

Abstract. In this paper, we propose a new hash function based on RC4 and we call it RC4-Hash. This proposed hash function produces variable length hash output from 16 bytes to 64 bytes. Our RC4-Hash has several advantages over many popularly known hash functions. Its efficiency is comparable with widely used known hash function (e.g., SHA-1). Seen in the light of recent attacks on MD4, MD5, SHA-0, SHA-1 and on RIPEMD, there is a serious need to consider other hash function design strategies. We present a concrete hash function design with completely new internal structure. The security analysis of RC4-Hash can be made in the view of the security analysis of RC4 (which is well studied) as well as the attacks on different hash functions. Our hash function is very simple and rules out all possible generic attacks. To the best of our knowledge, the design criteria of our hash function is different from all previously known hash functions. We believe our hash function to be secure and will appreciate security analysis and any other comments.

Keywords: Hash Function, RC4, Collision Attack, Preimage Attack.

1 Introduction

Hash functions are of fundamental importance in cryptographic protocols. They compress a string of arbitrary length to a string of fixed length. We know that digital signatures are very important in information security. The security of digital signatures depends on the cryptographic strength of the underlying hash functions. Other applications of hash functions in cryptography are data integrity, time stamping, password verification, digital watermarking, group signature, e-cash and in many other cryptographic protocols.

Hash functions are usually designed from scratch or made out of a block cipher in a black box manner. Some of the well studied hash functions constructed from scratch are SHA-family [31,9], MD4 [26], MD5 [27], RIPEMD [25], Tiger [1], HAVAL [39] etc. Whereas PGV hash function [24], MDC2 [6] etc. are designed in a black box manner.

Since among SHA-family SHA-0 [31], SHA-1 [9] were broken by Wang *et al.* [35,36], we can not be confident about the security of other algorithms in the

R. Barua and T. Lange (Eds.): INDOCRYPT 2006, LNCS 4329, pp. 80–94, 2006.
© Springer-Verlag Berlin Heidelberg 2006

SHA-family because their design principles are similar. Likewise MD4, MD5, RIPEMD and HAVAL were also broken [33,34,37,38]. So, we need to design new, variable length hash algorithms with different internal structures keeping security and efficiency in mind.

In response to the SHA-1 vulnerability [36] that was announced in Feb. 2005, NIST held a Cryptographic Hash Workshop on 2005 to solicit public input on its cryptographic hash function policy and standards. NIST continues to recommend a transition from SHA-1 to the larger approved hash functions (SHA-224, SHA-256, SHA-384, and SHA-512). In response to the workshop, NIST has also decided that it would be prudent in the long-term to develop an additional hash function through a public competition, similar to the development process for the block cipher in the Advanced Encryption Standard (AES).

It will be useful and interesting to propose some robust hash functions which are based on some well studied and structurally different from the broken class. In this direction we propose a hash function (RC4-Hash) whose basic structure is based on RC4. It also has the desirable advantage of variable length hash output. In fact our design provides hash output from 16 bytes to 64 bytes with little or no modification in the actual algorithm. It provides a wide range of security depending on the applications. In this context it may be noted that there are very few hash families providing variable size hash output. We provide security analysis against meaningful known attacks. We take care of the weakness of RC4 in a manner such that it will not affect the security of the Hash function. Many results on RC4 can be used to show the security of RC4-Hash against known attacks and importantly resistances against attacks by Wang *et al.* and Kelsey-Schneier second preimage attack [16]. Its efficiency is also comparable with SHA-1.

The rest of the paper is organized as follows. In Section 2 we give a simple description and some of the security analysis of RC4. We also give a short note on hash functions. RC4 based hash function is analyzed in Section 3 followed by a security/performance analysis of RC4-Hash in Section 4. We conclude in Section 5.

2   Preliminaries

We first describe the RC4 algorithm and its known security analysis which are relevant to this paper. Then we give a short note on hash functions. RC4 was designed by Ron Rivest in 1987 and kept as a trade secret until it leaked out in 1994. It consists of a table of all the 256 possible 8-bit words and two 8-bit pointers. Thus it has a huge internal state of $log_2(2^8! \times (2^8)^2) \approx 1700$ bits. For a detailed discussion on RC4 see Master's thesis of Itsik Mantin [18].

2.1   RC4 Algorithm

Let $[N] := [0, N-1] := \{0, 1, \cdots, N-1\}$ and $\mathsf{Perm}(A)$ be the set of all permutations on A. In this paper, we will be interested on $\mathsf{Perm}([N])$ (or we write Perm), where $N = 256 = 2^8$. For $S \in \mathsf{Perm}$, we denote $S[i]$ to the value of the permutation S at the position $i \in [N]$. In this paper, the addition modulo N is denoted by " + ", otherwise it will be stated clearly. The function $\mathsf{Swap}(S[i], S[j])$ means the swapping operation between $S[i]$ and $S[j]$. The key-scheduling algorithm and key-generation algorithm are defined in Figure 1.

RC4-KSA(K)	RC4-PRBG(S)

for i = 0 to N – 1
    S[i] = i;
j = 0;
for i = 0 to N – 1
    j = j + S[i] + K[i mod κ];
    Swap(S[i],S[j]);
	i = 0, j = 0;
	Pseudo-Random Bytes Generation:
	    i = (i + 1) mod N;
	    j = (j + S[i]) mod N;
	    Swap(S[i],S[j]);
	    out = S[(S[i] + S[j]) mod N];

Fig. 1. The Key Scheduling Algorithm (RC4-KSA) and Pseudo-Random Byte Generation Algorithm (RC4-PRBG or PRBG) in RC4. Here $K = K[0] \parallel \cdots \parallel K[\kappa - 1]$, $K[i] \in [N]$. and κ is the size of the secret key in bytes.

2.2   Some Relevant Security Analysis of RC4

In this section we briefly explain few attacks on RC4 which are important in this paper while considering the security analysis of RC4-Hash.

The Distribution After Key-Scheduling Algorithm (or RC4-KSA) Is Close to Uniform

RC4 can be viewed as a close approximation of exchange shuffle. In exchange shuffle, the value of j in Key-Scheduling Algorithm is chosen randomly (unlike RC4-KSA where it is updated recursively based on a secret key). Simion and Schmidt [30] studied the distribution of the permutation after exchange shuffle. Mironov [21] showed that the statistical distance between the output after t exchange shuffles and uniform distribution on permutations is close to $e^{\frac{-2t}{N}}$. Thus, when $t = N$, it has significant statistical distance which is e^{-2}. At the same time, if the number of random shuffle is large compared to N then the statistical distance is close to zero which means the two distributions are almost identical. Even though RC4-KSA is not the same as exchange shuffle, one can hope for a similar property. More precisely, we assume that if K is chosen randomly then the distribution of the pair (S, j) after the execution of RC4-KSA is close to uniform distribution i.e., $(S, j) = \mathrm{RC4\text{-}KSA}(K)$ is uniformly distributed on Perm $\times$ $[N]$ provided K is chosen uniformly.

The Distribution of RC4-PRBG Output Is Not Uniform

There are many observations [11,12,20,23] which proves that the distribution of RC4-PRBG(S) can not be uniform even if we assume that S is uniformly distributed. For example,

1. Mantin and Shamir [20] showed that the probability of second byte being zero is close to $\frac{2}{N}$ as compared to the probability $\frac{1}{N}$ in case of random Byte generation.

2. Paul and Preneel [23] showed that the probability that first two bytes are equal is close to $\frac{1}{N}(1 - \frac{1}{N})$.

3. Fluhrer and McGrew [11] computed probabilities for different possible outputs (e.g., the first two bytes are $(0,0)$ has probability close to $\frac{1}{N^2} + \frac{1}{N^3}$) and showed that the probability is not the same as that of uniform distribution.

4. A Finney [7] state at any stage i in RC4-PRBG is a pair $(S, j) \in$ Perm $\times [N]$ where $j = i + 1$ and $S[j] = 1$. One can check that if we have a Finney state in PRBG just before updating i then next state is also Finney. The converse is also true i.e., the Finney state should arise from a Finney state only. It is easy to see that if $(S, 1)$ is a Finney state at stage $i = 0$, then all N output from PRBG are distinct. Probability that a pair (S, j) chosen randomly for some i is a Finney state is $\frac{1}{N^2}$. One might expect that the output of PRBG is not uniform (as the output of PRBG with distinct bytes are more likely due to the Finney states).

5. Golic [12] proved the following result. Let the output n-bit word sequence of RC4 is $Z = (Z_t)_{t=1}^{t=\infty}$ and $z = (z_t)_{t=1}^{t=\infty}$ denote the least significant bit output sequence of RC4. Let $\ddot{z} = (\ddot{z}_t = z_t + z_{t+2})_{t=1}^{t=\infty}$ denotes the second binary derivative then $\ddot{z}$ is correlated to 1 with the correlation coefficient close to 15×2^{-3n} and output sequence length required to detect a statistical weakness is around $64^n/225$.

Besides these attacks there are some more attacks on RC4, for example, Fault analysis [2,15]. But those attacks are not meaningful in the contaxt of hash function cryptanalysis.

2.3 A Brief Note on Hash Function

A hash function is usually designed as follows : First a compression function $C : \{0,1\}^c \times \{0,1\}^a \rightarrow \{0,1\}^c$ is designed. We denote $C(h, x) = h'$ by $h \xrightarrow{x} h'$. Then given a message M such that $|M| < 2^{64}$, a pad is appended at the end of the message. For example, $\overline{M} := \mathsf{pad}(M) = M \parallel 10^k \parallel \mathsf{bin}_{64}(|M|)$, where $\mathsf{bin}_{64}(x)$ is the 64-bit binary representation of x and k is the least non-negative integer such that $|M| + k + 65 \equiv 0 \bmod a$. Now write $\overline{M} = M_1 \parallel \cdots \parallel M_t$ (for some $t > 0$) where $|M_i| = a$. We choose an initial value $\mathsf{IV} := h_0 \in \{0,1\}^c$ and then compute the hash values

$$h_0 \xrightarrow{M_1} h_1 \xrightarrow{M_2} \cdots \xrightarrow{M_{t-1}} h_{t-1} \xrightarrow{M_t} h_t$$

where h_t is the final hash value, i.e., $H(M) = h_t$ and $|h_i| = c$. The function C is known as the compression function and the iteration method is known as the classical iteration.

We have three most important notions of security in hash functions which we describe below. For more detail discussions one can see [32].

1. **Collision Attack:** Find $M_1 \neq M_2$, such that $H(M_1) = H(M_2)$.
2. **Preimage Attack:** Given a random $y \in \{0,1\}^c$, find M so that $H(M) = y$.
3. **Second Preimage Attack:** Given a message M_1, find M_2 such that $H(M_1) = H(M_2)$.

If it is hard to find any of the above attack (or attacks) then we say the hash function is resistant to these attacks. For example, if there is no efficient collision finding algorithm then the hash function is said to be collision resistant. For a c-bit hash function, exhaustive search requires $2^{c/2}$ complexity for collision and 2^c complexity for both preimage and second preimage both. In case of collision attack, birthday attack is popularly used exhaustive search. Recently, Kelsey-Schneier [16] has shown a generic attack for second preimage for classical hash function with complexity much less than 2^c.

Subsequently, a wide pipe hash design has been suggested [17]. In this design, there is an underlying function $C : \{0,1\}^w \times \{0,1\}^a \rightarrow \{0,1\}^w$, called *compression-like function* and a *post processing function* $g : \{0,1\}^w \rightarrow \{0,1\}^c$. Given a padded message $M = M_1 \parallel \cdots \parallel M_t$, with $|M_i| = a$, the hash value is computed as follows :

$$h_0 \xrightarrow{M_1} h_1 \xrightarrow{M_2} \cdots \xrightarrow{M_{t-1}} h_{t-1} \xrightarrow{M_t} h_t, \; H(M) = g(h_t).$$

If w (the intermediate state size) is very large compare to c (the final hash size), then the security of H may be assumed to be strong [17] even though there are some weakness in the compression-like function C. Kelsey-Schneier second preimage attack also will not work if $w > 2c$. The post processor g need not be very fast as it is applied once for each message. Thus, design of a wide pipe hash function has several advantages over other designs like classical hash functions. In this context, we would like to mention that, there are several other designs like prefix-free MD hash function [8], chop-MD [8], EMD [4] etc.

3 RC4-Hash Algorithm: RC4 Based Hash Function

Now we describe our newly proposed hash function based on RC4, RC4-Hash. This hash function has the following properties;

1. It is, in fact, a hash family denoted as RCH_ℓ, $16 \leq \ell \leq 64$ where RCH_ℓ : $\{0,1\}^{<2^{64}} \rightarrow \{0,1\}^{8\ell}$. Here $\{0,1\}^{<2^{64}}$ denotes the set of all messages whose length is at most $2^{64} - 1$ which is reasonable in all practical applications.
2. Our hash function is also a wide pipe hash function (see Section 2.3). Like other hash functions we will use an initial value and a variant of padding rule which provides a dynamic hash function (i.e., it produces independent hash outputs of different sizes for one message).

Algorithm RCH$_\ell(M)$

Padding Rule: We pad the message as follows : $\text{pad}(M) = \text{bin}_8(\ell) \parallel M \parallel 1 \parallel 0^k \parallel \text{bin}_{64}(|M|)$, where $\text{bin}_{64}(|M|)$ is the 64-bit binary representation of number of bits of M and k is the least non-negative integer such that $8 + |M| + 1 + k + 64 \equiv 0 \mod 512$. Write $\text{pad}(M) = M_1 \parallel \cdots \parallel M_t$ such that $|M_i| = 512$.

(2) Classical Iteration: Let $M_1 \parallel \cdots \parallel M_t$ be the padded message. Let $(S_0, j_0) := (S^{\mathsf{IV}}, 0)$ be an initial value (S^{IV} is given in Appendix). We invoke the compression-like function C (given in Figure 2) iteratively similar to the classical iteration as follows:

$$(S_0, j_0) \xrightarrow{M_1} (S_1, j_1) \xrightarrow{M_2} \cdots (S_{t-1}, j_{t-1}) \xrightarrow{M_t} (S_t, j_t) := C^+(M).$$

Recall that, $(S, j) \xrightarrow{X} (S^*, j^*)$ means that $C((S, j), X) = (S^*, j^*)$, where $C :$ Perm $\times [N] \times \{0, 1\}^{512} \to$ Perm $\times [N]$

(3) Post-processing: The post processing is divided into following steps. Let (S_t, j_t) be the internal state after the classical iteration i.e., $C^+(M) = (S_t, j_t)$.

1. Compute $S_{t+1} = S_0 \circ S_t$ and $j_{t+1} = j_t$.
2. We define the final hash value $\text{RCH}_\ell(M)$ by $\text{HBG}_\ell\big(\text{OWT}(S_{t+1}, j_{t+1})\big)$ (HBG_ℓ and OWT are given in Figure 2).

4 Security Analysis and Performance

In this section, we give security analysis against preimage, second preimage and collision attacks (see Section 2.3). We also compute the number of basic operations to compute the hash value such as table lookup and modular addition. We first explain the role of the each part of our hash function in view of the security analysis.

The role of OWT

First note that OWT is believed to be an one-way transformation since we define $\text{OWT}(S, j) = (S^*, j^*)$, where $S^* = \text{Temp1} \circ \text{Temp2} \circ \text{Temp1}$ (see the algorithm in Figure 2 for the definition of Temp1 and Temp2). One can easily invert from Temp2 to Temp1, but Temp1 would not be controlled as there is no choice of message in this part of the algorithm. Thus, it would be difficult to guess Temp2 such that $S^* = \text{Temp1} \circ \text{Temp2} \circ \text{Temp1}$.

It is also not easy to find fixed point with respect to the permutation (i.e. $\text{OWT}(S, j) = (S, j')$). This is why we define $\text{OWT}(S, j) = \text{Temp1} \circ \text{Temp2} \circ \text{Temp1}$ instead of any other composition. If we define, $\text{OWT}(S, j) = \text{Temp2} \circ \text{Temp1}$ then one can invert $\text{Temp2} = id$ (the identity permutation) to obtain Temp1. Then, it is easy to check that Temp1 is a fixed point for this definition of OWT. Similarly one can find a fixed point when we define $\text{OWT}(S, j) = \text{Temp2} \circ \text{Temp1} \circ \text{Temp2}$. In our definition this method does not work.

The Compression-like function C

The compression-like function C has output size about 1692 bits ($= 1700 - 8$ as 8-bit i is not a part of a state) which is much larger than three times of the size

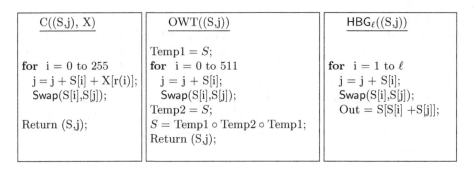

C((S,j), X)	OWT((S,j))	HBG$_\ell$((S,j))

for i = 0 to 255
  j = j + S[i] + X[r(i)];
  Swap(S[i],S[j]);
 	
Return (S,j);
	Temp1 = S;
	for i = 0 to 511
	  j = j + S[i];
	  Swap(S[i],S[j]);
	Temp2 = S;
	S = Temp1 ∘ Temp2 ∘ Temp1;
	Return (S,j);
		for i = 1 to ℓ
		  j = j + S[i];
		  Swap(S[i],S[j]);
		  Out = S[S[i] + S[j]];

Fig. 2. The Compression-like function C, OWT and HBG$_\ell$ in RC4-Hash. Here, $X = X[0] \parallel \cdots \parallel X[63]$, $|X[i]| = 8$ and ∘ means the composition of the permutations. The function $r : [256] \to [64]$ is known as reordering like in MD4 and MD5 (the function r is given in Appendix), that is the mappings restricted on $[0, 63]$, $[64, 127]$, $[128, 191]$ and $[192, 255]$ are injective.

of hash output (8ℓ-bits which is at most 512 bits). Thus, generic attacks such as Kelsey-Schneier [16] second-preimage attack does not work here.

The choice of Initial Value
We have chosen an initial value S^{IV} such that it is not b-conserving. Here we give a short note on b-conserving state and b-exact key. A b-exact key [10] can be considered as one of the weak keys of RC4 key scheduling algorithm. Much research has been devoted to find out several weak keys [29,10].

Definition 1. [10] (1) If $S[t] \equiv t \mod b$ for all t, the permutation S is said to be b-conserving. If $S(t) \equiv t \mod b$ for at least $N - 2$ values of t, then the permutation S is almost b-conserving.

(2) Let b and κ be two integers, and let K be an κ-byte key. Then K is called a b-exact key if for any index r, $K[r \mod \kappa] \equiv (1-r) \mod b$. Moreover, if $K[0] = 1$ and $msb(K[1]) = 1$ then K is called a special b-exact key where $msb(x)$ means the most significant bit of x.

The following result says that the permutation generated after key-scheduling algorithm is b-conserving with high probability if the key is a special b-exact key. Thus, one can use this to make a distinguishing attack as the distribution of the permutation is reduced on the set of all b-conserving permutations.

Theorem 1. [10] Let b be a power of 2 such that $b|\kappa$ and let K be a special b-exact key of κ bytes. Then the probability that the permutation generated after key-scheduling algorithm based on the key K is b-conserving, is at least $\frac{2}{5}$.

The reason why we exclude b-conserving S^{IV} for all b is that if the initial value (permutation) is b-conserving then the space of intermediate permutations can be reduced using Theorem 1 by choosing b-exact message block (here message block plays role of key).

In [14], a related key cryptanalysis has been provided where the the number of key bytes is very close to $N = 256$. This cryptanalysis does not work for smaller number of key bytes, (note that, in our RC4-Hash message blocks are of 64 bytes which is small enough). In [10], a key recovery attack is presented in RC4 where a known IV is appended with secret key. It is possible to reconstruct the secret key when different initial values are appended with the same secret key and few outputs of RC4-PRBG are known. The above cryptanalysis does not help directly to obtain an attack on our hash function.

The choice of Reordering: Reordering is also playing an important role to resist collision attack based on some internal collision patterns. We will give details of these attacks later when we study the collision resistance.

The padding rules makes it a Dynamic Hash Function: Here we use a slightly different padding rule than what in other known hash functions. We append the length representation of hash output size at the beginning so that we can produce different and independent looking hash values for different hash sizes of same message. If we do not pad the length of the hash size then for any message M, $\mathsf{RCH}_{\ell'}(M)$ is nothing but the truncation of $\mathsf{RCH}_\ell(M)$, where $\ell' < \ell$.

4.1 Preimage Resistance

Given a hash value of a message randomly chosen from a message space, we want to show the difficulty of finding its any preimage. Since we have a one-way transformation OWT, one can use "meet in the middle attack" just after invoking one way transformation and before invoking hash byte generation. More precisely, given a hash value $h = h_0||h_1|| \cdots ||h_{l-1}$ we first invert HBG (this is possible since hash byte generation algorithm is invertible) and store a set A of pairs (S, j) which outputs h after hash byte generation. Then we can choose message M randomly and compute $\mathsf{OWT}(C^+(M))$ and look for collision on the set A. But the complexity of this "meet in the middle attack" requires approximately $2^{1692/2} = 2^{846}$ queries (birthday attack on $\mathsf{Perm}[256]$ which is roughly 1692 bits). One can use a little different approach by using b-predictive a-state as explained below.

Preimage Attack based on Predictive RC4 states

Definition 2. (1) *An a-state is a partially specified RC4 state, that includes i, j, and a elements of S (not necessarily consecutive). More precisely, the tuple $p = (i, j, (i_1, \cdots, i_a), (j_1, \cdots, j_a))$ is said to be an a-state.*
(2) *An a-state $p = (i, j, (i_1, \cdots, i_a), (j_1, \cdots, j_a))$ is compatible with a RC4 state (i, j, S) if $S[i_k] = j_k$ for $1 \leq k \leq a$. We say that p predicts rth output if for all states compatible with p, produce the same output byte after r rounds. An a-state p is said to be b-predictive a-state if p predicts $r_1 < \cdots < r_b(\leq 2N)$ outputs.*

In [20], Mantin and Shamir have shown a distinguishing attack based on b-predictive a-state which requires $O(N^{2a-b+3})$ output bytes. Later, Paul and

Preneel [22] modified this definition by considering $1 = r_1 < \cdots < r_b \leq N$. According to this definition, they have shown that b-predictive a-state can exists only if $a \geq b$. In [11] total number of a special b-predictive b-state (known as fortuitous state where all b predicted states are consecutive) is given (see Table 1). Note that our hash output bytes are consecutive.

Table 1. The second column is the number of special b-predictive b states known as fortuitous states. Here, total states means the number of possible different choices of i, j and values of S in the corresponding b indices. For example, in the case of $b = 2$, the total states is $256 \times 256 \times 255 \times 256 \approx 2^{31.99}$. Thus, $516/2^{31.99} = 2^{-22.9}$ is the probability that a random state is one of the fortuitous state of length 2.

b	Number	Total states	Prob.
2	516	$2^{31.99}$	$2^{-22.9}$
3	290	$2^{39.98}$	$2^{-31.8}$
4	6540	$2^{47.97}$	$2^{-35.2}$
5	25,419	$2^{55.94}$	$2^{-41.3}$
6	101,819	$2^{63.92}$	$2^{-47.2}$

Suppose that we are given a hash value generated from a b-predictive b-state with a some choice of j. This means that the b-byte hash output is determined only by b elements of intermediate permutation S and j where $\mathsf{OWT}(C^+(M)) = (S, j)$. So any output of $\mathsf{OWT}(C^+(M))$ satisfying $b + 1$ conditions can become a preimage for a given hash value. Now the probability that a random message satisfies $b + 1$ conditions is $\frac{1}{N^2(N-1)\cdots(N-b+1)}$. The remaining has $l - b$ hash bytes will be same with probability $\frac{1}{N^{l-b}}$. Thus, the probability to get a preimage will be $\frac{1}{N^{l-b+2}(N-1)\cdots(N-b+1)}$. One can check that the probability is less than $\frac{1}{N^l}$ for $b \leq 64$. We give the probability for smaller values of ℓ in Table 1. Thus, the preimage attack based on fortuitous state does not help and it needs N^ℓ complexity.

4.2   Second Preimage Resistance

In [16], Kelsey and Schneier described a general second preimage attack which reduces the complexity from 2^n (trivial case for n-bit output) to about $2^{n/2}$. We can apply their attack to classical MD-construction which repeats compression function such that the length of intermediate value is same as that of hash output. Recently, Rivest [28] suggested the dithering method secure against Kelsey-Schneier second-preimage attack. Lucks [17] also suggested wide pipe hash, whose length of intermediate value (w-bit) is longer than that of hash output (n-bit). In case $w \geq 2n$, wide pipe hash is secure against Kelsey-Schneier second-preimage attack. The design principle of RCH_ℓ follows wide pipe hash. In case RCH_ℓ, w is about 1692 bits and hash output is less than 512 bits. Therefore, since the complexity of Kelsey-Schneier second preimage attack is about 2^{846}, we can say that RCH_ℓ is secure against Kelsey-Schneier second-preimage attack.

4.3   Collision Resistance

(1) Complexity of Birthday attack

We first state the result of Bellare and Kohno [3]. Let X and Y be two independent and identically distributed random variable taking values on a set $R = \{r_1, \cdots, r_L\}$. Let p_i be the probability that $X = r_i$ (which is also same for $Y = r_i$). It is easy to check that $\Pr[X = Y] = \sum_{i=1}^{L} p_i^2$. Now consider a function $f : D \rightarrow R$ and let x and y be chosen uniformly and independently from D. Then, $\Pr[f(x) = f(y)] = \sum_{i=1}^{L} p_i^2$. Thus, we need at least $1/(\sum_{i=1}^{L} p_i^2)^{1/2}$ many queries to obtain a collision by using birthday attack on f.

Now we consider $D = \mathsf{Perm} \times [N]$, $R = \{0,1\}^{8\ell}$ and $f : \mathsf{Perm} \times [N] \rightarrow \{0,1\}^{\ell}$ be HBG_ℓ function. HBG_ℓ is nothing but RC4-PRBG and hence we consider different distinguishing attack described in Section 2.2 to compute the birthday attack complexity.

(a) Mantin and Shamir's 2nd byte distinguishing attack: Let $y_1, \cdots, y_\ell$ be the bytes of PRBG output. It was shown that given $j = 0$ and S is chosen uniformly the probability that $y_2 = 0$ (zero byte) is close to $2/N$ (instead of $1/N$ for a true uniform distribution) [20]. Now there are $2^{8(\ell-1)}$ outputs which have 2nd byte 0. Assuming that all other remaining outputs are equally probable, we see that the birthday attack complexity is close to $q = 2^{4\ell} \times 2^{-.001} = 2^{4\ell-.001}$. Thus, the security is .001 bit less compare to the ideal situation. Moreover, here we assume that $j = 0$. Bias for the distinguishing attack is much less when we have a uniform distribution on j, which is more likely in our case.

(b) Paul and Preneel distinguishing attack: Let us study Paul and Preneel's [23] distinguishing attack in the view of Bellare-Kohno Birthday attack complexity. In this attack, it is proved that first two bytes are equal with probability close to $1/N(1 - 1/N)$. One can make similar calculation to see that the birthday attack complexity is very close to $2^{4\ell-.00000008}$.

One can make for similar analysis for other distinguishing attack given in [11,12]. Now we study the birthday attack in the view of Finney state.

(c) Finney State: Let (S_1, j_1) and (S_2, j_2) be chosen uniformly then the probability that the hash outputs are equal (i.e., $\mathsf{HBG}_\ell(S_1, j_1) = \mathsf{HBG}_\ell(S_2, j_2)$) is close to $\frac{1}{N^4 \times N(N-1)\cdots(N-\ell+1)} + (1 - \frac{1}{N^4})\frac{1}{N^\ell}$. This can be computed by conditioning on the event that both states are Finney state. Thus, the birthday attack complexity can be computed and which is approximately $N^{4\ell-.000001375}$.

Note that all these calculations are based on some assumptions. Actual birthday attack complexity may be different but it is not easy to calculate as the output distribution of HBG_ℓ is not known.

(2) Attack using characteristic for internal collision

In this section we write RCH to denote RCH_ℓ when the analysis does not depend on the choice of ℓ. Collision attack focuses on finding a characteristic with high probability. Recently, Wang *et al.* suggested new attack strategies to find collision-finding characteristics with high probability by using both addition and XOR difference. Especially, their attack method is deeply related to the properties of boolean functions used in each hash function. They also found collisions of MD4, MD5, HAVAL, SHA-0 and showed the complexity of finding a collision of SHA-1 is 2^{63} operations. Unlike MD4 style hash functions, RCH uses a nonlinear function, exchange shuffle. Since the exchange shuffle prevents the addition or XOR difference from being preserved and there is no boolean function, we can not apply Wang *et al.* attack method to RCH. Therefore we need different approach for security analysis of RCH. As the first step of security analysis, we consider two characteristics with small steps of RCH. Let x and x' be two message blocks and x_i and x'_i denote the message bytes at stage i.

First Example: For any i and $S[i] = a$, $S[i+1] = b$, $x_{i+1} = x_i$, if $i = j$ before updating j and $x_i + a \equiv 0 \bmod 256$ and $x_{i+1} + b \equiv 0 \bmod 256$, then final intermediate permutation S and j become same for x and x' such that $x'_i = x_i + 1$, $x'_{i+1} = x_{i+1}$ and $x'_{i+2} = x_{i+2} + 255$. Note that here we need three conditions to control the values of $S[i], S[i+1]$ and j.

Second Example: For any i and $S[i] = a$, $S[i+1] = b$, $x_i = x_{i+1} = x_{i+2} = x_{i+3} - 4$, if $i = j$ before updating j and $x_i + a \equiv 1 \bmod 256$ and $a \equiv b - 1 \bmod 256$, then final intermediate permutation S and j become same for x and x' such that $x'_i = x_i - 1$, $x'_{i+1} = x_{i+1}$, $x'_{i+2} = x_{i+2} - 1$, $x'_{i+3} = x_{i+3} + 2$ and $x'_{i+4} = x_{i+4} - 3$. Note that here we need three conditions to control the values of $S[i], S[i+1]$ and j.

First example is a 3-step characteristic with 3 conditions such that two message bytes are different for x and x'. The length of characteristic is same as the number of conditions. Second example is a 5-step characteristic with 3 conditions such that four message bytes are different for x and x'. We need more different message bytes for x and x' in order to get a long length of characteristic with few conditions. Since RCH uses each message byte four times with a reordering method, in case of using many different message bytes for x and x', an attacker has to make a complicated long characteristics for other rounds.

Here, we consider a specific attacker to try to construct characteristics such that each step has one condition. In this case, we can say security bound of attack complexity. If two messages differ in k_1 and k_2 positions and let i_1 and i_2 be its inverse with respect to the round function r_1 (say). Then we need to put conditions on $S[i_1], S[i_1 + 1] \cdots, S[i_2]$ and j. The reordering we have chosen have the property that for any k_1 and k_2,

$$\sum_{k=1}^{3} |r_k^{-1}(k_1) - r_k^{-1}(k_2)| \geq 24,$$

and hence the total number of conditions is at least 30. This is because we need $|r_k^{-1}(k_1) - r_k^{-1}(k_2)| + 2$ conditions for each round. Thus, we need 2^{240} queries to find the collision. Note that, this is a heuristic argument. Intuitively it is not possible to get a collision with above method within this complexity. In fact, it is not clear how to make a collision attack with this complexity.

(3) Attack using b-conserving property

Next, we consider the case of using b-conserving property. RCH_ℓ has a initial permutation S^{IV} such that there is no b-conserving property. Even though S^{IV} is not b-conserving, intermediate permutation can be b-conserving by applying a specific message. If intermediate permutation of each step is random, we can compute the probability that there exists b-conserving permutation in the intermediate value for each b as follows.

1. For $b = 2$, $(128!)^2/256! \approx 2^{-252}$
2. For $b = 4$, $(64!)^4/256! \approx 2^{-490}$
3. For $b = 8$, $(32!)^8/256! \approx 2^{-743}$
4. For $b = 16$, $(16!)^{16}/256! \approx 2^{-976}$

In order to get a 2-conserving intermediate permutation, we need 2^{252} queries of C and then we can choose message blocks such that all intermediate permutations onward are almost 2-conserving with probability $2/5$. Therefore, we can reduce the size of intermediate value from 1684-bit (corresponding to 256!) to at least 1432-bit (there are $128! \times 128!$ 2-conserving permutations) so that we can find a collision in intermediate value with complexity at least 2^{716} which is more than that of trivial collision attack with hash output less than 512-bit. As other cases have very small probabilities, we ignore them.

Performance

This hash function is based on the RC4 structure which itself is a very fast algorithm. For each 512 bit messages we need 1024 modulo sum and 1536 lookup (to compute $C^+(\cdot)$). The post processing is little bit costly but it would not matter if we hash long message as it is applied once for each message. In post-processing we have $512 + \ell$ addition and $2048 + 3\ell$ lookup. We have checked the performance with SHA-1 and we have noted that SHA-1 is roughly 1.5 times faster than our algorithm. We hope that this algorithm can be improved in near future.

5   Conclusion

In this paper we presented a new hash function RC4-Hash, and claim that it is secure as well as very fast. This hash function is based on the simple structure of RC4. This proposed hash function generate variable size hash outputs (like a family of hash functions e.g., SHA family). It's structure is different from that of many well known hash functions. Due to its completely new internal structure and huge size of internal state (approximately 1700 bits) it resists all generic attacks as well as path breaking attacks by Wang *et al.* It is very simple to implement and efficient in software and is compatible with different level of security. We hope that this new hash function will be found useful. Note, RC4 is based on 8 bit arithmetic, but there are RC4 like ciphers [5,13] exploiting 32/64 bit architecture of present day machines with enhanced speed. It may be a future work to design hash function based on the generalized RC4 with robust security and increased speed.

Acknowledgements. We wish to thank Dr. Pinakpani Pal for helping us in software implementation for checking the performance of RC4-Hash. We wish to thank Professor Rana Barua and the anonymous reviewers for their detailed comments that improved the technical quality and the editorial presentation of this paper. The first author was supported by the Korea Research Foundation Grant funded by the Korean Government (MOEHRD) (KRF-2005-213-C00005).

References

1. Ross J. Anderson and E. Biham. TIGER: A Fast New Hash Function. In *FSE'1996*, *Lecture Notes in Computer Science*, pages 89–97, Springer-Verlag, 1996.
2. E. Biham, L. Granboulan and P. Q. Nguyen. Impossible Fault Analysis of RC4 and Differential Falut Analysis of RC4. In *FSE'2005*, volume **3557** of *Lecture Notes in Computer Science*, pages 359–367, Springer-Verlag, 2005.
3. M. Bellare and T. Kohno. Hash Function Balance and Its Impact on Birthday Attacks. In *Advances in Cryptology-Eurocrypt'2004*, volume **3027** of *Lecture Notes in Computer Science*, pages 401–418, Springer-Verlag, 2004.
4. M. Bellare and T. Ristenpart. Multi-Property-Preserving Hash Domain Extension and the EMD Transform. *To appear in Asiacrypt'2006*. See at http://www-cse.ucsd.edu/users/tristenp/.
5. E. Biham, J. Seberry. **Py** (Roo): A Fast and Secure Stream Cipher using Rolling Arrays. eSTREAM, ECRYPT Stream Cipher Project, Report 2005/023, 2005.
6. B. O. Brachtl, D. Coppersmith, M. M. Hyden, S. M. Matyas, C. H. Meyer, J. Oseas, S. Pilpel, M. Schilling. *Data Authentication Using Modification Detection Codes Based on a Public One Way Encryption Function*. U.S. Patent Number 4,908,861, March 13, 1990.
7. H. Finney. An RC4 cycle that can't happen. *Post in sci.crypt*, September 1994.
8. J. S. Coron, Y. Dodis, C. Malinaud and P. Puniya. Merkle-Damgard Revisited: How to Construct a Hash Function. In *Advances in Cryptology-Crypto'2005*, volume **3621** of *Lecture Notes in Computer Science*, pages 430–448. Springer-Verlag, 2005.
9. FIPS 180-1. Secure Hash Standard, US Department of Commerce, Washington D. C, Springer Verlag, 1996.
10. S. Fluhrer, I. Mantin, A. Shamir. Weaknesses in the Key Scheduling Algorithm of RC4. In *SAC'2001*, volume **2259** of *Lecture Notes in Computer Science*, pages 1–24, Springer-Verlag, 2001.
11. S. Fluhrer and D. McGrew. Statistical Analysis of the Alleged RC4 Keystream Generator. In *FSE'2000*, volume **1978** of *Lecture Notes in Computer Science*, pages 19–30, Springer-Verlag, 2000.
12. J. Golic. Linear Statistical Weakness of Alleged RC4 Keystream Generator. In *Advances in Cryptology-Eurocrypt'1997*, volume **1233** of *Lecture Notes in Computer Science*, pages 226–238, Springer-Verlag, 1997.
13. G. Gong, K. C. Gupta, M. Hell and Y. Nawaz. Towards a General RC4-Like Keystream Generator In *CISC'2005*, volume **3822** of *Lecture Notes in Computer Science*, pages 162–174, Springer-Verlag, 2005.
14. A. Grosul and D. Wallach. A Related Key Cryptanalysis of RC4. *Department of Computer Science, Rice University, Technical Report TR-00-358*, June 2000.

15. J. J. Hoch, A. Shamir. Fault Analysis of Stream Ciphers. CHES: Cryptographic Hardware and Embedded Systems, CHES'04, *Lecture Notes in Computer Science*, pages 240–253, Springer-Verlag, 2004.

16. J. Kelsey, B. Schneier. Second Preimages on n-Bit Hash Functions for Much Less than 2^n Work. In *Advances in Cryptology-Eurocrypt'2005*, volume **3494** of *Lecture Notes in Computer Science*, pages 474–490, Springer-Verlag, 2005.

17. S. Lucks. A Failure-Friendly Design Principle for Hash Functions. In *Advances in Cryptology-Asiacrypt'2005*, volume **3788** of *Lecture Notes in Computer Science*, pages 474–494, Springer-Verlag, 2005.

18. I. Mantin. Analysis of the stream cipher RC4. Master's thesis, Weizmann Institute, Israel 2001.

19. I. Mantin. A Practical Attack on the Fixed RC4 in the WEP Mode. In *Advances in Cryptology-Asiacrypt'2005*, volume **3788** of *Lecture Notes in Computer Science*, pages 395–411, Springer-Verlag, 2005.

20. I. Mantin and A. Shamir. A Practical Attack on Broadcast RC4. In *FSE'2001*, volume **2355** of *Lecture Notes in Computer Science*, pages 152–164, Springer-Verlag, 2001.

21. I. Mironov. Not (So) Random Shuffle of RC4. In *Advances in Cryptology-Crypto'2002*, volume **2442** of *Lecture Notes in Computer Science*, pages 304–319, Springer-Verlag, 2002.

22. S. Paul and B. Preneel. Analysis of Non-fortuitous Predictive States of the RC4 Keystream Generator. In *Indocrypt'2003*, volume **2904** of *Lecture Notes in Computer Science*, pages 52–67, Springer-Verlag, 2003.

23. S. Paul and B. Preneel. A New Weakness in the RC4 Keystream Generator and an Approach to Improve the Security of the Cipher. In *FSE'2004*, volume **3017** of *Lecture Notes in Computer Science*, pages 245–259, Springer-Verlag, 2004.

24. B. Preneel, R. Govaerts and J. Vandewalle. *Cryptographically secure hash functions: an overview*. ESAT Internal Report, K. U. Leuven, 1989.

25. RIPE, Integrity Primitives for secure Information systems, Final report of RACE Integrity Primitive Evaluation (RIPE-RACE 1040) *Lecture Notes in Computer Science*, Springer-Verlag, 1995.

26. Ronald L. Rivest. The MD4 message-digest algorithm. In *Crypto'1990*, volume **537** of *Lecture Notes in Computer Science*, pages 303–311, Springer-Verlag, 1991.

27. Ronald L. Rivest. The MD5 message-digest algorithm. Request for comments (RFC 1320), Internet Activities Board, Internet Privacy Task Force, 1992.

28. Ronald L. Rivest. Abelian square-free dithering for iterated hash functions. In *First Hash Workshop by NIST*, October 2005.

29. A. Roos. A Class of Weak Keys in the RC4 Stream Cipher. *Post in sci.crypt*, September 1995.

30. F. Schmidt and R. Simion, Card shuffling and a transformation on Sn. Acquationes Mathematicae, vol. 44, pp. 11-34, 1992.

31. SHA-0, A federal standard by NIST, 1993.

32. D. R. Stinson. *Cryptography , Theory and Practice, Second Edition*. CRC Press, 2002.

33. X. Wang, X. Lai, D. Feng, H. Chen and X. Yu. Cryptanalysis of the Hash Functions MD4 and RIPEMD. In *Advances in Cryptology-Eurocrypt'2005*, volume **3494** of *Lecture Notes in Computer Science*, pages 1–18, Springer-Verlag, 2005.

34. X. Wang and H. Yu. How to Break MD5 and Other Hash Functions. In *Advances in Cryptology-Eurocrypt'2005*, volume **3494** of *Lecture Notes in Computer Science*, pages 19–35, Springer-Verlag, 2005.

35. X. Wang, H. Yu and Y. L. Yin. Efficient Collision Search Attacks on SHA-0. In *Advances in Cryptology-Crypto'2005*, volume **3621** of *Lecture Notes in Computer Science*, pages 1–16, Springer-Verlag, 2005.
36. X. Wang, Y. L. Yin and H. Yu. Finding Collisions in the Full SHA-1. In *Advances in Cryptology-Crypto'2005*, volume **3621** of *Lecture Notes in Computer Science*, pages 17–36, Springer-Verlag, 2005.
37. H. Yu, X. Wang, A. Yun and S. Park. Cryptanalysis of the Full HAVAL with 4 and 5 Passes. To appear in *FSE'2006*, Springer-Verlag, 2006.
38. H. Yu, G. Wang, G. Zhang and X. Wang. The Second-Preimage Attack on MD4. In *CANS'2005*, volume **3810** of *Lecture Notes in Computer Science*, pages 1–12, Springer-Verlag, 2005.
39. Y. Zheng, J. Pieprzyk and J. Seberry HAVAL - A One-Way Hashing Algorithm with Variable Length of Output In *ASIACRYPT 1992, Lecture Notes in Computer Science*, pages 83–104, Springer-Verlag, 1992.

Appendix

• Here we describe the reordering we are using in the hash algorithm. We use the identity function for r_0 and r_i's are defined as in below for $1 \leq i \leq 3$. Note that the function r restricted on $[64i, 64i + 63]$ is nothing but r_i, $0 \leq i \leq 3$

r_1 : 0, 55, 46, 37, 28, 19, 10, 1, 56, 47, 38, 29, 20, 11, 2, 57, 48, 39, 30, 21, 12, 3, 58, 49, 40, 31, 22, 13, 4, 59, 50, 41, 32, 23, 14, 5, 60, 51, 42, 33, 24, 15, 6, 61, 52, 43, 34, 25, 16, 7, 62, 53, 44, 35, 26, 17, 8, 63, 54, 45, 36, 27, 18, 9.

r_2: 0, 57, 50, 43, 36, 29, 22, 15, 8, 1, 58, 51, 44, 37, 30, 23, 16, 9, 2, 59, 52, 45, 38, 31, 24, 17, 10, 3, 60, 53, 46, 39, 32, 25, 18, 11, 4, 61, 54, 47, 40, 33, 26, 19, 12, 5, 62, 55, 48, 41, 34, 27, 20, 13, 6, 63, 56, 49, 42, 35, 28, 21, 14, 7.

r_3 : 0, 47, 30, 13, 60, 43, 26, 9, 56, 39, 22, 5, 52, 35, 18, 1, 48, 31, 14, 61, 44, 27, 10, 57, 40, 23, 6, 53, 36, 19, 2, 49, 32, 15, 62, 45, 28, 11, 58, 41, 24, 7, 54, 37, 20, 3, 50, 33, 16, 63, 46, 29, 12, 59, 42, 25, 8, 55, 38, 21, 4, 51, 34, 17.

• **The initial value permutation or** S^{IV} **is the following:**

145, 57, 133, 33, 65, 49, 83, 61, 113, 171, 63, 155, 74, 50, 132, 248, 236, 218, 192, 217, 23, 36, 79, 72, 53, 210, 38, 59, 54, 208, 185, 12, 233, 189, 159, 169, 240, 156, 184, 200, 209, 173, 20, 252, 96, 211, 143, 101, 44, 223, 118, 1, 232, 35, 239, 9, 114, 109, 161, 183, 88, 66, 219, 78, 157, 174, 187, 193, 199, 99, 52, 120, 89, 166, 18, 76, 241, 13, 225, 6, 146, 151, 207, 177, 103, 45, 148, 32, 29, 234, 7, 16, 19, 91, 108, 186, 116, 62, 203, 158, 180, 149, 67, 105, 247, 3, 128, 215, 121, 127, 179, 175, 251, 104, 246, 98, 140, 11, 134, 221, 24, 69, 190, 154, 253, 168, 68, 230, 58, 153, 188, 224, 100, 129, 124, 162, 15, 117, 231, 150, 237, 64, 22, 152, 165, 235, 227, 139, 201, 84, 213, 77, 80, 197, 250, 126, 202, 39, 0, 94, 42, 243, 228, 87, 82, 27, 141, 60, 160, 46, 125, 112, 181, 242, 167, 92, 198, 172, 170, 55, 115, 30, 107, 17, 56, 31, 135, 229, 40, 111, 37, 222, 182, 25, 43, 119, 244, 191, 122, 102, 21, 93, 97, 131, 164, 10, 130, 47, 176, 238, 212, 144, 41, 14, 249, 220, 34, 136, 71, 48, 142, 73, 123, 204, 206, 4, 216, 196, 214, 137, 255, 195, 26, 8, 51, 178, 2, 138, 254, 90, 194, 81, 245, 106, 95, 75, 86, 163, 205, 70, 226, 28, 147, 85, 5, 110.

Security of VSH in the Real World

Markku-Juhani O. Saarinen

Information Security Group
Royal Holloway, University of London
Egham, Surrey TW20 0EX, UK
m.saarinen@rhul.ac.uk

Abstract. In Eurocrypt 2006, Contini, Lenstra, and Steinfeld proposed a new hash function primitive, VSH, *very smooth hash*. In this brief paper we offer commentary on the resistance of VSH against some standard cryptanalytic attacks, including preimage attacks and collision search for a truncated VSH. Although the authors of VSH claim only collision resistance, we show why one must be very careful when using VSH in cryptographic engineering, where additional security properties are often required.

1 Introduction

Many existing cryptographic hash functions were originally designed to be *message digests* for use in digital signature schemes. However, they are also often used as building blocks for other cryptographic primitives, such as pseudorandom number generators (PRNGs), message authentication codes, password security schemes, and for deriving keying material in cryptographic protocols such as SSL, TLS, and IPSec.

These applications may use truncated versions of the hashes with an implicit assumption that the security of such a variant against attacks is directly proportional to the amount of entropy (bits) used from the hash result. An example of this is the $HMAC-n$ construction in IPSec [1]. Some signature schemes also use truncated hashes. Hence we are driven to the following slightly nonstandard definition of security goals for a hash function usable in practice:

1. **Preimage resistance.** For essentially all pre-specified outputs X, it is difficult to find a message Y such that $H(Y) = X$. The difficulty should be $\approx 2^l$ when there are l pre-specified bits in X.
2. **2nd-preimage resistance.** Given a pre-specified message X, it is difficult to find another message Y so that $H(X) = H(Y)$. The difficulty should be $\approx 2^l$ when there are l pre-specified bits that match in the hashes.
3. **Collision resistance.** It should require $\approx 2^{l/2}$ effort to find any two messages X and Y that produce a collision $H(X) = H(Y)$ in l pre-specified bits in the hashes.

In addition to the above three usual goals, we state a fourth, more informal goal – **pseudorandomness**. In essence, we would like a PRNG, stream cipher, or other derived design that relies on a hash function to have at least $\approx 2^{l/2}$ security, as if it was secured with a "real" pseudorandom function.

R. Barua and T. Lange (Eds.): INDOCRYPT 2006, LNCS 4329, pp. 95–103, 2006.
© Springer-Verlag Berlin Heidelberg 2006

Pseudorandomness implies that a hash has good statistical properties and resistance against a wide array of distinguishing attacks.

All of the mentioned desirable properties are difficult if not impossible to prove without nonstandard assumptions. We note that proofs based on assumptions are themselves assumptions, whether their origins are in the traditions of symmetric or asymmetric cryptanalysis. An assumption based on the sieving phase of the NFS factoring algorithm may seem like a "hard problem" to a researcher who has spent a lot of time tweaking the sieving phase of the NFS factoring algorithm. On the other hand, a researcher who has dedicated years of effort into symmetric cryptanalysis may feel that symmetric cryptography possesses equally well studied "hard problems", while also allowing more efficient overall implementation.

A "political" standardisation consideration is that (by definition) VSH has a backdoor in the secret factorisation of n. In the past it has been difficult to popularise cryptographic technologies that rely on trusted third parties.

In our opinion VSH is a simple, elegant design that is based on a plausible complexity-theoretic assumption (VSSR: Very Smooth number nontrivial modular Square Root). However, it should not be considered a general-purpose hash function as usually understood in security engineering.

On VSH Security Claims

> "VSH is not a Hash Function."
> – *Arjen K. Lenstra, Eurocrypt 2006* [1]

Collision resistance is the only property proven for VSH. In Section 3 of the VSH paper [2], short message inversion (equivalent to preimage resistance) is considered and one possible "solution" is provided. As will be shown in Section 2.1 of this paper, the solution is not adequate.

The authors therefore clearly expected VSH to exhibit some level of preimage and 2nd preimage resistance. These are standard requirements in the very definition of a "cryptographic hash function". The authors of VSH are very clear in that "VSH should not be used to model random oracles". Random oracle behaviour is not a standard hash function security requirement.

Some researchers tend to concentrate their efforts on showing that their hash functions provide collision resistance, while ignoring other security properties. However, it is well known that collision resistance does not imply preimage-resistance or other important hash function properties.

To illustrate this point, we present a classical counter-example. Consider an $l + 1$-bit hash $H'(x)$ that has been constructed from an l - bit hash H as follows:

$$\text{If } |x| < l - 1 \text{ then } H'(x) = x \,\|\, 1 \,\|\, 0\,0 \cdots 0.$$
$$\text{If } |x| \geq l - 1 \text{ then } H'(x) = H(x) \,\|\, 1.$$

---

[1] Quoted with permission. During the conference A.K. Lenstra used some of the results from this note in his presentation, with appropriate credit. This has led some people to mistakenly think that the results in this note were already contained in [2]. All cryptanalytic results presented in this paper are by the author; a draft was circulated with the authors of VSH before Eurocrypt 2006.

That is, if the message x is less than $l-1$ bits long, $H'(x)$ consists of the message itself, a single 1 bit and a padding of zero bits. If the message is $l-1$ bits or longer, the resulting hash consists of a (secure) hash of x, followed by a single 1 bit.

It is easy to show that H' is collision resistant if H is. It is also easy to see that H' is *not* preimage resistant for a large proportion of hash outputs, and that a slightly truncated version is *not* collision resistant.

2   The VSH Algorithm

We describe the VSH algorithm in its most basic form, essentially as it appears in the beginning section 3 of [2]. We note that the attacks can be extended to most of the variants given in the VSH paper, especially the Fast VSH variant in section 3.1 of [2]. [2]

Let $p_1 = 2, p_2 = 3, p_3 = 5, \ldots$ be the sequence of primes. Let n be a large RSA composite. Let k, the block length, be the largest integer such that $\prod_{i=1}^{k} p_i < n$. Let m be a be an l-bit message to be hashed, consisting of bits $m_1, m_2, \ldots, m_l$, and assume that $l < 2^k$. To compute the hash of m:

1. Let $x_0 = 1$.
2. Let $L = \lceil l/k \rceil$ the number of blocks. Let $m_i = 0$ for $l < i \leq Lk$ (padding).
3. Let $l = \sum_{i=1}^{k} l_i 2^{i-1}$ with $l_i \in \{0,1\}$ be the binary representation of the message length l and define $m_{Lk+i} = l_i$ for $1 \leq i \leq k$.
4. For $j = 0, 1, \ldots, L$ in succession compute

$$x_{j+1} = x_j^2 \prod_{i=1}^{k} p_i^{m_{(jk+i)}} \quad \bmod \ n.$$

5. Return x_{L+1}.

Selecting a 1024-bit modulus n has been suggested in the original paper, indicating 131-bit block size k.

2.1   Preimage Resistance

VSH is multiplicative: Let x, y, and z be three bit strings of equal length, where z consists only of zero bits and the strings satisfy $x \wedge y = z$. It is easy to see that

$$H(z)H(x \vee y) \equiv H(x)H(y) \ (\bmod \ n).$$

This multiplicative property is similar, although simpler, than the one used by Coppersmith to attack (then) Annex D of X.509 [3].

---

[2] There were many changes to VSH before its final publication, most recently in early March 2006 when message length padding was changed to be performed *after* the message been hashed, rather than at the beginning. Such small changes have significant implications on the development of practical attacks. Remarkably, the "security proof" required no modification. The attacks discussed in this paper apply only to the published Eurocrypt version of VSH; other attacks may be devised on other variants.

As a result VSH succumbs to a classical time-memory trade-off attack that applies to multiplicative and additive hashes. The attack is similar in many aspects to Shanks' baby-step giant-step algorithm for discrete logarithms [5].

We set the secret message m as $(x \lor y)$ and rewrite the equation as

$$H(y) = H(x)^{-1}H(z)H(m) \pmod{n}.$$

To solve the l-bit preimage m of $H(m)$:

1. Tabulate $H(x \,\|\, 00 \cdots 0)^{-1}H(z)H(m) \pmod{n}$ for $0 \le x < 2^{l/2}$.
2. Do table lookups for $H(00 \cdots 0 \,\|\, y)$ for $y = 0, 1, 2, \ldots$, looking for a match.

The algorithm terminates when $m = x \,\|\, y$, in other words before $y < 2^{l/2}$. A preimage attack on VSH therefore has $\approx 2^{l/2}$ complexity rather than $\approx 2^l$ as expected.

Final squarings proposed in section 3 of [2] under subtitle "short message inversion" do not protect against this attack.

This type of attack is extremely serious if VSH is used to secure passwords, a typical application for hash functions. Note that the complexity of attack does not depend on the modulus size n, but on the entropy of the password strings.

Example 1. VSH is being used to secure a 4 character lower case alphabetic password M, stored with ASCII encoding. For demonstration purposes we choose k = 32 and a 169-bit modulus n:

$$n = (2^{84} + 3)(2^{85} - 19)$$
$$= 748288838313422294120286382894166426220969123119047.$$

The hash of the secret is

$$H(m) = 168441206251546173371590624134667166930498668664325.$$

In this case $H(z) = 13$; the first iteration yields 1, and the second round 13, the sixth prime, as the length of the message is $2^5 = 32$ bits. We tabulate $H(x)^{-1}H(z)H(m)$ $\pmod{n}$ for $26^2 = 676$ values $T[0 \ldots 675]$:

```
x: aa..   Binary: 01100001 01100001 00000000 00000000
T[0] = 91345572106882035279752100576530653

x: ab..   Binary: 01100001 01100010 00000000 00000000
T[1] = 116156501606261492576199026944080853
              . . .
x: zz..   Binary: 01111010 01111010 00000000 00000000
T[675] = 384284712674090018973838770853950813384926485216514
```

In the second phase we run through the values of $H(y)$:

```
H(..aa) = 3904844677556216209933
H(..ab) = 3396095819174949308197
    . . .
```

A match is found after 83 steps at $H(\,.\,.\,\mathtt{df}) = 3020566045699958278116255949$, which matches with $T[18] = H(\mathtt{as}\,.\,.\,)^{-1}H(z)H(m) \pmod{n}$. Hence the secret password M is "$\mathtt{asdf}$".

Note that it is not necessary to store the entire value to the table $T[i]$; appropriate number of least significant bits usually suffices. When the table is indexed by, say, $T[i] \bmod 2^{32}$, search becomes an $O(1)$ operation.

This example illustrates that password cracking time is effectively "square-rooted" by this attack; l-character passwords offer a level of security expected from $l/2$-character passwords.

2.2   One-Wayness (of the "Cubing" Variant)

In section 3.4 of the VSH specification, a variant that uses cubing instead of squaring in its compression function is proposed. Using the Jacobi symbol, the compression function

$$x_{j+1} = x_j^3 \prod_{i=1}^{k} p_i^{m_i} \bmod n,$$

becomes

$$\left(\frac{x_{j+1}}{n}\right) = \left(\frac{x_j}{n}\right) \prod_{i=1}^{k} \left(\frac{p_i}{n}\right)^{m_i}.$$

We define a "binary" version of the Jacobi symbol:

$$j(c, n) = \frac{1}{2}\left(1 - \left(\frac{c}{n}\right)\right).$$

We now have a linear equation giving the parity of some message bits:

$$j(x_{j+1}, n) = j(x_j, n) + \sum_{i=1}^{k} j(p_i, n)m_i \pmod 2.$$

Note that the Jacobi symbol can be very efficiently computed and that $j(p_i, n)$ is essentially randomly 0 or 1 for each randomly generated composite n. If the same message has been hashed with k different moduli n, a system of k linear equations can be obtained, leading to disclosure of bits by solving the system of equations.

The same attack applies to the standard squaring version as well, but it only leaks information about the message length. This was not the case for VSH versions 3.57 and before (ePrint revisions of VSH published before March 2006), where information about the contents of the last message block could be obtained.

One-wayness is implied by the standard hash security requirement of preimage resistance. If one obtains some information about some of the preimage bits easily, one can find the rest faster in an exhaustive search, as the search space is smaller.

Example 2. Assume that a 64-bit password has been hashed with VSH. For demonstration purposes we define the modulus n to be equivalent to the RSA-1024 factoring challenge number $n = 1350..(300 \text{ digits})..7563$ [4].

The Jacobi symbols for the first small primes modulo n are:

$$\left(\frac{2}{n}\right) = -1 \ \left(\frac{3}{n}\right) = -1 \ \left(\frac{5}{n}\right) = -1 \ \left(\frac{7}{n}\right) = 1 \ \left(\frac{11}{n}\right) = 1 \ \left(\frac{13}{n}\right) = -1 \ \cdots$$

Since the length padding (last round) will simply consist of cubing the product of primes and multiplying that with length indicator $p_6 = 13$, we may write

$$\left(\frac{H(m)}{n}\right) = \left(\frac{13}{n}\right) \prod_{i=1}^{64} \left(\frac{p_i}{n}\right)^{m_i}.$$

Using the binary $j(c, n)$ function and knowledge of n, this can be further simplified into the following parity equation:

$$j(H(m), n) \equiv 1 + m_1 + m_2 + m_3 + m_6 + m_7 + m_{10} + m_{13} + m_{14} + m_{15} +$$
$$m_{16} + m_{17} + m_{22} + m_{24} + m_{25} + m_{26} + m_{27} + m_{28} + m_{29} +$$
$$m_{31} + m_{33} + m_{36} + m_{39} + m_{40} + m_{43} + m_{44} + m_{46} + m_{49} +$$
$$m_{51} + m_{52} + m_{57} + m_{59} + m_{61} + m_{64} \pmod{2}.$$

We can therefore speed up dictionary search against the password by a factor close to two as half of the password candidates can be rejected with simple bit shift, AND and XOR operations, rather than with computationally expensive modular arithmetic required to compute the full hash.

Note that if the same secret has been hashed with multiple different moduli n, the speedup grows almost exponentially; two distinct moduli yield a speedup factor close to 4 etc.

2.3   Collision Search for Truncated VSH Variants

VSH produces a very long hash (typically 1024 bits). There are no indications that a truncated VSH hash offers security that is commensurate to the hash length. This appears to rule out the applicability of VSH in digital signature schemes which produce signatures shorter than the VSH hash result, such as Elliptic Curve signature schemes.

To illustrate this point, we will describe give an attack on one truncated variant of VSH.

Partial Collision Attacks. We will first discuss a generic technique for turning a partial collision attack into a full collision attack.

Assume that there is a fast $O(1)$ mapping f that causes the hash result of an l-bit hash H to be in some smaller subset of possible outputs: $H(f(x)) \in S$, where $|S| < 2^l$. Typically f would be chosen in such a way that certain hash result bits are forced to have the same constant value. In other words, f forces partial collisions. Note that f itself should not produce too many collisions, i.e. $x_1 \neq x_2$ usually means that $f(x_1) \neq f(x_2)$.

If such an f can be found, and it is fast, the complexity of finding full collisions becomes $\approx \sqrt{|S|}$. Note that f does not need to be able to force the hash to S on each iteration, it is sufficient that it works with reasonable probability. The iteration in low-memory parallel collision search algorithm becomes $s_{i+1} = H(f(s_i))$, and generic parallel collision search algorithms such as those described in [6] can be used.

Attack on VSH Truncated to Least Significant 128 Bits. We will instantiate this attack on a VSH variant that only uses the least-significant 128 bits of the hash function result. For basic VSH (1024-bit n, k=131) the result of hashing a 128-bit message $m_1|m_2|\cdots|m_{128}$ can be simplified to:

$$x = \left(19\left(\prod_{i=1}^{128} p_i^{m_i}\right)^2 \bmod n\right) \bmod 2^{128}.$$

The constant $19 = p_8$ is caused by the length padding in the second (and final) round.

It is easy to see that modular reduction by n occurs in this case with less than 50% probability if m is random (or randomised) and its Hamming weight behaves accordingly. This is due to the fact that if only half of the bits in the message are ones, the product of corresponding small primes will be roughly the same bit size as $\sqrt{n}$. The square of this will still be less than n with a significant probability and hence there is no modular reduction by n. Hamming weight of a random bit string is binomially distributed. In practice the modular reduction happens in this case with roughly $P \approx 0.35$ probability. We get the following approximation that is valid with significant probability:

$$x = 19\left(\prod_{i=1}^{128} p_i^{m_i}\right)^2 \bmod 2^{128}.$$

Note that the iteration is independent of the RSA modulus n if there is no reduction.

Precomputation phase: For each of the 2^{41} bit strings r of length 41 we compute and store r into a lookup table, indexed by the product

$$\left(\prod_{i=2}^{42} p_i^{r_{i-1}}\right)^{-1} \bmod 2^{42}.$$

We will choose the f mapping as follows: Select message bits $m_{43}, m_{44}, \ldots, m_{128}$ from corresponding bits of s_i. Compute the partial product $\prod_{j=43}^{128} p_j^{m_j} \bmod 2^{42}$ and use that to select message bits $m_2, m_3, \ldots, m_{42}$ using the lookup table (m_1 is always set to zero).

This will often ($P \approx 0.5$) force the least significant 42 bits to a certain constant value, 19, on each iteration. Note that if the table lookup fails, we may select $m_2, m_3, \ldots, m_{42}$ to be some arbitrary deterministic value; one that satisfies $s_i \equiv 19 \pmod{2^l}$ for some $l < 42$ would be a good choice.

Hence we have can cause the iteration to run in a significantly smaller subset with essentially $O(1)$ effort (constant-factor increase), and collisions can be found significantly faster.

Example 3. We will start with $s_1 = 2^{42} + 19$, and try to produce a sequence satisfying $s_i \equiv 19 \pmod{2^{42}}$ for a significant portion of i.

The partial product $\prod_{i=43}^{128} p_i^{m_i} \bmod 2^{42}$ yields $p_{43} = 191$ for s_1. We will then perform a lookup in the precomputed table; it turns out that selecting message bits m_1 through m_{42} as

```
01110010 01010101 00000000 11100001 11110111 00
```

will force the product the desired subset, as the product of primes corresponding to those message bits is

$$3 \cdot 5 \cdot 7 \cdot 17 \cdot 29 \cdot 37 \cdot 43 \cdot 53 \cdot 97 \cdot 101 \cdot 103 \cdot 131 \cdot 137 \cdot 139 \cdot 149 \cdot 151 \cdot 163 \cdot 167 \cdot 173$$

$$= 116421357191179516863577800910009 5,$$

and this multiplied by the partial product satisfies

$$191 \cdot 116421357191179516863577800910009 5 \equiv 1 \pmod{2^{42}}.$$

Clearly squaring a number that is congruent to 1 mod 2^{42} maintains that property. The final multiplication by 19 results in that that the second element of the sequence satisfies the desired property $s_2 \equiv 19 \pmod{2^{42}}$. We have

$$s_2 = 19 \left(191 \cdot 116421357191179516863577800910009 5\right)^2 \bmod 2^{128}$$

$$= 79424F79408D6B27F52A500000000001 3_{16}$$

With this sequence we only need to rely on a birthday collision in the upper $128 - 42 = 86$ bits of the sequence. Roughly 2^{43} iterations are required with algorithms of [6] to achieve this.

Note that with some probability this algorithm will yield false collisions due to the fact that the inverse of the partial product is not always found in the lookup table. Modular reduction by n may also cause false collisions. This only results in a constant factor increase to the complexity of the algorithm, however; we only need to restart with different starting points until a proper collision is found.

Overall complexity. In essence, the complexity of this attack against VSH truncated to l bits is:

- Pre-computing the table offline: $\approx 2^{\frac{l}{3}}$ time and space.
- Finding collisions: $\approx 2^{\frac{l}{3}}$ iterations.
- Total cost: roughly $\approx 2^{\frac{l}{3}}$, rather than $\approx 2^{\frac{l}{2}}$ as expected from a hash function with good pseudorandomness properties.

We acknowledge that this represents just *one* way of truncating VSH – using, say, the most significant bits of the result would be an even worse option. Many other truncated variants can be attacked using a different f function.

2.4   Other Features of VSH

The authors of VSH do not explicitly note this, but the hash function result can be updated after small changes without computing the entire hash again. A "bit flip" in a message will always cause a predictable change in the message result (it becoming multiplied mod n by certain power of a small prime or its inverse). This is due to the highly algebraic nature of the hash.

We note such a property may be useful in some applications where rapid update of the hash is required, but it is undesirable in many more as it can facilitate adaptive attacks against some cryptographic protocols. Similar multiplicative property was sufficient for the X.509 Annex D hash function to be considered broken [3].

Acknowledgments

The author would like to thank Arjen K. Lenstra and other authors of VSH for encouragement. The paper wouldn't exist if Kenny Paterson wouldn't have pointed out that publication of relatively simple results is important for "real world" security engineers. Keith Martin, Daniel J. Bernstein and anonymous program committee members helped to make the paper significantly easier to read.

References

1. M. BELLARE, R. CANETTI AND H. KRAWCZYK. *HMAC: Keyed-Hashing for Message Authentication.* IETF RFC 2104, 1997.
2. S. CONTINI, A.K. LENSTRA AND R. STEINFELD. VSH, an efficient and provable collision resistant hash function. Advances in Cryptology – EUROCRYPT 2006, LNCS 4004, Springer-Verlag, 2006.
3. D. COPPERSMITH. Analysis of ISO/CCITT Document X.509 Annex D. IBM Research Division, Yorktown Heights, N.Y., 11 June 1989.
4. RSA LABORATORIES. RSA-1024 Factoring Challenge Number. Available from: http://www.rsasecurity.com/rsalabs/node.asp?id=2093
5. D. SHANKS. Class number, a theory of factorization and genera. Proc. Symp. Pure Math. pp. 415 – 550. AMS, Providence, R.I., 1979.
6. P. VAN OORSCHOT AND M. WIENER. Parallel collision search with cryptanalytic applications. Journal of Cryptology, 12 (1999), pp. 1 – 28, 1999.

Cryptanalysis of Two Provably Secure Cross-Realm C2C-PAKE Protocols

Raphael C.-W. Phan[1] and Bok-Min Goi[2,*]

[1] Information Security Research (iSECURES) Lab,
Swinburne University of Technology (Sarawak Campus), 93576 Kuching, Malaysia
rphan@swinburne.edu.my
[2] Centre for Cryptography & Information Security (CCIS), Faculty of Engineering,
Multimedia University, 63100 Cyberjaya, Malaysia
bmgoi@mmu.edu.my

Abstract. Password-Authenticated Key Exchange (PAKE) protocols allow parties to share secret keys in an authentic manner based on an easily memorizable password. Byun *et al.* first proposed a cross realm client-to-client (C2C) PAKE for clients of different realms (with different trusted servers) to establish a key. Subsequent work includes some attacks and a few other variants either to resist existing attacks or to improve the efficiency. However, all these variants were designed with heuristic security analysis despite that well founded provable security models already exist for PAKEs, e.g. the Bellare-Pointcheval-Rogaway model. Recently, the first provably secure cross-realm C2C-PAKE protocols were independently proposed by Byun *et al.* and Yin-Bao, respectively; i.e. security is proven rigorously within a formally defined security model and based on the hardness of some computationally intractable assumptions. In this paper, we show that both protocols fall to undetectable online dictionary attacks by any adversary. Further we show that malicious servers can launch successful man-in-the-middle attacks on the variant by Byun *et al.*, while the Yin-Bao variant inherits a weakness against unknown key-share attacks. Designing provably secure protocols is indeed the right approach, but our results show that such proofs should be interpreted with care.

Keywords: Password-authenticated key exchange, cross realm, client-to-client, cryptanalysis, provable security, security model.

1 Introduction

A 2-party password-based authenticated key exchange (PAKE) protocol establishes a shared secret key between two parties. Authentication of parties is based on knowledge of a shared low-entropy password. The first known PAKE is due to Bellovin and Merritt [9]. This concept has also been extended to 3 parties, e.g. two clients and a trusted server or key distribution center (KDC).

---

* The second author acknowledges the Malaysia IRPA grant (04-99-01-00003-EAR).

R. Barua and T. Lange (Eds.): INDOCRYPT 2006, LNCS 4329, pp. 104–117, 2006.

FORMAL SECURITY MODELS. The formal security model for 2-party PAKE proto-
cols was proposed by Bellare *et al.* [8] so called the Bellare-Pointcheval-Rogaway
(BPR2000) model, building on work by Bellare and Rogaway in [6,7]. Later,
Abdalla *et al.* [2] extended this model to the 3-party case.

One informal approach to designing security protocols is to list all known
attacks and argue why a protocol resists them. This list is clearly not exhaustive,
and sometimes fails to catch specific types of attacks. The main problem is that
this heuristic approach assumes the particular behaviour of the adversary, i.e. he
is assumed to attack in some way. History [8,21] has shown that this is not the
right approach, because intuitively an adversary behaves in any way he prefers
as long as he can break the system. Thus it is often that such a protocol is broken
and a minor fix proposed, etc. This cycle continues resulting in many slightly
different protocol variants because breaks and subsequent fixes are heuristically
done. There are many such instances but to be concise we only cite here a few
recent ones: [10,22,23,29].

In contrast the approach based on formal security models does not assume
on any specific attack method an adversary may use. Instead a communication
model is defined that describes how parties within the protocol, as well as an
adversary, communicate with each other, and what sort of information formalized via the notion of oracle queries, is available to or may be under the control
of the adversary. Then, security properties of a protocol are defined as one or
more games each intended to capture a security property, played by the adversary within the pre-defined communication model. A protocol is secure with
respect to the defined security properties if the adversary's advantage in winning the game(s) is negligible, and further that the task of an adversary winning
is reduced to computationally intractable assumption(s). This approach is also
known as provable security [26]. Once proven secure, a protocol is guaranteed to
resist attacks by any adversary who works within the communication model regardless of what specific attacks are mounted, as long as the assumptions remain
intractable.

However, defining an appropriate model is not a trivial task, because not
including some types of queries e.g. the Corrupt query [14,15], or improperly
defining the adversarial game [8] may result in a security proof that fails to
capture valid attacks (see [8,14,15] for more details).

PAKEs FOR CROSS REALMS. It is sometimes desirable that client parties from
different environments (realms) be able to establish shared secret keys. Byun *et
al.* [10] proposed a PAKE protocol that allows to achieve this, by using the KDCs
in the different realms as the go-between, i.e. to perform translation of encrypted
or blinded secrets in one realm to the other under passwords shared between the
KDCs. Such protocols are more popularly known as *cross-realm* C2C-PAKE
protocols. For ease of notation, we will simply call these C2C-PAKEs for the
rest of this paper.

Considering this cross realm setting, several additional security issues arise
that would otherwise not be relevant in a single realm setting, e.g. protecting
secrets of the client in one realm from a malicious server [13] or a malicious

client [22] in the other realm. For more details of the variants and analyses, see [10,13,27,22,24,12,28].

The original C2C-PAKE protocol [10] was shown to be secure by arguing that it resists specific attacks although the argument for one security property was related to a computationally intractable assumption. Nevertheless, the analysis is still adhoc and not done in a formal security model. Byun-Lee-Lim [12] and Yin-Bao [28] independently proposed the first provably secure C2C-PAKE protocols. The Byun-Lee-Lim variant is called EC2C-PAKE. We call the Yin-Bao variant as C2C-PAKE-YB.

OUR CONTRIBUTIONS. The main aim of this paper is to advocate that provable security is the right approach to analysis and design of C2C-PAKEs, and AKEs in general, but we caution that proving such formal security is not an easy task. Already, some provably secure protocols have been shown [14,15] to exhibit flaws because of subtle points missed out in the security model used to conduct the proofs. To demonstrate our point, we first show how any adversary can mount undetectable online dictionary attacks [18] on both provably secure EC2C-PAKE and C2C-PAKE-YB protocols to recover the password. A discussion of the relevance and significance of this sort of attack is given in Section 3.1. We then show how malicious servers can launch man-in-the-middle attacks on EC2C-PAKE and cause client parties to share different keys with the server. Even adding an extra mutual authenticator step [8] does not help. A further variant of the latter attack even causes a party to think it is sharing a key with a party different from the party it is supposed to share with: unknown key-share attack [17,20]. The existence of malicious servers acting as active adversaries is indeed considered by the security proof of EC2C-PAKE [12] but still it was not able to catch our attacks. Fortunately, the C2C-PAKE-YB model captures these later attacks because it disregards as impossible to obtain key privacy when active server adversaries exist. Nevertheless, C2C-PAKE-YB exhibits another weakness inherited from its predecessor [3,14] that allows an unknown key-share attack to be mounted by a malicious client insider. To draw lessons from these results, we discuss why the provable security proofs failed to capture flaws in the protocols that allowed our attacks to work.

2   Two Provably Secure C2C-PAKE Protocols

We now describe each of provably secure protocols [12,28] in turn. We will use the notations given in Table 1. Unless otherwise mentioned, all described operations are done modulo p, except operations in the exponents, and all protocols are based on Diffie-Hellman (DH) type assumptions.

2.1   Description of EC2C-PAKE

We give in Figure 1 a concise view of the EC2C-PAKE protocol proposed by Byun-Lee-Lim (henceforth simply Byun *et al.*) [12].

Table 1. Notations

A, B	The clients
ID_i	The identity of party i
KDC_i	The key distribution center which stores the identity (ID_i) and password (pw_i) of client i in its realm
pw_i	Client i's human-memorizable password shared with KDC_i
K	The symmetric secret key shared between different KDCs
$E_K(\cdot)$	Symmetric encryption using the secret key, K
$\mathcal{E}_{pw_i}(\cdot)$	Ideal cipher, which is a random 1-to-1 function, using the password (pw_i)
p	Sufficiently large prime
g	The generator of GF(p)
$\mathcal{H}, H_1, H_2, H_3$	Cryptographic hash functions
l_r	A security parameter
$Ticket_i$	Ticket for receiving party i, equal to $E_K(k, ID_j, ID_i, L)$ where k is a random element of Z_p^*, L is the lifetime of $Ticket_i$ and ID_j the identity of the sender party
$MAC_K(\cdot)$	A message authentication code using the secret key, K
$\|$	Message concatenation
$x \in_\$ Z_p^*$	Randomly choosing an element x of Z_p^*

1. Client A wishing to initiate a secret communication session by generating a secret session key sk with B in a different realm, randomly chooses a value $x \in Z_p^*$ and computes $E_x = \mathcal{E}_{pw_A}(g^x)$. Then, A sends $\langle E_x, ID_A, ID_B \rangle$ to KDC_A.

2. Based on the received ID_A, KDC_A retrieves pw_A from its database and uses this to decrypt E_x to recover g^x. KDC_A randomly chooses $y \in Z_p^*$ and computes $E_y = \mathcal{E}_{pw_A}(g^y)$ and $R = \mathcal{H}(g^{xy})$. It also randomly chooses $k \in Z_p^*$ and computes $E_R = E_R(k, ID_A, ID_B)$. It then computes $Ticket_B = E_K(k, ID_A, ID_B, L)$ where L is $Ticket_B$'s lifetime. Then, KDC_A replies $\langle E_y, E_R, Ticket_B \rangle$ to A.

3. Upon receiving the message, A computes the ephemeral R and decrypts E_R to obtain k, ID_A and ID_B. It checks that ID_A and ID_B are valid. A randomly chooses $a \in Z_p^*$ and computes $E_a = g^a \| MAC_k(g^a)$, and forwards $\langle ID_A, E_a, Ticket_B \rangle$ to B.

4. B randomly chooses $x' \in Z_p^*$ and computes $E_{x'} = \mathcal{E}_{pw_B}(g^{x'})$. Then, B sends $\langle E_{x'}, Ticket_B \rangle$ to KDC_B.

5. KDC_B decrypts $Ticket_B$ using K to obtain k, L and ID_A. It verifies that the lifetime L and ID_A are valid. KDC_B then randomly chooses $y' \in Z_p^*$ and computes $E_{y'} = \mathcal{E}_{pw_B}(g^{y'})$ and $E_{R'} = E_{R'}(k, ID_A, ID_B)$, where $R' = \mathcal{H}(g^{x'y'})$. It then sends $\langle E_{y'}, E_{R'} \rangle$ to B.

6. B decrypts $E_{y'}$, computes R', then uses this to decrypt $E_{R'}$ to obtain k. Using this, B checks the integrity of g^a by verifying the previously received E_a. It then randomly chooses $b \in Z_p^*$ and computes $sk = \mathcal{H}(ID_A \| ID_B \| g^a \| g^b \| g^{ab})$ and $E_b = g^b \| MAC_k(g^b)$ and sends E_b to A.

7. On receiving E_b, A checks the integrity of g^b and also computes sk.

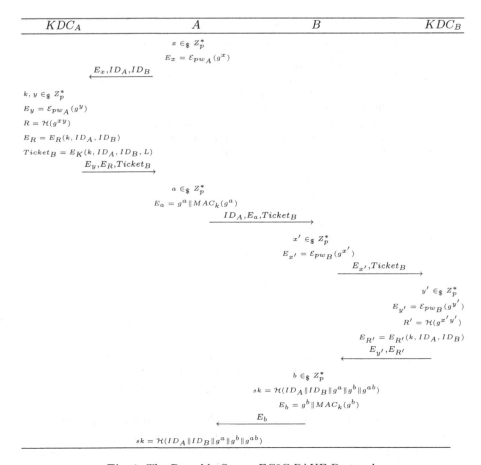

Fig. 1. The Provably Secure EC2C-PAKE Protocol

The authors further claim that mutual authentication can be provided by using an additional authenticator [8]. Let $sk = \mathcal{H}(ID_A \| ID_B \| g^a \| g^b \| g^{ab})$. Then at the end of the protocol, B sends $\mathcal{H}(sk\|1)$ to A, who verifies this by using his own computed sk, and then sends $\mathcal{H}(sk\|2)$ to B to be verified in turn. The final session key is computed by both parties as $sk' = \mathcal{H}(sk\|0)$.

2.2   Description of C2C-PAKE-YB

The C2C-PAKE-YB protocol proposed by Yin and Bao [28], as shown in Figure 2, is basically derived from the PAKE by Abdalla and Pointcheval in [3][1] by splitting the single server into two KDCs. The model used to prove the security of C2C-PAKE-YB is also based on [2,3]. Basically, the C2C-PAKE-YB protocol involves the following three steps:

---

[1] Note that a revised version of [3] appears in [4] but the C2C-PAKE-YB is based on the earlier version in [3].

1. Assume that clients A and B from different realms desire to establish a shared secret session key. In this step, A (resp. B) randomly chooses values x (resp. y) $\in Z_p^*$. Then, they respectively compute $X = g^x$ and $Y = g^y$, and then further compute blinded values as $X^* = X \cdot H_1(pw_A)$ and $Y^* = Y \cdot H_1(pw_B)$. Finally, they send the blinded values together with their identities to their corresponding KDCs. Note that both clients can perform this step simultaneously and independently.

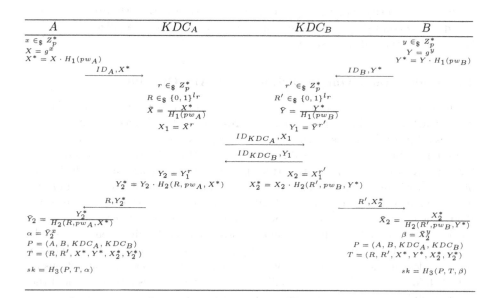

Fig. 2. The Provably Secure C2C-PAKE-YB Protocol

2. Upon receiving the messages from the clients, KDC_A (resp. KDC_B) randomly chooses $r \in Z_p^*$ and $R \in \{0,1\}^{l_r}$ (resp. r' and R'), where l_r is a security parameter, then by using pw_A (resp. pw_B) in its database, it computes $\bar{X} = \frac{X^*}{H_1(pw_A)}$ (resp. $\bar{Y} = \frac{Y^*}{H_1(pw_B)}$) and reblinds it as $X_1 = \bar{X}^r$ (resp. $Y_1 = \bar{Y}^{r'}$). KDC_A and KDC_B exchange $\langle ID_{KDC_A}, X_1 \rangle$ and $\langle ID_{KDC_B}, Y_1 \rangle$ secretly[2]. Then, KDC_A (resp. KDC_B) computes $Y_2 = Y_1^r$ and $Y_2^* = Y_2 \cdot H_2(R, pw_A, X^*)$ (resp. $X_2 = X_1^{r'}$ and $X_2^* = X_2 \cdot H_2(R', pw_B, Y^*)$). Finally, KDC_A returns $\langle R, Y_2^* \rangle$ to A (resp. KDC_B returns $\langle R', X_2^* \rangle$ to B).

3. Both clients compute their shared session key sk, i.e. A (resp. B) computes $\bar{Y}_2 = \frac{Y_2^*}{H_2(R, pw_A, X^*)}$ and $\alpha = \bar{Y}_2^x$ (resp. $\bar{X}_2 = \frac{X_2^*}{H_2(R', pw_B, Y^*)}$ and $\beta = \bar{X}_2^y$) and the secret key $sk = H_3(P, T, \alpha)$ (resp. $sk = H_3(P, T, \beta)$) where the transcript $T = (R, R', X^*, Y^*, X_2^*, Y_2^*)$ and party list $P = (A, B, KDC_A, KDC_B)$.

---

[2] This *authenticated* private communication channel can be established by the pre-agreed key shared between them or their public keys in PKI [28].

3   Cryptanalysis of the Provably Secure EC2C-PAKE

The EC2C-PAKE is interesting because it is claimed to be the first provably secure C2C-PAKE protocol and based on computationally intractable assumptions, when previous variants were based on heuristic design principles and underwent adhoc security analyses. EC2C-PAKE [12] is claimed to securely protect the established session key sk, to provide forward secrecy and to be secure against malicious servers.

Provable security alone intrigued us to study it in detail. Unfortunately, we found attacks that disprove the security claims in [12]. We categorize our attacks based on the type of adversaries.

3.1   Undetectable Online Dictionary by Any Outsider

We first describe an *undetectable online dictionary* attack [18]. Recall a dictionary attack is only interesting for password-based AKEs, since it exhaustively guesses all possible values of the secret and verifies if the guess is correct. Thus dictionary attacks are only feasible in the case of PAKEs since passwords have low entropy. An *offline* dictionary attack is where the adversary can do all his verifications without needing to interact with legitimate protocol parties. An *online* dictionary attack requires to interact with the parties in protocol sessions to verify the password guesses. One way to prevent online dictionary attacks is to trigger an alarm if the number of detected unsuccessful password login attempts exceeds a certain threshold [19,2,3,4,16]; but firstly this is typically outside the scope of a protocol design and more of an implementation detail. Secondly, assumptions in [19,2,3,4,16] for guarding against online dictionary attacks is only possible if the unsuccessful attempts are *detectable* [18,16]. But existing security models do not properly handle detecting this (see [16] for some discussion of this).

Ding and Horster [18] showed some *undetectable* online dictionary attacks. Basically, this means the adversary is interacting with legitimate parties in protocol sessions (thus it is online), but the parties are unaware that the adversary is present but instead think they are interacting with legitimate honest counterparts (thus undetectable). Therefore such attacks are hard to trace and legitimate parties become oracles to the adversary without noticing anything wrong. Indeed a party could keep a counter of how many different sessions that another party is interacting with it and refuse to further communicate if the counter exceeds a certain threshold, but there is a possibility [18] that it came from an honest party hence a denial of service occurs, which is clearly not desirable. The main point is the inability to differentiate between interactions with another legitimate honest party and interactions with the malicious adversary that makes undetectable online dictionary attacks so subtle.

Our attack exploits the fact that a key distribution center KDC_A for party A simply acts as an oracle in Step (2) to perform encryptions under the password pw_A shared with A. Recall that encryptions under the low-entropy password need to be used carefully to avoid being exploited for dictionary attack verifications [13,22,27]. The attack proceeds as follows:

An adversary A' chooses any $x^\star \in Z_p^*$ and computes $g^{x^\star}$. Then for every guess $pw_A^\star$ of pw_A:

1. Computes $E_{x^\star} = \mathcal{E}_{pw_A^\star}(g^{x^\star})$ and sends $\langle E_{x^\star}, ID_A, ID_B \rangle$ to KDC_A.
2. KDC_A decrypts $E_{x^\star}$ with pw_A and obtains $z_1 = \mathcal{E}_{pw_A}^{-1}(E_{x^\star}) = \mathcal{E}_{pw_A}^{-1}(\mathcal{E}_{pw_A^\star}(g^{x^\star}))$. Then, it randomly chooses $y \in Z_p^*$ and computes $E_y = \mathcal{E}_{pw_A}(g^y)$ and $R^\star = \mathcal{H}(z_1^y)$. It also randomly chooses $k \in Z_p^*$ and computes $E_{R^\star} = E_{R^\star}(k, ID_A, ID_B)$, and then computes $Ticket_B = E_K(k, ID_A, ID_B, L)$. Then, KDC_A replies $\langle E_y, E_{R^\star}, Ticket_B \rangle$ to A.
3. This is received by A' who decrypts E_y by computing $z_2 = \mathcal{E}_{pw_A^\star}^{-1}(E_y) = \mathcal{E}_{pw_A^\star}^{-1}(\mathcal{E}_{pw_A}(g^y))$. It then computes $R^{\star\star} = \mathcal{H}(z_2^{x^\star})$ and decrypts $E_{R^\star}$ by computing $E_{R^{\star\star}}^{-1}(E_{R^\star}) = E_{R^{\star\star}}^{-1}(E_{R^\star}(k, ID_A, ID_B))$. It checks if ID_A and ID_B are valid. If so, the guess of $pw_A^\star$ is correct. Else, it guesses a new one and repeats.

Our attack on EC2C-PAKE does not require knowledge of any secrets and can be mounted by any outsider, in contrast to some attacks on previous C2C-PAKE variants [13,27,22][3]. It similarly applies to the original C2C-PAKE in [10]. What is more, there is no distinction between offline and undetectable online dictionary attacks in the security models for protocols because the adversary when mounting an undetectable online dictionary attack appears as a legitimate party so other parties are unable to distinguish between the adversary and any other honest party.

3.2   Two Attacks by a Malicious Server Insider

Our second type of attack exploits the fact that a participating KDC_I (for $I \in \{A, B\}$) knows the MAC key k to be used between both parties A and B. KDC_I can then successfully launch a man-in-the-middle attack and end up sharing different keys with A and B respectively, while they would think they are sharing the same key with each other, as follows:

1. The protocol steps proceed as normal, but in Step (3) KDC_I replaces E_a sent by A to B with $E_{a'} = g^{a'} \| MAC_k(g^{a'})$, for any $a' \in Z_p^*$ chosen by KDC_I.
2. The rest of the steps proceed as normal, until Step (6) where KDC_I replaces E_b sent by B to A with $E_{b'} = g^{b'} \| MAC_k(g^{b'})$, for any $b' \in Z_p^*$ chosen by KDC_I.
3. A computes $sk = \mathcal{H}(ID_A \| ID_B \| g^a \| g^{b'} \| g^{a'})$ while B computes $sk^* = \mathcal{H}(ID_A \| ID_B \| g^{a'} \| g^b \| g^{a'b})$. Even if the extra mutual authenticator as described in Section 3.1 of [12] is added, KDC_I still succeeds in sharing sk with A and sharing sk^* with B because KDC_I knows both keys sk and sk^* and can take part in the mutual authenticator steps.

---

[3] We remark that the Denning-Sacco attack on [10] by Kim *et al.* in [22] is flawed because they wrongly assume the insider adversary A knows the ephemeral Diffie-Hellman key R' shared by client B and KDC_B. This contradicts the DH assumption.

A further variant of the above attack can be launched by KDC_B to cause B to think that it is $I \neq A$ who wishes to establish a key with it. This unknown key-share attack [17,20] is as follows:

1. The steps are as normal until Step (3) where KDC_B replaces the $\langle ID_A, E_a \rangle$ sent from A to B with $\langle ID_I, E_{a'} \rangle$, where $E_{a'} = g^{a'} \| MAC_k(g^{a'})$, for any $a' \in Z_p^*$ chosen by KDC_B.
2. Then in Step (5) instead of computing $E_{R'}(k, ID_A, ID_B)$, KDC_B computes $E_{R'}(k, ID_I, ID_B)$ and sends this back to B.
3. In Step (6), KDC_B replaces E_b sent by B to A with $E_{b'} = g^{b'} \| MAC_k(g^{b'})$, for any $b' \in Z_p^*$ chosen by KDC_B.
4. A computes $sk = \mathcal{H}(ID_A \| ID_B \| g^a \| g^{b'} \| g^{ab'})$ while B computes $sk^* = \mathcal{H}(ID_I \| ID_B \| g^{a'} \| g^b \| g^{a'b})$. Even if the extra mutual authenticator as described in Section 3.1 of [12] is added, KDC_B will still succeed in sharing sk with A while A thinks it is sharing with B; and KDC_B will also share sk^* with B while B thinks it is sharing with I, because KDC_B knows both keys sk and sk^* and can easily take part in the mutual authenticator.

Indeed, our results confirm the fact [2,16] that it is difficult to guard against malicious servers since they know the password of clients in the same realm and thus also know the MAC key k used by parties A and B. This fact however was overlooked by the EC2C-PAKE designers [12]. As a side note, this attack also applies to the variant in [22].

On the other hand, it is fortunate that attacks by malicious servers described in this subsection do not apply to C2C-PAKE-YB because its security model reasonably assumes that key privacy cannot be achieved against malicious servers acting as active adversaries.

4   Cryptanalysis of the Provably Secure C2C-PAKE-YB

C2C-PAKE-YB is equally interesting. In fact, with the right tools of which the provable security approach is one, properly designed protocols with expected security properties can be obtained. However, provably secure protocols should be scrutinized to ensure they are properly designed and their security models do not miss catching known security flaws.

4.1   Undetectable Online Dictionary Attack by Any Outsider

We show an undetectable online dictionary attack on C2C-PAKE-YB by any adversary, making use of only the SendServer() query defined in the C2C-PAKE-YB security model.

1. Choose $x', y' \in Z_p^*$ and compute $X' = g^{x'}, Y' = g^{y'}$.
2. For all guesses of pw'_A and pw'_B:
    (a) Compute $X^* = X' \cdot H_1(pw'_A)$ and $Y^* = Y' \cdot H_1(pw'_B)$.

(b) Send $\langle ID_A, X^{*'} \rangle$ to KDC_A and $\langle ID_B, Y^{*'} \rangle$ to KDC_B; i.e. obtain Send-Server($KDC_A, \langle ID_A, X^{*'} \rangle$) and SendServer($KDC_B, \langle ID_B, Y^{*'} \rangle$) queries.

(c) KDC_A computes $\bar{X}' = \frac{X^{*'}}{H_1(pw_A)} = \frac{X' \cdot H_1(pw'_A)}{H_1(pw_A)}$ and KDC_B computes $\bar{Y}' = \frac{Y^{*'}}{H_1(pw_B)} = \frac{Y' \cdot H_1(pw'_B)}{H_1(pw_B)}$.

(d) KDC_A computes $X'_1 = \bar{X}'^r$ and KDC_B computes $Y'_1 = \bar{Y}'^r$.

(e) KDC_A sends $\langle ID_{KDC_A}, X'_1 \rangle$ to KDC_B and KDC_B sends $\langle ID_{KDC_B}, Y'_1 \rangle$ to KDC_A.

(f) KDC_B computes $X'_2 = X'^{r'}_1 = g^{x' \cdot r \cdot r'}$ and KDC_A computes $Y'_2 = Y'^{r'}_1 = g^{y' \cdot r \cdot r'}$.

(g) KDC_B computes $X_2^{*'} = X'_2 \cdot H_2(R', pw_B, Y^*)$ and KDC_A computes $Y_2^{*'} = Y'_2 \cdot H_2(R, pw_A, X^*)$.

(h) KDC_A outputs $\langle R, Y_2^{*'} \rangle$ and KDC_B outputs $\langle R', X_2^{*'} \rangle$.

(i) Compute $\bar{Y}_2 = \frac{Y_2^{*'}}{H_2(R, pw'_A, X^*)}$ and $\bar{X}_2 = \frac{X_2^{*'}}{H_2(R', pw'_B, Y^*)}$.

(j) Compute $z = \bar{Y}_2^{(y'^{-1})}$ and $z' = \bar{X}_2^{(x'^{-1})}$.

(k) Check if $z = z'$.

The intuition behind this attack is that the servers KDC_A (resp. KDC_B) act as oracles that perform unmasking by dividing with a function of the password pw_A (resp. pw_B) and then remasking by multiplying with another function of the password pw_A (resp. pw_B) on whatever X^* (resp. Y^*) values that are supplied by A (resp. B). KDC_A (resp. KDC_B) then outputs Y_2^* (resp. X_2^*) that the adversary attempts to unmask by dividing with a function of the guessed password pw'_A (resp. pw'_B). If the guess is correct, then the adversary obtains the correct $\bar{Y}_2$ and $\bar{X}_2$ values that can each be used to compute $g^{r \cdot r'}$. Checking that they match will allow the password guess to be verified.

Discussion. It is worthwhile to discuss why C2C-PAKE-YB falls to dictionary attacks even though this appears to be treated in their Theorem 2 [28]. Firstly, they considered curbing *detectable* online dictionary attacks by limiting the number q_s of guessing attempts made via queries to SendServer() and SendClient(). However, as remarked in [18,16] and in Section 3.1, this cannot prevent *undetectable* online dictionary attacks since the server cannot differentiate between successful or failed login attempts. Thus, since a SendServer() oracle is used to model *active* attacks by an adversary, the inability to detect the adversary's attack attempts means the undetectable attempts may not even be recorded in the counter for q_s. Secondly, recall that our attack only makes use of the SendServer() query defined in the C2C-PAKE-YB security model. In their proof, Yin-Bao considered security against dictionary attacks by a malicious client and by a malicious server, respectively, as independent cases. Although a provable security model allows to capture all possible attacks as long as the polynomial time adversary is constrained to the defined oracle queries, however, when arguing within a security proof, one needs to be careful not to limit the consideration to specific adversarial behaviours or to specific approaches that an adversary may use to attack. The flaw in the proof argument of Yin-Bao lies in that they considered a malicious client mounting attacks as one legitimate client party, whereas our attacks does not require the adversary to be a legitimate party and instead mounts attacks on both sides of the client parties, i.e. A and B. Therefore, although the goal that our attack achieves is indeed considered by Theorem 2, the argument for Theorem 2 was too specific to catch our attack.

Another point is that since undetectable online dictionary attacks are caused by parties in general not being able to distinguish between interactions with other honest parties or with the adversary; therefore it may be interesting to consider incorporating security against undetectable online dictionary attacks directly into the security model of a PAKE protocol.

4.2   Unknown Key-Share Attack by a Malicious Client Insider

Finally, we observe that the C2C-PAKE-YB model in [28] does not include Corrupt queries. It is now known that exclusion of this gives rise to attacks that cannot be captured by a proof in the security model, as demonstrated by Choo et al. [14] on the 3PAKE protocol by Abdalla and Pointcheval [3] whose security is proven in the BPR2000 model [8,2]. Abdalla and Pointcheval later [4] revised their protocol to prevent this.

Since C2C-PAKE-YB is based on 3PAKE [3], we remark that it directly inherits the weakness of 3PAKE that allowed the Choo et al. attack. To elaborate, Choo et al. showed that by corrupting another legitimate client C, then an adversary $\mathcal{A}$ can end up sharing a session key with client A but with A thinking it is sharing with client B who is not sharing any key with A or C. Furthermore, note that corruption of a client C is essentially equivalent to having a malicious client insider C as the adversary. But while C2C-PAKE-YB claims security against the latter, the former is surprisingly not considered. In more detail, the attack follows:

1. The protocol steps proceed as normal with A sending to KDC_A the value X^* and the identities ID_A, ID_B notifying KDC_A that it wishes to initiate a session with B. Similarly B sends to KDC_B the value Y^* and identities ID_B, ID_A notifying KDC_B that it wishes to initiate a session with A.
2. C randomly chooses $e \in Z_p^*$ and computes $E^* = g^e \cdot H_1(pw_C)$. C then replaces $\langle ID_A, ID_B, X^* \rangle$ with $\langle ID_A, ID_C, X^* \rangle$ causing KDC_A to believe the recipient client is C and not B; and replaces $\langle ID_B, ID_A, Y^* \rangle$ with $\langle ID_C, ID_A, E^* \rangle$ causing KDC_C to believe C wishes to initiate a session with A.
3. The rest of the steps proceed normally, but where KDC_C is involved instead of KDC_B; and E^* and pw_C are used instead of Y^* and pw_B.
4. KDC_A outputs $\langle R, E_2^* \rangle$ to A and KDC_C outputs $\langle R', X_2^* \rangle$ to C.
5. A will compute $\bar{Y}_2 = g^{err'}$, $\alpha = g^{xerr'}$ and the secret key $sk = H_3(P, T, \alpha)$. C will compute $\bar{X}_2 = g^{xrr'}$, $\beta = g^{xerr'} = \alpha$ and the same secret key $sk = H_3(P, T, \beta)$, where $P = \langle A, B, KDC_A, KDC_B \rangle$.

A ends up thinking it is sharing a key with B when it is actually sharing with C and C knows what this key sk is. To fix this, Choo et al. [14] informally suggest to include the identities of both clients A and B into the computation of the password-based blinding factors, and Abdalla-Pointcheval's full version [4] and appendix proofs do now take this into consideration. The flaw by Yin-Bao was to base their C2C-PAKE-YB and corresponding security model on the earlier variant in [3], and to not consider Corrupt queries.

5   Concluding Remarks

Byun *et al.* [12] and Yin-Bao [28] respectively proposed first known provably secure C2C-PAKE protocols, with security based on computationally intractable assumptions and further the Yin-Bao variant has security properties proven in a formal security model. These are nice results. But, provable security proofs for PAKE protocols should be done carefully to avoid miscatching known attacks. A recent example is the first provably secure n-party PAKE protocol in the DPWA setting [11] that was shown in [25] to fall to attacks that it was designed to resist. Other examples are in [14,15]. Though the responsibility rests on protocol designers to carefully define adequate security models and check the correctness of their security proofs, the community in particular protocol implementers should exercise caution when interpreting provable security proofs. Experience in the analysis and design of security protocols [1,5,8,14,15] has shown that even seemingly sound designs may exhibit problems, and though provable security is the right approach, years of public scrutiny should still complement the process before a protocol is deemed secure. Instead, adhoc protocol designs with heuristic security proofs should be discouraged.

Acknowledgement

The first author thanks Liqun Chen for discussions on and a copy of [13]. We thank Colin Boyd for previously suggesting at ACNS '06 that we check what problems in the security proof of a provably secure protocol that cause it to fall to attacks. We thank God for His many blessings. We dedicate this work to our daughters Rachel; and Yue Tian, Yue Chen, for their cheekiness and keeping our ageing minds constantly at work.

References

1. M. Abadi. Explicit Communication Revisited: Two New Attacks on Authentication Protocols. *IEEE Transactions on Software Engineering*, Vol. 23, No. 3, pp. 185-186, 1997.
2. M. Abdalla, P.-A. Fouque, and D. Pointcheval. Password-Based Authenticated Key Exchange in the Three-Party Setting. *Proc. PKC '05*, LNCS 3386, pp. 65-84, 2005.
3. M. Abdalla and D. Pointcheval. Interactive Diffie-Hellman Assumptions with Applications to Password-Based Authentication. *Proc. FC '05*, LNCS 3570, pp. 341-356, 2005.

4. M. Abdalla and D. Pointcheval. Interactive Diffie-Hellman Assumptions with Applications to Password-Based Authentication. Full version of [3], available online at http://www.di.ens.fr/~pointche/pub.php?reference=AbPo05.

5. R. Anderson. Security Engineering − A Guide to Building Dependable Distributed Systems. Wiley, USA, 2001.

6. M. Bellare and P. Rogaway. Entity Authentication and Key Distribution. *Advances in Cryptology - Crypto '93*, LNCS 773, pp. 232-249, 1993.

7. M. Bellare and P. Rogaway. Provably Secure Session Key Distribution: the Three Party Case. *Proc. ACM STOC '95*, pp. 57-66, 1995.

8. M. Bellare, D. Pointcheval, and P. Rogaway. Authenticated Key Exchange Secure against Dictionary Attacks. *Advances in Cryptology - Eurocrypt '00*, LNCS 1807, pp. 139-155, 2000.

9. S. Bellovin and M. Merritt. Encrypted Key Exchange: Passwords based Protocols Secure against Dictionary Attacks. *Proc. IEEE Symposium on Security & Privacy '92*, pp. 72-84, 1992.

10. J.W. Byun, I.R. Jeong, D.H. Lee, and C.S. Park. Password-Authenticated Key Exchange between Clients with Different Passwords. *Proc. ICICS '02*, LNCS 2513, pp. 134-146, 2002.

11. J.W. Byun and D.H. Lee. N-Party Encrypted Diffie-Hellman Key Exchange Using Different Passwords. *Proc. ACNS '05*, LNCS 3531, pp. 75-90, 2005.

12. J.W. Byun, D.H. Lee, and J. Lim. Efficient and Provably Secure Client-to-Client Password-Based Key Exchange Protocol. *Proc. APWeb '06*, LNCS 3841, pp. 830-836, 2006.

13. L. Chen. A Weakness of the Password-Authenticated Key Agreement between Clients with Different Passwords Scheme. *Circulated for consideration at the 27th SC27/WG2 meeting in Paris, France*, ISO/IEC JTC 1/SC27 N3716, 2003-10-20.24, 2003.

14. K.-K.R. Choo, C. Boyd, and Y. Hitchcock. Examining Indistinguishability-Based Proof Models for Key Establishment Protocols. *Advances in Cryptology - Asiacrypt '05*, LNCS 3788, pp. 585-604, 2005.

15. K.-K.R. Choo, C. Boyd, and Y. Hitchcock. Errors in Computational Complexity Proofs for Protocols. *Advances in Cryptology - Asiacrypt '05*, LNCS 3788, pp. 624-643, 2005.

16. Y. Cliff, Y.S.T. Tin, and C. Boyd. Password Based Server Aided Key Exchange. *Proc. ACNS '06*, LNCS 3989, pp. 146-161, 2006.

17. W. Diffie, P.C. van Oorschot, and M.J. Wiener. Authentication and Authenticated Key Exchanges. *Design, Codes and Cryptography*, Vol. 2, No. 2, pp. 107-125, 1992.

18. Y. Ding and P. Horster. Undetectable On-line Password Guessing Attacks. *ACM Operating Systems Review*, Vol. 29, No. 4, pp. 77-86, 1995.

19. Y. Hitchcock, Y.S.T. Tin, J.M. Gonzalez Nieto, C. Boyd, and P. Montague. A Password-Based Authenticator: Security Proof and Applications. *Progress in Cryptology - Indocrypt '03*, LNCS 2904, pp. 388-401, 2003.

20. B.S. Kaliski Jr. An Unknown Key-Share Attack on the MQV Key Agreement Protocol. *ACM TISSEC*, Vol. 4, No. 3, pp. 275-288, 2001.

21. S. Katzenbeisser. On the Integration of Watermarks and Cryptography. *Proc. IWDW '03*, LNCS 2939, pp. 50-60, 2003.

22. J. Kim, S. Kim, J. Kwak, and D. Won. Cryptanalysis and Improvement of Password-Authenticated Key Exchange Scheme between Clients with Different Passwords. *Proc. ICCSA '04*, LNCS 3043, pp. 895-902, 2004.

23. S. Kim, H. Lee and H. Oh. Enhanced ID-based Authenticated Key Agreement Protocols for a Multiple Independent PKG Environment. *Proc. ICICS '05*, LNCS 3783, pp. 323-336, 2005.
24. R.C.-W. Phan and B.-M. Goi. Cryptanalysis of an Improved Client-to-Client Password-Authenticated Key Exchange (C2C-PAKE) Scheme. *Proc. ACNS '05*, LNCS 3531, pp. 33-39, 2005.
25. R.C.-W. Phan and B.-M. Goi. Cryptanalysis of the N-Party Encrypted Diffie-Hellman Key Exchange Using Different Passwords. *Proc. ACNS '06*, LNCS 3989, pp. 226-238, 2006.
26. J. Stern. Why Provable Security Matters? *Advances in Cryptology - Eurocrypt '03*, LNCS 2656, pp. 449-461, 2003.
27. S. Wang, J. Wang, and M. Xu. Weaknesses of a Password-Authenticated Key Exchange Protocol between Clients with Different Passwords. *Proc. ACNS '04*, LNCS 3089, pp. 414-425, 2004.
28. Y. Yin and L. Bao. Secure Cross-Realm C2C-PAKE Protocol. *Proc. ACISP '06*, LNCS 4058, pp. 395-406, 2006.
29. E.-J. Yoon and K.-Y. Yoo. Cryptanalysis of Two User Identification Schemes with Key Distribution Preserving Anonymity. *Proc. ICICS '05*, LNCS 3783, pp. 315-322, 2005.

Efficient and Provably Secure Generic Construction of Three-Party Password-Based Authenticated Key Exchange Protocols*

Weijia Wang and Lei Hu

State Key Laboratory of Information Security,
Graduate School of Chinese Academy of Sciences, Beijing 100049, China
{wwj, hu}@is.ac.cn

Abstract. Three-party password-based authenticated key exchange (3-party PAKE) protocols make two communication parties establish a shared session key with the help of a trusted server, with which each of the two parties shares a predetermined password. Recently, with the first formal treatment for 3-party PAKE protocols addressed by Abdalla et al., the security of such protocols has received much attention from cryptographic protocol researchers. In this paper, we consider the security of 3-party PAKE protocols against undetectable on-line dictionary attacks which are serious and covert threats for the protocals. We examine two 3-party PAKE schemes proposed recently by Abdalla et al. and reveal their common weakness in resisting undetectable on-line dictionary attacks. With reviewing the formal model for 3-party PAKE protocols of Abdalla et al. and enhancing it by adding the authentication security notion for the treatment of undetectable attacks, we then present an efficient generic construction for 3-party PAKE protocols, and prove it enjoys both the semantic security and the authentication security.

Keywords: password, authenticated key exchange, key distribution, multi-party protocol.

1   Introduction

Three-party password-based authenticated key exchange protocols (3-party PAKE or 3PAKE) enable communicating parties within a large network, who only share a weak (low entropy) password with a trusted server respectively, to authenticate each other with the help of the trusted server and establish a strong session key for protecting their subsequent communications over the public channel. In this solution, a communicating party who wants to build secure communications with other parties does not need to remember so many passwords whose number would be large linearly in the number of all possible partners, instead it only holds a password shared with a trusted server. Due to this advantage, these protocols are particularly appealing for those real-world applications in which communication parties are human beings who are equipped

---

* This work was supported by NSFC(60373041,60573053).

R. Barua and T. Lange (Eds.): INDOCRYPT 2006, LNCS 4329, pp. 118–132, 2006.

with lightweight or mobile client machines that can not afford a heavyweight infrastructure such as public key infrastructure (PKI) and common secrets with every party.

Unlike public-key based key exchange protocols which rely on the existence of PKI, password-based authenticated key exchange protocols, in which secret keys shared among communication parties are not distributed over a large space, but are rather drawn from a small set of values, have a challenge from so-called exhaustive dictionary attacks. Generally, we can divide such attacks into the following three classes [13]:

1. Off-line dictionary attacks: Only by using the eavesdropped information, an attacker guesses a password and verifies its guess off-linely. No participation of the honest client or the server is required, so these attacks can not be noticed.
2. Undetectable on-line dictionary attacks: An attacker tries to verify a password guess in an on-line transaction. However, a failed guess can not be detected by the honest client or the server, since one of them is not able to distinguish a malicious request from an honest one.
3. Detectable on-line dictionary attacks: Similar to above, an attacker attempts to use a guessed password in an on-line transaction. Using the response from the honest client or the server, it verifies the correctness of its guess. But a failed guess can be detected by the honest client or the server.

Among these attacks, detectable on-line dictionary attacks are unavoidable and should be handled by taking additional precautions such as logging failed protocol attempts and invalidating the use of the password after a certain number of failures. However, both off-line and undetectable on-line dictionary attacks are serious attacks against password-based settings so that a secure password-based protocol should ideally resist the two types of attacks. Nevertheless, undetectable on-line dictionary attacks are always more difficult to be found than off-line ones in the design of password-based protocols, especially in that of 3-party PAKE cases so that some 3-party PAKE protocols are still susceptible to the undetectable attacks even if they are claimed to be provably secure [1, 2].

Our contribution. In this paper, we study the design of 3-party PAKE protocols resisting dictionary attacks, especially against both off-line and undetectable on-line dictionary attacks. Two formal treatments for 3-party PAKE protocols are proposed recently by Abdalla et al. [1, 2]. Unfortunately, these two schemes still suffer from undetectable on-line dictionary attacks due to the attacks being out of the scope of the security model [1] considered. In section 2, we will briefly describe such attacks against the above two schemes. Then, we review the formal model for 3-party PAKE protocols provided by Abdalla et al. [1] and enhance it by adding the authentication security notion for the treatment of undetectable attacks. Finally, we present a new generic construction scheme for the 3-party PAKE protocols. Compared with the resolution proposed by Abdalla et al. [1], our scheme is not only more efficient, but also resistant to both off-line and undetectable on-line dictionary attacks.

Related work. Password-based authenticated key exchange protocols are attractive due to their simplicity and convenience, and have received much interest in the research community. Many password-based protocols and their various formal treatments were presented in the last few years [4, 5, 3, 7, 8, 9, 10, 11, 12, 13, 14, 15, 16, 17]. Most of them considered different aspects of password-based protocols in the 2-party setting (2-party PAKE or 2PAKE), while only a few of them [13, 16, 17] dealt with 3-party password-based authenticated key exchange protocols.

However, more recently, the importance of 3-party PAKE protocols has been realized by protocol researchers, especially followed by an increasing recognition that precise definitions and formalization were needed. The first formal treatment for 3-party PAKE protocols was provided recently by Abdalla et al. [1]. In their paper, they presented the first formal security model for 3-party PAKE protocols, based on that of Bellare and Rogaway [5] for key distribution schemes and that of Bellare, Pointcheval, and Rogaway [3] for password-based authenticated key exchange. At the same time, they also gave a generic construction of 3-party PAKE protocols, and proved its security under their security model. Subsequently, based on the two-party encrypted key exchange protocol of Bellovin and Merritt [6], Abdalla et al. [2] designed an efficient 3-party PAKE protocol and put forth its security proof by using their security model. To the best of our knowledge, only the two schemes mentioned above are 3-party PAKE schemes with provable security.

2   Undetectable On-line Dictionary Attacks

2.1   Attacks on the General Construction of Abdalla et al.

As shown in Figure 1, the generic construction (referred as GPAKE) of Abdalla et al. [1] for 3-party PAKE protocols is essentially a compiler, which consists of a secure 2-party PAKE protocol, a secure key distribution (KD), and a secure message authentication code (MAC) scheme.

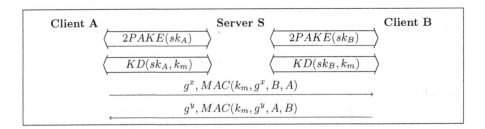

Fig. 1. GPAKE: a general construction of 3PAKE protocols provided by Abdalla et al. [1]

In this construction, since only the semantic security of the 2-party PAKE protocol used in it is required, which means an adversary can not obtain any information about a free session key, but no unilateral authentication security from clients to the sever is considered, an inside attacker can still impersonate the other legitimate party and mount undetectable on-line dictionary attacks against the sever in the 3-party PAKE protocol, when the latter is implemented by using a 2PAKE scheme without the authentication security from clients to the server such as OMDHKE protocol [9].

More specifically, assume user B is an inside attacker and it masquerades as user A by guessing password pw'_A to initiate the protocol designed as above. After finishing the first step of the protocol, B can build a session key sk_B with the server S by using its own legal password pw_B, and also obtain a test session key sk'_A by the guessing password pw'_A. Upon receiving the messages $KD(sk_B, k_m)$ and $KD(sk_A, k_m)$, the adversary can compute k_m and k'_m by using the real session key sk_B and the test one sk'_A, respectively. Finally, by checking $k'_m = k_m$, the malicious adversary B can verify its guessing password. If the above equation holds, it shows the guessing password is correct and the inside adversary B succeeds. Otherwise the adversary B continues to initiate a new instance of the protocol with the server S who can not detect such attacks since there is no unilateral authentication from the client to the server. As the distribution space of the password pw_A of the user A is small, it is easy for the adversary to guess and obtain the correct password after a certain number of on-line attacks.

2.2 Attacks on Another 3PAKE Protocol of Abdalla et al.

After proposing GPAKE, Abdalla et al. [2] also present a new efficient 3-party PAKE protocol, which is based on the encrypted key exchange protocols of Bellovin and Merrit [6]. Moreover, under the so-called Chosen-basis Decisional Diffie-Hellman assumptions (CDDH), which is stronger than the standard Decisional Diffie-Hellman (DDH) assumption, they proved the security of this 3-party PAKE protocol in their formal model for 3-party schemes [1]. Unfortunately, similar as their GPAKE construction, the protocol still suffers from undetectable on-line password guessing attacks.

We depict the protocol in Figure 2, where $\mathbb{G}$ is a group of a prime order p and g is a generator of $\mathbb{G}$; l_r is a security parameter; and H_1, H_2 and H_3 are Hash functions that are regarded as random oracles.

Similarly to the analysis of section 2.1, we assume that the user B is the malicious inside attacker and masquerades as user A by guessing password pw'_A to initiate the protocol. The inside attacker firstly chooses random numbers x and y, and then sends to the server S $X^{*'}$ and Y^*, which are computed by using the guessing password pw'_A and its own legal password pw_B, respectively. Upon receiving the two pieces of message, the server generates and responds to the inside attacker B with $(R, \overline{Y}^*)$ and $(R, \overline{X}^{*'})$ according to the protocol, using the passwords of users A and B. After getting the above message, the malicious user B can verify whether its guessing password is correct or not by comparing the two session keys obtained finally from normal computation. If the two keys are equal, it shows the guessing password is correct and the inside attacker B

$$
\begin{array}{ccc}
\textbf{Client A} & \textbf{Server S} & \textbf{Client B}
\end{array}
$$

Client A	Server S	Client B
$x \leftarrow_R Z_p; X \leftarrow g^x$	$r \leftarrow_R Z_p; R \leftarrow_R \{0,1\}^{l_r}$	$y \leftarrow_R Z_p; Y \leftarrow g^y$
$pw_{A,1} \leftarrow H_1(pw_A)$		$pw_{B,1} \leftarrow H_1(pw_B)$
$X^* \leftarrow X \cdot pw_{A,1}$		$Y^* \leftarrow Y \cdot pw_{B,1}$

$$\xrightarrow{X^*} \qquad\qquad \xleftarrow{Y^*}$$

$$
\begin{aligned}
pw_{A,1} &\leftarrow H_1(pw_A)\\
pw_{B,1} &\leftarrow H_1(pw_B)\\
X &\leftarrow X^*/pw_{A,1}\\
Y &\leftarrow Y^*/pw_{B,1}\\
\overline{X} &\leftarrow X^r\\
\overline{Y} &\leftarrow Y^r\\
pw_{A,2} &\leftarrow H_2(R, pw_A, X^*)\\
pw_{B,2} &\leftarrow H_2(R, pw_B, Y^*)\\
\overline{Y}^* &\leftarrow \overline{Y} \cdot pw_{A,2}\\
\overline{X}^* &\leftarrow \overline{X} \cdot pw_{B,2}
\end{aligned}
$$

$$\xleftarrow{R, \overline{Y}^*} \qquad\qquad \xrightarrow{R, \overline{X}^*}$$

Client A	Client B
$pw_{A,2} \leftarrow H_2(R, pw_A, X^*)$	$pw_{B,2} \leftarrow H_2(R, pw_B, Y^*)$
$\overline{Y} \leftarrow \overline{Y}^*/pw_{A,2}; K \leftarrow \overline{Y}^x$	$\overline{X} \leftarrow \overline{X}^*/pw_{B,2}; K \leftarrow \overline{X}^y$
$T \leftarrow R, X^*, Y^*, \overline{X}^*, \overline{Y}^*$	$T \leftarrow R, X^*, Y^*, \overline{X}^*, \overline{Y}^*$
$SK \leftarrow H_3(A, B, S, T, K)$	$SK \leftarrow H_3(A, B, S, T, K)$

Fig. 2. An efficient 3-party PAKE protocol provided by Abdalla et al. [2]

succeeds. Otherwise the adversary continues a new round of undetectable attacks on the server S until it finds the correct password pw_A.

Remark 1. In 2-party PAKE protocols, to resist undetectable on-line dictionary attacks, we can modify and extend those protocols only having semantic security by using generic transformations similar to those of Bellare et al. [3] for mutual authentication between two communicating parties. However, for 3-party PAKE protocols, only adding mutual authentication between two communicating parties in the end can not enhance those protocols to be resistant to undetectable on-line dictionary attacks. The difference is that there are inside attackers in the 3-party scenario, who themselves can play the legal role of one of the involved client users and impersonate the other client party by guessing the value of its password. After finishing the protocol with the trusted server, insider attackers can verify whether or not a password guess is correct by comparing the session keys obtained from legal and impersonating identifications respectively. So if only communicating parties authenticate each other, inside attackers can still guess the correct password by keeping on-line interacting with the trusted server which cannot detect such attacks.

3   Security Model of 3-Party PAKE Protocols

In this section, we review the main points of the formal security model of 3-party password authenticated key exchange protocols [1], which is a generalization of that for 2-party authenticated key exchange protocols, and in which a new oracle is introduced to represent the trusted server.

The reader is assumed to be familiar with the security model of 2-party PAKE protocols [3]. Hence, we do not present its definition here, and refer the reader to [3, 1] for more details. However, it should be noted that for simplicity of the proof of GPAKE, Abdalla et al. provide a new model [1] for the 2-party PAKE protocols by modifying the previous one [3] slightly and call it Real-Or-Random (ROR) model for the 2-party case, in which *Reveal* queries are no longer allowed and the adversary is allowed to ask as many *Test* queries as it wants. This property is inherited by the ROR model of the 3-party case, which will be presented in the description below.

Additionally, to consider the security against undetectable on-line dictionary attacks, we add the authentication security notion as an extension for the ROR model of 3PAKE protocols provided by Abdalla et al. [1].

3.1 Communication Model

Protocol participants. The participants in a 3-party PAKE setting consist of two sets: $\mathcal{U}$, the set of all client users and $\mathcal{S}$, the set of trusted servers. The set $\mathcal{S}$ is assumed to involve only a single trusted server for the simplicity of the proof, which can be easily extended to the case considering multiple servers. Here we further divide the set $\mathcal{U}$ into two disjoint subsets: $\mathcal{C}$, the set of honest clients and $\mathcal{E}$, the set of malicious clients. That is, the set of all users $\mathcal{U}$ is the union $\mathcal{C} \bigcup \mathcal{E}$. The malicious set $\mathcal{E}$ corresponds to the set of inside attackers, who exist only in the 3-party setting.

Long-lived keys. Each client user $U \in \mathcal{U}$ holds a password pw_U. The single server $S \in \mathcal{S}$ holds a vector $pw_S = \langle pw_S[U] \rangle_{U \in \mathcal{U}}$ with an entry for each client in which $pw_S[U]$ may be equal to pw_U in symmetric model or a transformation of pw_U as defined in [3]. The set of passwords pw_E, where $E \in \mathcal{E}$, is assumed to be held by the inside attackers.

Protocol execution. In the model, it is assumed that an adversary $\mathcal{A}$ has full control over the communication channels and can create several concurrent instances of the protocol. During the execution of the protocol, the interaction between an adversary and the protocol participants occurs only via oracle queries, which model the adversary capabilities in a real attack. These queries are as follows, where U^i (S^j, respectively) denotes the i-th (j-th, respectively) instance of a participant U (S, respectively):

1. $Execute(U_1^{i_1}, S^j, U_2^{i_2})$: This query models passive attacks, where the attacker gets access to honest executions among the client instances $U_1^{i_1}$ and $U_2^{i_2}$ and trusted server instance S^j by eavesdropping. The output of this query consists of the message that was exchanged during the honest execution of the protocol.
2. $SendClient(U^i, m)$: This query models an active attack against clients, in which the adversary sends a message to the client instance U^i. The output of this query is the message that client instance U^i would generate upon receipt of message m.

3. $SendServer(S^j, m)$: This query models an active attack against the server, in which the adversary sends a message to server instance S^j. It outputs the message which server instance S^j would generate upon receipt of message m.
4. $Reveal(U^i)$: This query models the misuse of session keys by clients. Only if the session key of the client instance U^i is defined, the query is available and returns to the adversary the session key.
5. $Test(U^i)$: This query is used to measure the semantic security of the session key of the client instance U^i. If the session key is not defined, it returns $\perp$. Otherwise, it returns either the session key held by the client instance U^i if $b = 0$ or a random number of the same size if $b = 1$, where b is the hidden bit selected at random prior to the first call.

3.2   Security Definitions

Notation. As in [1], which in turn build on [5,3], an instance U^i is said to be *opened* if the query $Reveal(U^i)$ has been made by the adversary. We say an instance U^i is *unopened* if it is not *opened*. An instance U^i is said to be *accepted* if it goes into an accept state after receiving the last expected protocol message.

Partnering. Our definitional approach of partnering is from [3], which uses the notion of session identifications (*sid*). More specifically, we say two instances U_1^i and U_2^j are partners if the following conditions are met: (1) Both U_1^i and U_2^j accept; (2) Both U_1^i and U_2^j share the same *sid*; (3) The partner identification for U_1^i is U_2^j and vice-versa; and (4) No instance other than U_1^i and U_2^j accepts with a partner identification equal to U_1^i or U_2^j.

Freshness. If an instance U^i has been accepted, both the instance and its partner are unopened and they are both instances of honest clients, we say the instance U^i is *fresh*.

Semantic security in the ROR model. The security notion is defined in the context of executing a 3-party PAKE protocol P in the presence of an adversary $\mathcal{A}$. During executing the protocol, the adversary $\mathcal{A}$ is allowed to send multiple queries to the *Execute*, *SendClient*, *SendServer*, and *Test* oracles and asks at most one *Test* query to each fresh instance of each honest client, while it is no longer allowed to ask *Reveal* queries. Finally $\mathcal{A}$ outputs its guess b' for the bit b hidden in the Test oracle. An adversary $\mathcal{A}$ is said to be successful if $b' = b$. We denote this event by *Succ*. Provided that passwords are drawn from dictionary $\mathcal{D}$, we define the advantage of $\mathcal{A}$ in violating the semantic security of the protocol P and the advantage function of the protocol P, respectively, as follows:

$$Adv_{P,\mathcal{D}}^{ror-ake}(\mathcal{A}) = 2 \cdot Pr[Succ] - 1,$$
$$Adv_{P,\mathcal{D}}^{ror-ake}(t, R) = \max_{\mathcal{A}}\{Adv_{P,\mathcal{D}}^{ror-ake}(\mathcal{A})\},$$

where the *maximum* is taken over all $\mathcal{A}$ with time-complexity at most t and using resources at most R (such as the number of oracle queries).

We say a 3-party PAKE protocol P is semantically secure if the advantage $Adv_{P,\mathcal{D}}^{ror-ake}(t, R)$ is only negligibly larger than $kn/|\mathcal{D}|$, where n is number of active sessions and k is a constant. Certainly, one can hope for the best scenario in which $k = 1$ and an adversary has an advantage of $n/|\mathcal{D}|$ since it simply guesses a password in each of the active sessions.

Authentication security. To measure the security of a 3PAKE protocol resisting the above focused undetectable on-line dictionary attacks, in this paper we consider the unilateral authentication from the client to the trusted server, wherein the adversary may be an inside attacker and impersonates another client user. We denote by $Succ_P^{auth(C \to S)}(\mathcal{A})$ the probability that an adversary $\mathcal{A}$ successfully impersonates a client instance during executing the protocol P while the trusted server does not detect it. Further, $Succ_P^{auth(C \to S)}(t, R) = \max_{\mathcal{A}}\{Succ_P^{auth(C \to S)}(\mathcal{A})\}$ is defined as the maximum over all $\mathcal{A}$ running in time at most t and using resources at most R. We say a 3-party PAKE protocol P is client-to-server authentication secure if $Succ_P^{auth(C \to S)}(t, R)$ is negligible in the security parameter.

4   General Construction of 3-Party PAKE Protocols

In this section, we present a new generic construction (referred as NGPAKE) of 3-party password-based key exchange protocol, by which we can create a series of provably secure 3-party PAKE protocols. Similarly to the construction of Abdalla et al. [1], our construction is essentially a form of compiler transforming any secure 2-party PAKE protocol into a secure 3-party PAKE protocol. It consists of two components: a 2-party password-based key exchange and a 2-party MAC-based key exchange protocol. Compared with the construction provided by Abdalla et al. [1], our construction not only avoids the use of a secure component (i.e. Key Distribution) and adds no additional burden of communication and computation, but also implements stronger security—both the semantic security and the client-to-server authentication security.

4.1   Scheme Description

Assume that two client users A and B want to establish a secure session key with the help of the trusted server S, based on their passwords pw_A and pw_B stored in the server. Firstly, the users A and B build two secure high-entropy session key sk_A and sk_B with the trusted server S, respectively, by using any semantic secure 2-party PAKE protocol. Secondly, using the session keys generated in the first step as the MAC key, A and B can concurrently authenticate and send their respective temporary Diffie-Hellman public keys to the server S. Thirdly, upon receiving and confirming the temporary public keys from the clients A and B, the server S authenticates and transfers temporary public keys of A and B to B and A, respectively, similarly by using the same MAC scheme with the session keys sk_A and sk_B. In this manner, A and B finally finish establishing a session

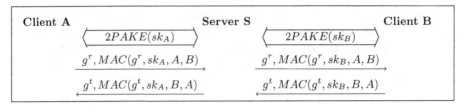

Fig. 3. NGPAKE: our new generic construction of 3PAKE protocol

key in an authenticated way, with the cooperation of the trusted server S. See Figure 3 for more details.

4.2 Building Blocks

We briefly describe the two cryptographic primitives used as building blocks in our scheme. Refer to [1] for more details.

Decisional Diffie-Hellman assumption. Let us consider two experiments: $Exp_{\mathbb{G}}^{ddh-real}$ and $Exp_{\mathbb{G}}^{ddh-rand}$. In the former, g^x, g^y and g^{xy} are given to $\mathcal{A}$, and in the latter g^x, g^y and g^z are provided, where x, y and z are drawn at random from $\{1, ..., |\mathbb{G}|\}$. Set

$$Adv_{\mathbb{G}}^{ddh}(t) = \max_{\mathcal{A}}\{|Pr[Exp_{\mathbb{G}}^{ddh-real}(\mathcal{A}) = 1] - Pr[Exp_{\mathbb{G}}^{ddh-rand}(\mathcal{A}) = 1]|\},$$

where the maximum is over all adversaries $\mathcal{A}$ running in time at most t. The DDH assumption in $\mathbb{G}$ holds if $Adv_{\mathbb{G}}^{ddh}(t)$ is a negligible function of t.

Message authentication codes. A Message authentication code scheme can be written as MAC = (Tag; Ver), where Tag is a MAC generation algorithm and Ver is a MAC verification one. It is existential unforgeability under chosen-message attacks (EUF-CMA) if adversaries can not create a new valid message-tag pair, even after obtaining many valid message-tag pairs. Namely, the maximal value, $Adv_{MAC}^{euf-cma}(t, q_g, q_v)$, of advantages of all $\mathcal{A}$ in violating EUF-CMA with at most t time complexity and at most q_g and q_v queries to its MAC generation and verification oracles, respectively, is a negligible function of the parameters above.

4.3 Security of Our Construction

We prove our new construction of 3-party PAKE protocols satisfies the semantic security and the client-to-server authentication security provided the Decisional Diffie-Hellman assumption holds in $\mathbb{G}$ and the underlying primitives it uses are secure.

Theorem 1. *Let 2PAKE be a semantic secure 2-party PAKE protocol and MAC be a secure MAC algorithm. Let q_{exe} and q_{test} denote the numbers of queries to Execute and Test oracles, and q_{send}^A, q_{send}^B, and q_{ake} be the numbers of queries to the SendClient and SendServer oracles with respect to each of the two 2PAKE protocols and the final two authenticated key exchange protocols. Then,*

$$Adv_{NGPAKE,\mathcal{D}}^{ror-ake}(t, q_{exe}, q_{test}, q_{send}^A, q_{send}^B, q_{ake}) \leq$$
$$2 \cdot Adv_{2PAKE,\mathcal{D}}^{ror-ake}(t, q_{exe}, q_{exe} + q_{send}^A, q_{send}^A)$$
$$+ 2 \cdot Adv_{2PAKE,\mathcal{D}}^{ror-ake}(t, q_{exe}, q_{exe} + q_{send}^B, q_{send}^B)$$
$$+ 2 \cdot q_{ake} \cdot Adv_{MAC}^{euf-cma}(t, 2, 0)$$
$$+ 2 \cdot Adv_{\mathbb{G}}^{ddh}(t + 8(q_{exe} + q_{ake})\tau_{\mathbb{G}})$$

and

$$Succ_{NGPAKE}^{auth(C \to S)}(t, q_{exe}, q_{test}, q_{send}^A, q_{send}^B, q_{ake}) \leq$$
$$Adv_{2PAKE,\mathcal{D}}^{ror-ake}(t, q_{exe}, q_{exe} + q_{send}^A, q_{send}^A)$$
$$+ Adv_{2PAKE,\mathcal{D}}^{ror-ake}(t, q_{exe}, q_{exe} + q_{send}^B, q_{send}^B)$$
$$+ q_{ake} \cdot Adv_{MAC}^{euf-cma}(t, 2, 0),$$

where $\tau_{\mathbb{G}}$ denotes the exponentiation computational time in $\mathbb{G}$.

Proof. We follow the original proof of Abdalla et al. [1]. For simplicity, the set of honest users is assumed to involve only users A and B. It can be easily extended to the multiple-party case. Let $\mathcal{A}$ be an adversary attacking NGPAKE in the Real-Or-Random model with time-complexity at most t, and asking at most q_{exe} queries to its *Execute* oracle, q_{test} queries to its *Test* oracle, q_{send}^A queries to *SendClient* and *SendServer* oracles corresponding to the 2PAKE protocol between A and the trusted server S, q_{send}^B queries to the oracles corresponding to the protocol between B and S, q_{ake}^{AS} queries to *SendClient* and *SendSever* oracles corresponding to the authenticated key exchange (AKE) protocol between A and S, and q_{ake}^{BS} queries to the oracles corresponding to the protocol between B and S. We incrementally define a sequence of games starting at the real game G_0 and ending up at the game G_6 in which the advantage of the adversary is zero. We define $Succ_i$ as the event that $\mathcal{A}$ guess b hidden in the *Test* oracles correctly in game G_i.

Furthermore, we assume that when the game below aborts or stops with no answer for b hidden in the *Test* oracles from $\mathcal{A}$, we guess a random bit for b, in which the success probability of the adversary is straightforwardly $1/2$.

Game G_0: This game represents the real attack game, where all the instances of clients A and B and the trusted server S modeled as oracles are available to the adversary. By definition, we have

$$Adv_{NGPAKE,\mathcal{D}}^{ror-ake}(\mathcal{A}) = 2 \cdot Pr[Succ_0] - 1. \tag{1}$$

Game G_1: In this game, we modify the simulation of the oracles by using a random session key sk_A', instead of the session key sk_A, as the MAC key in all of the sessions between A and S. As the following lemma shows, the difference between the success probabilities of the adversary between the current and previous game is at most the probability of breaking the security of the underlying 2PAKE protocol between A and S.

Lemma 1. $|Pr[Succ_1] - Pr[Succ_0]| < Adv_{2PAKE,\mathcal{D}}^{ror-ake}(t, q_{exe}, q_{exe} + q_{send}^A, q_{send}^A)$.

Proof. Given an adversary $\mathcal{A}_1$ and a distinguisher for G_0 and G_1, we can construct an active adversary $\mathcal{A}_{2PAKE}$ against the indistinguishability of the 2PAKE protocol between A and S as follows. In the initialization, $\mathcal{A}_{2PAKE}$ chooses a bit b randomly, selects passwords uniformly from $\mathcal{D}$ for all clients in the system except A and provides $\mathcal{A}_1$ with those for all the malicious users. Next, it runs $\mathcal{A}_1$ and simulates oracles as follows.

Consider a query *SendSever* or *SendClient* from $\mathcal{A}_1$. If the query corresponds to an instance of 2PAKE protocol between B and S, $\mathcal{A}_{2PAKE}$ can make reply by using the password of B. If the query corresponds to an instance of 2PAKE protocol between A and S, $\mathcal{A}_{2PAKE}$ can make reply by asking its own *Send* oracles. Once this query triggers the acceptance of the given instance of client A or S, which is also fresh, $\mathcal{A}_{2PAKE}$ asks a *Test* query to that instance and treats its output as the session key shared between A and S. For all remaining *SendClient* and *SendServer* queries from $\mathcal{A}_1$, $\mathcal{A}_{2PAKE}$ can easily simulate the corresponding oracles by either raising its *Send* query to the given 2PAKE protocol or using the secrets held preliminarily. Additionally, since *Execute* queries essentially consist of *Send* ones, $\mathcal{A}_{2PAKE}$ can simulate the corresponding oracles as above. With regard to *Test* queries by $\mathcal{A}_1$, $\mathcal{A}_{2PAKE}$ answers it according to the value of b, namely the real session key if 1 or a random key, otherwise. Finally, $\mathcal{A}_1$ outputs its guess b', and then $\mathcal{A}_{2PAKE}$ outputs 1 if $b' = b$ or 0, otherwise.

It is clear that the probability that $\mathcal{A}_1$ succeeds in G_0 is exactly the probability that $\mathcal{A}_{2PAKE}$ outputs 1 when its *Test* query returns the real session key. Similarly, the probability that $\mathcal{A}_1$ succeeds in G_1 is exactly that of the $\mathcal{A}_{2PAKE}$ outputs 1 when its *Test* query returns a random key. As a result, the lemma follows easily with $\mathcal{A}_{2PAKE}$ running at most time-complexity t and asking at most $q_{exe} + q_{send}^A$ queries to its *Test* oracle, at most q_{exe} queries to its Execute oracle, and at most q_{send}^A queries to its *Send* oracle. □

Game G_2: In this game, we employ a random session key sk'_B, instead of the session key sk_B agreed by B and S, as the MAC key in all of the sessions between B and S, similar as the change from G_0 to G_1. With the similar arguments, one can prove the following lemma.

Lemma 2. $|Pr[Succ_2] - Pr[Succ_1]| < Adv_{2PAKE,\mathcal{D}}^{ror-ake}(t, q_{exe}, q_{exe} + q_{send}^B, q_{send}^B)$. □

Game G_3: This game is modified as follows. If the adversary asks a *SendClient* or *SendServer* query for AKE between A and S involving a new pair message-tage not previously generated by an oracle, then we consider the MAC tag invalid and abort the game. According to the following lemma, the difference of the success probabilities of the adversary between in the current game and in the previous one should be negligible in use of a secure MAC scheme.

Lemma 3. $|Pr[Succ_3] - Pr[Succ_2]| \leq q_{ake}^{AS} \cdot Adv_{MAC}^{euf-cma}(t, 2, 0)$.

Proof. We use the so-called "hybrid arguments" method, in which q_{ake}^{AS} hybrids are defined as follows. In hybrid E_i, where $0 \leq i \leq q_{ake}^{AS}$, *SendClient* or *SendServer* queries for AKE between A and S in the first i sessions are treated as in game G_3 and the remaining $q_{ake}^{AS} - i$ ones are treated as in game G_2.

Given $\mathcal{A}_3^i$ be a distinguisher for hybrids E_i and E_{i-1}, by which we can build an adversary $\mathcal{A}_{mac}^i$ against the security of MAC scheme. Let F be the event in which a message-tag pair generated by $\mathcal{A}_3^i$ is considered invalid in hybrid E_i but valid in hybrid E_{i-1}. Since $Pr[F]$ is at most the probability that a new message-tag pair is forged by $\mathcal{A}_{mac}^i$ under a chosen-message attack, with time-complexity t and making at most two queries to its MAC generation oracle (to answer the $SendClient$ and $SendServer$ queries from $\mathcal{A}_3^i$) and no queries to its verification oracle, we have that $Pr[F] \leq Adv_{MAC}^{euf-cma}(t, 2, 0)$. Unless F occurs, hybrids E_i and E_{i-1} are identical, hence we have $Pr[Succ_{E_i} \wedge \neg F] = Pr[Succ_{E_{i-1}} \wedge \neg F]$ and then obtain $|Pr[Succ_{E_i}] - Pr[Succ_{E_{i-1}}]| \leq Adv_{MAC}^{euf-cma}(t, 2, 0)$. The lemma follows from the fact that there are at most q_{ake}^{AS} hybrids. $\square$

Game G_4: The treatments for $SendClient$ or $SendServer$ queries for AKE between B and S are modified similarly as in the previous game. By the similar arguments, one can prove the following lemma.

Lemma 4. $|Pr[Succ_4] - Pr[Succ_3]| \leq q_{ake}^{BS} \cdot Adv_{MAC}^{euf-cma}(t, 2, 0).$ $\square$

Notice that the proof on the authentication security in Theorem 1 is finished by combining the previous lemmas.

Game G_5: In this game, it is assumed that our simulator is initially given a random DDH triple $(X; Y; Z)$, where $X = g^x$, $Y = g^y$, and $Z = g^{xy}$. When processing the query $SendClient(A^i, start)$, the simulator selects two random values x_0 and a_0 in Z_p, computes $X_0 = X^{a_0} g^{x_0}$, and stores (a_0, x_0, X_0) in a list Λ_A. For $SendClient(B^j, start)$ in the same session, the simulator selects b_0 and y_0, computers Y_0 and stores them in a list Λ_B in the same measure. Upon receipt of both $SendSever(A^i, (X_0, m_A))$ and $SendServer(B^j, (Y_0, m_B))$ of the same session, the simulator checks the existence of X_0 and Y_0 by using Λ_A and Λ_B, respectively. If their existence is exact, it computes $Z_0 = Z^{a_0 b_0} \times Y_0^{x_0 b_0} \times X^{a_0 y_0} \times g^{x_0 y_0}$ in preparation for answering the $Test$ query. For other case or query, the simulator processes them as the previous game.

Since the case of MAC forgeries is excluded from the previous game and one set of random variables is in fact replaced by another set of identically distributed random variable, this game is equivalent to the previous one. So, we have $Pr[Succ_4] = Pr[Succ_5]$.

Game G_6: In this game, we take a random triple $(g^x; g^y; g^z)$, intead of a DDH triple, as the given triple for the simulator. As the following lemma shows, the current and previous games are indistinguishable since DDH is hard in $\mathbb{G}$.

Lemma 5. $|Pr[Succ_6] - pr[Succ_5]| \leq Adv_{\mathbb{G}}^{ddh}(t + 8(q_{exe} + q_{ake})\tau_{\mathbb{G}}).$

Proof. Assume $\mathcal{A}$ is a distinguisher for the current and previous games, by which we build an adversary $\mathcal{A}^{ddh}$ against the DDH problem in $\mathbb{G}$ as follows. Initially, with a triple $(X; Y; Z)$ as input, $\mathcal{A}^{ddh}$ selects a bit b at random and then starts running $\mathcal{A}$. By using the input triple, $\mathcal{A}^{ddh}$ can easily deal with *SendClient, SendServer, Execute,* or *Test* query from $\mathcal{A}$ as in the previous game. Finally, $\mathcal{A}$ output a bit b'. If $b' = b$, then $\mathcal{A}^{ddh}$ returns 1 or 0, otherwise. As a result, one can easily see that $\mathcal{A}^{ddh}$ runs $\mathcal{A}$ exactly as in game G_5 if the triple $(X; Y; Z)$ is a true DDH triple, and as in game G_6 if it is a random triple. Hence, the probabilities that $\mathcal{A}^{ddh}$ outputs 1 in the former and in the latter are exactly $Pr[Succ_5]$ and $Pr[Succ_6]$, respectively. The lemma follows from the fact that $\mathcal{A}^{ddh}$ has time-complexity at most $t + 8(q_{exe} + q_{ake})\tau_{\mathbb{G}}$, where $q_{ake} = q_{ake}^{AS} + q_{ake}^{BS}$, due to the additional time for the computations of the random self-reducibility.   □

Thus far, since no information on the bit b in the *Test* oracle is leaked to the adversary, we have $Pr[Succ_6] = 1/2$. Combining with the previous lemmas, one gets the result on the semantic security in Theorem 1.   □

Remark 2. As a matter of fact, most of the existing provably secure 2-party PAKE protocols, especially those based on Diffie-Hellman key exchange, can be represented as the following generic form, which is shown in figure 4. Firstly, the client C sends its temporary public key pk_C to the server S under the mask of the shared password pw. Upon receiving the message from C, S gets pk_C by using pw, computes the session key sk with pk_C and generates its authenticator $Auth_S$. Then it sends $Auth_S$ and its masked temporary public key pk_S to C. After obtaining this message, C checks $Auth_S$. If it is invalid, C aborts. Otherwise C computes the session key sk and its own authenticator $Auth_C$ and then return $Auth_C$ to S. Finally, S confirms sk by authenticating $Auth_C$.

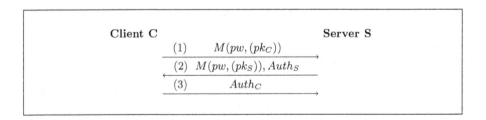

Fig. 4. The generic form of the present provably secure 2PAKE protocols

However, due to considering the efficiency, the last message, namely the authentication from clients to the server, is omitted in some 2-party PAKE protocols such as PPK protocol [14] and OMDHKE protocol [9], which still keep semantic security. If these protocols are used as a secure block in building a 3-party PAKE protocol under the generic construction of Abdalla et al. [1], the resulting protocol will suffer from the undetectable on-line dictionary attacks as mentioned above. On the contrary, if these protocols are used in our construction, both the efficiency and the authentication security for clients will still be held in the resulting scheme.

Hence, as a generic construction scheme of 3-party PAKE protocols, NGPAKE has both better adaptability and higher efficiency than GPAKE [1] does. Furthermore, when a 2-party PAKE protocol with the mutual authentication is employed in the instantiation of our scheme, the authenticators from clients to the server in the 2PAKE component can be removed without any effect on the security of the whole protocol, if one is eager to capture a higher efficiency. On the other hand, it seems that GPAKE are more easily scaled up to an N-party version than NGPAKE, but its forementioned security problem still exits in an N-party case. Therefore, how to construct a secure and efficient general construction for N-party protocol is another interesting challenge.

Remark 3. With no requirement for Random Oracle (RO) model, our generic construction can be instantiated by using any present 2-party PAKE protocol enjoying the semantic security in the standard model such as KOY protocol [12] so as to generate a secure 3-party PAKE protocol in the standard model. Certainly, one can also use any 2-party scheme with the semantic security in RO model such as the PAK suite protocols [14] and OMDHKE protocol [9] in the instantiation of our scheme, and then gets a secure 3PAKE protocol in RO model.

Remark 4. Though we do not consider a security of key privacy as in [1], the sub-protocol executed by both clients in the last stage of our scheme is substantially an authenticated Diffie-Hellman key exchange as in GPAKE, so it is apparent that NGPAKE also holds key privacy.

5   Conclusion

Absence of the authentication from clients to the trusted server is the common cause that the two 3-party PAKE schemes provided by Abdalla et al. [1] [2] are subject to undetectable on-line dictionary attacks. For this, by fully considering the generic construction of most present provably secure 2PAKE protocols, we propose a new generic construction (NGPAKE) for 3-party PAKE protocols, which is to a certain degree superior in adaptability, security and efficiency to GPAKE provided by Abdalla et al. [1].

References

1. M. Abdalla, P.-A. Fouque, and D. Pointcheval. Password-based authenticated key exchange in the three-party setting. In S. Vaudenay, editor, *Public Key Cryptography - PKC 2005*, volume 3386 of *LNCS*, pages 65–84. Springer, 2005.
2. M. Abdalla and D. Pointcheval. Interactive diffie-hellman assumptions with applications to password-based authentication. In A. S. Patrick and M. Yung, editors, *Financial Cryptography and Data Security - FC 2005*, volume 3570 of *LNCS*, pages 341–356. Springer, 2005.
3. M. Bellare, D. Pointcheval, and P. Rogaway. Authenticated key exchange secure against dictionary attacks. In B. Preneel, editor, *Advances in Cryptology - EUROCRYPT 2000*, volume 1807 of *LNCS*, pages 139–155. Springer, 2000.

4. M. Bellare and P. Rogaway. Entity authentication and key distribution. In D. R. Stinson, editor, *Advances in Cryptology - CRYPTO 1993*, volume 773 of *LNCS*, pages 232–249. Springer, 1993.

5. M. Bellare and P. Rogaway. Provably secure session key distribution: the three party case. In *ACM Symposium on Theory of Computing - STOC 1995*, pages 57–66. ACM, 1995.

6. S. M. Bellovin and M. Merritt. Encrypted key exchange: Password-based protocols secure against dictionary attacks. In *Proceedings of the 1992 IEEE Symposium on Security and Privacy*, pages 72–84. IEEE Computer Society Press, 1992.

7. V. Boyko, P. D. MacKenzie, and S. Patel. Provably secure password-authenticated key exchange using diffie-hellman. In B. Preneel, editor, *Advances in Cryptology - EUROCRYPT 2000*, volume 1807 of *LNCS*, pages 156–171. Springer, 2000.

8. E. Bresson, O. Chevassut, and D. Pointcheval. Security proofs for an efficient password-based key exchange. In Sushil Jajodia, Vijayalakshmi Atluri, and Trent Jaeger, editors, *ACM Conference on Computer and Communications Security, CCS 2003*, pages 241–250. ACM, 2003.

9. E. Bresson, O. Chevassut, and D. Pointcheval. New security results on encrypted key exchange. In F. Bao, R. H. Deng, and J. Zhou, editors, *Public Key Cryptography - PKC 2004*, volume 2947 of *LNCS*, pages 145–158. Springer, 2004.

10. O. Goldreich and Y. Lindell. Session-key generation using human passwords only. In J. Kilian, editor, *Advances in Cryptology - CRYPTO 2001*, volume 2139 of *LNCS*, pages 408–432. Springer, 2001.

11. D. P. Jablon. Strong password-only authenticated key exchange. *ACM Computer Communication Review.*, 26:5–26, October 1996.

12. J. Katz, R. Ostrovsky, and M. Yung. Efficient password-authenticated key exchange using human-memorable passwords. In B. Pfitzmann, editor, *Advances in Cryptology - EUROCRYPT 2001*, volume 2045 of *LNCS*, pages 475–494. Springer, 2001.

13. C.-L. Lin, H.-M. Sun, M. Steiner, and T. Hwang. Three-party encrypted key exchange without server public-keys. *IEEE Communication Letters*, 5(12):497–499, 2001.

14. P. D. MacKenzie. The pak suite: Protocols for password-authenticated key exchange. In *Submission to IEEE P1363.2*, 2002.

15. P. D. MacKenzie, S. Patel, and R. Swaminathan. Password-authenticated key exchange based on rsa. In T. Okamoto, editor, *Advances in Cryptology - ASIACRYPT 2000*, volume 1976 of *LNCS*, pages 599–613. Springer, 2000.

16. M. Steiner, G. Tsudik, and M. Waidner. Refinement and extension of encrypted key exchange. *Operating Systems Review*, 29(3):22–30, 1995.

17. H.-T. Yeh, H.-M. Sun, and T. Hwang. Efficient three-party authentication and key agreement protocols resistant to password guessing attacks. *J. Inf. Sci. Eng.*, 19(6):1059–1070, 2003.

On the Importance of Public-Key Validation in the MQV and HMQV Key Agreement Protocols

Alfred Menezes and Berkant Ustaoglu

Department of Combinatorics & Optimization, University of Waterloo
{ajmeneze, bustaoglu}@uwaterloo.ca

Abstract. HMQV is a hashed variant of the MQV key agreement protocol proposed by Krawczyk at CRYPTO 2005. In this paper, we present some attacks on HMQV and MQV that are successful if public keys are not properly validated. In particular, we present an attack on the two-pass HMQV protocol that does not require knowledge of the victim's ephemeral private keys. The attacks illustrate the importance of performing some form of public-key validation in Diffie-Hellman key agreement protocols, and furthermore highlight the dangers of relying on security proofs for discrete-logarithm protocols where a concrete representation for the underlying group is not specified.

1 Introduction

Public-key validation is a process whose purpose is to verify that a public key possesses certain arithmetic properties. Public-key validation is especially important in Diffie-Hellman protocols where a party $\hat{B}$ derives a secret session key K by combining his private key with a public key received from a second party $\hat{A}$ and subsequently uses K in some symmetric-key protocol (e.g., encryption or message authentication) with $\hat{A}$. A dishonest party $\hat{A}$ might select an invalid public key in such a way that the use of K reveals information about $\hat{B}$'s private key. Lim and Lee [18] demonstrated the importance of public-key validation by presenting *small-subgroup attacks* on some discrete logarithm key agreement protocols that are effective if the receiver of a group element does not verify that the element belongs to the desired group of high order (e.g., a prime-order DSA-type subgroup of $\mathbb{F}_p^*$). In [5,3], *invalid-curve attacks* were designed that are effective on elliptic curve protocols if the receiver of a point does not verify that the point indeed lies on the chosen elliptic curve. Kunz-Jacques et al. [15] showed that the zero-knowledge proof proposed in [4] for proving possession of discrete logarithms in groups of unknown order can be broken if a dishonest verifier selects invalid parameters during its interaction with the prover. More recently, Chen, Cheng and Smart [7] illustrated the importance of public-key validation in identity-based key agreement protocols that use bilinear pairings.

The MQV protocols [16] are a family of authenticated Diffie-Hellman protocols that have been widely standardized [1,2,9,27]. In the two-pass and three-pass versions of the protocol, the communicating parties $\hat{A}$ and $\hat{B}$ exchange static

R. Barua and T. Lange (Eds.): INDOCRYPT 2006, LNCS 4329, pp. 133–147, 2006.
© Springer-Verlag Berlin Heidelberg 2006

(long-term) and ephemeral (short-term) public keys, and thereafter derive a secret key from these values. In the one-pass version, only one party contributes an ephemeral public key. In 2005, Krawczyk [12,13] presented the HMQV protocols, which are hashed variants of the MQV protocols. The primary advantages of HMQV over MQV are better performance and a rigorous security proof. The improved performance of HMQV is a direct consequence of not requiring the validation of ephemeral and static public keys — unlike with MQV where these operations are mandated. Despite the omission of public-key validation, Krawczyk was able to devise proofs that the HMQV protocols are secure in the random oracle model assuming the intractability of the computational Diffie-Hellman problem (and some variants thereof) in the underlying group.

Menezes [19] identified some flaws in the HMQV security proofs and presented small-subgroup attacks on the protocols. The attacks exploit the omission of validation for both ephemeral and static public keys, and allow an adversary to recover the victim's static private key. The attacks on the one-pass protocol are the most realistic, while the attacks on the two-pass and three-pass protocols are harder to mount in practice because the adversary needs to learn some of the victim's ephemeral private keys.

In this paper, we further investigate the effects of omitting public-key validation in HMQV and MQV. For the most part, we will only consider the two-pass HMQV protocol (which we call *the* HMQV protocol), which is the core member of the HMQV family. We identify a subtle flaw in the HMQV security proof which leads to an attack that does not require knowledge of ephemeral private keys, thereby contradicting the claim made in [13] that the HMQV protocol (without public-key validation) is provably secure if the adversary never learns any ephemeral private keys. We also consider the vulnerability of HMQV and MQV if only static public keys are validated, or if only ephemeral public keys are validated. These hypothetical scenarios are worth investigating because the reasons for omitting public-key validation can be different for ephemeral and static keys — validation of ephemeral public keys may be omitted for performance reasons, while validation of static public keys may be omitted because the certification authority may not be configured to perform such tests [13].

We emphasize that many of the attacks described in this paper cannot be mounted in realistic settings. For example, the aforementioned attack on HMQV that does not require knowledge of ephemeral private keys is described in certain underlying groups that have never been proposed for practical use. Moreover, this attack fails if the underlying group is a DSA-like group or a prime-order subgroup of an elliptic curve group as proposed for standardization in [14]. We also caution against inferring from our work that one must necessarily (fully) validate public keys in all Diffie-Hellman key agreement protocols. For example, the version of HMQV proposed in [14] only requires that a few simple and efficient checks be performed on static and ephemeral public keys. Moreover, even in the situation where one is concerned that ephemeral private keys might be leaked, [14] only requires that ephemeral and static public keys be *jointly* validated, thus saving a potentially expensive validation step (cf. §6).

The remainder of this paper is organized as follows. The MQV and HMQV protocols are reviewed in §2. The new attack on HMQV that does not require knowledge of ephemeral private keys is presented in §3, and the associated flaw in the HMQV security proof is identified. In §4 we present attacks on HMQV in the case where only ephemeral public keys are validated. In §5 it is shown that MQV is insecure if validation of ephemeral public keys is omitted. In §6 we describe the approach taken in [14] to guard against the kinds of attacks discussed in this paper. An example is presented in §7 to illustrate the potential pitfalls if public keys are not completely validated. The paper concludes with some remarks in §8.

2   The HMQV Key Agreement Protocol

Let $G = \langle g \rangle$ be a multiplicatively-written cyclic group of prime order q, and let 1 denote the identity element in G. Let H be a hash function, and let $\overline{H}$ be an l-bit hash function where $l = (\lfloor \log_2 q \rfloor + 1)/2$. Party $\hat{A}$'s static private key is an integer $a \in_R [0, q-1]$, while her static public key is the group element $A = g^a$. Similarly, party $\hat{B}$ has a static key pair (B, b) where $b \in_R [0, q-1]$ and $B = g^b$.

2.1   Description of HMQV

In the (two-pass) HMQV protocol as presented in [12,13], parties $\hat{A}$ and $\hat{B}$ establish a secret session key as follows:

1. $\hat{A}$ selects an ephemeral private key $x \in_R [0, q-1]$ and computes her ephemeral public key $X = g^x$. $\hat{A}$ then sends $(\hat{A}, \hat{B}, X)$ to $\hat{B}$.
2. Upon receiving $(\hat{A}, \hat{B}, X)$, $\hat{B}$ checks that $X \neq 0$,[1] selects an ephemeral key pair (Y, y), and sends $(\hat{B}, \hat{A}, Y)$ to $\hat{A}$. $\hat{B}$ proceeds to compute $s_B = y + eb \bmod q$ and $\sigma = (XA^d)^{s_B}$ where $d = \overline{H}(X, \hat{B})$ and $e = \overline{H}(Y, \hat{A})$.[2]
3. Upon receiving $(\hat{B}, \hat{A}, Y)$, $\hat{A}$ checks that $Y \neq 0$, and computes $s_A = x + da \bmod q$ and $\sigma = (YB^e)^{s_A}$ where again $d = \overline{H}(X, \hat{B})$ and $e = \overline{H}(Y, \hat{A})$.
4. The secret session key is $K = H(\sigma) = H(g^{s_A s_B})$.

The messages transmitted in steps (1) and (2) may include certificates for the static public keys A and B, respectively. Note that HMQV does not mandate that static and ephemeral public keys be validated, i.e., verified as being non-identity elements of G.

---

[1] Note that $0 \notin G$. The check $X \neq 0$ (and $Y \neq 0$) makes sense in some settings, e.g., when G is a multiplicative subgroup of a finite field; in this case 0 is the additive identity of the field.

[2] The HMQV papers [12,13] do not explicitly state that s_A (and s_B) should be computed modulo q. The attacks in this paper can still be launched if s_A (and s_B) are not reduced modulo q.

2.2   Description of MQV

The three essential differences between the MQV protocol (as standardized in [27]) and the HMQV protocol are the following:

1. Static and ephemeral public keys must be validated in MQV.[3]
2. In MQV, the integers d and e are derived from the group elements X and Y, respectively. For example, if G is a group of elliptic curve points, then d and e are derived from the l least significant bits of the x-coordinates of X and Y respectively.
3. The secret session key is $K = H(\sigma, \hat{A}, \hat{B})$.[4]

2.3   Security Proofs

Krawczyk [12,13] provided a very extensive analysis of HMQV. He proved that the protocol satisfies the Canetti-Krawczyk definition [6] for secure key agreement, under the assumptions that H and $\overline{H}$ are random oracles and that the computational Diffie-Hellman (CDH) problem[5] in G is intractable. The Canetti-Krawczyk security definition is a very strong one in that the adversary controls all communications between parties and its goal is very modest — distinguishing a target session key from a purely random key. The protocol remains secure even if the adversary is allowed to learn session keys different from the target session key. Krawczyk also proved that the protocol is resistant to attacks when the adversary is permitted to learn the ephemeral private keys of sessions; for this property the 'Gap Diffie-Hellman' and 'KEA1' assumptions about G are needed.

2.4   An Attack

We describe the attack on HMQV that was presented in [19]. The attack exploits the omission of public-key validation for ephemeral and static public keys, and also the ability of the adversary to learn the victim's ephemeral private keys.

We suppose that there is an algebraic structure R (e.g., a field, ring, or group) such that:

1. The elements of R are represented in the same format as elements of G (e.g., bitstrings of the same length).
2. The group operation for G is defined on elements of R.

For the attack in this section, we further assume that there is a subset G' of R such that:

3. G' is a cyclic group with respect to the operation defined on G.
4. G' has order t where $t = 2^r$ for some small r (e.g., $r = 4$).

---

[3] Actually 'embedded' validation may be performed on ephemeral public keys. The details are not relevant to the attacks presented in this paper.

[4] The identities $\hat{A}$, $\hat{B}$ are included in the derivation of K from σ in order to thwart Kaliski's unknown-key share attack [10].

[5] The CDH problem in $G = \langle g \rangle$ (with respect to g) is that of computing $\text{CDH}(X, Y) = X^y = Y^x$ given g, $X = g^x$ and $Y = g^y$.

For example, if G is a multiplicative subgroup of order q of $\mathbb{F}_p^*$ (the integers modulo p) and $t \mid (p-1)/q$, then we can take $R = \mathbb{F}_p$ and G' to be the unique subgroup of $\mathbb{F}_p^*$ of order t. Note that elements of G and R have the same representation (integers modulo p), and the common operation is multiplication modulo p. As a second example, suppose that $G = E(\mathbb{F}_p)$ where $E : V^2 = U^3 + \alpha U + \beta$ is an (additively-written) elliptic curve defined over $\mathbb{F}_p$, and let $E' : V^2 = U^3 + \alpha U + \beta'$ be another elliptic curve defined over $\mathbb{F}_p$ such that $t \mid \#E'(\mathbb{F}_p)$. Then we can take $R = E'(\mathbb{F}_p)$ and G' to be a subgroup of $E'(\mathbb{F}_p)$ of order t. Again, the elements of G and R have the same representation (pairs of integers modulo p), and the group law for E and E' are the same since the usual chord-and-tangent laws for E and E' do not (explicitly) use the coefficients β and β' (see §4).

The attack proceeds as follows. The adversary $\hat{A}$ chooses an element $\gamma \in G'$ of order $t = 2^r$, selects $a, x \in [1, t-1]$, computes $A = \gamma^a$ and $X = \gamma^x$, and sends $(\hat{A}, \hat{B}, X)$ to $\hat{B}$. While $\hat{B}$ is computing the session key $K = H((XA^d)^{s_B})$, the adversary learns $\hat{B}$'s ephemeral private key y. Let $\beta = XA^d = \gamma^{x+da}$, so $K = H(\beta^{s_B})$. $\hat{A}$ then learns the session key K.[6] Now $\hat{A}$ computes $K' = H(\beta^c)$ for $c = 0, 1, 2, \ldots, t-1$ until $K' = K$, in which case $\hat{A}$ has determined $s_B \bmod t$. After repeating this procedure a few times, $\hat{A}$ can use the Leadbitter-Smart lattice attack [17] to find the l most significant bits of b. The remaining l bits of b can thereafter be determined in $O(q^{1/4})$ time using Pollard's lambda method [24].

3   No Ephemeral Private Key Leakage

The adversary in the attack of §2.4 requires knowledge of the victim's ephemeral private keys. While resistance to ephemeral private key leakage is a desirable attribute of a key establishment protocol[7], it is arguably not a fundamental security requirement. In [13] it is claimed that the HMQV protocol is *provably secure* if the adversary does not learn any ephemeral private keys. In this section we demonstrate that this claim is false.

3.1   A New Attack

Suppose that $G = \langle g \rangle$ is a multiplicatively-written group of prime order q, and suppose that the CDH problem in G is intractable. We further assume that R is a ring such that:

1. The elements of R are represented in the same format as elements of G (e.g., bitstrings of the same length).
2. The multiplication operation for R is defined in the same way as the operation for G. In particular, G is a subgroup of the group of units $U(R)$ of R.

---

[6] Suppose, for example, that $\hat{B}$ sends $\hat{A}$ an authenticated message $(m, \tau = \mathrm{MAC}_K(m))$. Then $\hat{A}$ can learn K by computing $\tau' = \mathrm{MAC}_{K'}(m)$ where $K' = H(\beta^c)$ for $c = 0, 1, 2, \ldots, t-1$ until $\tau' = \tau$.

[7] In [12,13], resistance of Diffie-Hellman protocols to damage from the disclosure of ephemeral private keys is described as a 'prime security concern'.

3. There exists an element $T \in R$, $T \neq 0$, such that $T^2 = 0$ (where "0" denotes the additive identity element in R).

The new attack on HMQV assumes that parties do not validate ephemeral public keys. The adversary $\hat{C}$ intercepts the message $(\hat{A}, \hat{B}, X)$ sent by $\hat{A}$ and replaces it with $(\hat{A}, \hat{B}, T)$. Similarly, $\hat{C}$ intercepts $\hat{B}$'s response $(\hat{B}, \hat{A}, Y)$ and forwards $(\hat{B}, \hat{A}, T)$ to $\hat{A}$. If R is commutative then, assuming that $s_A \geq 2$ and $s_B \geq 2$, it is easy to see that both $\hat{A}$ and $\hat{B}$ compute the session key $K = H(0)$. Of course, $\hat{C}$ can also compute this key.

If R is not commutative, then the value of the session key depends on the particular exponentiation method used by the parties. Suppose that $\hat{A}$ determines the session key by first calculating $t_A = es_A \bmod q$ and then using simultaneous multiple exponentiation [20, Algorithm 14.88] to compute $\sigma = T^{s_A} B^{t_A}$ and $K = H(\sigma)$. This algorithm first computes TB and initializes an accumulator to 1. It then repeatedly examines the bits of s_A and t_A from left to right. During each iteration, either 1, T, B or TB is multiplied into the accumulator which is then squared. Now, if the most significant bits of s_A and t_A are 1 and 0, respectively, then the accumulator takes on the values 1, T, $T^2, \ldots$. Hence $\hat{A}$ computes $\sigma = 0$. Similarly, $\hat{B}$ may compute $\sigma = 0$, in which case $\hat{C}$ also learns the session key $K = H(0)$.

3.2   Examples of Groups

We give two examples of groups that satisfy the conditions of §3.1. These examples do not have any immediate practical relevance since such groups are not being deployed in practice. Nonetheless, they serve to refute the claim made in [12,13] that HMQV is provably secure regardless of the representation used for the elements of the G (subject to the constraint that the CDH problem in G be intractable).

A Commutative Example. Let p be a 1024-bit prime such that $p - 1$ has a 160-bit prime divisor q. Consider the commutative ring $R = \mathbb{Z}_{p^2}$. Then $U(R)$ is cyclic and $\#U(R) = p(p - 1)$. Let G be the order-q cyclic subgroup of $U(R)$. The CDH problem in G is believed to be intractable. The element $T = p \in \mathbb{Z}_{p^2}$ satisfies $T \neq 0$ and $T^2 = 0$.

A Non-commutative Example. Again, let p be a 1024-bit prime such that $p - 1$ has a 160-bit prime divisor q. Consider the non-commutative ring R of 2×2 matrices over $\mathbb{F}_p$. Then $\#U(R) = (p^2 - 1)(p^2 - p)$. Let $g \in U(R)$ be an element of order q, and let $G = \langle g \rangle$. The CDH problem in G is equivalent to the CDH problem in the order-q subgroup of $\mathbb{F}_p^*$ (see [21]) and is therefore believed to be intractable. The element

$$T = \begin{bmatrix} 0 & 0 \\ 1 & 0 \end{bmatrix}$$

satisfies $T \neq 0$ and $T^2 = 0$.

3.3   Flaw in the HMQV Proof

The HMQV security proof in [13] has two main steps. First, an 'exponential challenge-response' signature scheme XCR is defined and proven secure in the random oracle model under the assumption that the CDH problem in G is intractable. Second, the security of XCR (actually a 'dual' version of XCR) is proven to imply the security of HMQV.

In the XCR signature scheme, a verifier $\hat{A}$ selects $x \in_R [0, q - 1]$ and sends the challenge $X = g^x$ and a message m to the signer $\hat{B}$. $\hat{B}$ responds by selecting $y \in_R [0, q - 1]$ and sending the signature $(Y = g^y, \sigma = X^{s_B})$ to $\hat{A}$ where $s_B = y + eb \bmod q$, $e = \overline{H}(Y, m)$, and (B, b) is $\hat{B}$'s static key pair. The signature is accepted by $\hat{A}$ provided that $Y \neq 0$ and $\sigma = (YB^e)^x$. XCR signatures are different from ordinary digital signatures — $\hat{A}$ cannot convince a third party that $\hat{B}$ generated a signature (Y, σ) for message m and challenge X because $\hat{A}$ could have generated this signature herself.

The XCR security proof in [12,13] uses the forking lemma of Pointcheval and Stern [23]. The proof hypothesizes the existence of a forger who, on input $B, X_0 \in_R G$ and a signing oracle for $\hat{B}$, produces a message m_0 and a valid signature (Y_0, σ) for m_0;[8] that is $Y_0 \neq 0$ and $\sigma = (Y_0 B^e)^{x_0}$ where $e = \overline{H}(Y_0, m_0)$ and $X_0 = g^{x_0}$.[9] Now, in order to compute $\mathrm{CDH}(B, X_0)$, the forger is run twice with input B, X_0. The forger's hash function and signature queries are suitably answered so that the two invocations of the forger eventually produce valid forgeries (m_0, Y_0, σ) and (m_0, Y_0, σ') where $e = \overline{H}(Y_0, m_0)$, $e' = \overline{H}'(Y_0, m_0)$, and $e \not\equiv e' \pmod{q}$. To conclude the argument, one notes that

$$\frac{\sigma}{\sigma'} = \frac{(Y_0 B^e)^{x_0}}{(Y_0 B^{e'})^{x_0}} = (B^{x_0})^{e-e'} \tag{1}$$

whence $\mathrm{CDH}(B, X_0) = (\sigma/\sigma')^{(e-e')^{-1}}$ can be efficiently computed.

The flaw in this argument is the assumption that the Y_0 terms in (1) can be cancelled under the sole condition that $Y_0 \neq 0$. While the cancellation in (1) is valid if $Y_0 \in G$ (which is the case if Y_0 has been validated), in general one needs to make additional assumptions including that Y_0 is invertible. Thus, since the description of XCR does not mandate that the verifier validate Y, the XCR security proof in [12,13] is incorrect.

This flaw in the XCR security proof accounts for the following attack on XCR. Let R and T be as defined in §3.1, and suppose for the sake of concreteness that R is commutative. A forger can respond to $\hat{A}$'s challenge (X, m) with the signature $(Y = T, \sigma = 0)$. The signature is accepted by $\hat{A}$ since $T \neq 0$ and $(TB^e)^x = 0$. This attack on XCR in turn explains why the attack described in §3.1 can be launched on HMQV.

---

[8] There is also the requirement that (m_0, Y_0) did not appear in any of $\hat{B}$'s responses to the forger's signature queries.

[9] The XCR security definition in [12,13] incorrectly states that the forger's output (m_0, Y_0, σ) should satisfy $Y_0 \neq 0$ and $\sigma = X_0^{y_0+eb}$ where $Y = g^{y_0}$. The latter condition is *not* equivalent to the condition $\sigma = (Y_0 B^e)^{x_0}$ in the case where $Y_0 \notin G$ — indeed y_0 is not even defined when $Y_0 \notin G$.

4   No Static Public-Key Validation

We describe an attack on HMQV in the hypothetical situation where ephemeral public keys are validated but static public keys aren't. As mentioned in the Introduction, this situation is worth investigating because validation for static public keys may be omitted if a certification authority is not configured to perform such tests. We describe attacks that can be mounted in the realistic setting where G is a DSA-type group or an elliptic curve group.

4.1   DSA-Type Groups

We suppose that G is the order-q subgroup of $\mathbb{F}_p^*$, and that $t = 2^r$ is a divisor of $(p-1)/q$. Let $\gamma \in \mathbb{F}_p^*$ be an element of order t. Using the notation introduced in §2.4, we have $R = \mathbb{F}_p$ and $G' = \langle \gamma \rangle$.

The adversary $\hat{A}$ selects a valid $X \in G$ and computes $d = (\overline{H}(X, \hat{B}))^{-1} \bmod q$ and $A = \gamma X^{-d^{-1} \bmod q}$. She certifies A as her (invalid) static public key and sends X to $\hat{B}$ who computes $\beta = XA^d = \gamma^d$ and $K = H(\beta^{s_B})$. As in the attack described in §2.4, $\hat{A}$ learns y, K, and $s_B \bmod t$; repeating this procedure yields half the bits of b.

4.2   Elliptic Curves Groups

We suppose that $G = E(\mathbb{F}_p)$ where $E : V^2 = U^3 + \alpha U + \beta$ is an elliptic curve of prime order defined over the prime field $\mathbb{F}_p$. Let $P_1 = (u_1, v_1)$ and $P_2 = (u_2, v_2)$ be two finite points in $E(\mathbb{F}_p)$ with $P_1 \neq -P_2$, and let $P_3 = (u_3, v_3) = P_1 + P_2$. The usual formulae for computing P_3 are:

$$u_3 = \lambda^2 - u_1 - u_2, \tag{2}$$
$$v_3 = \lambda(u_1 - u_3) - v_1, \tag{3}$$

where

$$\lambda = \frac{v_2 - v_1}{u_2 - u_1} \text{ or } \lambda = \frac{3u_1^2 + \alpha}{2v_1},$$

depending on whether $P_1 \neq P_2$ or $P_1 = P_2$. Note that the formulae do not (explicitly) depend on the coefficient β.

The adversary $\hat{A}$'s goal is to select two points $A, X \in \mathbb{F}_p \times \mathbb{F}_p$ such that (i) X is valid, i.e., $X \in E(\mathbb{F}_p)$, $X \neq \infty$; and (ii) $T = X + dA$ is a point of order 16 on some curve $E' : V^2 = U^3 + \alpha U + \beta'$ defined over $\mathbb{F}_p$, where $d = \overline{H}(X, \hat{B})$ and $X + dA$ is computed using the formulas for $E(\mathbb{F}_p)$. Using the notation introduced in §2.4, we have $R = E'(\mathbb{F}_p)$, $G' = \langle T \rangle$, and $t = 16$. The adversary then certifies A as her (invalid) static public key and sends X to $\hat{B}$, who computes $K = H(s_B T)$. As in the attack described in §2.4, $\hat{A}$ learns y, K, and $s_B \bmod t$; repeating this procedure yields half the bits of b.

The adversary can proceed to determine A and X as follows. She first selects an arbitrary finite point $X = (u_2, v_2) \in E(\mathbb{F}_p)$ such that $d = \overline{H}(X, \hat{B})$ is odd.

Now let $A = (z, 0)$, where $z \in \mathbb{F}_p$ is an indeterminate whose value will be specified later. Since d is odd, application of the group law for E yields $dA = A$. The coordinates (u_3, v_3) of $T = X + dA$ are then derived using (2) and (3):

$$u_3 = \left(\frac{v_2}{u_2 - z}\right)^2 - z - u_2 \quad \text{and} \quad v_3 = \frac{v_2}{u_2 - z}(z - u_3). \tag{4}$$

Define

$$\beta' = v_3^2 - u_3^3 - \alpha u_3, \tag{5}$$

so that $T = (u_3, v_3)$ is an $\mathbb{F}_p$-point on the elliptic curve

$$E' : V^2 = U^3 + \alpha U + \beta'. \tag{6}$$

We next show how division polynomials can be used to select $z \in \mathbb{F}_p$ so that T has order 16. The following result is well known (e.g., see [25]).

Theorem 1. Consider the division polynomials $\Psi_k(U, V) \in \mathbb{F}_p[U, V]$ associated with an elliptic curve $E/\mathbb{F}_p : V^2 = U^3 + \alpha U + \beta$ and defined recursively as follows:

$$\Psi_1(U, V) = 1$$
$$\Psi_2(U, V) = 2V$$
$$\Psi_3(U, V) = 3U^4 + 6\alpha U^2 + 12\beta U - \alpha^2$$
$$\Psi_4(U, V) = 4V(U^6 + 5\alpha U^4 + 20\beta U^3 - 5\alpha^2 U^2 - 4\alpha\beta U - 8\beta^2 - \alpha^3)$$
$$\Psi_{2k+1}(U, V) = \Psi_{k+2}\Psi_k^3 - \Psi_{k+1}^3\Psi_{k-1} \text{ for } k \geq 2$$
$$\Psi_{2k}(U, V) = \Psi_k(\Psi_{k+2}\Psi_{k-1}^2 - \Psi_{k-2}\Psi_{k+1}^2)/2V \text{ for } k \geq 3.$$

Let Ψ_k' be the polynomial obtained by repeatedly replacing occurrences of V^2 in Ψ_k by $U^3 + \alpha U + \beta$, and define

$$f_k = \begin{cases} \Psi_k'(U, V), & \text{if } k \text{ is odd,} \\ \Psi_k'(U, V)/V, & \text{if } k \text{ is even.} \end{cases}$$

Then in fact $f_k \in \mathbb{F}_p[U]$. Moreover, if $P = (u, v) \in E(\overline{\mathbb{F}}_p)$ such that $2P \neq \infty$, then $kP = \infty$ if and only if $f_k(u) = 0$.

It follows from Theorem 1 that the roots of the polynomial

$$g(U) = \frac{f_{16}(U)}{f_8(U)}$$

are precisely the U-coordinates of points of order 16 in $E(\overline{\mathbb{F}}_p)$, and hence $\deg(g) = 96$.

Now to determine T, the adversary computes $h(z) = g(u_3)$, where $g(U)$ is associated with $E' : V^2 = U^3 + \alpha U + \beta'$, and where u_3 and β' are defined in (4) and (5). It can be seen that $h(z) = h_1(z)/h_2(z)$, where $h_1, h_2 \in \mathbb{F}_p[z]$ and

$\deg(h_1) = 288.$[10] If the polynomial h_1 has a root z in $\mathbb{F}_p$, then the associated point T is guaranteed to have order 16 in $E'(\mathbb{F}_p)$. Since X can be chosen uniformly at random from $E(\mathbb{F}_p)$, it is reasonable to make the heuristic assumption that h_1 is a "random" degree-288 polynomial over $\mathbb{F}_p$. The following result ensures that there is a very good chance that h_1 will indeed have a root in $\mathbb{F}_p$. The result is well known (e.g., see Exercise 1 in §4.6.2 of [11]), but we include its proof for the sake of completeness.

Lemma 1. For $p \gg n \geq 10$, the proportion of degree-n polynomials over $\mathbb{F}_p$ that have at least one root in $\mathbb{F}_p$ is approximately $(1 - \frac{1}{e}) \approx 0.632$.

Proof. It suffices to consider monic polynomials over $\mathbb{F}_p$.

The generating function for the number of monic polynomials over $\mathbb{F}_p$ with respect to degree is

$$\Phi(x) = \sum_{i \geq 0} p^i x^i = \frac{1}{1 - px}. \tag{7}$$

Let $L(n, p)$ denote the number of degree-n monic irreducible polynomials over $\mathbb{F}_p$. Since every monic polynomial can be written as a product of monic irreducible polynomials, the generating function $\Phi(x)$ can be written as

$$\Phi(x) = \prod_{i \geq 1} \left(\frac{1}{1 - x^i} \right)^{L(i,p)}. \tag{8}$$

Now, the generating function for monic polynomials with no linear factors (i.e., no roots in $\mathbb{F}_p$) is

$$\tilde{\Phi}(x) = \prod_{i \geq 2} \left(\frac{1}{1 - x^i} \right)^{L(i,p)}. \tag{9}$$

Multiplying (8) by $(1 - x)^{L(1,p)} = (1 - x)^p$ yields

$$\tilde{\Phi}(x) = \frac{(1 - x)^p}{1 - px}. \tag{10}$$

Letting $[\cdot]$ denote the coefficient operator, it follows from (10) that the number $R(n, p)$ of monic polynomials of degree n over $\mathbb{F}_p$ that have at least one root in $\mathbb{F}_p$ is

$$R(n, p) = p^n - [x^n]\tilde{\Phi}(x) = p^n - \sum_{i=0}^{n} \binom{p}{i}(-1)^i p^{n-i}.$$

For $p \gg n \geq 10$, we have

$$R(n, p) \approx p^n \sum_{i=1}^{n} \frac{(-1)^{i-1}}{i!} \approx p^n \sum_{i \geq 1} \frac{(-1)^{i-1}}{i!} = p^n \left(1 - \frac{1}{e} \right).$$

---

[10] More generally, if $t = 2^r$ then $\deg(g) = 3 \cdot 2^{2r-3}$ and $\deg(h_1) = 9 \cdot 2^{2r-3}$.

Example 1. (determination of A, X, T and E') Consider the NIST-recommended elliptic curve [8] defined by the equation $E : V^2 = U^3 - 3U + \beta$ over $\mathbb{F}_p$, where $p = 2^{192} - 2^{64} - 1$ and

$$\beta = 2455155546008943817740293915197451784769108058161191238065.$$

Suppose that we select

$$X = (6020462823756886567582134805875261119166989766368846884818,$$
$$1740503322936220314048575522802194103640234889273866650641)$$

in $E(\mathbb{F}_p)$, and

$$A = (2664590514587922359853612655516270937783866981812798250851, 0).$$

Then the point $T = X + A$ computed using the group law for $E(\mathbb{F}_p)$ is

$$T = (5350077178842604929587851454217201721791103389533004256989,$$
$$4170329249603673452251890924513609385018269372344921771517).$$

T is a point of order 16 on $E' : y^2 = x^3 - 3x + \beta'$, where

$$\beta' = 2271835836669632292423953498680460143165540922751246538627.$$

5   No Ephemeral Public-Key Validation

In this section we consider attacks in the hypothetical situation where static public keys are validated but ephemeral public keys aren't. We don't know of any attacks on HMQV in the case where the underlying group G is a DSA-type group or an elliptic curve group (cf. §4.1 and §4.2). In particular, we don't know how to extend the attacks described in §4.1 and §4.2 to this setting. The difficulty is in part because of the complicated relationship between X and $d = \overline{H}(X, \hat{A})$ whereby d is not determined until X has been fixed.

However, we observe that attacks can be launched on MQV if ephemeral public keys are not validated. Suppose that $G = E(\mathbb{F}_p)$ where $E/\mathbb{F}_p : V^2 = U^3 + \alpha U + \beta$ is an elliptic curve of prime order. The adversary $\hat{C}$, who wishes to impersonate $\hat{A}$ to $\hat{B}$, selects $u_1 \in_R \mathbb{F}_p$ and sets $X = (u_1, z)$ where z is an indeterminate. Since in MQV d depends only on u_1, $\hat{C}$ can then compute $\tilde{A} = dA$, where A is $\hat{A}$'s (valid) static public key. Using the method of §4.2, $\hat{C}$ can use the t-th division polynomial (for some small t) to determine $z, \beta' \in \mathbb{F}_p$ so that $T = X + \tilde{A}$ has order t on $E' : V^2 = U^3 + \alpha U + \beta'$. The adversary sends X to $\hat{B}$ who computes the session key $K = H(T^{sB}, \hat{A}, \hat{B})$. Now $\hat{C}$ can guess the session key with non-negligible success probability $\frac{1}{t}$. Alternatively, if $\hat{C}$ can learn $\hat{B}$ ephemeral's private keys y, then $\hat{C}$ can determine $\hat{B}$'s static private key b as in §2.4.

6   Partial Validation

It may be possible to circumvent the attacks described in the preceding sections without performing (full) public-key validation on static and ephemeral public keys. For example, consider the version of HMQV that has recently been proposed for standardization by the IEEE 1363 standards group [14]. This proposal specifies HMQV in the concrete setting of a DSA-type group G, i.e., G is the order-q subgroup of the multiplicative group $\mathbb{F}_p^*$ of a prime field. The only checks required on ephemeral and static public keys is that they be integers in the interval $[2, p-1]$. In [14] it is claimed that this instantiation of HMQV is provably secure (under the assumptions that CDH in G is intractable, and that the employed hash functions are random functions) as long as ephemeral private keys are never leaked. Moreover, in order to resist attacks that may be mounted in the face of ephemeral private key leakage, the recipient of an ephemeral key X and static key A only needs to verify that $Z^q = 1$ and $Z \neq 1$ where $Z = XA^d$. Such a check is more efficient that separately verifying that $A^q = 1$ and $X^q = 1$. Again, [14] claims that this version of HMQV is provably secure even if the adversary is able to learn some ephemeral private keys.

7   Almost Validation

A public key X is said to have been *almost validated* if it has been verified that $X \in G$ but not necessarily that $X \neq 1$. Protocol descriptions sometimes inadvertently omit the condition $X \neq 1$; see for example the 'G-tests' in [13]. Performing almost validation instead of full validation of public keys may lead to new vulnerabilities. This section gives an example of this likelihood.

In the one-pass HMQV protocol [13], only the initiator contributes an ephemeral public key. The initiator $\hat{A}$ sends $(\hat{A}, \hat{B}, X)$ to $\hat{B}$ and computes the session key $K = H(B^{s_A})$ where $s_A = x + da \mod q$ and $d = \overline{H}(X, \hat{A}, \hat{B})$. The receiver $\hat{B}$ verifies that $X \neq 0$ and computes $K = H((XA^d)^b)$.

In [19] it was shown that the one-pass HMQV protocol succumbs to a Kaliski-style unknown-key share attack [10] even if public keys are (fully) validated. The attack is 'on-line' in the sense that the adversary needs to have her static public key certified during the attack. We next present an 'off-line' Kaliski-style attack on the one-pass HMQV protocol which succeeds if ephemeral public are (fully) validated but static public keys are only almost validated.

The adversary $\hat{C}$ registers in advance the static public key $C = 1$ with the certification authority. Now, when $\hat{A}$ sends $(\hat{A}, \hat{B}, X)$, $\hat{C}$ replaces this message with $(\hat{A}, \hat{C}, T)$ where $T = XA^d$ and $d = \overline{H}(X, \hat{A}, \hat{B})$. Note that T is valid, whereas C is only partially valid. The recipient $\hat{B}$ computes $d' = \overline{H}(T, \hat{C}, \hat{B})$ and

$$K = H((TC^{d'})^b) = H(T^b) = H((XA^d)^b).$$

Thus $\hat{A}$ and $\hat{B}$ have computed the same session key, but $\hat{B}$ mistakenly believes that the key is shared with $\hat{C}$.

8   Concluding Remarks

The attacks on HMQV presented in §2.4, §3.1 and §4 are also effective on MQV if validation of static or ephemeral public keys is omitted. The attacks are summarized in Table 1. While these attacks are not necessarily practical and may

Table 1. Attacks on HMQV (and MQV without validation) described in this paper. † The attack of §5 applies to MQV only.

Static public keys validated?	Ephemeral public keys validated?	Ephemeral private keys secure?	Attacks
$\sqrt{}$	$\sqrt{}$	$\sqrt{}$	No attack known
$\sqrt{}$	$\sqrt{}$	$\times$	No attack known
$\times$	$\sqrt{}$	$\sqrt{}$	No attack known
$\sqrt{}$	$\times$	$\sqrt{}$	§3.1, §5†
$\sqrt{}$	$\times$	$\times$	§3.1, §5†
$\times$	$\times$	$\sqrt{}$	§3.1, §5†
$\times$	$\sqrt{}$	$\times$	§4.1, §4.2
$\times$	$\times$	$\times$	§2.4, §3.1, §4.1, §4.2, §5†

not be a threat in real-world settings, they nonetheless illustrate the importance of performing some form of validation for static and ephemeral public keys in Diffie-Hellman key agreement protocols. Furthermore, the attacks highlight the danger of relying on security proofs for discrete-logarithm protocols where a concrete representation for the underlying group is not specified. In particular, since public keys in HMQV are not necessarily valid, the security of HMQV depends on several aspects of the representation for the underlying group G including the manner in which the group operation is performed, and the particular algorithm chosen for computing $(XA^d)^{s_B}$ and $(YB^e)^{s_A}$. For other examples of the pitfalls when relying on security proofs where a concrete representation of the underlying group is not specified, see [22] and [26].

Acknowledgements

We would like to thank Daniel Panario for his help with the proof of Lemma 1. We also thank Darrel Hankerson for his comments on earlier drafts of this paper.

References

1. ANSI X9.42, *Public Key Cryptography for the Financial Services Industry: Agreement of Symmetric Keys Using Discrete Logarithm Cryptography*, American National Standards Institute, 2003.
2. ANSI X9.63, *Public Key Cryptography for the Financial Services Industry: Key Agreement and Key Transport Using Elliptic Curve Cryptography*, American National Standards Institute, 2001.

3. A. Antipa, D. Brown, A. Menezes, R. Struik and S. Vanstone, "Validation of elliptic curve public keys", *Public Key Cryptography – PKC 2003*, Lecture Notes in Computer Science, 2567 (2003), 211-223.

4. E. Bangerter, J. Camenisch and U. Maurer, "Efficient proofs of knowledge of discrete logarithms and representations in groups with hidden order", *Public Key Cryptography – PKC 2005*, Lecture Notes in Computer Science, 3386 (2005), 154-171.

5. I. Biehl, B. Meyer and V. Müller, "Differential fault analysis on elliptic curve cryptosystems", *Advances in Cryptology – CRYPTO 2000*, Lecture Notes in Computer Science, 1880 (2000), 131-146.

6. R. Canetti and H. Krawczyk, "Analysis of key-exchange protocols and their use for building secure channels", *Advances in Cryptology – EUROCRYPT 2001*, Lecture Notes in Computer Science, 2045 (2001), 453-474. Full version available at http://eprint.iacr.org/2001/040/.

7. L. Chen, Z. Cheng and N. Smart, "Identity-based key agreement protocols from pairings", Cryptology ePrint Archive: Report 2006/199. Available at http://eprint.iacr.org/2006/199.

8. FIPS 186-2, *Digital Signature Standard (DSS)*, Federal Information Processing Standards Publication 186-2, National Institute of Standards and Technology, 2000.

9. IEEE Std 1363-2000, *Standard Specifications for Public-Key Cryptography*, 2000.

10. B. Kaliski, "An unknown key-share attack on the MQV key agreement protocol", *ACM Transactions on Information and System Security*, 4 (2001), 275-288.

11. D. Knuth, *Seminumerical Algorithms*, vol. 2 of *Art of Computer Programming*, 3rd ed., Addison-Wesley, 1997.

12. H. Krawczyk, "HMQV: A high-performance secure Diffie-Hellman protocol", *Advances in Cryptology – CRYPTO 2005*, Lecture Notes in Computer Science, 3621 (2005), 546-566.

13. H. Krawczyk, "HMQV: A high-performance secure Diffie-Hellman protocol", Full version of [12], available at http://eprint.iacr.org/2005/176/.

14. H. Krawczyk, "HMQV in IEEE P1363", submission to the IEEE P1363 working group, July 7 2006. Available at http://grouper.ieee.org/groups/1363/ P1363-Reaffirm/submissions/krawczyk-hmqv-spec.pdf.

15. S. Kunz-Jacques, G. Martinet, G. Poupard and J. Stern, "Cryptanalysis of an efficient proof of knowledge of discrete logarithm", *Public Key Cryptography – PKC 2006*, Lecture Notes in Computer Science, 3958 (2006), 27-43.

16. L. Law, A. Menezes, M. Qu, J. Solinas and S. Vanstone, "An efficient protocol for authenticated key agreement", *Designs, Codes and Cryptography*, 28 (2003), 119-134.

17. P. Leadbitter and N. Smart, "Analysis of the insecurity of ECMQV with partially known nonces", *Information Security – ISC 2003*, Lecture Notes in Computer Science, 2851 (2003), 240-251.

18. C. Lim and P. Lee, "A key recovery attack on discrete log-based schemes using a prime order subgroup", *Advances in Cryptology – CRYPTO '97*, Lecture Notes in Computer Science, 1294 (1997), 249-263.

19. A. Menezes, "Another look at HMQV", *Journal of Mathematical Cryptology*, to appear. Available at http://eprint.iacr.org/2005/205/.

20. A. Menezes, P. van Oorschot and S. Vanstone, *Handbook of Applied Cryptography*, CRC Press, 1996.

21. A. Menezes and Y.-H. Wu, "The discrete logarithm problem in $GL(n,q)$, *Ars Combinatoria*, 47 (1998), 23-32.

22. D. Naccache, N. Smart, and J. Stern, "Projective coordinates leak", *Advances in Cryptology – EUROCRYPT 2004*, Lecture Notes in Computer Science, 3027 (2004), 257-267.

23. D. Pointcheval and J. Stern, "Security arguments for digital signatures and blind signatures", *Journal of Cryptology*, 13 (2000), 361-396.

24. J. Pollard, "Monte Carlo methods for index computation mod p", *Mathematics of Computation*, 32 (1978), 918-924.

25. R. Schoof, "Elliptic curves over finite fields and the computation of square roots mod p", *Mathematics of Computation*, 44 (1985), 483-494.

26. N. Smart, "The exact security of ECIES in the generic group model", *Cryptography and Coding*, Lecture Notes in Computer Science, 2260 (2001), 73-84.

27. SP 800-56A *Special Publication 800-56A, Recommendation for Pair-Wise Key Establishment Schemes Using Discrete Logarithm Cryptography*, National Institute of Standards and Technology, March 2006.

Another Look at "Provable Security". II

Neal Koblitz[1] and Alfred Menezes[2]

[1] Department of Mathematics, University of Washington
koblitz@math.washington.edu
[2] Department of Combinatorics & Optimization, University of Waterloo
ajmeneze@uwaterloo.ca

Abstract. We discuss the question of how to interpret reduction arguments in cryptography. We give some examples to show the subtlety and difficulty of this question.

1   Introduction

Suppose that one wants to have confidence in the security of a certain cryptographic protocol. In the "provable security" paradigm, the ideal situation is that one has a tight reduction (see §4 for a definition and discussion of tightness) from a mathematical problem that is widely believed to be intractable to a successful attack (of a prescribed type) on the protocol. This means that an adversary who can attack the system must also be able to solve the (supposedly intractable) problem in essentially the same amount of time with essentially the same probability of success. Often, however, the best that researchers have been able to achieve falls short of this ideal. Sometimes reductionist security arguments have been found for modified versions of the protocol, but not for the actual protocol that is used in practice; or for a modified version of the type of attack, but not for the security definition that people really want; or based on a somewhat contrived and unnatural modified version of the mathematical problem that is believed to be hard, but not based on the actual problem that has been extensively studied. In other cases, an asymptotic result is known that cannot be applied to specific parameters without further analysis. In still other cases, one has a reduction, but one can show that there cannot be (or is unlikely to be) a tight reduction.

In this paper we give examples that show the subtle questions that arise when interpreting reduction arguments in cryptography.

2   Equivalence But No Reductionist Proof

In [13], Boneh and Venkatesan showed that an efficient reduction from factoring to the RSA problem (the problem of inverting the function $y = x^e \bmod N$) is unlikely to exist. More precisely, they proved that for small encryption exponent e the existence of an efficient "algebraic" reduction would imply that factoring is easy.

R. Barua and T. Lange (Eds.): INDOCRYPT 2006, LNCS 4329, pp. 148–175, 2006.

The paper [13] appeared at a time of intense rivalry between RSA and elliptic curve cryptography (ECC). As enthusiastic advocates of the latter, we were personally delighted to see the Boneh–Venkatesan result, and we welcomed their interpretation of it — that, in the words of their title, "breaking RSA may not be equivalent to factoring" — as another nail in the coffin of RSA.

However, to be honest, another interpretation is at least as plausible. Both factoring and the RSA problem have been studied intensively for many years. In the general case no one has any idea how to solve the RSA problem without factoring the modulus. Just as our experience leads us to believe that factoring (and certain other problems, such as the elliptic curve discrete logarithm problem) are hard, so also we have good reason to believe that, in practice, the RSA problem *is* equivalent to factoring. Thus, an alternative interpretation of the Boneh–Venkatesan result is that it shows the limited value of reduction arguments, and an alternative title of the paper [13] would have been "Absence of a reduction between two problems may not indicate inequivalence."

Which interpretation one prefers is a matter of opinion, and that opinion may be influenced, as in our own case, by one's biases in favor of or against RSA.

3   Results That Point in Opposite Directions

3.1   Reverse Boneh–Venkatesan

A recent result [16] by D. Brown can be seen as giving support to the alternative interpretation of Boneh–Venkaesan that we described at the end of §2. For small encryption exponents e,[1] Brown proves that if there is an efficient program that, given the RSA modulus N, constructs a straight-line program that efficiently solves the RSA problem,[2] then the program can also be used to efficiently factor N. This suggests that for small e the RSA problem may very well be equivalent to factoring. If one believes this interpretation, then one might conclude that small e are more secure than large e. In contrast, the result of Boneh–Venkatesan could be viewed as suggesting that large values of e are more secure than small ones.

As Brown points out in §5 of [16], his result does not actually contradict Boneh–Venkatesan. His reduction of factoring to a straight-line program for finding e-th roots does not satisfy the conditions of the reductions treated in [13]. His use of the e-th root extractor cannot be modeled by an RSA-oracle, as required in [13], because he applies the straight-line program to ring extensions of $\mathbb{Z}/N\mathbb{Z}$.[3]

Brown's choice of title is a helpful one: "Breaking RSA may be as difficult as factoring." All one has to do is put it together in a disjunction with the title of [13], and one has a statement that cannot lead one astray, and accurately summarizes what is known on the subject.

---

[1] Brown's result actually applies if e just has a small prime factor.

[2] This essentially means that it constructs a polynomial that inverts the encryption function.

[3] For example, when $e = 3$ the polynomial that inverts cube roots is applied to the ring $\mathbb{Z}/N\mathbb{Z}[X]/(X^2 - u)$, where the Jacobi symbol $\left(\frac{u}{N}\right) = -1$.

3.2   Random Padding Before or After Hashing?

When comparing ElGamal-like signature schemes, one finds that some, such as Schnorr signatures [35], append a random string to the message before evaluating the hash function; and some, such as the Digital Signature Algorithm (DSA) and the Elliptic Curve Digital Signature Algorithm (ECDSA), apply the hash function before the random padding. Is it more secure to do the padding before or after hashing? What do the available "provable security" results tell us about this question?

As we discussed in §5.2 of [27], the proof that forgery of Schnorr signatures is equivalent to solving the discrete log problem (see the sketch in §5.1 of [27] and §8.3 below, and the detailed proof in [33,34]) relies in an essential way on the fact that an attacker must choose the random r before making his hash query. For this reason, the proof does not carry over to DSA, where only the message m and not r is hashed. In §5.2 of [27] we commented that

> ...replacing $H(m, r)$ by $H(m)$ potentially gives more power to a forger, who has control over the choice of k (which determines r) but no control over the (essentially random) hash value. If H depends on r as well as m, the forger's choice of k must come before the determination of the hash value, so the forger doesn't "get the last word."

That was our attempt to give an intuitive explanation of the circumstance that in the random oracle model Schnorr signatures, unlike the closely related DSA signatures, have been tied to the discrete logarithm problem (DLP) through a reduction argument. One could conclude from our comment that it's more secure to do the padding before hashing.

However, we were very much at fault in misleading the reader in this way. In fact, there is another provable security result, due to D. Brown [14,15], that points in the opposite direction. It says: *If the hash function and pseudorandom bit generator satisfy certain reasonable assumptions, then ECDSA is secure against chosen-message attack by a universal forger*[4] *provided that the "adaptive semi-logarithm problem" in the elliptic curve group is hard.*[5] Brown comments in [15] that his security reduction would not work for a modification of ECDSA in which r as well as the message m is hashed. Brown does not claim that the modified version is therefore less secure than the original version of ECDSA with only the message hashed. However, in an informal communication [17] he explained how someone might make such a claim: namely, the inclusion of a random r along with m in the input could be viewed as "giving an attacker extra play

---

[4] A forger is *universal* (or *selective* in Brown's terminology) if it can forge an arbitrary message that it is given.

[5] A semi-logarithm of a point Q with respect to a basepoint P of prime order p is a pair (t, u) of integers mod p such that $t = f(u^{-1}(P + tQ))$, where the "conversion function" f is the map from points to integers mod p that is used in ECDSA. The adaptive semi-logarithm problem is the problem of finding a semi-logarithm of Q to the base P given an oracle that can find a semi-logarithm of Q to any base of the form eP with $e \neq 1$.

with the hash function," and this could lead to a breach. (But note that both the results in [33,34] and in [14,15] assume that the hash function is strong.)

Once again we have provable security results that suggest opposite answers to a simple down-to-earth question. Is it better to put in the random padding before or after evaluating the hash function? As in the case of the question in §3.1, both answers "before" and "after" can be supported by reduction arguments.

In §8 we shall discuss another question — whether or not forgery of Schnorr-type signatures is equivalent to the DLP — for which different provable security results give evidence for opposite answers.

4   Non-tightness in Reductions

We first give an informal definition of tightness of a reduction. Suppose that we have an algorithm for solving problem A that takes time at most T and is successful for a proportion at least ϵ of the instances of A, where T and ϵ are functions of the input length. A reduction from a problem B to A is an algorithm that calls upon the algorithm for A a certain number of times and solves B in time T' for at least a proportion ϵ' of the instances of B. This reduction is said to be *tight* if $T' \approx T$ and $\epsilon' \approx \epsilon$. Roughly speaking, it is *non-tight* if $T' \gg T$ or if $\epsilon' \ll \epsilon$.

Suppose that researchers have been able to obtain a highly non-tight reduction from a hard mathematical problem to breaking a protocol. There are various common ways to respond to this situation:

1. Even a non-tight reduction is better than nothing at all. One should regard the cup as half-full rather than half-empty, derive some reassurance from what one has, and try not to think too much about what one wishes one had.[6]
2. Even though the reduction is not tight, it is reasonable to expect that in the future a tighter reduction will be found.
3. Perhaps a tight reduction cannot be found for the protocol in question, but a small modification of the protocol can be made in such a way as to permit the construction of a tight reduction — and we should regard this reduction as a type of assurance about the original protocol.
4. A tight reduction perhaps can be obtained by relaxing the underlying hard problem (for example, replacing the computational Diffie–Hellman problem by the decision Diffie–Hellman problem).
5. Maybe the notion of security is too strict, and one should relax it a little so as to make possible a tight reduction.

---

[6] We are reminded of the words of the popular song
   If you can't be with the one you love,
   Love the one you're with,
   (*Stephen Stills*, 1970). The version for cryptographers is:
   If you can't prove what you'd love to prove,
   Hype whatever you prove.

6. Perhaps the protocol is secure in practice, even though a tight reduction may simply not exist.

7. Perhaps the protocol is in fact insecure, but an attack has not yet been discovered.

These seven points of view are not mutually exclusive. In fact, protocol developers usually adopt some combination of the first six interpretations — but generally not the seventh.

4.1   Insecure But Provably Secure: An Example

We now give an example that is admittedly somewhat artificial. Let us step into a time machine and go back about 25 years to a time when naive index-calculus was pretty much the best factoring algorithm. Let us also suppose that 2^{2a} operations are feasible, but $2^{(2\sqrt{2})a}$ operations are not.

Let N be a c-bit RSA modulus, and let r be an a-bit integer. Let $F = \{p_1, \ldots, p_r\}$ be a factor base consisting of the first r primes. Let 2^b be the expected time needed before a randomly selected $x \bmod N$ has the property that $x^2 \bmod N$ is p_r-smooth (this means that it has no prime factors greater than p_r). The usual estimate is that $2^b \approx u^u$, where $u = c/a$. (Actually, it's more like $u = c/(a + \log(a \ln 2))$, where log denotes $\log_2$, but let's ignore second-order terms.)

If x has the property that $x^2 \bmod N$ is p_r-smooth, then by its "exponent-vector" we mean the vector in $\mathbb{F}_2^r$ whose components ϵ_i are the exponents of p_i in the squarefree part of $x^2 \bmod N$.

The basic (naive) index-calculus algorithm involves generating roughly r such x values and then solving an $r \times r$-matrix over $\mathbb{F}_2$. The first part takes roughly $r2^b \approx 2^{a+b}$ operations, and the second part takes roughly 2^{2a} operations. So one usually chooses $b \approx a$. However, in our protocol, in order to be able to give a "proof" of security we'll optimize slightly differently, taking $b \approx 2a$.

Note that for fixed c, the value of a chosen with $b \approx 2a$ is different from the optimal value a' that one would choose to factor N. In the former case one sets $2^{2a} \approx u^u$ (where $u = c/a$) — that is, $2a \approx \frac{c}{a} \log u$ — and in the latter case one sets $a' \approx \frac{c}{a'} \log u'$ (where $u' = c/a'$). Since u' is of the same order of magnitude as u, by dividing these two equations we get approximately $a' \approx \sqrt{2}a$. This leads to the estimate $2^{(2\sqrt{2})a}$ for the number of operations needed to factor N.

We now describe the protocol. Alice wants to prove her identity to Bob, i.e., prove that she knows the factors of her public modulus N. Bob sends her a challenge that consists of s linearly independent vectors in $\mathbb{F}_2^r$, where $0 \leq s \leq r - 1$. Alice must respond with an x such that $x^2 \bmod N$ is p_r-smooth and such that its exponent-vector is not in the subspace S spanned by Bob's challenge vectors. (The idea is to prevent an imposter from giving a correct response by combining earlier responses of Alice; thus, in practice Bob would be sure to include the exponent-vectors of Alice's earlier responses among his challenge vectors.) Alice can do this quickly, because it is easy to find square roots modulo N if one knows the factorization of N.

We now reduce factoring to impersonating Alice. Let IO be the impersonator-oracle. To factor N, we make r calls to IO (where each time our challenge vectors consist of the exponent-vectors of all the earlier responses of IO) to get a set of relations whose exponent-vectors span $\mathbb{F}_2^r$. After that we merely have to find k more randomly generated x with p_r-smooth x^2 mod N in order to have probability $1 - 2^{-k}$ of factoring N. Finding these x's takes time about $k2^b$. Since we have to solve a matrix each time, the time is really $k(2^b + 2^{2a})$. If a call to IO on average takes time T, then the total time to factor N is $T' \approx k(2^b + 2^{2a}) + rT \approx k2^{2a+1} + 2^a T$ since $b = 2a$ and $r \approx 2^a$. We are assuming that factoring N requires $2^{(2\sqrt{2})a}$ operations, and so we obtain the nontrivial lower bound $T \geq 2^{(2\sqrt{2}-1)a}$. Whenever one is able to prove a lower bound for an adversary's running time that, although far short of what one ideally would want, is highly nontrivial and comes close to the limits of practical feasibility, such a result can be viewed as reassuring (see also Remark 2 below).

However, the protocol is insecure, because it can be broken in time roughly $2^b = 2^{2a}$.

This example is unrealistic not only because we're supposing that naive index-calculus is the best factoring algorithm, but also because it should have been obvious from the beginning that the protocol is insecure. We thus state as an open problem:

Problem. Find an example of a natural and realistic protocol that has a plausible (non-tight) reductionist proof of security, and is also insecure when used with commonly accepted parameter sizes.

Remark 1. Either success or failure in solving this problem would be of interest. If someone finds a (non-tightly) provably secure but insecure protocol, then the importance of the tightness question in security reductions will be clearer than ever. On the other hand, if no such example is found after much effort, then practitioners might feel justified in doubting the need for tightness in reductions.

Remark 2. It should be noted that something like this has already been done in the context of symmetric–key message authentication codes (MAC's). In [18] Cary and Venkatesan presented a MAC scheme for which they had a security proof (it was not actually a reductionist proof). Their scheme depended on a parameter l, and for the practical value $l = 32$ their proof showed that a collision cannot be found without at least 2^{27} MAC queries. Even though this figure falls far short of what one ideally would want — namely, 64 bits of security — it could be viewed as providing some assurance that the scheme does in fact have the desired security level. However, in [8] Blackburn and Paterson found an attack that could find a collision using $2^{48.5}$ MAC queries and a forgery using 2^{55} queries. This example shows that the exact guarantees implied by a proof have to be taken seriously, or else one might end up with a cryptosystem that is provably secure and also insecure.

4.2   Coron's Result for RSA Signatures

We first recall the basic RSA signature scheme with full-domain hash function. Suppose that a user Alice with public key (N, e) and secret exponent d wants to sign a message m. She applies a hash function $H(m)$ which takes values in the interval $0 \leq H(m) < N$, and then computes her signature $s = H(m)^d$ mod N.

When Bob receives the message m and the signature s, he verifies the signature by computing $H(m)$ and then s^e mod N. If these values are equal, he is satisfied that Alice truly sent the message (because presumably only Alice knows the exponent d that inverts the exponentiation $s \mapsto s^e$) and that the message has not been tampered with (because any other message would presumably have a different hash value).

We now describe a classic reductionist security argument for this signature scheme [6]:

Reductionist security claim. If the problem of inverting $x \mapsto x^e$ mod N is intractable, then the RSA signature with full-domain hash function is secure in the random oracle model from chosen-message attack by an existential forger.

Argument. Suppose that we are given an arbitrary integer y, $0 \leq y < N$, and asked to find x such that $y = x^e$ mod N. The claim follows if we show how we could find x (with high probability) if we had a forger that can mount chosen-message attacks.

So suppose that we have such a forger. We give it Alice's public key (N, e) and wait for its queries. In all cases but one, we respond to the hash query for a message m_i by randomly selecting $x_i \in \{0, 1, \ldots, N - 1\}$ and setting the hash value h_i equal to x_i^e mod N. For just one value m_{i_0} we respond to the hash query by setting $h_{i_0} = y$ (recall that y is the integer whose inverse under the map $x \mapsto x^e$ mod N we are required to find). We choose i_0 at random and hope that $m = m_{i_0}$ happens to be the message whose signature will be forged by our existential forger. Any time the forger makes a signature query for a message m_i with $i \neq i_0$, we send x_i as its signature. Notice that this will satisfy the forger, since $x_i^e \equiv h_i \pmod{N}$. If the forger ends up outputting a valid signature s_{i_0} for m_{i_0}, that means that we have a solution $x = s_{i_0}$ to our original equation $y = x^e$ mod N with unknown x. If we guessed wrong and m_{i_0} was not the message that the forger ends up signing, then we won't be able to give a valid response to a signature query for m_{i_0}. The forger either will fail or will give us useless output, and we have to start over again. Suppose that q_h is a bound on the number of queries of the hash function. If we go through the procedure k times, the probability that every single time we fail to solve $y = x^e$ mod N for x is at most $(1 - 1/q_h)^k$. For large k, this approaches zero; so with high probability we succeed. This completes the argument.

Notice that the forgery program has to be used roughly $O(q_h)$ times (where q_h is the number of hash queries) in order to find the desired e-th root modulo

N. A result of Coron [19] shows that this can be improved to $O(q_s)$, where q_s denotes a bound on the number of signature queries.[7] (Thus, $q_h = q_s + q'_h$, where q'_h is a bound on the number of hash function queries that are not followed later by a signature query for the same message.)

Moreover, in a later paper [20] Coron essentially proves that his result cannot be improved to give a tight reduction argument; $O(q_s)$ is a lower bound on the number of calls on the forger needed to solve the RSA problem.

From the standpoint of practice (as emphasized, for example, in [5]) this non-tightness is important. What it means is the following. Suppose that you anticipate that a chosen-message attacker can get away with making up to 2^{20} signature queries. You want your system to have 80 bits of security; that is, you want a guarantee that such a forger will require time at least 2^{80}. The results of [19,20] mean that you should use a large enough RSA modulus N so that you're confident that e-th roots modulo N cannot be found in fewer than $2^{100} = 2^{20} \cdot 2^{80}$ operations. Thus, you should use a modulus N of about 1500 bits.

4.3   The Implausible Magic of One Bit

We now look at a construction of Katz and Wang [25], who show that by adding only a single random bit to a message, one can achieve a tight reduction.[8] To sign a message m Alice chooses a random bit b and evaluates the hash function H at m concatenated with b. She then computes $s = (H(m, b))^d \bmod N$; her signature is the pair (s, b). To verify the signature, Bob checks that $s^e = H(m, b) \bmod N$.

Remarkably, Katz and Wang show that the use of a single random bit b is enough to get a tight reduction from the RSA problem to the problem of producing a forgery of a Katz–Wang signature. Namely, suppose that we have a forger in the random oracle model that asks for the signatures of some messages and then produces a valid signature of some other message. Given an arbitrary integer y, the simulator must use the forger to produce x such that $y = x^e \bmod N$. Without loss of generality we may assume that when the forger asks for the hash value $H(m, b)$, it also gets $H(m, b')$ (where b' denotes the complement of b). Now when the forger makes such a query, the simulator selects a random bit c and two random integers t_1 and t_2. If $c = b$, then the simulator responds with $H(m, b) = t_1^e y$ and $H(m, b') = t_2^e$; if $c = b'$, it responds with $H(m, b) = t_2^e$ and $H(m, b') = t_1^e y$. If the forger later asks the simulator to sign the message m, the simulator responds with the corresponding value of t_2. At the end the forger outputs a signature that is either an e-th root of t_2^e or an e-th root of $t_1^e y$ for some t_1 or t_2 that the simulator knows. In the latter case, the simulator has succeeded in its task. Since this happens with probability $1/2$, the simulator is almost certain — with probability $1 - 2^{-k}$ — to find the desired e-th root

---

[7] In the above argument, instead of responding only to the i_0-th hash query with $h_{i_0} = y$, Coron's idea was to respond to a certain optimal number $i_0, i_1, \ldots$ with $h_{i_j} = yz_j^e$ with z_j random.

[8] We shall describe a slightly simplified version of the Katz–Wang scheme. In particular, we are assuming that Alice never signs the same message twice.

after running the forger k times. This gives us a tight reduction from the RSA problem to the forgery problem.

From the standpoint of "practice-oriented provable security" the Katz–Wang modification provides a much better guarantee than did the RSA signature without the added bit. Namely, in order to get 80 bits of security one need only choose N large enough so that finding e-th roots modulo N requires 2^{80} operations — that is, one needs roughly a 1000-bit N. Thus, the appending of a random bit to the message allows us to shave 500 bits off our modulus!

This defies common sense. How could such a "magic bit" have any significant impact on the true security of a cryptosystem, let alone such a dramatic impact? This example shows that whether or not a cryptographic protocol lends itself to a tight security reduction argument is not necessarily related to the true security of the protocol.

Does tightness matter in a reductionist security argument? Perhaps not, if, as in this case, a protocol with a non-tight reduction can be modified in a trivial way to get one that has a tight reduction. On the other hand, the example in §4.1 shows that in some circumstances a non-tight reduction might be worthless. Thus, the question of how to interpret a non-tight reductionist security argument has no easy answer.

One interpretation of Coron's lower bound on tightness is that if the RSA problem has s_1 bits of security and if we suppose that an attacker could make 2^{s_2} signature queries, then RSA signatures with full-domain hash have only $s_1 - s_2$ bits of security. However, such a conclusion seems unwarranted in light of the Katz–Wang construction. Rather, it is reasonable to view Coron's lower bound on tightness as a result that casts doubt not on the security of the basic RSA signature scheme, but rather on the usefulness of reduction arguments as a measure of security of a protocol. This point of view is similar to the alternative interpretation of Boneh–Venkatesan's result that we proposed in §2.

5   Equivalence But No Tight Reduction

Let $\mathcal{P}$ denote a presumably hard problem underlying a cryptographic protocol; that is, solving an instance of $\mathcal{P}$ will recover a user's private key. For example, the RSA version of factorization is the problem $\mathcal{P}$ whose input is a product N of two unknown k-bit primes and whose output is the factorization of N.

Let $\mathcal{P}_m$ denote the problem whose input is an m-tuple of distinct inputs for $\mathcal{P}$ of the same bitlength and whose output is the solution to $\mathcal{P}$ for any one of the inputs. In the cryptographic context, m might be the number of users. In that case, solving $\mathcal{P}_m$ means finding the private key of any one of the users, while solving $\mathcal{P}$ means finding the private key of a specified user. We call the former "existential key recovery" and the latter "universal key recovery." A desirable property of a cryptosystem is that these two problems be equivalent — in other words, that it be no easier to recover the private key of a user of the attacker's choice than to recover the private key of a user that is specified to the attacker.

To see how this issue might arise in practice, let's suppose that in a certain cryptosystem a small proportion — say, 10^{-5} — of the randomly assigned private keys are vulnerable to a certain attack. From the standpoint of an individual user, the system is secure: she is 99.999% sure that her secret is safe. However, from the standpoint of the system administrator, who is answerable to a million users, the system is insecure because an attacker is almost certain (see below) to eventually obtain the private key of one or more of the users, who will then sue the administrator. Thus, a system administrator has to be worried about existential key recovery, whereas an individual user might care only about universal key recovery.

5.1   The RSA Factorization Problem

In the case of RSA, is $\mathcal{P}_m$ equivalent to $\mathcal{P}$? (For now we are asking about algorithms that solve all instances of a problem; soon we shall consider algorithms that solve a non-negligible proportion of all instances.) It is unlikely that there is an efficient reduction from $\mathcal{P}$ to $\mathcal{P}_m$. Such a reduction would imply that the following cannot be true: for every k there are a small number $r_k < m$ of moduli N that are much harder to factor than any other $2k$-bit N. On the other hand, all of our knowledge and experience with factoring algorithms support the belief that, in fact, these two problems are in practice equivalent, and that RSA does enjoy the property that existential and universal private key recovery are equivalent.

When studying the security of a protocol, one usually wants to consider algorithms that solve only a certain non-negligible proportion of the instances.[9] In this case there is an easy reduction from $\mathcal{P}$ to $\mathcal{P}_m$: given an input to $\mathcal{P}$, randomly choose $m - 1$ other inputs to form an input to $\mathcal{P}_m$. One can check that this transforms an algorithm that solves a non-negligible proportion of instances of $\mathcal{P}_m$ to one that solves a non-negligible proportion of instances of $\mathcal{P}$.

However, the proportion of instances solved can be dramatically different. An algorithm $\mathcal{A}$ that solves ϵ of the instances of $\mathcal{P}$, where ϵ is small but not negligible, gives rise to an algorithm $\mathcal{A}_m$ that solves $\nu = 1 - (1 - \epsilon)^m$ of the instances of $\mathcal{P}_m$ (this is the probability that at least one of the m components of the input can be solved by $\mathcal{A}$). For small ϵ and large m, $\nu \approx 1 - e^{-\epsilon m}$. For example, if $\epsilon = 10^{-5}$ and $m = 10^6$, then ν is greater than 99.99%. Thus, from a theoretical point of view there seems to be a significant distance between universal private key recovery $\mathcal{P}$ and existential private key recovery $\mathcal{P}_m$ for many systems such as RSA. In other words, we know of no reductionist argument to show that if RSA is secure from the standpoint of an individual user, then it must also be secure from the standpoint of the system administrator.

---

[9] In this section probabilities are always taken over the set of problem instances (of a given size), and not over sets of possible choices (coin tosses) made in the execution of an algorithm. If for a given problem instance the algorithm succeeds for a non-negligible proportion of sequences of coin tosses, then we suppose that the algorithm is iterated enough times so that it is almost certain to solve the problem instance.

But once again, all of our experience and intuition suggest that there is no real distance between the two versions of the RSA factoring problem. This is because for all of the known subexponential-time factoring algorithms, including the number field sieve, the running time is believed not to be substantially different for (a) a randomly chosen instance, (b) an instance of average difficulty, and (c) a hardest possible instance. No one knows how to prove such a claim; indeed, no one can even give a rigorous proof of the $L_{1/3}$ running time for the number field sieve. And even if the claim could be proved for the current fastest factoring algorithm, we would be very far from proving that there could never be a faster algorithm for which there was a vast difference between average-case and hardest-case running times. This is why there is no hope of proving the tight equivalence of universal and existential private key recovery for RSA.

5.2   A Non-cryptographic Example

Consider the problem $\mathcal{P}$ of finding all the prime factors of an arbitrary integer N. Let us say that N is "k-easy" if it has at most one prime divisor greater than 2^k. If k is small, then $\mathcal{P}$ in that case can be solved efficiently by first using trial division, perhaps in conjunction with the Lenstra elliptic curve factoring algorithm, to pull out the prime factors $< 2^k$, and then applying a primality test to what's left over if it's greater than 1.

It is not hard to see that the proportion ϵ of n-bit integers N that are k-easy is at least k/n. Namely, for $1 \le j < 2^k$ consider N that are of the form pj for primes p. The number of such n-bit integers is asymptotic to

$$\frac{2^{n-1}}{j \ln(2^n/j)} > \frac{2^{n-1}}{\ln 2^n} \frac{1}{j}.$$

Thus, the proportion of n-bit integers that are k-easy is greater than

$$\frac{1}{\ln 2^n} \sum_{1 \le j < 2^k} \frac{1}{j} \approx \frac{\ln 2^k}{\ln 2^n} = \frac{k}{n}.$$

As an example, let's take $n = 2000$, $k = 20$. Then $\epsilon \ge 0.01$. We saw that for $m = 1000$ more than 99.99% of all instances of $\mathcal{P}_m$ can be quickly solved. In contrast, a significant proportion of the instances of $\mathcal{P}$ are outside our reach. Obviously, it is not feasible to factor a 2000-bit RSA modulus. But there is a much larger set of 2000-bit integers that cannot be completely factored with current technology. Namely, let $S_{\ge 1}$ denote the set of integers that have at least one prime factor in the interval $[2^{300}, 2^{500}]$ and at least one prime factor greater than 2^{500}. At present a number in $S_{\ge 1}$ cannot feasibly be factored, even using a combination of the elliptic curve factorization method and the number field sieve; and a heuristic argument, which we now give, shows that at least 25% of all 2000-bit integers N lie in $S_{\ge 1}$.

To see this, let S_k denote the set of integers that have exactly k prime factors in $[2^{300}, 2^{500}]$ and at least one prime factor greater than 2^{500}. Writing a 2000-bit

$N \in S_1$ in the form $N = lm$ with l a prime in $[2^{300}, 2^{500}]$ and $m \in S_0$, we see that the number of such N is equal to

$$\sum_{l \text{ prime in } [2^{300}, 2^{500}]} \# \left(S_0 \cap \left[\frac{1}{l} 2^{1999}, \frac{1}{l} 2^{2000} \right] \right).$$

The probability that an integer in the latter interval satisfies the two conditions defining S_0 is at least equal to

$$\text{Prob(not divisible by any prime } p \in [2^{300}, 2^{500}]) - \text{Prob}(2^{500} - \text{smooth})$$

$$\approx \prod_{p \in [2^{300}, 2^{500}]} \left(1 - \frac{1}{p} \right) - u^{-u},$$

where $u = (2000 - \log_2 l)/500 \geq 3$. The product is equal to $\exp \sum \ln(1 - \frac{1}{p}) \approx \exp \sum (-1/p) \approx \exp(-\ln\ln 2^{500} + \ln\ln 2^{300}) = 0.6$, and so the probability that an integer in $[\frac{1}{l} 2^{1999}, \frac{1}{l} 2^{2000}]$ lies in S_0 is greater than 50%. Thus, the proportion of 2000-bit integers N that lie in $S_{\geq 1} \supset S_1$ is at least

$$\frac{1}{2} \sum_{l \text{ prime in } [2^{300}, 2^{500}]} \frac{1}{l} \approx \frac{1}{2} (\ln\ln 2^{500} - \ln\ln 2^{300}) = \frac{1}{2} \ln(5/3) \approx 0.25,$$

as claimed.

This problem $\mathcal{P}$ does not seem to have any cryptographic significance: it is hard to imagine a protocol whose security is based on the difficulty of completely factoring a randomly chosen integer. Rather, its interest lies in the fact that, despite its apparent resemblance to the RSA factoring problem, it spectacularly fails to have a certain property — tight equivalence of existential and universal solvability — that intuitively seems to be a characteristic of RSA factoring. This example also suggests that it is probably hopeless to try to *prove* that universal and existential private key recovery are tightly equivalent for RSA.

5.3   Use Different Elliptic Curves or the Same One?

Let us look at universal versus existential private key recovery in the case of elliptic curve cryptography (ECC). Suppose that each user chooses an elliptic curve E over a finite field $\mathbb{F}_q$, a subgroup of $E(\mathbb{F}_q)$ whose order is a k-bit prime p, a basepoint P in the subgroup, and a secret key $x \bmod p$; the public key is $Q = xP$. Let $\mathcal{P}$ denote the elliptic curve discrete logarithm problem (ECDLP), that is, the problem of recovering the secret key x from the public information. Let $\mathcal{P}_m$ denote the problem whose input is an m-tuple of ECDLP inputs with distinct orders p of the subgroups and whose output is any one of the m discrete logarithms. Once again, it seems intuitively clear that $\mathcal{P}_m$ is as hard as $\mathcal{P}$, although it is very unlikely that a tight reduction from $\mathcal{P}$ to $\mathcal{P}_m$ could be found.

In contrast, suppose that everyone uses the same elliptic curve group, and only the private/public key pairs (x, Q) differ. In that case ECC *provably* enjoys the property of tight equivalence of existential and universal private key recovery. The reason is that the ECDLP on a fixed group is "self-reducible." That means that, given an instance we want to solve, we can easily create an m-tuple of distinct random instances such that the solution to any one of them gives us the solution to the problem we wanted to solve. Namely, given an input Q, we randomly choose m distinct integers y_i modulo p and set $Q_i = y_i Q$. A $\mathcal{P}_m$-oracle will solve one of the ECDLP instances with input Q_i. Once we know its discrete log x_i, we immediately find $x = y_i^{-1} x_i \bmod p$. This shows that for the ECDLP on a fixed curve the universal private key recovery problem $\mathcal{P}$ reduces (tightly) to the existential private key recovery problem $\mathcal{P}_m$.

Thus, if we want a cryptosystem with the *provable* security property of tight equivalence of existential and universal private key recovery, then we should not only choose ECC in preference to RSA, but also insist that all users work with the same elliptic curve group.

Needless to say, we are not suggesting that this would be a good reason to choose one type of cryptography over another. On the contrary, what this example shows is that it is sometimes foolish to use the existence or absence of a tight reductionist security argument as a guide to determine which version of a cryptosystem is preferable.

Remark 3. We should also recall the problematic history of attempts to construct cryptosystems whose security is based on a problem for which the average cases and the hardest cases are *provably* equivalent.[10] This was finally done by Ajtai and Dwork [2] in 1997. However, the following year Nguyen and Stern [30] found an attack that recovers the secret key in the Ajtai–Dwork system unless parameters are chosen that are too large to be practical (see also [31]).

6   Pseudorandom Bit Generators

A pseudorandom bit generator G is a function — actually, a family of functions parameterized by n and $M \gg n$ — that takes as input a random sequence of n bits (called the "seed") and outputs a sequence of M bits that appear to be random. More precisely, G is said to be *asymptotically secure in the sense of indistinguishability* if there is no polynomial time statistical test that can distinguish (by a non-negligible margin) between its output and random output. An alternative and at first glance weaker notion of security is that of the "next bit" test: that there is no value of j for which there exists a polynomial time algorithm that, given the first $j - 1$ bits, can predict the j-th bit with greater than $\frac{1}{2} + \epsilon$ chance of success (where ϵ is non-negligible as a function of n). A theorem of Yao (see [26], pp. 170-171) shows that these two notions of security

---

[10] Discrete-log-based systems do not have this property because the underlying problem is self-reducible only after the group has been fixed; there is clearly no way to reduce one instance to another when the groups have different orders.

are equivalent. However, that theorem is non-tight in the sense that ϵ-tolerance for the next bit test corresponds only to $(M\epsilon)$-tolerance for indistinguishability.

If one wants to analyze the security of a pseudorandom bit generator more concretely, one has to use a more precise definition than the asymptotic one. Thus, for given values of n and M, G is said to be (T, ϵ)-*secure in the sense of indistinguishability* if there is no algorithm (statistical test) with running time bounded by T such that the probability of a "yes" answer in response to the output of G and the probability of a "yes" answer in response to a truly random sequence of M bits differ in absolute value by at least ϵ. The relation between indistinguishability and the next bit test is that we have to know that our generator is $(T, \epsilon/M)$-secure in the next bit sense in order to conclude that it is (T, ϵ)-secure in the sense of indistinguishability.

6.1   The Blum–Blum–Shub Generator

Let N be an n-bit product of two large primes that are each $\equiv 3 \pmod 4$ (such an N is called a "Blum integer"), and choose a (small) integer j. The Blum–Blum–Shub (BBS) pseudorandom bit generator G takes a random $x \bmod N$ and produces $M = jk$ bits as follows. Let $x_0 = x$, and for $i = 1, \ldots, k$ let[11]

$$x_i = \min\{x_{i-1}^2 \bmod N,\ N - (x_{i-1}^2 \bmod N)\}.$$

Then the output of G consists of the j least significant bits of x_i, $i = 1, \ldots, k$.

Obviously, the larger j is, the faster G generates M bits. However, the possibility of distinguishing the generated sequence from a truly random sequence becomes greater as j grows. In [41] and [3] it was proved that $j = O(\log \log N)$ bits can be securely extracted in each iteration, under the assumption that factoring is intractable.

This asymptotic result was used to justify recommended values of j. For example, in 1994 the Internet Engineering Task Force [21] made the following recommendation (in this and the following quote the modulus is denoted by n rather than N):

> Currently the generator which has the strongest public proof of strength is called the Blum Blum Shub generator... If you use no more than the $\log_2(\log_2(s_i))$ low order bits, then predicting any additional bits from a sequence generated in this manner is provable [sic] as hard as factoring n.

This recommendation has been repeated more recently, for example, in the book by Young and Yung ([43], p. 68):

> The Blum–Blum–Shub PRBG is also regarded as being secure when the $\log_2(\log_2(n))$ least significant bits...are used (instead of just the least significant bit). So, when n is a 768-bit composite, the 9 least significant bits can be used in the pseudorandom bit stream.

---

[11] The original generator described in [9] has $j = 1$ and $x_i = x_{i-1}^2 \bmod N$.

Let us compare this recommendation with the best security bounds that are known. In what follows we set

$$L(n) \approx 2.8 \cdot 10^{-3} \exp\left(1.9229(n \ln 2)^{1/3}(\ln(n \ln 2))^{2/3}\right),$$

which is the heuristic expected running time for the number field sieve to factor a random n-bit Blum integer (here the constant $2.8 \cdot 10^{-3}$, which is taken from [40], was obtained from the reported running time for factoring a 512-bit integer), and we assume that no algorithm can factor such an integer in expected time less than $L(n)$.

For the $j = 1$ version of Blum–Blum–Shub the best concrete security result (for large n) is due to Fischlin and Schnorr [22], who showed that the BBS generator is (T, ϵ)-secure in the sense of indistinguishability if

$$T \leq \frac{L(n)(\epsilon/M)^2}{6n \log n} - \frac{2^7 n(\epsilon/M)^{-2} \log(8n(\epsilon/M)^{-1})}{\log n}, \tag{1}$$

where log denotes $\log_2$ here and in the sequel.

For $j > 1$ the Fischlin–Schnorr inequality (1) was generalized by Sidorenko and Schoenmakers [40], who showed that the BBS generator is (T, ϵ)-secure if

$$T \leq \frac{L(n)}{36n(\log n)\delta^{-2}} - 2^{2j+9}n\delta^{-4}, \tag{2}$$

where $\delta = (2^j - 1)^{-1}(\epsilon/M)$. For large n this is an improvement over the inequality

$$T \leq \frac{L(n)(\epsilon/M)^8}{2^{4j+27}n^3}, \tag{3}$$

which is what follows from the security proof in [3].

Returning to the parameters recommended in [21] and [43], we take $n = 768$ and $j = 9$. Suppose we further take $M = 10^7$ and $\epsilon = 0.01$. According to inequality (2), the BBS generator is secure against an adversary whose time is bounded by -2^{192}. (Yes, that's a negative sign!) In this case we get a "better" result from inequality (3), which bounds the adversary's time by 2^{-264}. (Yes, that's a negative exponent!) These less-than-reassuring security guarantees are not improved much by changing M and ϵ. For example, if $M = 2^{15}$ and $\epsilon = 0.5$, we get $T \leq -2^{136}$ and $T \leq 2^{-134}$ from (2) and (3), respectively. Thus, depending on whether we use (2) or (3), the adversary's running time is bounded either by a negative number or by 10^{-40} clock cycles!

Nor does the recommendation in [21] and [43] fare well for larger values of n. In Table 1, the first column lists some values of n; the second column gives $L(n)$ to the nearest power of 2 (this is the bound on the adversary's running time that would result from a tight reduction); the third column gives the corresponding right-hand side of inequality (2); and the fourth column gives the right-hand side of (3). Here we are taking $j = \lfloor \log n \rfloor$, $M = 10^7$, and $\epsilon = 0.01$.

Table 1. The BBS generator: bounds on the adversary's running time with $j = \lfloor \log n \rfloor$

n	$L(n)$	Bound from (2)	Bound from (3)
1024	2^{78}	-2^{199}	2^{-258}
2048	2^{108}	-2^{206}	2^{-235}
3072	2^{130}	-2^{206}	2^{-215}
7680	2^{195}	-2^{213}	2^{-158}
15360	2^{261}	-2^{220}	2^{-99}

Thus, the asymptotic result in [3,41], which seemed to guarantee that we could securely extract $j = \lfloor \log n \rfloor$ bits in each iteration, does not seem to deliver in practice what it promises in theory.

Suppose that we retreat from the idea of getting $j = \lfloor \log n \rfloor$ bits from each iteration, and instead use the BBS generator to give just $j = 1$ bit per iteration. Now the security guarantees given by the inequalities (1) and (3) are better, but not by as much as one might hope. Table 2 gives the corresponding right-hand sides of (1) (in the third column) and (3) (in the fourth column) for $j = 1$, $M = 10^7$, and $\epsilon = 0.01$.

Table 2. The BBS generator: bounds on the adversary's running time with $j = 1$

n	$L(n)$	Bound from (1)	Bound from (3)
1024	2^{78}	-2^{79}	2^{-222}
2048	2^{108}	-2^{80}	2^{-194}
3072	2^{130}	-2^{80}	2^{-175}
7680	2^{195}	2^{115}	2^{-114}
15360	2^{261}	2^{181}	2^{-51}

The cross-over point at which the Fischlin–Schorr inequality starts to give a meaningful security guarantee is about $n = 5000$ (for which the right-hand side of (1) is roughly 2^{84}). Unfortunately, it is not very efficient to have to perform a 5000-bit modular squaring for each bit of the pseudorandom sequence.

Remark 4. The recommended value $j = \log(\log N)$ in [21] and [43] was obtained by taking the asymptotic result $j = O(\log(\log N))$ and setting the implied constant C in the big-O equal to 1. The choice $C = 1$ is arbitrary. In many asymptotic results in number theory the implicit constant is much greater, so with equal justification one might decide to take $C = 100$. It is amusing to note that if one did that with 1000-bit N, one would get a completely insecure BBS generator. Since $j = 100 \log(\log N) = 1000$, one would be using all the bits of x_i. From the output an attacker could easily determine N (by setting $N_1 = x_2 \pm x_1^2$, $N_i = \gcd(N_{i-1}, x_{i+1} \pm x_i^2)$, so that $N_i = N$ for $i \geq i_0$ for quite small i_0), after which the sequence would be deterministic for the attacker.

6.2   The Gennaro Generator

Let p be an n-bit prime of the form $2q + 1$ with q prime, and let c be an integer such that $c \gg \log n$. Let g be a generating element of $\mathbb{F}_p^*$. The Gennaro pseudorandom bit generator G takes a random $x \bmod p - 1$ and produces $M = (n - c - 1)k$ bits as follows (see [23]). Let $x \mapsto \tilde{x}$ be the function on n-bit integers $x = \sum_{l=0}^{n-1} s_l 2^l$ given by $\tilde{x} = s_0 + \sum_{l=n-c}^{n-1} s_l 2^l$. Let $x_0 = x$, and for $i = 1, \ldots, k$ let $x_i = g^{\tilde{x}_{i-1}} \bmod p$. Then the output of G consists of the 2nd through $(n - c)$-th bits of x_i, $i = 1, \ldots, k$ (these are the bits that are ignored in $\tilde{x}_i$).

In comparison with the BBS generator, each iteration of the exponentiation $x_i = g^{\tilde{x}_{i-1}} \bmod p$ takes longer than modular squaring. However, one gets many more bits each time. For example, with the parameters $n = 1024$ and $c = 160$ that are recommended in [24] each iteration gives 863 bits.

In [24], Howgrave-Graham, Dyer, and Gennaro compare the Gennaro generator (with $n = 1024$ and $c = 160$) with a SHA-1 based pseudorandom bit generator (namely, the ANSI X9.17 generator) that lacks a proof of security:

> ...SHA-1 based pseudorandom number generation is still considerably faster than the one based on discrete logarithms. However, the difference, a factor of less than 4 on this hardware, may be considered not too high a price to pay by some who wish to have a "provably secure," rather than a "seemingly secure" (i.e., one that has withstood cryptographic attack thus far) system for pseudorandom number generation.

The proof of security for the Gennaro generator is given in §4 of [23]. Interestingly, Gennaro uses the next bit test rather than the indistinguishability criterion to derive his results. However, it is the latter criterion rather than the next bit test that is the widely accepted notion of security of a pseudorandom bit generator. As mentioned above, to pass from the next bit test to indistinguishability, one must replace ϵ by ϵ/M in the inequalities. One finds [39] that Gennaro's proof then gives the following inequality for the adversary's time:

$$T \leq \frac{L(n)(n - c)^3}{16c(\ln c)(M/\epsilon)^3}. \tag{4}$$

For $n = 1024$, $c = 160$, $M = 10^7$, and $\epsilon = 0.01$, the right-hand side of (4) is 18. Thus, the security guarantees that come with the Gennaro generator are not a whole lot more reassuring than the ones in §6.1.

We conclude this section by repeating the comment we made in §5.5 of [27]:

> Unfortunately, this type of analysis [incorporating the measure of non-tightness into recommendations for parameter sizes] is generally missing from papers that argue for a new protocol on the basis of a "proof" of its security. Typically, authors of such papers trumpet the advantage that their protocol has over competing ones that lack a proof of security (or that have a proof of security only in the random oracle model), then give a non-tight reductionist argument, and at the end give key-length recommendations that would make sense if their proof had been

tight. They fail to inform the potential users of their protocol of the true security level that is guaranteed by the "proof" if, say, a 1024-bit prime is used. It seems to us that cryptographers should be consistent. If one really believes that reductionist security arguments are very important, then one should give recommendations for parameter sizes based on an honest analysis of the security argument, even if it means admitting that efficiency must be sacrificed.

7   Short Signatures

In the early days of provable security work, researchers were content to give asymptotic results with polynomial-time reductions. In recent years, they have increasingly recognized the importance of detailed analyses of their reductions that allow them to state their results in terms of specified bounds, probabilities, and running times.

But regrettably, they often fail to follow through with interpretations in practical terms of the formulas and bounds in their lemmas and theorems. As a result, even the best researchers sometimes publish results that, when analyzed in a concrete manner, turn out to be meaningless in practice. In this section we give an example of this.

First we recall that when analyzing the security of a signature scheme against chosen-message attack in the random oracle model, one always has two different types of oracle queries — signature queries and hash function queries — each with a corresponding bound on the number of queries that an attacker can make.[12] In practice, since signature queries require a response from the target of the attack, to some extent they can be limited. So it is reasonable to suppose that the bound q_s is of the order of a million or a billion. In contrast, a query to the hash oracle corresponds in practice to simply evaluating a publicly available function. There is no justification for supposing that an attacker's hash queries will be limited in number by anything other than her total running time. Thus, to be safe one should think of q_h as being 2^{80}, or at the very least 2^{50}.

We now give an overview of three signature schemes proposed by Boneh-Lynn-Shacham [12] and Boneh-Boyen [11]. All three use bilinear pairings to obtain short signatures whose security against chosen-message attack is supported by reductionist arguments. Let k denote the security parameter; in practice, usually $k \approx 80$. For efficient implementation it is generally assumed that the group order q is approximately 2^{2k}, which is large enough to prevent squareroot attacks on the discrete log problem.

In the Boneh-Lynn-Shacham (BLS) signature scheme the signatures then have length only about $2k$. In [12] this scheme is shown to be secure against chosen-message attack in the random oracle model if the Computational Diffie-Hellman problem is hard.

---

[12] We shall continue to use the notation q_s and q_h for these bounds, even though we are also using q to denote the prime group order. We apologize to the reader for our over-use of the letter q.

In [11] Boneh and Boyen propose two alternatives to the BLS scheme. The first one (referred to below as the "BB signature scheme") has roughly twice the signature length of BLS, namely, $4k$, but it can be proven secure against chosen-message attack without using the random oracle model, assuming that the so-called Strong Diffie-Hellman problem (SDH) is intractable. The second signature scheme proposed in the paper (the "BB hash-signature scheme") is a variant of the first one in which the message must be hashed. Its proof of security uses the random oracle assumption. Like the BLS scheme, the BB hash-signature scheme has signature length roughly $2k$ rather than $4k$; moreover, it has the advantage over BLS that verification is roughly twice as fast.

The proofs in [11] are clear and readable, in part because the authors introduce a simplified version of the BB scheme (the "basic" BB scheme) in order to formulate an auxiliary lemma (Lemma 1) that is used to prove the security of both the full BB scheme (without random oracles) and the BB hash-signature scheme (with random oracles). What concerns us is the second of these results (Theorem 2).

We now describe our reason for doubting the value of that result. We shall give Lemma 1 and Theorem 2 of [11] in a slightly simplified form where we omit mention of the probabilities ϵ and ϵ', which are not relevant to our discussion. The underlying hard problem SDH for both BB schemes is parameterized by an integer that we shall denote q'_s.

Lemma 1. Suppose that q'_s-SDH cannot be solved in time less than t'. Then the basic signature scheme is secure against a weak chosen-message attack by an existential forger whose signature queries are bounded by q''_s and whose running time is bounded by t'', provided that

$$q''_s < q'_s \qquad \text{and} \qquad t'' \le t' - \Theta(q'^2_s T),$$

where T is the maximum time for a group exponentiation.

Theorem 2. Suppose that the basic signature scheme referred to in Lemma 1 is existentially unforgeable under a weak chosen-message attack with bounds q''_s and t''. Then the corresponding hash-signature scheme is secure in the random oracle model against an adaptive chosen-message attack by an existential forger whose signature queries are bounded by q_s, whose hash queries are bounded by q_h, and whose running time is bounded by t, provided that

$$q_s + q_h < q''_s \qquad \text{and} \qquad t \le t'' - o(t'').$$

Casual readers are likely to view this theorem as a fairly precise and definitive security guarantee, especially since the authors comment: "Note that the security reduction in Theorem 2 is tight... Proofs of signature schemes in the random oracle model are often far less tight." Readers are not likely to go to the trouble of comparing the statement of the theorem with that of Lemma 1, particularly since in [11] several pages of text separate the lemma from the theorem. But such a comparison must be made if we want to avoid ending up in the embarrassing situation of the previous section (see Tables 1 and 2), where the adversary's running time was bounded by a negative number.

If we put the two statements side by side and compare them, we see that in order for the bound on the adversary's running time to be a positive number it is necessary that

$$q_h^2 < t' \approx 2^k,$$

where k is the security parameter. In practice, this means that we need $q_h \ll 2^{40}$.[13] Thus, there is no security guarantee at all for the hash-signature scheme in Theorem 2 unless one assumes that the adversary is severely limited in the number of hash values she can obtain.

The conclusion of all this is not, of course, that the signature scheme in Theorem 2 of [11] is necessarily insecure, but rather that the provable security result for it has no meaning if parameters are chosen for efficient implementation.

8   The Paillier–Vergnaud Results for Schnorr Signatures

In [32] Paillier and Vergnaud prove that it is unlikely that a reduction — more precisely, an "algebraic" reduction — can be found from the Discrete Logarithm Problem (DLP) to forging Schnorr signatures. After describing this result and its proof, we compare it with various positive results that suggest equivalence between forgery of Schnorr-type signatures and the DLP.

8.1   Schnorr Signatures

We first recall the Schnorr signature scheme [35].

Schnorr key generation. Let q be a large prime, and let p be a prime such that $p \equiv 1 \pmod{q}$. Let g be a generator of the cyclic subgroup G of order q in $\mathbb{F}_p^*$. Let H be a hash function that takes values in the interval $[1, q-1]$. Each user Alice constructs her keys by selecting a random integer x in the interval $[1, q-1]$ and computing $y = g^x \bmod p$. Alice's public key is y; her private key is x.

Schnorr signature generation. To sign a message m, Alice must do the following:

1. Select a random integer k in the interval $[1, q-1]$.
2. Compute $r = g^k \bmod p$, and set $h = H(m, r)$.
3. Set $s = k + hx \bmod q$.

The signature for the message is the pair of integers (h, s).

Schnorr signature verification. To verify Alice's signature (h, s) on a message m, Bob must do the following:

1. Obtain an authenticated copy of Alice's public key y.
2. Verify that h and s are integers in the interval $[0, q-1]$.

---

[13] If we had a 160-bit group order and took $q_h = 2^{50}$, then Theorem 2 and Lemma 1 would give us the bound $t \leq -2^{100}$ for the adversary's running time.

3. Compute $u = g^s y^{-h} \bmod p$ and $v = H(m, u)$.
4. Accept the signature if and only if $v = h$.

8.2 Paillier–Vergnaud

Before giving the Paillier–Vergnaud result, we need some preliminaries. First, suppose that we have a group G generated by g. By the "discrete log" of $y \in G$ we mean a solution x to the equation $g^x = y$. In [32] the "one-more DLP" problem, denoted n-DLP, is defined as follows.

n-*DLP.* Given $r_0, r_1, \ldots, r_n \in G$ and a discrete log oracle $DL(\cdot)$ that can be called upon n times, find the discrete logs of all $n + 1$ elements r_i.

Second, by an "algebraic" reduction $\mathcal{R}$ from the DLP to forgery, Paillier and Vergnaud mean a reduction that is able to perform group operations but is not able to use special features of the way that group elements are represented. In addition, they suppose that the choices made while carrying out $\mathcal{R}$ are accessible to whomever is running the reduction algorithm (in the proof below this is the n-DLP solver). With these definitions, they prove the following result.

Theorem. Suppose that G is a group of order q generated by g. Suppose that $\mathcal{R}$ is an algebraic reduction from the DLP to universal forgery with a key-only attack that makes n calls to the forger. Then n-DLP is easy.

Proof. Let $r_0, r_1, \ldots, r_n \in G$ be an instance of n-DLP. We are required to find all $n + 1$ discrete logs, and we can call upon the oracle $DL(\cdot)$ n times. The reduction $\mathcal{R}$ will find the discrete logarithm of any element if it is given a forger that will break n different instances (chosen by $\mathcal{R}$) of the Schnorr signature scheme. We ask $\mathcal{R}$ to find the discrete log of r_0. Then n times the reduction algorithm produces a Schnorr public key y_i and a message m_i. Each time we simulate the forger by choosing $r = r_i$, computing the hash value $h_i = H(m_i, r_i)$, and then setting s_i equal to the discrete log of $r_i y_i^{h_i}$, which we determine from the oracle:

$$s_i = DL(r_i y_i^{h_i}).$$

We send (h_i, s_i), which is a valid signature for m_i with public key y_i, to $\mathcal{R}$. Finally, $\mathcal{R}$ outputs the discrete log x_0 of r_0.

In order to compute the public key y_i, $\mathcal{R}$ must have performed group operations starting with the only two group elements that it was given, namely, g and r_0. Thus, for some integer values α_i and β_i that are accessible to us, we have $y_i = g^{\alpha_i} r_0^{\beta_i}$. Once we learn x_0 (which is the output of $\mathcal{R}$), we can compute

$$x_i = s_i - h_i(\alpha_i + x_0 \beta_i) \bmod q,$$

which is the discrete logarithm of r_i, $i = 1, \ldots, n$. We now know the discrete logs of all the $n + 1$ values $r_0, \ldots, r_n$. This completes the proof.

Paillier and Vergnaud proved similar results for other signature schemes based on the DLP, such as DSA and ECDSA. In the latter cases they had to modify the n-DLP slightly: the discrete log oracle is able to give the queried discrete logs to different bases g_i.

Intuitively, the "one-more DLP" problem seems to be equivalent to the DLP, even though there is an obvious reduction in just one direction. Thus, the Paillier–Vergnaud results can be paraphrased as follows: *A reduction from the DLP to forgery is unlikely unless the DLP is easy.* In this sense the above theorem has the same flavor as the result of Boneh and Venkatesan [13] discussed in §2. As in that case, one possible interpretation of Paillier–Vergnaud is that there might be a security weakness in Schnorr-type signatures. Indeed, that interpretation is suggested by the title "Discrete-log-based signatures may not be equivalent to discrete log" and by the claim in the Introduction that "our work disproves that Schnorr, ElGamal, DSA, GQ, etc. are maximally secure."[14]

On the other hand, as in §2, an alternative explanation is that their work gives a further illustration of the limitations of reduction arguments. It is instructive to compare the negative result of Paillier–Vergnaud concerning the existence of reductions with the following two positive reductionist security results for Schnorr-type signature schemes.

8.3   Random Oracle Reductions

Reductionist security claim. In the Schnorr signature scheme, if the hash function is modeled by a random oracle, then the DLP reduces to universal forgery.

Argument. Suppose that the adversary can forge a signature for m. After it gets $h = H(m, r)$, suppose that it is suddenly given a second hash function H'. Since a hash function has no special properties that the forger can take advantage of, whatever method it used will work equally well with H replaced by H'. In other words, we are using the random oracle model for the hash function. So the forger uses $h' = H'(m, r)$ as well as $h = H(m, r)$ and produces two valid signatures (h, s) and (h', s') for m, with the same r but with $h' \neq h$. Note that the value of k is the same in both cases, since r is the same. By subtracting the two values $s \equiv k + xh$ and $s' \equiv k + xh'$ (mod q) and then dividing by $h' - h$, one can use the forger's output to immediately find the discrete log x.[15]

The above argument is imprecise. Strictly speaking, we should allow for the possibility that a forger gets $H(m, r)$ for several different values of r and signs only one of them. In that case we guess which value will be signed, and run the forger program several times with random guesses until our guess is correct. We described a rigorous argument (for a stronger version of the above claim) in §5 of [27], and full details can be found in [33,34].

Note that the need to run the forger many times leads to a non-tight reduction. In [34] it is shown that it suffices to call on the forger approximately q_h times, where q_h is a bound on the number of hash function queries. In [32] Paillier and Vergnaud prove that, roughly speaking, an algebraic reduction in the random oracle model cannot be tighter than $\sqrt{q_h}$. Much as Coron did in the case

---

[14] Paillier and Vergnaud do acknowledge, however, that their work leads to "no actual attack or weakness of either of these signature schemes."

[15] Note that one does not need to know k.

of RSA signatures, Paillier and Vergnaud establish a lower bound on tightness of the reduction.

What do we make of the circumstance that, apparently, no tight reduction from the DLP to forgery is possible in the random oracle model, and no reduction at all is likely in a standard model? As usual, several interpretations are possible. Perhaps this shows that reductions in the random oracle model are dangerous, because they lead to security results that cannot be achieved in a standard model. On the other hand, perhaps we can conclude that the random oracle model should be used, because it can often come closer to achieving what our intuition suggests should be possible. And what about the non-tightness? Should we ignore it, or should we adjust our recommendations for key sizes so that we have, say, 80 bits of security after taking into account the non-tightness factor?

8.4 Brown's Result for ECDSA

Finally, we discuss another positive result that concerns ECDSA. We shall state without proof an informal version of a theorem of D. Brown [14,15].

Theorem. Suppose that the elliptic curve is modeled by a generic group. Then the problem of finding a collision for the hash function reduces to forgery of ECDSA signatures.

Brown's theorem falls outside the framework of the results in [32]. It is a reduction not from the DLP to forgery, but rather from collision finding to forgery. And it is a tight reduction. By making the generic group assumption, one is essentially assuming that the DLP is hard (see [36]). If the hash function is collision-resistant, then the assumed hardness of the DLP (more precisely, the generic group assumption) implies hardness of forgery. However, in [14] there is no reduction from the DLP to forgery.

Both Brown and Paillier–Vergnaud make similar assumptions about the group. The latter authors implicitly assume that n-DLP is hard, and they assume that a reduction uses the group in a "generic way," that is, computes group operations without exploiting any special features of the encodings of group elements. Similarly, Brown assumes that the elliptic curve group is for all practical purposes like a generic group, and, in particular, the DLP is hard.

But their conclusions are opposite one another. Paillier and Vergnaud prove that no reduction is possible in the standard model, and no tight reduction is possible in the random oracle model. Brown gives a tight reduction — of a different sort than the ones considered in [32] — which proves security of ECDSA subject to his assumptions.

So *is* forgery of Schnorr-type signatures equivalent to the DLP? The best answer we can give is to quote a famous statement by a recent American president: it all depends on what the definition of "is" is.[16]

---

[16] The context was an explanation of his earlier statement that "there *is* no sexual relationship with Ms. Lewinsky." A statement to the effect that "there *is* no relationship of equivalence between the DLP and forgery of discrete-log-based signatures" is, in our judgment, equally implausible.

9   Conclusions

In his 1998 survey article "Why chosen ciphertext security matters" [37], Shoup explained the rationale for attaching great importance to reductionist security arguments:

> This is the preferred approach of modern, mathematical cryptography. Here, one shows with mathematical rigor that any attacker that can break the cryptosystem can be transformed into an efficient program to solve the underlying well-studied problem (e.g., factoring large numbers) that is widely believed to be very hard. Turning this logic around: if the "hardness assumption" is correct as presumed, the cryptosystem is secure. This approach is about the best we can do. If we can prove security in this way, then we essentially rule out all possible shortcuts, even ones *we have not yet even imagined*. The only way to attack the cryptosystem is a full-frontal attack on the underlying hard problem. Period. (p. 15; emphasis in original)

Later in [37] Shoup concluded: "Practical cryptosystems that are provably secure are available, and there is very little excuse for not using them." One of the two systems whose use he advocated because they had proofs of security was RSA-OAEP [7].

Unfortunately, history has not been kind to the bold opinion quoted above about the reliability of provable security results. In 2001, Shoup himself [38] found a flaw in the purported proof of security of general OAEP by Bellare and Rogaway. The same year, Manger [29] mounted a successful chosen-ciphertext attack on RSA-OAEP. Interestingly, it was not the flaw in the Bellare–Rogaway proof (which was later patched for RSA-OAEP) that made Manger's attack possible. Rather, Manger found a shortcut that was "not yet even imagined" in 1998, when Shoup wrote his survey.

It is often difficult to determine what meaning, if any, a reductionist security argument has for practical cryptography. In recent years, researchers have become more aware of the importance of concrete analysis of their reductions. But while they often take great pains to prove precise inequalities, they rarely make any effort to explain what their mathematically precise security results actually mean in practice.

For example, in [1] the authors construct a certain type of password-based key exchange system and give proofs of security in the random oracle model based on hardness of the computational Diffie–Hellman (CDH) problem. Here is the (slightly edited) text of their basic result (Corollary 1 of Theorem 1, pp. 201-202 of [1]) that establishes the relation between the "advantage" of an adversary in breaking their SPAKE1 protocol and the advantage of an adversary in solving the CDH:

> *Corollary 1.* Let G be a represent group of order p, and let $\mathcal{D}$ be a uniformly distributed dictionary of size $|\mathcal{D}|$. Let SPAKE1 be the above password-based encrypted key exchange protocol associated with these primitives. Then for any numbers t, q_{start}, q_{send}^A, q_{send}^B, q_H, q_{exe},

$$\text{Adv}^{\text{ake}}_{\text{SPAKE},\mathcal{D}}(t, q_{\text{start}}, q^A_{\text{send}}, q^B_{\text{send}}, q_H, q_{\text{exe}})$$

$$\leq 2 \cdot \left(\frac{q^A_{\text{send}} + q^B_{\text{send}}}{|\mathcal{D}|} + \sqrt[6]{\frac{2^{14}}{|\mathcal{D}|^2} \text{Adv}^{\text{CDH}}_G(t') + \frac{2^{15} q^4_H}{|\mathcal{D}|^2 p}} \right)$$

$$+ 2 \cdot \left(\frac{(q_{\text{exe}} + q_{\text{send}})^2}{2p} + q_H \text{Adv}^{\text{CDH}}_G(t + 2q_{\text{exe}}\tau + 3\tau) \right),$$

where q_H represents the number of queries to the H oracle; q_{exe} represents the number of queries to the Execute oracle; q_{start} and q^A_{send} represent the number of queries to the Send oracle with respect to the initiator A; q^B_{send} represents the number of queries to the Send oracle with respect to the responder B; $q_{\text{send}} = q^A_{\text{send}} + q^B_{\text{send}} + q_{\text{start}}$; $t' = 4t + O((q_{\text{start}} + q_H)\tau)$; and τ is the time to compute one exponentiation in G.

The paper [1] includes a proof of this bewildering and rather intimidating inequality. But the paper gives no indication of what meaning, if any, it would have in practice. The reader who might want to use the protocol and would like to find parameters that satisfy security guarantees and at the same time allow a reasonably efficient implementation is left to fend for herself.

In the provable security literature the hapless reader is increasingly likely to encounter complicated inequalities involving more than half a dozen variables. (For other examples, see Theorem 5 in [28] and Theorems 2 and 3 in [4].) The practical significance of these inequalities is almost never explained. Indeed, one has to wonder what the purpose is of publishing them in such an elaborate, undigested form, with no interpretation given. Whatever the authors' intent might have been, there can be little doubt that the effect is not to enlighten their readers, but only to mesmerize them.

* * *

Embarking on a study of the field of "provable security," before long one begins to feel that one has entered a realm that could only have been imagined by Lewis Carroll, and that the Alice of cryptographic fame has merged with the heroine of Carroll's books:

> Alice felt dreadfully puzzled. The Hatter's remark seemed to her to have no sort of meaning in it, and yet it was certainly English. (*Alice's Adventures in Wonderland* and *Through the Looking-Glass*, London: Oxford Univ. Press, 1971, p. 62.)

The Dormouse proclaims that his random bit generator is provably secure against an adversary whose computational power is bounded by a negative number. The Mad Hatter responds that he has a generator that is provably secure against an adversary whose computational resources are bounded by 10^{-40} clock cycles. The White Knight is heralded for blazing new trails, but upon further examination one notices that he's riding backwards. The Program Committee is made up of Red Queens screaming "Off with their heads!" whenever authors submit a paper with no provable security theorem.

Lewis Carroll's Alice wakes up at the end of the book and realizes that it has all been just a dream. For the cryptographic Alice, however, the return to the real world might not be so easy.

Acknowledgments

We would like to thank Andrey Sidorenko for his valuable comments on pseudorandom bit generators and Bart Preneel for answering our queries about the provable security of MAC algorithms. We also wish to thank Ian Blake and Dan Brown for reading and commenting on earlier drafts of the paper. Needless to say, all the opinions expressed in this article are the sole responsibility of the authors.

References

1. M. Abdalla and D. Pointcheval, Simple password-based encrypted key exchange protocols, *Topics in Cryptology – CT-RSA 2005*, LNCS 3376, Springer-Verlag, 2005, pp. 191-208.
2. M. Ajtai and C. Dwork, A public-key cryptosystem with worst-case/average-case equivalence, *Proc. 29th Symp. Theory of Computing*, A.C.M., 1997, pp. 284-293.
3. W. Alexi, B. Chor, O. Goldreich, and C. P. Schnorr, RSA and Rabin functions: Certain parts are as hard as the whole, *SIAM J. Computing*, **17** (1988), pp. 194-209.
4. P. Barreto, B. Libert, N. McCullagh, and J.-J. Quisquater, Efficient and provably-secure identity-based signatures and signcryption from bilinear maps, *Advances in Cryptology – Asiacrypt 2005*, LNCS 3788, Springer-Verlag, 2005, pp. 515-532.
5. M. Bellare, Practice-oriented provable-security, *Proc. First International Workshop on Information Security (ISW '97)*, LNCS 1396, Springer-Verlag, 1998, pp. 221-231.
6. M. Bellare and P. Rogaway, Random oracles are practical: a paradigm for designing efficient protocols, *Proc. First Annual Conf. Computer and Communications Security*, ACM, 1993, pp. 62-73.
7. M. Bellare and P. Rogaway, Optimal asymmetric encryption — how to encrypt with RSA, *Advances in Cryptology – Eurocrypt '94*, LNCS 950, Springer-Verlag, 1994, pp. 92-111.
8. S. Blackburn and K. Paterson, Cryptanalysis of a message authentication code due to Cary and Venkatesan, *Fast Software Encryption 2004*, LNCS 3017, Springer-Verlag, 2004, pp. 446-453.
9. L. Blum, M. Blum, and M. Shub, A simple unpredictable pseudo-random number generator, *SIAM J. Computing*, **15** (1986), pp. 364-383.
10. M. Blum and S. Micali, How to generate cryptographically strong sequences of pseudo-random bits, *SIAM J. Computing*, **13** (1984), pp. 850-864.
11. D. Boneh and X. Boyen, Short signatures without random oracles, *Advances in Cryptology – Eurocrypt 2004*, LNCS 3027, Springer-Verlag, 2004, pp. 56-73.
12. D. Boneh, B. Lynn, and H. Shacham, Short signatures from the Weil pairing, *Advances in Cryptology – Asiacrypt 2001*, LNCS 2248, Springer-Verlag, 2001, pp. 514-532.

13. D. Boneh and R. Venkatesan, Breaking RSA may not be equivalent to factoring, *Advances in Cryptology – Eurocrypt '98*, LNCS 1233, Springer-Verlag, 1998, pp. 59-71.

14. D. Brown, Generic groups, collision resistance, and ECDSA, *Designs, Codes and Cryptography*, **35** (2005), pp. 119-152.

15. D. Brown, On the provable security of ECDSA, in I. Blake, G. Seroussi, and N. Smart, eds., *Advances in Elliptic Curve Cryptography*, Cambridge University Press, 2005, pp. 21-40.

16. D. Brown, Breaking RSA may be as difficult as factoring, http://eprint.iacr.org/2005/380

17. D. Brown, unpublished communication, February 2006.

18. M. Cary and R. Venkatesan, A message authentication code based on unimodular matrix groups, *Advances in Cryptology – Crypto 2003*, LNCS 2729, Springer-Verlag, 2003, pp. 500-512.

19. J.-S. Coron, On the exact security of full domain hash, *Advances in Cryptology – Crypto 2000*, LNCS 1880, Springer-Verlag, 2000, pp. 229-235.

20. J.-S. Coron, Optimal security proofs for PSS and other signature schemes, *Advances in Cryptology – Eurocrypt 2002*, LNCS 2332, Springer-Verlag, 2002, pp. 272-287.

21. D. Eastlake, S. Crocker, and J. Schiller, RFC 1750 – Randomness Recommendations for Security, available from http://www.ietf.org/rfc/rfc1750.txt

22. R. Fischlin and C. P. Schnorr, Stronger security proofs for RSA and Rabin bits, *J. Cryptology*, **13** (2000), pp. 221-244.

23. R. Gennaro, An improved pseudo-random generator based on the discrete log problem, *J. Cryptology*, **18** (2005), pp. 91-110.

24. N. Howgrave-Graham, J. Dyer, and R. Gennaro, Pseudo-random number generation on the IBM 4758 Secure Crypto Coprocessor, *Workshop on Cryptographic Hardware and Embedded Systems* (CHES 2001), LNCS 2162, Springer-Verlag, 2001, pp. 93-102.

25. J. Katz and N. Wang, Efficiency improvements for signature schemes with tight security reductions, *10th ACM Conf. Computer and Communications Security*, 2003, pp. 155-164.

26. D. Knuth, *Seminumerical Algorithms*, vol. 2 of *Art of Computer Programming*, 3rd ed., Addison-Wesley, 1997.

27. N. Koblitz and A. Menezes, Another look at "provable security," to appear in *J. Cryptology*; available from http://eprint.iacr.org/2004/152.

28. P. Mackenzie and S. Patel, Hard bits of the discrete log with applications to password authentication, *Topics in Cryptology – CT-RSA 2005*, LNCS 3376, Springer-Verlag, 2005, pp. 209-226.

29. J. Manger, A chosen ciphertext attack on RSA Optimal Asymmetric Encryption Padding (OAEP) as standardized in PKCS #1 v2.0, *Advances in Cryptology – Crypto 2001*, LNCS 2139, Springer-Verlag, 2001, pp. 230-238.

30. P. Q. Nguyen and J. Stern, Cryptanalysis of the Ajtai–Dwork cryptosystem, *Advances in Cryptology – Crypto '98*, LNCS 1462, Springer-Verlag, 1998, pp. 223-242.

31. P. Q. Nguyen and J. Stern, The two faces of lattices in cryptology, *Cryptography and Lattices – Proc. CALC 2001*, LNCS 2146, Springer-Verlag, 2001, pp. 146-180.

32. P. Paillier and D. Vergnaud, Discrete-log-based signatures may not be equivalent to discrete log, *Advances in Cryptology – Asiacrypt 2005*, LNCS 3788, Springer-Verlag, 2005, pp. 1-20.

33. D. Pointcheval and J. Stern, Security proofs for signature schemes, *Advances in Cryptology – Eurocrypt '96*, LNCS 1070, Springer-Verlag, 1996, pp. 387-398.

34. D. Pointcheval and J. Stern, Security arguments for digital signatures and blind signatures, *J. Cryptology*, **13** (2000), pp. 361-396.

35. C. P. Schnorr, Efficient signature generation for smart cards, *J. Cryptology*, **4** (1991), pp. 161-174.

36. V. Shoup, Lower bounds for discrete logarithms and related problems, *Advances in Cryptology – Eurocrypt '97*, LNCS 1233, Springer-Verlag, 1997, pp. 256-266.

37. V. Shoup, Why chosen ciphertext security matters, IBM Research Report RZ 3076 (#93122) 23/11/1998.

38. V. Shoup, OAEP reconsidered, *Advances in Cryptology – Crypto 2001*, LNCS 2139, Springer-Verlag, 2001, pp. 239-259.

39. A. Sidorenko, unpublished communication, March 2006.

40. A. Sidorenko and B. Schoenmakers, Concrete security of the Blum–Blum–Shub pseudorandom generator, *Cryptography and Coding 2005*, LNCS 3796, Springer-Verlag, 2005, pp. 355-375.

41. U. V. Vazirani and V. V. Vazirani, Efficient and secure pseudo-random number generation, *Proc. IEEE 25th Annual Symp. Foundations of Computer Science*, 1984, pp. 458-463.

42. A. Yao, Theory and applications of trapdoor functions, *Proc. IEEE 23rd Annual Symp. Foundations of Computer Science*, 1982, pp. 80-91.

43. A. Young and M. Yung, *Malicious Cryptography: Exposing Cryptovirology*, Wiley, 2004.

Efficient CCA-Secure Public-Key Encryption Schemes from RSA-Related Assumptions

Jaimee Brown, Juan Manuel González Nieto, and Colin Boyd

Information Security Institute
Queensland University of Technology, Australia
{j2.brown, j.gonzaleznieto, c.boyd}@qut.edu.au

Abstract. We build new RSA-based encryption schemes secure against adaptive chosen-ciphertext attack (CCA-secure) without random oracles. To do this, we first define a new general RSA-related assumption, the Oracle RSA-type assumption, and give two specific instances of this assumption. Secondly, we express RSA-based encryption schemes as tag-based encryption schemes (TBE), where the public exponent is the tag. We define selective-tag weak chosen-ciphertext security for the special RSA-based case and call it selective-exponent weak chosen-ciphertext security. RSA-based schemes secure in this sense can be used as a building block for the construction of chosen-ciphertext secure encryption schemes using a previous technique. We build two concrete CCA-secure encryption schemes whose security is based on the two concrete Oracle RSA-type assumptions respectively, and whose efficiency is comparable to the most efficient CCA-secure schemes known.

Keywords: chosen-ciphertext security, public key encryption, RSA assumptions.

1 Introduction

Indistinguishability against adaptive chosen ciphertext attack (IND-CCA), where an adversary is given the capability to decrypt ciphertexts of his choice, with the exception of a target ciphertext, is considered to be the correct notion of security for general-purpose public key encryption schemes. We refer to such schemes as CCA-secure schemes. In the literature, there are a number of approaches for obtaining encryption schemes that are CCA-secure. Much of this work, however, has been only achieved with proofs in the random oracle model, the most famous being OAEP [3]. When, in practice, these random oracles are replaced by hash functions, the security argument becomes heuristic only and does not guarantee security against all attacks under the standard assumptions.

In the standard model three main techniques have been proposed for constructing CCA-secure encryption schemes. The first approach, from Naor and Yung [21] and subsequently Dolev, Dwork and Naor [14], builds CCA-secure schemes from any chosen-plaintext secure scheme (CPA-secure) and any non-interactive zero knowledge (NIZK) proof system. The resulting schemes, however, are too inefficient for practical use, since they use expensive NIZK proofs.

R. Barua and T. Lange (Eds.): INDOCRYPT 2006, LNCS 4329, pp. 176–190, 2006.

Cramer and Shoup [11] proposed the first encryption scheme that was simultaneously practical and CCA-secure in the standard model. Cramer and Shoup [12] later generalised their encryption scheme by defining *hash proof systems* (HPS) and giving a framework for constructing CCA-secure encryption schemes using a HPS constructed from a general subset membership problem. Kurosawa and Desmedt [18] later showed how to obtain CCA-secure hybrid encryption schemes using a HPS as a building block, and in particular described an efficient hybrid encryption scheme based on the Cramer-Shoup cryptosystem.

More recently, Canetti, Halevi and Katz [8] proposed a framework (CHK) for constructing chosen-ciphertext secure encryption schemes from ID-based encryption (IBE) schemes secure against selective-identity chosen-plaintext attack. Boneh and Katz [7] improved the efficiency of the CHK construction, and the two related works were combined in a later paper [6]. The resulting schemes are both simple and efficient, with proofs of security in the standard model. Interestingly, the authors note that the resulting schemes seem to achieve chosen-ciphertext security using a different approach to the two previous techniques that use NIZK proofs and HPS. More precisely, the schemes do not use a *proof of well-formedness* as in the two previous approaches and hence do not fall within the general paradigm for chosen-ciphertext encryption described by Elkind and Sahai [15]. Kiltz [17] showed that tag-based encryption (TBE) is a more general case of IBE and can in fact be used as the building block for the CHK framework in place of the IBE. Kiltz defined selective-tag security for TBE and showed that a TBE secure in this sense is a sufficient building block for the CHK transformation and then proposed a new TBE that can be used to construct a reasonably efficient CCA-secure scheme from the Decisional Linear Assumption [5].

It is worth noting that despite these three approaches to building provably CCA-secure encryption schemes, there has yet to emerge an efficient CCA-secure encryption scheme based on RSA or related assumptions in the standard model.

1.1 Our Contributions

NEW RSA-RELATED ASSUMPTION. We define a new general RSA-related assumption, namely the *Oracle RSA-type assumption*, which will be used to prove the security of the new schemes we introduce later in the paper. The Oracle RSA-type assumption is a variant of the general *Decisional RSA-type assumption* which is a decisional RSA-based assumption of a specific form. The Oracle RSA-type assumption can be viewed as the analog of the Oracle Diffie-Hellman assumption [2] for an RSA context. We give concrete examples of Oracle RSA-type assumptions derived from previously studied decisional RSA-based assumptions.

SELECTIVE-EXPONENT SECURITY FOR RSA-BASED ENCRYPTION. We observe that an RSA-based encryption scheme can be considered as a TBE where the exponent e is the tag. We redefine the notion of selective-tag weak CCA security [17] for this special case of RSA-based TBEs, and call it *selective-exponent weak chosen-ciphertext security*, or more simply *selective-exponent security*. In a selective-exponent attack, an adversary is given access to a decryption oracle from which he can obtain any decryptions of ciphertexts with respect to any (public) exponent except the target (public key) exponent. We show how we can build schemes that are secure against this attack based on our newly defined Oracle RSA-type assumption. Moreover, we describe two concrete schemes based on the two concrete examples given.

NEW EFFICIENT CCA-SECURE RSA-BASED SCHEMES. Using the Kiltz generalised transformation of the CHK transformation we show how to obtain RSA-based encryption schemes secure against chosen-ciphertext attack. More precisely, using the efficiency improvement of Boneh and Katz [7], a CCA-secure RSA-based scheme can be constructed from a selective-exponent secure RSA-based scheme, a message authentication code and a secure encapsulation scheme. The efficiency of the resulting scheme is essentially the same as the selective-exponent secure RSA-based scheme. We give two concrete instantiations of the construction to yield two new chosen-ciphertext secure encryption schemes based on RSA-related assumptions, whose efficiencies are comparable to the most efficient known chosen-ciphertext secure schemes.

2   Preliminaries

2.1   Notation

For a set S, the notation $|S|$ is used to denote the cardinality of S. For a string or number n, the notation $|n|$ is used to denote the bit length of n. For a set S, $x \in_R S$ denotes selecting an element x uniformly at random from S.

The term PPT is used to describe an algorithm that is probabilistic and runs in polynomial time.

Algorithms which operate on some group G will be given string representations of elements of G, so a map from G to $\{0,1\}^{len}$ is implicitly assumed. Also, we implicitly assume that numbers are encoded in binary. The notation $a \circ b$ is denotes the concatenation of strings a, b, or if a, b are numbers, the concatenation of their bit-string representations.

3   RSA-Related Oracle Assumptions

We propose a variant of a general RSA-based assumption. The following discussion is for general RSA-based problems, but specific instances are discussed later in the section. Let $f_e : \mathbb{Z}_N^* \to \mathbb{Z}_N^*$ be the RSA trapdoor one-way function for the public exponent e, that is, $f_e(s) = s^e \mod N$. Also let $g : \mathbb{Z}_N^* \to S$ be a one-way function such that the distribution of $(f_e(s), g(s))$ for a randomly generated s is difficult to distinguish from $(f_e(s), g(t))$ for random s, t. (The function g may also be determined by the exponent e, but we will leave our notation as g for the general case.) The trapdoor information for inverting any f_e is the factorisation of the RSA modulus $N = pq$. Moreover, this trapdoor information enables one to decide whether a given (A, B) is from the distribution $(f_e(s), g(s))$ or from a random distribution. We call the assumption that $(f_e(s), g(s))$ is difficult to distinguish from random the Decisional RSA-type assumption. This assumption is described more precisely in the following definition.

Definition 1. [**D-RSA-type**] *For a security parameter k, let N be an RSA modulus, e a public exponent. The Decisional RSA-type assumption is that, for all PPT adversaries $\mathcal{A}$, the advantage of $\mathcal{A}$ in distinguishing an RSA-type tuple from random is negligible in k. More precisely,*

$$Adv_{k,N,e}^{drsa-type}(\mathcal{A}) = \quad | \Pr[s \in_R \mathbb{Z}_N^* : \mathcal{A}(k, f_e(s), g(s)) = 1]$$
$$- \Pr[s, t \in_R \mathbb{Z}_N^* : \mathcal{A}(k, f_e(s), g(t)) = 1]|$$

is negligible in k.

Now consider the distribution of $(f_e(s), H(g(s))$ where H is a hash function, and f_e, g are as before. The following Hash RSA-type assumption roughly states that $H(g(s))$ looks like a random string, even if $f_e(s)$ is also known.

Definition 2. [**H-RSA-type**] *For a security parameter k, let N be an RSA modulus and e a public exponent. Also let len be a number, and $H : \{0,1\}^* \to \{0,1\}^{len}$ be a hash function. The Hash RSA-type assumption is that for all PPT adversaries $\mathcal{A}$, the advantage of $\mathcal{A}$ in distinguishing a Hash RSA-type tuple from random is negligible in k. More precisely,*

$$Adv_{k,N,e,H}^{hrsa-type}(\mathcal{A}) = \quad | \Pr[s \in_R \mathbb{Z}_N^* : \mathcal{A}(k, f_e(s), H(g(s))) = 1]$$
$$- \Pr[s \in_R \mathbb{Z}_N^*; r \in_R \{0,1\}^{len} : \mathcal{A}(k, f_e(s), r) = 1]|$$

is negligible in k.

Now consider giving the adversary $\mathcal{A}$ access to an oracle $\mathcal{O}(\bar{e}, X)$ which computes $\mathcal{O}(\bar{e}, X) = H(g(f_{\bar{e}}^{-1}(X)))$ for any given $\bar{e} \neq e$. The Oracle RSA-type (O-RSA-type) assumption states that such an oracle does not help $\mathcal{A}$ to break the H-RSA-type assumption.

Definition 3. [**O-RSA-type**] *For a security parameter k, let N be an RSA modulus and let e be a public exponent. Also let len be a number, and $H : \{0,1\}^* \to \{0,1\}^{len}$ be a hash function. Also define oracle $\mathcal{O}$ which computes $\mathcal{O}(\bar{e}, X) = H(g(f_{\bar{e}}^{-1}(X)))$. The Oracle RSA-type assumption is that, for all PPT adversaries $\mathcal{A}$, the advantage of $\mathcal{A}$ in distinguishing a Hash RSA-type tuple from random is negligible in k, even when $\mathcal{A}$ has access to $\mathcal{O}$. More precisely,*

$$Adv_{N,e,H}^{orsa-type}(\mathcal{A}) = \quad | \Pr[s \in_R \mathbb{Z}_N^* : \mathcal{A}^{\mathcal{O}(.,.)}(f_e(s), H(g(s))) = 1]$$
$$- \Pr[s \in_R \mathbb{Z}_N^*; r \in_R \{0,1\}^{len} : \mathcal{A}^{\mathcal{O}(.,.)}(f_e(s), r) = 1]|$$

is negligible in k, when $\mathcal{A}$ is restricted to oracle queries $\mathcal{O}(\bar{e}, .)$ for $\bar{e} \neq e$.

These RSA-type assumptions are general for f_e, g functions described. The strength of the actual assumptions will depend on the hash function H, and the particular f_e, g. Section 3.2 discusses the choices for H, and section 3.3 discusses some particular choices of f_e, g. For the general case, we propose that if the underlying Decisional RSA-type assumption is valid, and an appropriate hash function H is chosen, then the Oracle RSA-type assumption derived from it will also be valid, albeit a stronger assumption.

3.1 Analogy with Oracle Diffie-Hellman Assumptions

We now demonstrate that the Oracle variant of an RSA-based assumption is analogous to the Oracle Diffie-Hellman [1,2] (ODH) assumption, a variant of the Decisional Diffie-Hellman assumption. Essentially, the ODH assumption states that, for a hash function $H : \{0,1\}^* \rightarrow \{0,1\}^\ell$ it is hard to distinguish $(g^u, g^v, H(g^{uv}))$ from (g^u, g^v, r) for $r \in_R \{0,1\}^\ell$ when given access to an oracle $\mathcal{O}_v(.)$ that computes $\mathcal{O}_v(X) = H(X^v)$ for any $X \neq g^u$. We point out that the steps in constructing an Oracle RSA-type assumption from a Decisional RSA-type assumption are the same as those used in constructing the ODH assumption from the Decisional Diffie-Hellman assumption. Table 1 attempts to illustrate the parallel between the various assumption variants for the two general problems.

Table 1. Comparing the RSA-type assumption variants to Diffie-Hellman assumption variants

	RSA-type in $\mathbb{Z}_N^*$ (given $e, f_e(s)$)	Diffie-Hellman in G (given g^u, g^v)
Decisional Problem	Distinguish $g(s)$ from random $\mathbb{Z}_N^*$ element	Distinguish g^{uv} from random G element
Hash Variant	Distinguish $H(g(s))$ from random string	Distinguish $H(g^{uv})$ from random string
Oracle Variant	Distinguish $H(g(s))$ from random string given $\mathcal{O}(\bar{e}, X) = H(g(f_{\bar{e}}^{-1}(X)))$	Distinguish $H(g^{uv})$ from random string given $\mathcal{O}_v(X) = H(X^v)$

3.2 Choice of Hash Function

It is very important that the hash function used in our Hash and Oracle RSA-type assumption are appropriately chosen so as to maximise our confidence in the assumptions. The requirements of the hash functions for our new RSA-type assumptions are very similar to those required by the Hash and Oracle variants of Diffie-Hellman. For the Oracle Diffie-Hellman problem, it was suggested in [1] that H be derived from some cryptographic hash function like SHA-1, in some suitable but unspecified way. The problem with this recommendation is that it is unclear exactly how to derive this H, and it is unclear what assumptions need to be made in order for the requirements of H to be satisfied. Moreover, SHA-1 was not constructed with this application in mind and so it would seem a leap of faith to assume that it is sufficient without giving fair justification.

We recommend H to be a function suitable for key derivation. Typically, this is achieved by choosing H_k at random from a family of universal hash functions [9], or almost-universal hash functions [26][1]. The function has a known key k which defines the function, and which should be chosen at random. The Leftover Hash Lemma [16] (adapted to almost-universal hash functions in [13]) roughly states that if the input distribution has min-entropy that is (sufficiently) larger than the length of the range of H_k, then the output distribution of H_k will be close to uniform on the range. More precisely, if x is from an input distribution with min-entropy γ and k is chosen uniformly at random from the set of keys, then $H_k(x)$ will be 2^{-e}-close to uniform for $e \leq \frac{\gamma-m}{2}$, where m is the bitlength of the range of H. For our Hash RSA-type assumption, the input to H_k is $g(s)$ for $s \in_R \mathbb{Z}_N^*$, so if $g(s)$ has high output entropy, which by the D-RSA-type assumption we are assuming, and the output length is sufficiently less than the input, the output will be a random looking string. If more random output bits than H_k offers are required, we can use $H_k(g(s))$ to key a pseudorandom generator (PRG), which will output a value computationally indistinguishable from a random string. Thus, we have $H(x) = PRG(H_k(x))$.

Note that besides a random looking output, we also require that H be chosen such that given an output $H(x)$ for a specific x, an adversary cannot recover x. If this is not satisfied, then the adversary may be able to distinguish Hash RSA-type tuples from random. By choosing a universal hash function that satisfies the Leftover Hash Lemma for sufficiently large e, we also ensure that the set of preimages for $H(x)$ is large. Thus, even if an adversary could find the set of preimages for $H(x)$, the probability of guessing the particular x is very small.

3.3   Instances of Oracle RSA-Type Assumptions

We describe two examples of Oracle RSA-type assumptions derived from Decisional RSA-type assumptions which have been studied previously in the literature.

INSTANCE 1. The Decisional Dependent-RSA (DDRSA) assumption, described by Pointcheval [23], considers $f_e(s) = s^e \mod N$ and $g(s) = (s+1)^e \mod N$, and states that $(f_e(s), g(s))$ is difficult to distinguish from randomly distributed $(f_e(s), r)$. The oracle variant of this Decisional RSA-type assumption, namely Oracle Dependent-RSA assumption, is defined as follows:

Definition 4. [ODRSA] *Let N be an RSA modulus, let e be a public exponent and let $\mathcal{A}$ be an adversary. Also let len be a number, and $H : \{0,1\}^* \to \{0,1\}^{len}$ be a hash function. Also define oracle $\mathcal{O}(\bar{e}, X) = H((X^{\bar{e}^{-1}} + 1)^{\bar{e}})$ to which $\mathcal{A}$ can query for any $\bar{e} \neq e$. The advantage of $\mathcal{A}$ in violating the ODRSA assumption is*

---

[1] There are a number of constructions of universal hash functions featured in the literature (see [25] section 6.7 for simple constructions, and [22] for a performance comparison of several constructions).

$$Adv_{N,e,H}^{odrsa}(\mathcal{A}) = \quad |\Pr[s \in_R \mathbb{Z}_N^* : \mathcal{A}^{\mathcal{O}(\cdot,\cdot)}(s^e, H((s+1)^e)) = 1]$$
$$- \Pr[s \in_R \mathbb{Z}_N^*; r \in_R \{0,1\}^{len} : \mathcal{A}^{\mathcal{O}(\cdot,\cdot)}(s^e, r) = 1]|$$

It has been stated in section 3.2 that given $H(x)$, it must be hard to find the exact value of x. To see this, suppose that for fixed e, $\mathcal{A}$ is given as input (α, β) and it is able to recover the correct x such that $H(x) = \beta$. $\mathcal{A}$ can then check whether β is from the ODRSA distribution by checking if $H(x^{e'})$ is equal to $\mathcal{O}(ee', \alpha^{e'}) = H((\alpha^d + 1)^{ee'})$.

INSTANCE 2. Catalano et al. [10] proposed the Decisional Small e-Residues (DSeR) assumption which roughly states that $s^e \mod N^2$ for $s \in_R \mathbb{Z}_N$ cannot be distinguished from random elements of $\mathbb{Z}_{N^2}^*$. They also showed that an equivalent assumption is that it is difficult to distinguish (s_a, s_b), where $s_a + s_b N = s^e \mod N^2$ for some $s \in \mathbb{Z}_N$, from a randomly distributed pair in $\mathbb{Z}_N \times \mathbb{Z}_N$. In terms of our RSA-type assumptions, $f_e(s) = s^e \mod N$ and $g(s) = \frac{(s^e \mod N^2) - (s^e \mod N)}{N}$. We call this assumption the Decision Small e-Residues-Related (DSeRR) assumption. The oracle variant of this assumption, namely the Oracle Small e-Residues-Related (OSeRR) assumption is defined as follows:

Definition 5. [OSeRR] *Let N be an RSA modulus, let e be a public exponent and let len be a number, and $H : \{0,1\}^* \to \{0,1\}^{len}$ be a hash function. Also define oracle $\mathcal{O}(\bar{e}, X) = H(\frac{(X^{\bar{e}^{-1}} \mod N)^{\bar{e}} \mod N^2 - X}{N})$ to which $\mathcal{A}$ can query for any $\bar{e} \neq e$. The advantage of $\mathcal{A}$ in violating the OSeRR assumption is*

$$Adv_{N,e,H}^{oserr}(\mathcal{A}) = |\Pr[s \in_R \mathbb{Z}_N^* : \mathcal{A}^{\mathcal{O}(\cdot,\cdot)}(s^e \mod N, H(\frac{(s^e \mod N^2) - (s^e \mod N)}{N})) = 1]$$
$$- \Pr[s \in_R \mathbb{Z}_N^*; r \in_R \{0,1\}^{len} : \mathcal{A}^{\mathcal{O}(\cdot,\cdot)}(s^e \mod N, r) = 1]|$$

4   RSA-Based Encryption is Tag-Based Encryption

A tag-based encryption scheme [19] (TBE) differs from a general public key encryption scheme in that the encryption and decryption algorithms take as input an additional public string argument called a *tag*.

Let S $= (\mathcal{G}, \mathcal{E}, \mathcal{D})$ be an RSA-based public-key encryption scheme where the public key PK is the RSA modulus $N = pq$, and the secret key SK its factorisation p, q. Usually, the exponents e, d such that $ed \equiv 1 \mod \phi(N)$ are included in the public and private key, respectively. However, if we instead include the exponent e as input to both the encryption and decryption algorithms, such an RSA-based scheme fits in with the definition of a tag-based encryption scheme, where the exponent e is the tag. The main difference will be that the decryption algorithm must first derive the private exponent d from the exponent/tag e, and then the decryption can proceed as before.

4.1   Selective-Exponent Security for RSA-Based Encryption

We now restate Kiltz's definition of selective-tag weak chosen-ciphertext security [17] for the case of general RSA-based schemes where the exponent is the tag, which we refer to as selective-exponent weak chosen-ciphertext security, or more simply selective-exponent security[2]. As usual, the adversary has a find stage and a guess stage. The selective-exponent attack allows the adversary to obtain decryptions of its choice of ciphertexts for any other public exponent than the target exponent e. In other words, an adversary can decrypt any ciphertext with respect to $\bar{d} = \bar{e}^{-1} \mod \phi(N)$ for any $\bar{e} \neq e$. The definition is as follows:

Definition 6. [Sel-Exp weak-CCA] *An RSA-based public key encryption scheme* $\mathcal{S} = (\mathcal{G}, \mathcal{E}, \mathcal{D})$ *is secure against selective-exponent weak chosen-ciphertext attack (or equivalently selective-exponent secure) if the advantage of all* PPT *adversaries* $\mathcal{A}$ *is negligible in the security parameter* k *in the following game:*

1. $\mathcal{A}$ *outputs target exponent* e
2. $(PK, SK) = \mathcal{G}(1^k)$. *Adversary* $\mathcal{A}$*'s find stage is given* 1^k *and* PK.
3. $\mathcal{A}$ *may make polynomially-many queries to a decryption oracle* $\mathcal{D}_{SK}(\bar{e}, .)$ *as long as* $\bar{e} \neq e$.
4. $\mathcal{A}$*'s find stage outputs two messages* m_0, m_1 *and internal state information* I_S. *A bit* $b \in_R \{0, 1\}$ *is selected and ciphertext* $C \leftarrow \mathcal{E}_{PK}(e, m_b)$ *and* I_S *is input to* $\mathcal{A}$*'s guess stage.*
5. $\mathcal{A}$ *can continue making queries to a decryption oracle* $\mathcal{D}_{SK}(\bar{e}, .)$ *as long as* $\bar{e} \neq e$.
6. $\mathcal{A}$ *outputs guess bit* b'.

We say that $\mathcal{A}$ *succeeds if* $b' = b$. *We denote the probability of this as* $\Pr_{\mathcal{A},\mathcal{S}}[Succ]$. *The adversary's advantage is defined as* $|\Pr_{\mathcal{A},\mathcal{S}}[Succ] - 1/2|$.

If we take away the adversary's capability to perform decryption queries, then we have a chosen-plaintext attack. Thus, an encryption scheme secure against selective-exponent weak CCA is also secure against chosen-plaintext attacks.

Selective-exponent weak CCA security differs from (full) chosen-ciphertext security in the type of decryptions the oracle allows. In a CCA attack, the exponent e is fixed, and decryptions are allowed for any ciphertext different from the target ciphertext. In a selective-exponent attack, any ciphertext can be decrypted but only using an exponent different from the target exponent.

Note that in the above definition, the target exponent is output prior to the generation of the keys, which may seem counter-intuitive for our RSA-based case. This is to be consistent with the more general selective-tag definition, where the tag is output at the beginning of the game. However, it would also make sense to allow the adversary to output the target exponent after being given PK, but before asking any decryption queries.

---

[2] We use a different term to describe selective-tag security for this special case of RSA-based TBE where the exponent is the tag so as to distinguish it from a RSA-based TBE where the exponent is fixed and part of the public key, and the tag is an additional parameter.

4.2   Constructing Selective-Exponent Secure Schemes

Catalano *et al.* [10] noted that given a trapdoor one-way function f, and function g such that $(f(s), g(s))$ cannot be distinguished from a randomly distributed pair $(f(s), r)$, the encryption scheme $\mathcal{E}(m) = (f(s), mg(s))$ will be semantically secure. Using our terminology from section 3 for the RSA trapdoor function, the encryption scheme $\mathcal{E}(m) = (f_e(s), mg(s))$ is semantically secure under the Decisional RSA-type assumption. We have observed that such an encryption scheme is a tag-based encryption scheme where the tag is the exponent e. Moreover, we now show that the encryption scheme $\mathcal{E}(m, e) = (f_e(s), m \oplus H(g(s)))$ is selective-exponent weak CCA-secure based on the Oracle RSA-type assumption. We define the scheme $\mathcal{S} = (\mathcal{G}, \mathcal{E}, \mathcal{D})$ as follows

Key Generation $\mathcal{G}(1^k)$: Generate an RSA modulus $N = pq$ and output the public key $PK = N$ and the secret key $SK = (p, q)$.

Encryption $\mathcal{E}_{PK}(e, m)$: Choose s uniformly at random from $\mathbb{Z}_N^*$ and compute $(A, B) = (f_e(s), m \oplus H(g(s)))$. The ciphertext is $C = (A, B)$

Decryption $\mathcal{D}_{SK}(e, C)$: Compute $s = f_e^{-1}(A)$ and then $m = B \oplus H(g(s))$

Theorem 1. *The above encryption scheme $\mathcal{S}$ is secure against selective-exponent weak chosen ciphertext attack under the Oracle RSA-type assumption.*

Proof (Proof sketch.). If there exists a selective-exponent adversary $\mathcal{A}$ that has non-negligible advantage in attacking $\mathcal{S}$, then we can build an adversary $\mathcal{B}$ which uses $\mathcal{A}$ to gain non-negligible advantage against the Oracle RSA-type assumption. The input to $\mathcal{B}$ is the RSA modulus N, the target exponent e (which we assume has already been output), and the pair (α, β). We suppose that $\mathcal{B}$ has access to an oracle $\mathcal{O}(.,.)$ that computes $\mathcal{O}(\bar{e}, X) = H(g(f_{\bar{e}}^{-1}(X)))$ for $\bar{e} \neq e$. The goal is to output 1 if (α, β) is from the distribution of $(f_e(s), H(g(s)))$ and 0 otherwise. We define $\mathcal{B}(N, e, \alpha, \beta)$ as follows

1. Run $\mathcal{A}(1^k, N)$
2. When $\mathcal{A}$ makes a decryption oracle query $\mathcal{D}(\bar{e}, \bar{A}, \bar{B})$, answer queries as follows:
   (a) If $\bar{e} = e$ output $\perp$
   (b) Else query oracle $Y = \mathcal{O}(\bar{e}, \bar{A})$
   (c) $m = \bar{B} \oplus Y$
   (d) return m
3. At some point, $\mathcal{A}$ outputs m_0, m_1. Select $b \in_R \{0, 1\}$
4. $A = \alpha, B = m_b \oplus \beta$. Send $c = (A, B)$ to $\mathcal{A}$
5. $\mathcal{A}$ continues to make decryption queries and they are answered as above.
6. $\mathcal{A}$ outputs guess bit b'. Return 1 if $b = b'$ or else 0.

Firstly, it can be seen that $\mathcal{B}$ is a legal adversary against the Oracle RSA-type assumption. In particular the oracle $\mathcal{O}$ is never queried for target exponent e. It is also evident that $\mathcal{B}$ provides a perfect simulation for the decryption queries made by $\mathcal{A}$.

Now, when (α, β) is from the Oracle RSA-type distribution, (A, B) will be a valid encryption of m_b. When $\mathcal{A}$ guesses correctly, which $\mathcal{A}$ does with non-negligible advantage, $\mathcal{B}$ will output 1 (correctly). On the other hand, when (α, β) comes from the random distribution, the ciphertext (A, B) will neither be a valid encryption of m_0 nor m_1, but rather the encryption of a random message. More precisely, $m_0 \oplus \beta$ will be indistinguishable from $m_1 \oplus \beta$ and the adversary cannot do better than guessing b with probability $1/2$. Therefore, $\mathcal{B}$ is a distinguisher for the Oracle RSA-type assumption.

4.3   Concrete RSA-Based Schemes Secure Against Selective-Exponent Attack

In section 3.3 we defined Oracle RSA-type assumptions derived from previously studied Decisional RSA-type assumptions. We can obtain concrete selective-exponent secure encryption schemes based on these assumptions by applying them to the simple construction above.

SCHEME 1. The encryption scheme $\mathcal{E}(e, m) = (s^e, m \oplus H((s+1)^e))$ is secure against selective-exponent CCA under the Oracle Dependent-RSA assumption.

SCHEME 2. The encryption scheme $\mathcal{E}(e, m) = (s^e \mod N,$
$m \oplus H(\frac{(s^e \mod N^2) - (s^e \mod N)}{N}))$ is secure against selective-exponent CCA under the Oracle Small e-Residues-Related (OSeRR) assumption.

5   Chosen-Ciphertext Security from RSA-Based Assumptions

The following construction is the Kiltz TBE generalisation of the Boneh and Katz [7] efficiency improvement on the CHK transformation, instantiated for the special case of an RSA-based encryption scheme. The construction builds a chosen-ciphertext secure encryption scheme from a selective-exponent secure RSA-based scheme, a message authentication code and a secure encapsulation scheme[3].

Let $\mathcal{S}' = (\mathcal{G}', \mathcal{E}', \mathcal{D}')$ be a selective-exponent weakly chosen-ciphertext secure RSA-based encryption scheme. Let $\Omega = (\mathsf{Init}, \mathcal{C}, \mathcal{R})$ be a secure encapsulation scheme and let $\Delta = (\mathsf{Mac}, \mathsf{Vrfy})$ be a message authentication code. We define the following encryption scheme $\mathcal{S} = (\mathcal{G}, \mathcal{E}, \mathcal{D})$:

Key Generation $\mathcal{G}(1^k)$: Input security parameter k

1. Run $\mathcal{G}'(1^k)$ to get (PK', SK')
2. Run $\mathsf{Init}(1^k)$ to get pub.
3. Output the public key $PK = (PK', \mathsf{pub})$ and the secret key $SK = SK'$.

---

[3] We refer the reader to the Boneh and Katz [7] paper for the definition of secure encapsulation schemes.

Encryption $\mathcal{E}(PK, m)$: Input message $m \in \mathbb{Z}_N^*$, public key PK

1. Run $\mathcal{C}(1^k, \mathsf{pub})$ to get $(r, \mathsf{com}, \mathsf{dec})$.
2. Encrypt $m \circ \mathsf{dec}$ under public key com; $c = \mathcal{E}'_{PK'}(\mathsf{com}, m \circ \mathsf{dec})$.
3. Compute $\mathsf{tag} = \mathsf{Mac}_r(c)$.
4. Ciphertext $C = (\mathsf{com}, c, \mathsf{tag})$

Decryption $\mathcal{D}(SK, C)$: Input ciphertext $C = (\mathsf{com}, c, \mathsf{tag})$, secret key SK

1. Decrypt c by computing $m \circ \mathsf{dec} = \mathcal{D}'_{SK}(\mathsf{com}, c)$
2. Run $\mathcal{R}(\mathsf{pub}, \mathsf{com}, \mathsf{dec})$ to obtain r. If $r \neq \perp$ and $\mathsf{Vrfy}_r(c, \mathsf{tag}) = 1$, output m. Otherwise output $\perp$.

Theorem 2. *If $\mathcal{S}'$ is an RSA-based encryption scheme secure against selective-exponent weak chosen-ciphertext attacks, Ω is a secure encapsulations scheme and Δ is a strong, one-time message authentication code, then $\mathcal{S}$ is a public key encryption scheme secure against chosen-ciphertext attack.*

The proof of this theorem follows from the proof of the more general construction from selective-tag weakly CCA-secure TBE to CCA-secure PKE of Kiltz [17] and will be given in the full version.

5.1   Efficient Instantiations

To instantiate the construction in the previous section, we must specify instantiations of a message authentication code, an encapsulation scheme and a RSA-based encryption scheme.

Message authentication code. Several efficient message authentication codes are known. For example, we could use a CBC-MAC with 128-bit AES as underlying block cipher.

Encapsulation scheme. The Encapsulation Scheme was discussed Boneh and Katz [7], and their scheme is also sufficient for our purposes.

Let $H : \{0,1\}^{448} \rightarrow \{0,1\}^{128}$ be a second preimage-resistant hash function. For example, H might be constructed by modifying the output length of SHA-1. Alternatively, H may be chosen from a family of universal one-way hash functions (UOWHFs)[4] and the key included in pub below.

- Init chooses a hash function h from a family of pairwise-independent hash functions mapping 448-bit string to 128-bit strings. Output $\mathsf{pub} = h$.
- $\mathcal{C}(\mathsf{pub})$ chooses x uniformly at random in $\{0,1\}^{448}$ and outputs $(r = h(x), \mathsf{com} = H(x), \mathsf{dec} = x)$.
- $\mathcal{R}(\mathsf{pub}, \mathsf{com}, \mathsf{dec})$ outputs $h(\mathsf{dec})$ if $H(\mathsf{dec}) = \mathsf{com}$ and $\perp$ otherwise.

Informally, binding is satisfied as long as H is second-preimage resistant. The proof of statistical hiding of the scheme follows from [7].

---

[4] Not to be confused with universal hash functions as discussed in section 3.2. UOWHFs were proposed by Naor and Yung [20].

RSA-Based Encryption Scheme. We apply, in turn, the selective-exponent secure encryption schemes described in section 4.3. For the scheme based on the ODRSA assumption, the following concrete CCA-secure scheme is obtained:

ODRSA-based Encryption Scheme. Let $H : \{0,1\}^{448} \rightarrow \{0,1\}^{128}$ be a second preimage-resistant hash function. Let PRG be a pseudorandom generator.

Key Generation $\mathcal{G}(1^k)$: An RSA modulus $N = pq$ is generated, a hash function h is chosen from a family of pairwise-independent hash functions, and a function $\tilde{H}$ is chosen from a family of (almost) universal hash functions. The public key is $PK = (N, h, \tilde{H})$. The secret key is $SK = (p, q)$.

Encryption $\mathcal{E}(PK, m)$: To encrypt a message $m \in \mathbb{Z}_N^*$, choose $x \in_R \{0,1\}^{448}$, and set $r = h(x)$ and $e = H(x)$. Next choose $s \in_R \mathbb{Z}_N^*$ and set $A = s^e, B = (m \circ x) \oplus PRG(\tilde{H}((s+1)^e))$. Then compute the MAC $\mathsf{tag} = \mathsf{Mac}_r((A, B))$. The ciphertext is $C = (e, A, B, \mathsf{tag})$.

Decryption $\mathcal{E}(SK, C)$: To decrypt $C = (e, A, B, \mathsf{tag})$, first compute $d = e^{-1}$ mod $\phi(N)$ and then compute $s = A^d$. The encrypted data $m \circ x$ can be obtained by computing $B \oplus PRG(\tilde{H}((s+1)^e))$. Now set $r = h(x)$ and check that both $\mathsf{Vrfy}_r((A, B), \mathsf{tag}) = 1$ and $e = H(x)$. If so output m otherwise output $\bot$.

OSeRR-based Encryption Scheme. Let $H : \{0,1\}^{448} \rightarrow \{0,1\}^{128}$ be a second preimage-resistant hash function. Let PRG be a pseudorandom generator.

Key Generation $\mathcal{G}(1^k)$: An RSA modulus $N = pq$ is generated, a hash function h is chosen from a family of pairwise-independent hash functions, and a function $\tilde{H}$ is chosen from a family of (almost) universal hash functions. The public key is $PK = (N, h, \tilde{H})$. The secret key is $SK = (p, q)$.

Encryption $\mathcal{E}(PK, m)$: To encrypt a message $m \in \mathbb{Z}_N^*$, a choose $x \in_R \{0,1\}^{448}$ is selected, and set $r = h(x)$ and $e = H(x)$. Next choose $s \in_R \mathbb{Z}_N^*$ and set $x = s^e$ mod N^2. Now set $A = x$ mod $N, B = (m \circ x) \oplus PRG(\tilde{H}(\frac{(x-A)}{N}))$. Then compute the MAC $\mathsf{tag} = \mathsf{Mac}_r((A, B))$. The ciphertext is $C = (e, A, B, \mathsf{tag})$.

Decryption $\mathcal{E}(SK, C)$: To decrypt $C = (e, A, B, \mathsf{tag})$, first compute $d = e^{-1}$ mod $\phi(N)$ and then compute $s = A^d$. The encrypted data $m \circ x$ can be obtained by computing $B \oplus PRG(\tilde{H}(\frac{s^e \bmod N^2 - A}{N}))$. Now set $r = h(x)$ and check that both $\mathsf{Vrfy}_r((A, B), \mathsf{tag}) = 1$ and $e = H(x)$. If so output m otherwise output $\bot$.

5.2 Comparing Efficiency with Previous Schemes

We now compare the new schemes with a number of other previous efficient, CCA-secure schemes: the instantiations of the BK construction with the two IBE schemes from Boneh and Boyen [4] (BK1, BK2), the recent CCA-schemes from Kiltz [17] (a BK construction and a hybrid scheme), and the Kurosawa-Desmedt [18] (KD) hybrid scheme. In table 2 we compare the schemes in terms

Table 2. Comparison of CCA-secure schemes. We denote $|p|$ as the bit-length of group elements, and $|N|$ as the bitlength of RSA modulus N. For a typical example, $|N| = |p| = 1024$ bits. Also, f-exps denotes exponentiations with a fixed base, where precomputations can improve efficiency, S-exp denotes an exponentiation with a short exponent (eg 128 bits), L-exp denotes an exponentiation with a long exponent ($\approx |N|$ bits). Note that Chinese Remainder Theorem optimisation can be applied here. A multi-exponentiation is counted as 1.5 exponentiations.

	Encryption	Decryption	Ciphertext overhead
ODRSA-based scheme	2 S-exp	1 S-exp + 1 L-exp	$|N| + 704$
OSeRR-based scheme	1 S-exp mod N^2	1 L-exp mod N + 1 S-exp mod N^2	$|N| + 704$
Kiltz	6 f-exps	1.5 exp	$4|p| + 256$
Kiltz hybrid	6 f-exps	1.5 exp	$4|p| + 128$
BK1	3.5 f-exps	1.5 exps+ 1 pairing	$2|p| + 704$
BK2	3.5 f-exps	1 f-exp + 1 pairing	$2|p| + 704$
KD	3.5 f-exps	1.5 exps	$2|p| + 128$

of exponentiations, as they dominate the computation in all these schemes for encryption and decryption, and ciphertext overhead, which is the ciphertext length minus the message size.

In encryption, all of the previous schemes use several fixed-base exponentiations. The online computation may be sped up by precomputing powers of the bases offline such that the online cost per exponentiation is approximately 0.2 of an exponentiation. Such precomputation is not possible for our new schemes. However, in particular applications that do not permit the storage of precomputed values, our schemes would seem to have advantage. Between our two new schemes, the ODRSA-based scheme is more efficient in encryption.

In decryption, the two Kiltz schemes and the KD scheme are the fastest, requiring only one multi-exponentiation. The BK schemes are slowed down by the pairing operation, which is approximately equivalent to 5 exponentiations [24]. Between our two new schemes, the ODRSA-based scheme is again the more efficient in terms of decryption.

For equal group element lengths, which is reasonable in typical examples, our new schemes give the shortest ciphertexts of all the schemes examined.

6   Conclusion and Future Work

We have shown how to build CCA-secure public key encryption schemes from RSA-related assumptions. We have given two concrete schemes which are quite efficient in terms of computation and ciphertext size. Although these schemes are based on new RSA-related assumptions, we believe that this work provides a useful and interesting approach to developing RSA-based schemes secure against adaptive chosen-ciphertext attacks without random oracles.

An obvious direction of future work would be to investigate in more detail the validity of the Oracle RSA-type assumptions for specific choices of H. Indeed, the assumptions presented in this paper are quite strong, but we should not dismiss the possibility that specific instantiations could be of comparable validity to standard assumptions.

References

1. M. Abdalla, M. Bellare, and P. Rogaway. DHIES: An Encryption Scheme Based on the Diffie-Hellman Problem, 2001. http://www.cs.ucsd.edu/users/mihir/papers/dhaes.pdf.
2. M. Abdalla, M. Bellare, and P. Rogaway. The Oracle Diffie-Hellman Assumptions and an Analysis of DHIES. In *CT-RSA 2001*, volume 2020 of *LNCS*, pages 143–158. Springer-Verlag, 2001.
3. M. Bellare and P. Rogaway. Optimal Asymmetric Encryption - How to Encrypt with RSA. In *EUROCRYPT '94*, volume 839 of *LNCS*, pages 93–111. Springer-Verlag, 1994.
4. D. Boneh and X. Boyen. Efficient Selective-ID Secure Identity-Based Encryption Without Random Oracles. In *EUROCRYPT '04*, volume 3027 of *LNCS*, pages 223–238. Springer-Verlag, 2004.
5. D. Boneh, X. Boyen, and H. Shacham. Short Group Signatures. In *CRYPTO '04*, volume 3152 of *LNCS*, pages 41–55. Springer-Verlag, 2004.
6. D. Boneh, R. Canetti, S. Halevi, and J. Katz. Chosen-Ciphertext Security from Identity-Based Encryption. http://crypto.stanford.edu/~dabo/papers/ccaibejour.pdf, 2005. Journal submission.
7. D. Boneh and J. Katz. Improved Efficiency for CCA-Secure Cryptosystems Built Using Identity-Based Encryption. In *CT-RSA 2005*, volume 3376 of *LNCS*, pages 87–103. Springer-Verlag, 2005.
8. R. Canetti, S. Halevi, and J. Katz. Chosen-Ciphertext Security from Identity-Based Encryption. In *EUROCRYPT '04*, volume 3027 of *LNCS*, pages 207–222. Springer-Verlag, 2004.
9. J. L. Carter and M. N. Wegman. Universal Classes of Hash Functions. *JCSS*, 18(2):143–154, April 1979.
10. D. Catalano, R. Gennaro, N. Howgrave-Graham, and P. Q. Nguyen. Paillier's cryptosystem revisited. In *CCS*, pages 206–214. ACM Press, 2001.
11. R. Cramer and V. Shoup. A Practical Public Key Cryptosystem Provably Secure Against Adaptive Chosen Ciphertext Attack. In *CRYPTO '98*, volume 1462 of *LNCS*, pages 13–25. Springer-Verlag, 1998.
12. R. Cramer and V. Shoup. Universal Hash Proofs and a Paradigm for Adaptive Chosen Ciphertext Secure Public-Key Encryption. In *EUROCRYPT '02*, volume 2332 of *LNCS*, pages 45–64. Springer-Verlag, 2002.
13. Y. Dodis, R. Gennaro, J. Håstad, H. Krawczyk, and T. Rabin. Randomness Extraction and Key Derivation Using the CBC, Cascade and HMAC Modes. In *CRYPTO '04*, volume 3152 of *LNCS*, pages 494–510. Springer-Verlag, 2004.
14. D. Dolev, C. Dwork, and M. Naor. Non-malleable cryptography. In *STOC*. ACM Press, 1991.

15. E. Elkind and A. Sahai. A Unified Methodology For Constructing Public-Key Encryption Schemes Secure Against Adaptive Chosen-Ciphertext Attack. Cryptology ePrint Archive, Report 2002/042, 2002. http://eprint.iacr.org/.
16. R. Impagliazzo, L. A. Levin, and M. Luby. Pseudo-random Generation from One-Way Functions. In *STOC*, pages 12–24. ACM press, 1989.
17. E. Kiltz. Chosen-Ciphertext Security from Tag-Based Encryption. In *TCC 2006*, volume 3876 of *LNCS*, pages 581–600. Springer-Verlag, 2006.
18. K. Kurosawa and Y. Desmedt. A New Paradigm of Hybrid Encryption Scheme. In *CRYPTO '04*, volume 3152 of *LNCS*, pages 426–442. Springer-Verlag, 2004.
19. P. D. MacKenzie, M. K. Reiter, and K. Yang. Alternatives to Non-malleability: Definitions, Constructions, and Applications (Extended Abstract). In *TCC 2004*, volume 2951 of *LNCS*, pages 171–190. Springer-Verlag, 2004.
20. M. Naor and M. Yung. Universal One-Way Hash Functions and their Cryptographic Applications. In *STOC*, pages 33–43, 1989.
21. M. Naor and M. Yung. Public-Key Cryptosystems Provably Secure Against Chosen Ciphertext Attacks. In *STOC*, pages 427–437. ACM Press, 1990.
22. W. Nevelsteen and B. Preneel. Software Performance of Universal Hash Functions. In *EUROCRYPT '99*, volume 1592 of *LNCS*, pages 24–41. Springer-Verlag, 1999.
23. D. Pointcheval. New Public Key Cryptosystems Based on the Dependent-RSA Problems. In *EUROCRYPT '99*, volume 1592 of *LNCS*, pages 239–254. Springer-Verlag, 1999.
24. M. Scott. Faster Pairings using an Elliptic Curve with an Efficient Endomorphism. Cryptology ePrint Archive, Report 2005/252, 2005. http://eprint.iacr.org/.
25. V. Shoup. *A Computational Introduction to Number Theory and Algebra*. Cambridge University Press, 2005. Available at http://shoup.net/ntb/.
26. D. R. Stinson. Universal Hashing and Authentication Codes. In *CRYPTO '91*, volume 576 of *LNCS*, pages 74–85. Springer-Verlag, 1991.

General Conversion for Obtaining Strongly Existentially Unforgeable Signatures

Isamu Teranishi[1,2], Takuro Oyama[2], and Wakaha Ogata[2]

[1] NEC Corporation
1753, Shimonumabe, Nakahara-Ku, Kawasaki, Kanagawa, 211-8666, Japan
[2] Tokyo Institute of Technology
2-12-1 Ookayama, Meguro-ku Tokyo, 152-8550, Japan
teranisi@ah.jp.nec.com, taku-zy@crypt.ss.titech.ac.jp,
wakaha@mot.titech.ac.jp

Abstract. We say that a signature scheme is strongly existentially unforgeable if no adversary, given message/signature pairs adaptively, can generate a new signature on either a signature on a new message or a new signature on a previously signed message. Strongly existentially unforgeable signature schemes are used to construct many applications, such as an IND-CCA2 secure public-key encryption scheme and a group signature scheme.

We propose two general and efficient conversions, both of which transform a secure signature scheme to a strongly existentially unforgeable signature scheme. There is a tradeoff between the two conversions. The first conversion requires the random oracle, but the signature scheme transformed by the first conversion has shorter signature length than the scheme transformed by the second conversion. The second conversion does not require the random oracle. Therefore, if the original signature scheme is of the standard model, the strongly existentially unforgeable property of the converted signature scheme is proved also in the standard model.

Both conversions ensure tight security reduction to the underlying security assumptions. Moreover, the transformed schemes by the first or second conversion satisfy the on-line/off-line property. That is, signers can precompute almost all operations on the signing before they are given a message.

Keywords: signature scheme, strong unforgeability, standard model.

1 Introduction

Strong existential unforgeability (SEU) is a stronger variant of the usual security notion, existential unforgeability, of a signature scheme. Ordinary existential unforgeability prohibits an adversary from forging a valid signature on a message which a signer has not signed. However, it does not prohibit an adversary from forging a new valid signature on a message which a signer has already signed. That is, the adversary, by giving a message/signature pair (M, σ), may be able to forge a new valid signature $\sigma' \neq \sigma$ on M. SEU is a security notion which ensures

R. Barua and T. Lange (Eds.): INDOCRYPT 2006, LNCS 4329, pp. 191–205, 2006.

not only existential unforgeability but also that no adversary can execute the type of forgery mentioned above.

SEU is useful in constructing many applications, such as IND-CCA2 secure public-key encryption schemes [DDN00, CHK04] and a group signature scheme [BBS04]. We review how SEU signatures are used in such applications. In the encryption schemes [DDN00, CHK04], an SEU signature σ is used as one part of a ciphertext. It is a signature on the other part C of the ciphertext. The SEU property ensures the IND-CCA2 security of these schemes. Indeed, if the signature scheme is not SEU, an adversary may be able to obtain a new ciphertext (C, σ') by modifying the signature of another ciphertext (C, σ). This means that the encryption is malleable [DDN00], and hence is not IND-CCA2 secure.

In group signature schemes [BBS04], an authority issues a signature σ on a user's secret key x in advance. The signature will be used as an ID of the user. Hence, if the user succeeds in forging a signature, he also succeeds in forging his ID. Therefore, no signature should be able to be forged, especially, a new signature $\sigma' \neq \sigma$ on the user's secret key x. Therefore, we require not only the usual existential unforgeability but SEU property.

1.1   Our Contributions

We propose two general and efficient conversions, both of which transform a secure signature scheme to a SEU signature scheme. There is tradeoff between the two conversions. The first conversion requires the random oracle [BR93], but the signature scheme transformed by the first conversion has shorter signature length than the scheme transformed by the second conversion.

The second conversion does not require the random oracle. Therefore, if the original signature scheme is of the standard model, the SEU property of the converted signature scheme is proved also in the standard model. The scheme transformed by the first conversion has SEU property, if the original scheme is existentially unforgeable, and the discrete logarithm problem is hard to solve. The scheme transformed by the second conversion has SEU property, if the above two assumptions hold and the collision resistance of a hash function holds.

Both conversions ensure the tight security reduction to the underlying security assumptions. That is, if there exists an adversary who succeeds in breaking the SEU property of the converted scheme with probability ε' within t' steps, there exists an adversary who can break at least one of assumptions mentioned above with probability $\varepsilon \simeq \varepsilon'$ within $t \simeq t'$ steps.

Moreover, the schemes transformed by the first or second conversion satisfy the on-line/off-line property. That is, signers can precompute almost all operations on the signing before they are given a message. Therefore, the signer can generate signatures quite efficiently.

1.2   Previous Work

In PKC 2006, Boneh, Shen, and Waters [BSW06] proposed a SEU signature scheme by modifying the Waters signature scheme [W05]. They also showed that their modification is applicable to not only the Waters scheme but also any existentially unforgeable signature schemes satisfying the *partitioned* property [BSW06]. However, there are a lot of non partitioned signature schemes, such as DSS. Moreover, the modified scheme does not satisfy the on-line/off-line property. Our conversions are the first one that can convert *any* signature scheme, and are also the first one that ensures the on-line/off-line property.

1.3   Idea Behind Construction

The idea behind two conversions is the same as that of the previous conversion [BSW06]. Therefore, we first review the naive idea of [BSW06]. A signature on the converted schemes is a pair (σ, r) satisfying the tricky property $\sigma = \mathsf{Sig}_{\mathsf{sk}}(C(\sigma\|M; r))$, where σ is a signature on the original scheme and $C(\sigma\|M; r)$ is the commitment of σ generated by using the random r. Since $\sigma = \mathsf{Sig}_{\mathsf{sk}}(C(\sigma\|M; r))$ holds, we can recognize σ as "the signature on (the commitment of) the signature itself". Therefore, in order to forge a new signature (σ', r') of the converted scheme on a message M, an adversary has to forge a signature (that is, $\sigma' = \mathsf{Sig}_{\mathsf{sk}}(C(\sigma'\|M; r'))$) of the original scheme on a *new* message $C(\sigma'\|M; r')$. However, it is impossible because the original scheme is existentially unforgeable. Therefore, the converted scheme is SEU secure.

Of course, the singer cannot compute (σ, r) satisfying such property, if we use an ordinally commitment scheme. Therefore, we use the chameleon commitment [KR97, KR00] as a function C. Here the chameleon commitment is the commitment such that the committer can change the committed value if he knows the secret key. If we use the chameleon commitment as C, the signer can compute such σ as follows: compute $\sigma = \mathsf{Sig}_{\mathsf{sk}}(C(m'; r'))$ on the random $C(m'; r')$ and then change the committed value to $\sigma\|M$ by using the secret key.

However, the idea mentioned above does not work generally. Recall that, when we prove the security of the converted scheme, the simulator cannot use the secret key of the signer. Moreover, recall that one can change the committed value of only if he knows the secret key. Hence the simulator cannot change the committed value and therefore cannot simulate the signing oracle. (Therefore, the the original scheme has to be partitioned in the previous paper [BSW06] in order to simulate the signing oracle).

In order to enable simulator to simulate the signing oracle, we introduce new ideas. In the first conversion, we introduce the random oracle and use not $\sigma = \mathsf{Sig}_{\mathsf{sk}}(C(\sigma\|M; r))$ but $\sigma = \mathsf{Sig}_{\mathsf{sk}}(C(\mathcal{H}(\sigma)\|M; r))$. The simulator simulates the signing oracle by setting the hash table of $\mathcal{H}$ appropriately. In the second conversion, we introduce a *chameleon commitment with two trapdoors*. The new chameleon commitment satisfies the property that one can change the committed value if he knows one of two trapdoors. Therefore, even if the simulator does not know the secret key, it can change the committed value by using the other trapdoor.

1.4   Related Work

Independently and concurrently, Steinfeld, Pieprzyk, and Wang [S97] propose a similar conversion to our second conversion. The key idea behind the constructions of theirs and ours are the same, but the details of the constructions are different, and there are tradeoff between two conversions. The difference is how "chameleon commitment with two trapdoors" are realized. We realize it based on the discrete logarithm, that is, we realize it by modifying the discrete logarithm based chameleon commitment $g^{H(x)}h^r$ [KR97, KR00] to $g^{H(x)}h_1{}^r h_2{}^s$, where H is the hash function, x is committed value and r and s are random numbers. In contrast, Steinfeld et.al. realize it based on two general chameleon commitments C_1 and C_2, that is, they sets $C(x)$ to $C_2(C_1(x; r); s)$. Therefore, if one sets C_1 and C_2 to the discrete logarithm based chameleon commitment, $C(x) = g_2{}^{H(g_1{}^{H(x)}h_1{}^r)}h_2{}^s$ holds.

In the efficiency point of view, our conversion is better than the conversion of [S97], (if C_1 and C_2 are the discrete logarithm based chameleon commitment). First, one can compute $g^{H(x)}h_1{}^r h_2{}^s$ by computing only one exponentiation by using simultaneous exponentiation technique [MOV96], although $g_2{}^{H(g_1{}^{H(x)}h_1{}^r)}h_2{}^s$ requires two exponentiation even if one uses the technique. Second, the public key length of our converted scheme is shorter than their converted scheme, because $g^{H(x)}h_1{}^r h_2{}^s$ only requires three public group elements g, h_1, and h_2 although $g_2{}^{H(g_1{}^{H(x)}h_1{}^r)}h_2{}^s$ requires four public group elements.

In contrast, their conversion has advantage that one can use any kind of chameleon commitment. This means that one can make a conversion, the security of which is based on something other than the discrete logarithm problem. This fact has not only theoretical interest but has the practical interest also. For instance, if one set C_1 and C_2 to the factoring base chameleon commitment [ST01, CLS06], the verification cost of the converted scheme becomes smaller, (although the signature length becomes larger).

2   Preliminary

Definition 1. (Existential Unforgeability [GMR88], Strong Existential Unforgeability (SEU) [ADR02]) Let κ be a security parameter, Σ=(Gen, Sig, Ver) be a signature scheme, and $\mathcal{A}$ be an adversary. Let $\mathcal{O}_{\mathsf{sk}}^{\mathsf{sig}}$ be an oracle named *signing oracle* which, on inputting a message M, outputs a signature σ on M. We consider the following game:

$$(\mathsf{pk}, \mathsf{sk}) \leftarrow \mathsf{Gen}(1^\kappa),$$
$$(M_0, \sigma_0) \leftarrow \mathcal{A}^{\mathcal{O}_{\mathsf{sk}}^{\mathsf{sig}}}(\mathsf{pk})$$
If $\mathsf{Ver}_{\mathsf{pk}}(M_0, \sigma_0) = \text{reject}$, return 0
Return 1.

We set (M_i, σ_i) to the pair of i-th signing query of $\mathcal{A}$ and the corresponding answer. We say that $\mathcal{A}$ *wins* if the output of the above game is 1 and $\mathcal{A}$ has not made the query M_0 to the signing oracle. We also say that $\mathcal{A}$ *wins strongly* if the output of the above game is 1 and $(M_0, \sigma_0) \neq (M_i, \sigma_i)$ holds for any i.

Let $t = t(\kappa)$, $q_S = q_S(\kappa)$, and $\varepsilon = \varepsilon(\kappa)$ be non negative valued functions. We say that $\Sigma = (\mathsf{Gen}, \mathsf{Sig}, \mathsf{Ver})$ is (t, q_S, ε)-*existentially unforgeable* (resp. (t, q_S, ε)-strongly existentially unforgeable (SEU)) if for any adversary $\mathcal{A}$ such that it terminates within t steps and has made at most q_S queries to the signing oracle, the probability that $\mathcal{A}$ will win (resp. strongly win) is less than ε.

If $\Sigma = (\mathsf{Gen}, \mathsf{Sig}, \mathsf{Ver})$ is a signature scheme in the random oracle model [BR93], we say that Σ is $(t, q_S, q_H, \varepsilon)$-*existentially unforgeable* (resp. $(t, q_S, q_H, \varepsilon)$-strongly existentially unforgeable (SEU)) if for any adversary $\mathcal{A}$ such that it terminates within t steps and has made at most q_S queries to the signing oracle and at most q_H queries to the random oracle, the probability that $\mathcal{A}$ will win (resp. strongly win) the above game is less than ε.

Definition 2 (Collision Resistant Hash Function). Let κ be a security parameter and Let $\{H_\kappa\}$ be a family of functions $H = H_\kappa : \{0,1\}^* \rightarrow \{0,1\}^\kappa$ named *hash functions*. Let $t = t(\kappa)$ and $\varepsilon = \varepsilon(\kappa)$ be non negative valued functions. We say that $\{H_\kappa\}$ is (t, ε)-*collision resistant* if any adversary $\mathcal{A}$, who terminates within t steps, satisfies $\Pr((m_0, m_1) \leftarrow \mathcal{A}(1^\kappa) : H(m_0) = H(m_1)) < \varepsilon$.

Definition 3 (Discrete Logarithm Assumption). Let κ be a security parameter, and $\{\mathcal{G}_\kappa\}$ be a family of cyclic groups $\mathcal{G} = \mathcal{G}_\kappa$ with the prime order $q = q_\kappa$. We say that (t, ε)-*discrete logarithm assumption* holds in $\{\mathcal{G}_\kappa\}$ if any adversary $\mathcal{A}$, who terminates within t steps, satisfies $\Pr(h \leftarrow \mathcal{G}, z \leftarrow \mathbb{Z}_q, g \leftarrow h^z, u \leftarrow \mathcal{A}(g, h) : z = u) < \varepsilon$.

3   Proposed Conversion in the Random Oracle Model

We construct a general and efficient conversion in the random oracle model, such that the conversion transforms a secure signature scheme to an SEU signature scheme. Let κ be a security parameter, $\mathcal{G}$ be a cyclic group with order q. Let $\mathcal{H} : \{0,1\}^* \rightarrow \mathbb{Z}_q$ be a hash function, which we will replace with the random oracle model when we prove the security of the converted scheme. Let $\Sigma = (\mathsf{Gen}, \mathsf{Sig}, \mathsf{Ver})$ be a signature scheme. Our conversion transforms the scheme Σ to the signature scheme $\Sigma' = (\mathsf{Gen}', \mathsf{Sig}', \mathsf{Ver}')$ described in Fig. 1. We note that we here use the chameleon commitment $C = g^x h^r$ [KR97, KR00].

The converted scheme satisfies the on-line/off-line property. More precisely, a signer can precompute x^{-1}, C and σ before it is given the message M. By precomputing C and σ, a signer can generate a signature only by computing one hash value $\mathcal{H}(M\|\sigma)$ and one multiplication $r = (t - m)x^{-1} \bmod q$.

Theorem 1. *Let S' be the signing cost of Σ' and E be the exponentiation cost on $\mathcal{G}$. Suppose that there exists an adversary that can break $(t', q_S, q_H, \varepsilon')$-SEU property of the signature scheme $\Sigma' = (\mathsf{Gen}', \mathsf{Sig}', \mathsf{Ver}')$ in the random oracle model. Then there exists an adversary that can break either the (t, q_S, ε)-existential unforgeability of the underlying signature scheme $\Sigma = (\mathsf{Gen}, \mathsf{Sig}, \mathsf{Ver})$, or the (t, ε)-discrete logarithm problem in $\mathcal{G}$. Here*

$$\begin{cases} t = t' + q_S(S' + E) + (\text{lower terms}), \\ \varepsilon = \frac{\varepsilon'}{9} - \frac{(q_H + q_S)q_S}{3q} - (\text{lower terms}). \end{cases}$$

—Gen$'(1^\kappa)$—
$(\mathsf{pk}, \mathsf{sk}) \leftarrow \mathsf{Gen}(1^\kappa)$, $g \leftarrow \mathcal{G}$, $x \leftarrow \mathbb{Z}_q$, $h \leftarrow g^x$.
$\mathsf{pk}' \leftarrow (\mathsf{pk}, g, h)$, $\mathsf{sk}' \leftarrow (\mathsf{sk}, x)$.
Output $(\mathsf{pk}', \mathsf{sk}')$.

—Sig$'_{\mathsf{sk}'}(M)$—
$t \leftarrow \mathbb{Z}_q$, $C \leftarrow g^t$. $\sigma \leftarrow \mathsf{Sig}_{\mathsf{sk}}(C)$, $m \leftarrow \mathcal{H}(M||\sigma)$.
Select $r \in \mathbb{Z}_q$ satisfying $m + rx = t \bmod q$.
$\sigma' \leftarrow (\sigma, r)$. Output σ'.

—Ver$'_{\mathsf{pk}'}(M, \sigma')$—
Parse σ' as (σ, r). $m \leftarrow \mathcal{H}(M||\sigma)$, $C \leftarrow g^m h^r$.
If $\mathsf{Ver}_{\mathsf{pk}}(C, \sigma) = $ accept then return accept.
Otherwise return reject.

Fig. 1. Proposed Conversion in the Random Oracle Model

Proof. Let $\mathcal{A}$ be an adversary against the $(t', q_S, q_H, \varepsilon')$-SEU property of Σ'. The adversary $\mathcal{A}$ is first given a public key $\mathsf{pk}' = (\mathsf{pk}, g, h)$. $\mathcal{A}$ makes queries $M_1, \ldots, M_{q_S}$ to the signing oracle $\mathcal{O}_{\mathsf{sk}'}^{\mathsf{Sig}'}$ adaptively, and receives the signatures $\sigma_1' = (\sigma_1, r_1), \ldots, \sigma_{q_S}' = (\sigma_{q_S}, r_{q_S})$ on these messages as the answers from $\mathcal{O}_{\mathsf{sk}'}^{\mathsf{Sig}'}$. $\mathcal{A}$ finally outputs a message M and a signature $\sigma' = (\sigma, r)$. We let m_i, m, C_i, and C be $\mathcal{H}(M_i||\sigma_i)$, $\mathcal{H}(M||\sigma)$, $g^{m_i} h^{r_i}$, and $g^m h^r$.

Let ε_1, ε_2, and ε_3 be the probability that $\mathcal{A}$ will break the SEU property and the following (1), (2), and (3) will hold:

(1) $C \neq C_i$ holds for any i.

(2) $C = C_i$ holds for some i. Moreover, there is k such that, when the signing oracle computes $\sigma_k = \mathsf{Sig}_{\mathsf{sk}}(C_k)$, $M_k||\sigma_k$ has already been queried to the random oracle by the signing oracle or the adversary.

(3) $C = C_i$ holds for some i. Moreover, there exists no such k as described in (2).

Note that the latter condition of (2) means that the equality $M_k||\sigma_k = M_j||\sigma_j$ holds for some $j < k$, or $\mathcal{A}$ succeeds in predicting σ_k and making query $M_k||\sigma_k$ to the random oracle before the signing oracle computes σ_k.

Clearly, at least one of ε_1, ε_2, or ε_3 is not less than $\varepsilon'/3$. By using $\mathcal{A}$ as a subroutine, we will construct three machines $\mathcal{B}_1$, $\mathcal{B}_2$, and $\mathcal{B}_3$ and will show the following facts:

1. If $\varepsilon_i \geq \varepsilon'/3$ holds for at least one of $i = 1, 2$, then $\mathcal{B}_i$ succeeds in breaking the $(t, q_S, 3\varepsilon)$-exisitential unforgeablilty of Σ.

2. If $\varepsilon_3 \geq \varepsilon'/3$ holds, then $\mathcal{B}_3$ succeeds in breaking $(t, 3\varepsilon)$-discrete logarithm assumption in $\mathcal{G}$ respectively.

A simulator flips a coin at the beginning of the simulation and executes one of the algorithms $\mathcal{B}_1$, $\mathcal{B}_2$, or $\mathcal{B}_3$. Clearly, the simulator will succeed in guessing the type of $\mathcal{A}$ with probability $1/3$. This means that the theorem holds.

The Case where $\varepsilon_1 \geq \varepsilon'/3$ holds: Let pk_* be a randomly selected public key of the signature scheme $\Sigma = (\mathsf{Gen}, \mathsf{Sig}, \mathsf{Ver})$. By using $\mathcal{A}$ as a subroutine, we construct a machine $\mathcal{B}_1$ that can break the (t, q_S, ε)-existensial unforgeability of Σ. $\mathcal{B}_1$ runs $\mathcal{A}$ as follows:

Setup: $\mathcal{B}_1$ executes the same procedure as Gen', except that $\mathcal{B}_1$ sets pk to pk_*. More precisely, $\mathcal{B}_1$ selects $g \in \mathcal{G}$ and $x \in \mathbb{Z}_q$ randomly and sets $h = g^x$ and $\mathsf{pk}' = (\mathsf{pk}_*, g, h)$. Then $\mathcal{B}_1$ provides pk' to $\mathcal{A}$.

Random Oracle Simulation: Let X be a query of $\mathcal{A}$. If $\mathcal{H}(X)$ has already been determined, $\mathcal{B}_1$ sends $\mathcal{H}(X)$ back to $\mathcal{A}$. Otherwise, $\mathcal{B}_1$ selects $m \in \mathbb{Z}_q$ randomly, sets $\mathcal{H}(X)$ to m, and sends $m = \mathcal{H}(X)$ back to $\mathcal{A}$.

Signing Oracle Simulation: Let M_i be the i-th queried message of $\mathcal{A}$. $\mathcal{B}_1$ selects $t_i \in \mathbb{Z}_q$ randomly, and sets $C_i = g^{t_i}$. Then $\mathcal{B}_1$ makes the query C_i to its signing oracle, and receives a signature σ_i on C_i as the answer.

$\mathcal{B}_1$ determines $m_i = \mathcal{H}(M_i \| \sigma_i)$ as in the case of the random oracle simulation. That is, if $\mathcal{H}(M_i \| \sigma_i)$ has not been determined yet, $\mathcal{B}_1$ takes m_i randomly and sets $m_i = \mathcal{H}(M_i \| \sigma_i)$. Otherwise, $\mathcal{B}_1$ obtains the hash value m_i from the hash table.

Then $\mathcal{B}_1$ selects $r_i \in \mathbb{Z}_q$ satisfying $m_i + r_i x = t_i \bmod q$, and sets $\sigma_i' = (\sigma_i, r_i)$. (Note that there is no such r_i in the case where $x = 0$ holds. However, it occurs only with negligible probability $1/q$). One can easily show that σ_i' is a valid signature on M_i. $\mathcal{B}_1$ finally sends σ_i' to $\mathcal{A}$.

Extraction: Suppose that $\mathcal{A}$ succeeds in forging a message/signature pair (M, σ'). That is, suppose that $\mathcal{A}$ outputs a message M and a valid signature $\sigma' = (\sigma, r)$ on M, such that $(M, \sigma') \neq (M_i, \sigma_i')$ holds for any i.

$\mathcal{B}_1$ determines $m = \mathcal{H}(M \| \sigma)$ as in the case of the random oracle simulation. That is, if $\mathcal{H}(M \| \sigma)$ has not been determined yet, $\mathcal{B}_1$ takes m_i randomly and sets $m = \mathcal{H}(M \| \sigma)$. Otherwise, $\mathcal{B}_1$ obtains the hash value m from the hash table.

Then $\mathcal{B}_1$ computes $C = g^m h^r$. Since $\sigma' = (\sigma, r)$ is a valid signature on M, σ is a valid signature on C.

Let $C_1, \ldots, C_{q_S}$ be the signing queries by $\mathcal{A}$. Recall that $\mathcal{B}_1$ is not allowed to output a message/signature pair such that $\mathcal{B}_1$ has sent the message as a query to the signing oracle. Recall also that $\mathcal{B}_1$ makes no signing query other than $C_1, \ldots, C_{q_S}$. Therefore, if $C \neq C_i$ holds for any i, $\mathcal{B}_1$ outputs (C, σ). Otherwise, the simulation fails.

The number of steps until $\mathcal{B}_1$ terminates is clearly not more than $t' + q_S E + 2E \leq t' + (S' + E)q_S +$ (lower terms). We estimate the success probability of $\mathcal{B}_1$. Recall that $\mathcal{B}_1$ succeeds in forging a signature if $C \neq C_i$ holds for any i. From the definition of ε_1, $\mathcal{A}$ succeeds in forging (M, σ') and $C \neq C_i$ holds for any i with probability ε_1. Since we supposed that $\varepsilon_1 \geq \varepsilon'/3$ holds, the probability that $\mathcal{B}_1$ will succeed in forging a signature is at least $\varepsilon_1 \geq (\varepsilon'/3) - ((q_H + q_S)q_S/q)$.

The Case where $\varepsilon_2 \geq \varepsilon'/3$ holds: By using $\mathcal{A}$ as a subroutine, we construct a machine $\mathcal{B}_2$ that can break the (t, q_S, ε)-existensial unforgeability of Σ.

$\mathcal{B}_2$ executes the same procedure as Gen', except that $\mathcal{B}_2$ sets pk to pk_*. That is, $\mathcal{B}_2$ selects $g \in \mathcal{G}$ and $x \in \mathbb{Z}_q$ randomly, sets $h = g^x$, and $\mathsf{pk}' = (\mathsf{pk}_*, g, h)$.

Then $\mathcal{B}_2$ provides pk$'$ to $\mathcal{A}$ and runs $\mathcal{A}$. $\mathcal{B}_2$ simulates the random oracle similarly to $\mathcal{B}_1$, maintaining a hash table.

For each k, when $\mathcal{A}$ makes k-th signing query M_k, $\mathcal{B}_2$ simulates the first step of the singing oracle. That is, $\mathcal{B}_2$ selects $t_i \in \mathbb{Z}_q$ randomly and sets $C_k = g^{t_k}$. (We stress that $\mathcal{B}_2$ simulates only the first step of the signing oracle. Therefore, $\mathcal{B}_2$ does not make query C_k to the signing oracle).

Then $\mathcal{B}_2$ finds $M||\sigma$ satisfying $\mathsf{Ver}_{\mathsf{pk}'}(C_k, \sigma) = \mathsf{accept}$ from the hash table. If there is such σ, (in this case, we call the current k an *expected k*), $\mathcal{B}_2$ checks that $C_k = C_i$ holds for some $i = 1, \ldots, k-1$, where C_i is the i-th signing query $\mathcal{B}_2$ made. If $C_k = C_i$ holds for some i, the simulation fails and $\mathcal{B}_2$ outputs a symbol fail_1 and terminates. Otherwise, $\mathcal{B}_2$ outputs (C_k, σ) as a forged pair.

In the case that k is not an expected one, that is, there is no σ in the hash table satisfying the above condition, $\mathcal{B}_2$ determines (σ_k, r_k) similarly to $\mathcal{B}_1$, and continues the simulation. (Remember that $\mathcal{B}_2$ makes a signing query C_k.)

If $\mathcal{A}$ outputs a forged pair before $\mathcal{B}_2$ found expected k, $\mathcal{B}_2$ outputs a symbol fail_2 and terminates.

The number of steps until $\mathcal{B}_2$ terminates is clearly not more than $t' + q_S E + 2E \leq t' + (S' + E)q_S + (\text{lower terms})$. We estimate the probability that $\mathcal{B}_2$ suceeds in forging a signature. From the definition of ε_2, the probaility that $\mathcal{B}_2$ outputs fail_2 is at most $1 - \varepsilon_2$. Moreover, the equality $C_k = C_i$ holds for the expected k and for some i with probability at most $(k-1)/q \leq q_S/q$, since $C_k = g^{t_k}$ distributes unigormly on $\mathcal{G}$. Therefore, $\mathcal{B}_2$ succeeds in forging a new message/signature pair (C_k, σ) with probabilily $\varepsilon_2 - (q_S/q) \geq (\varepsilon'/3) - ((q_H + q_S)q_S/q)$.

The Case where $\varepsilon_3 \geq \varepsilon'/3$ holds: By using $\mathcal{A}$ as a subroutine, we construct a machine $\mathcal{B}_3$ that can break the (t, ε)-discrete logarithm problem in $\mathcal{G}$. Let $(h_*, g_*) \in \mathcal{G}^2$ be an instance of the discrete logarithm problem in $\mathcal{G}$. The aim of $\mathcal{B}_3$ is to obtain $z_* \in \mathbb{Z}_q$ satisfying $g_* = h_*^{z_*}$. $\mathcal{B}_3$ runs $\mathcal{A}$ as follows:

Setup: $\mathcal{B}_3$ executes the same procedure as Gen$'$, except that $\mathcal{B}_3$ sets (g, h) to (g_*, h_*). More precisely, $\mathcal{B}_3$ executes $\mathsf{Gen}(1^\kappa)$, obtains (pk, sk) as the output of $\mathsf{Gen}(1^\kappa)$, and sets $(g, h) = (g_*, h_*)$ and pk$' = (\text{pk}, g, h)$. Then $\mathcal{B}_3$ provides pk$'$ to $\mathcal{A}$.

Random Oracle Simulation: Let X be a query of $\mathcal{A}$. If $\mathcal{H}(X)$ is already determined, $\mathcal{B}_3$ sends $\mathcal{H}(X)$ back to $\mathcal{A}$. Otherwise, $\mathcal{B}_3$ selects $m \in \mathbb{Z}_q$ randomly, sets $\mathcal{H}(X)$ to m, and sends $m = \mathcal{H}(X)$ back to $\mathcal{A}$.

Signing Oracle Simulation: Let M_i be the i-th queried message of $\mathcal{A}$. $\mathcal{B}_3$ selects $m_i, r_i \in \mathbb{Z}_q$ randomly, and sets $C_i = g^{m_i} h^{r_i}$. By using the secret key sk, $\mathcal{B}_3$ computes $\sigma_i = \mathsf{Sig}_{\mathsf{sk}}(C_i)$. If the hash value corresponding to $M_i||\sigma_i$ has already been determined, then the simulation fails and $\mathcal{B}_3$ outputs a symbol fail_1 and terminates. Otherwise, $\mathcal{B}_3$ sets the hash value $\mathcal{H}(M_i||\sigma_i)$ to m_i. One can easily show that $\sigma_i' = (\sigma_i, r_i)$ is a valid signature on M_i. $\mathcal{B}_3$ finally sends σ_i' to $\mathcal{A}$.

Extraction: Suppose that $\mathcal{A}$ outputs a message M and a valid signature $\sigma' = (\sigma, r)$ on M, such that $(M, \sigma') \neq (M_i, \sigma_i')$ holds for any i. If there is no i satisfying $C = C_i$, the simulation fails and $\mathcal{B}_3$ outputs a symbol fail_2 and terminates.

We consider the case where $C = C_i$ holds for some i. Let m and m_i be $\mathcal{H}(M\|\sigma)$ and $\mathcal{H}(M_i\|\sigma_i)$. Since the data $C = g^m h^r = g_*{}^m h_*{}^r$ is equal to $C_i = g^{m_i} h^{r_i} = g_*{}^{m_i} h_*{}^{r_i}$, the equation $h_*{}^{r-r_i} = g_*{}^{m_i-m}$ holds. If $r - r_i \neq 0$ holds, $\mathcal{B}_3$ succeeds in computing the discrete logarithm $z_* = (m_i - m)/(r - r_i) \bmod q$ of g_* based on h_*. Otherwise, the simulation fails and $\mathcal{B}_3$ outputs a symbol fail_3 and terminates.

The number of steps until $\mathcal{B}_3$ terminates is clearly not more than $t' + 2E + (S + 2E)q_S = t' + (S' + E)q_S + (\text{lower terms})$, where S is the signing cost of Σ.

We next estimate the probability that $\mathcal{B}_3$ will succeed in obtaining z_*. We suppose the following three events occur: $\mathcal{A}$ succeeds in forging a signature, $C = C_i$ holds for some i, and there is no k such that, when the signing oracle computes $\sigma_k = \mathsf{Sig}_{\mathsf{sk}}(C_k)$, $M_k\|\sigma_k$ has already been written in the hash table. From the definition of ε_3, these three events occur with probability at least $\varepsilon_3 \geq \varepsilon'/3$. Therefore, from the above assumption, $\mathcal{B}_3$ does not output fail_1 or fail_2.

We next estimate the probability that $\mathcal{B}_3$ will output fail_3. Recall that $\mathcal{A}$ and the signing oracle $\mathcal{O}_{\mathsf{sk}'}^{\mathsf{Sig}'}$ makes at most q_H and q_S hash queries respectively. Recall also that hash values are randomly taken from $\mathbb{Z}_q$. Therefore, the probability that

$$\exists \ell,\ \exists X \in (\text{hash table})\ :\ X \neq M_\ell\|\sigma_\ell \wedge \mathcal{H}(X) = \mathcal{H}(M_\ell\|\sigma_\ell)$$

will hold is at most $(q_H + q_S)q_S/q$.

We consider the case where there exists no such (ℓ, X). We show that $\mathcal{B}_3$ does not output fail_3 in this case. Let us make a contradictory supposition that $\mathcal{B}_3$ outputs fail_3. Then the equality $r = r_i \bmod q$ holds for some i. Recall that $C = C_i$ holds. Hence, it follows that $g^m h^r = C = C_i = g^{m_i} h^{r_i} = g^{m_i} h^r$. Therefore, $g^m = g^{m_i}$ holds. This means that $m = m_i \bmod q$ holds. Therefore, the equality $\mathcal{H}(M\|\sigma) = m = m_i = \mathcal{H}(M_i\|\sigma_i)$ holds. Since there is no X satisfying $X \neq M_i\|\sigma_i$ and $\mathcal{H}(X) = \mathcal{H}(M_i\|\sigma_i)$, the equality $(M, \sigma) = (M_i, \sigma_i)$ has to hold. From the definition of SEU, $(M, \sigma') = (M, (\sigma, r))$ is not equal to $(M_i, \sigma_i') = (M_i, (\sigma_i, r_i))$. Therefore, $r \neq r_i \bmod q$ has to hold, and this contradicts the assumption that $r = r_i \bmod q$ holds. This means that $\mathcal{B}_3$ does not output fail_3 in this case.

From the above discussion, the probability that $\mathcal{B}_3$ will succeed in obtaining the discrete logarithm is at least $(\varepsilon'/3) - (q_H + q_S)q_S/q$. $\qquad\square$

We finally estimate the security of our scheme more intuitively. In order to do it, we introduce the notions, which we call *difficulty*. For an adversary $\mathcal{X}$ against (t, ε)-discrete logarithm problem in $\mathcal{G}$, we let $\mathcal{X}^*$ be an adversary which executes $\mathcal{X}$ until $\mathcal{X}$ succeeds in solving the problem. Then $\mathcal{X}^*$ solves the discrete logarithm problem with t/ε steps on average and with probability 1. This means that we can use the value t/ε in order to estimate how difficult one solves the discrete logarithm problem. Therefore, we say that the discrete logarithm problem has *difficulty* T if there is no (t, ε)-adversary $\mathcal{X}$ satisfying $t/\varepsilon < T$. We also define the difficulties of the existential unforgeability and the SEU property similarly.

We will estimate the difficulty of the SEU property of the our proposed scheme.

Corollary 2. *Let S' be the signing cost of Σ', and E be the exponentiation cost on $\mathcal{G}$, Suppose that the existential unforgeability of Σ and the discrete logarithm problem in $\mathcal{G}$ have difficulty T_1 and T_2 respectively. We let C_0 denote $20(1+E/S)$. Then the SEU property of the proposed scheme Σ' has difficulty T'. Here*

$$T' \geq \min\{T_1, T_2\}/C_0 + \text{(lower terms)}.$$

Proof. We will show that there exists an adversary $\mathcal{A}_0$ which can break the SEU property of Σ' within $2E2^{\kappa/2} + \text{(lower terms)}$ step and with the probability 1. Therefore, by substituting $\mathcal{A}$ to $\mathcal{A}_0$ (if we need), we can assume that $t'/\varepsilon' \leq 2^{\kappa/2} + \text{(lower terms)}$ holds.

From the definition of q_S and q_H, the inequalities $Sq_S \leq t'$ and $q_H \leq t'$ hold. Since $S' = S + E + \text{(lower terms)}$ holds, it follows

$$t \leq t' + q_S(S' + E) \leq (1 + (S' + E)/S)t' = (1 + (S + 2E)/S)t' = (2 + 2E/S)t',$$

$$(q_H + q_S)q_S/(3q\varepsilon') \leq (t' + t'/S)(t'/S)/(3q\varepsilon') \simeq t'^2/(3qS\varepsilon') \leq (t'/\varepsilon')^2/(3qS)$$
$$\leq (2^{\kappa/2})^2/(3 \cdot 2^{\kappa-1}S) = 2/3S,$$

$$\varepsilon = \varepsilon'/9 - (q_H + q_S)q_S/(3q) = \varepsilon' \cdot (1/9 - (q_H + q_S)q_S/(3q\varepsilon')) \geq \varepsilon' \cdot (1/9 - 2/3S) \geq \varepsilon'/10$$

holds, (because $S \gg 0$ holds if $\kappa \gg 0$). Hence, it follows that

$$\min\{T_1, T_2\} \leq t/\varepsilon \leq (2 + 2E/S)t'/(\varepsilon'/10) = 20(1 + E/S) \cdot (t'/\varepsilon') = C_0 \cdot (t'/\varepsilon').$$

Hence $\min\{T_1, T_2\} \leq C_0 \cdot T'$ holds. Therefore, $T' \leq \min\{T_1, T_2\}/C_0$ holds.

We finally construct $\mathcal{A}_0$. $\mathcal{A}_0(\mathsf{pk}')$ computes the discrete logarithm x of (g, h) by using the Baby Step and Giant Step (BSGS) algorithm [BSS99], sends an arbitrarily message M to the signing oracle as a query, receives the answer $\sigma' = (\sigma, r)$, and computes $m = \mathcal{H}(M||\sigma)$ and $C = g^m h^r$. Then $\mathsf{Ver}_{\mathsf{pk}}(C, \sigma) = \text{accept}$ holds. $\mathcal{A}_0$ then selects an arbitrarily message $M_0 \neq M$, computes $m_0 = \mathcal{H}(M_0||\sigma)$, selects $r_0 \in \mathbb{Z}_q$ satisfying $m_0 + r_0 x = m + rx \bmod q$, sets $\sigma_0 = (\sigma, r_0)$, and outputs (M_0, σ_0). One can easily show that σ_0 is a valid signature on $M_0 \neq M$. Since BSGS algorithm requires $2^{\kappa/2} + \text{(lower terms)}$ steps, the number of steps of $\mathcal{A}_0$ is also $2^{\kappa/2} + \text{(lower terms)}$. □

4   Proposed Conversion in the Standard Model

By modifying the conversion of the last section, we construct a conversion in the standard model. That is, we construct a conversion such that the SEU property of the converted scheme can be proved without exploiting the random oracle. Fig. 2 describes the signature scheme transformed by our conversion. Here κ is a security parameter, $\mathcal{G}$ is a cyclic group with the prime order q, $\Sigma = (\mathsf{Gen}, \mathsf{Sig}, \mathsf{Ver})$ is a signature scheme, and $\{H_\kappa\}$ is a family of collision resistant hash functions $H = H_\kappa : \{0,1\}^* \to \mathbb{Z}_q$.

As in the case of the other conversion, the converted scheme satisfies the on-line/off-line property. More precisely, a signer can precompute x^{-1}, C and σ before it is given the message M. By precomputing C and σ, a signer can generate a signature only by computing one hash value $H(M||\sigma)$ and one multiplication $r = (t - m - sy)x^{-1} \bmod q$.

—Gen$'(1^\kappa)$—
$(\mathsf{pk}, \mathsf{sk}) \leftarrow \mathsf{Gen}(1^\kappa)$,
$g \leftarrow \mathcal{G}$,$x, y \leftarrow \mathbb{Z}_q$, $(h_1, h_2) \leftarrow (g^x, g^y)$.
$\mathsf{pk}' \leftarrow (\mathsf{pk}, g, h_1, h_2)$, $\mathsf{sk}' \leftarrow (\mathsf{sk}, x, y)$.
Output $(\mathsf{pk}', \mathsf{sk}')$.

—Sig$'_{\mathsf{sk}'}(M)$—
$t \leftarrow \mathbb{Z}_q$, $C \leftarrow g^t$. $\sigma \leftarrow \mathsf{Sig}_{\mathsf{sk}}(C)$, $m \leftarrow H(M||\sigma)$.
Randomly selects $r, s \in \mathbb{Z}_q$ satisfying $m + rx + sy = t \bmod q$.
$\sigma' \leftarrow (\sigma, r, s)$. Output σ'.

—Ver$'_{\mathsf{pk}'}(M, \sigma')$—
Parse σ' as (σ, r, s).
$m \leftarrow H(M||\sigma)$, $C \leftarrow g^m h_1{}^r h_2{}^s$.
If $\mathsf{Ver}_{\mathsf{pk}}(C, \sigma) = $ accept then return accept.
Otherwise return reject.

Fig. 2. Proposed Conversion in the Standard Model

Theorem 3. *Let S' be the signing cost of Σ'. Suppose that there exists a $(t', q_S, q_H, \varepsilon')$-adversary against the SEU property of the signature scheme $\Sigma' = (\mathsf{Gen}', \mathsf{Sig}', \mathsf{Ver}')$. Then there exists an adversary that can break either the (t, q_S, ε)-existential unforgeability of the underlying signature scheme $\Sigma = (\mathsf{Gen}, \mathsf{Sig}, \mathsf{Ver})$, the (t, ε)-discrete logarithm problem in $\mathcal{G}$, or (t, ε)-collision resistant of H. Here*

$$t = t' + q_S S' + \text{(lower terms)},$$
$$\varepsilon = \tfrac{\varepsilon'}{4} - \text{(lower terms)}.$$

Proof. Let $\mathcal{A}$ be an adversary that breaks the (t', q_S, ε')-SEU property of Σ'. The adversary $\mathcal{A}$ is first given a public key $\mathsf{pk}' = (\mathsf{pk}, g, h_1, h_2)$. $\mathcal{A}$ makes queries $M_1, \ldots, M_{q_S}$ to the signing oracle $\mathcal{O}^{\mathsf{Sig}'}_{\mathsf{sk}'}$ adaptively, and receives the signatures $\sigma'_1 = (\sigma_1, r_1, s_1), \ldots, \sigma'_{q_S} = (\sigma_{q_S}, r_{q_S}, s_{q_S})$ on these messages as the answers from $\mathcal{O}^{\mathsf{Sig}'}_{\mathsf{sk}'}$. $\mathcal{A}$ finally outputs a message M and a signature $\sigma' = (\sigma, r, s)$. We let m_i, m, C_i, and C be $H(M_i||\sigma_i)$, $H(M||\sigma)$, $g^{m_i} h_1{}^{r_i} h_2{}^{s_i}$, and $g^m h_1{}^r h_2{}^s$.

We distinguish among four types of forgeries:

Type 1: A forgery where $C \neq C_i$ for any i.
Type 2: A forgery where $(C, r, s) = (C_i, r_i, s_i)$ for some i.
Type 3A: A forgery where $C = C_i$ and $r \neq r_i$ for some i.
Type 3B: A forgery where $C = C_i$ and $s \neq s_i$ for some i.

By using $\mathcal{A}$ as a subroutine, we will construct four machines $\mathcal{B}_1$, $\mathcal{B}_2$, and $\mathcal{B}_{3A}$, $\mathcal{B}_{3B}$. We will show the following facts:

- If $\mathcal{A}$ is a Type 1 adversary, $\mathcal{B}_1$ succeeds in breaking the (t, q_S, ε')-existential unforgeability of Σ.
- If $\mathcal{A}$ is a Type 2 adversary, $\mathcal{B}_2$ succeeds in breaking (t, ε')-collision resistance of H.

– If $\mathcal{A}$ is a Type 3A or Type 3B adversary, $\mathcal{B}_{3A}$ or $\mathcal{B}_{3B}$ respectively succeeds in breaking the (t, ε')-discrete logarithm assumption in $\mathcal{G}$.

A simulator flips a coin at the beginning of the simulation to guess which type of forgery $\mathcal{A}$ will produce. Then the simulator executes an algorithm $\mathcal{B}_1$, $\mathcal{B}_2$, $\mathcal{B}_{3A}$, or $\mathcal{B}_{3B}$. Clearly, the simulator will succeed in guessing the type of $\mathcal{A}$ with probability $1/4$. This means that the theorem holds.

Type 1 adversary: Suppose that $\mathcal{A}$ is a Type 1 adversary that breaks the (t', q_S, ε')-SEU property of Σ'. Let pk_* be a randomly selected public key of the signature scheme $\Sigma = (\mathsf{Gen}, \mathsf{Sig}, \mathsf{Ver})$. By using $\mathcal{A}$ as a subroutine, we construct a machine $\mathcal{B}_1$ that can break the (t, q_S, ε')-existential unforgeability of Σ. $\mathcal{B}_1$ runs $\mathcal{A}$ as follows:

Setup: $\mathcal{B}_1$ firstly executes the same procedure as Gen', except that $\mathcal{B}_1$ sets pk to pk_*. More precisely, $\mathcal{B}_1$ selects $g \in \mathcal{G}$ and $x, y \in \mathbb{Z}_q$ randomly, sets $(h_1, h_2) = (g^x, g^y)$, and $\mathsf{pk} = \mathsf{pk}_*$ and $\mathsf{pk}' = (\mathsf{pk}, g, h_1, h_2)$. Then $\mathcal{B}_1$ provides pk' to $\mathcal{A}$.

Signing Oracle Simulation: Let M_i be the i-th queried message of $\mathcal{A}$. $\mathcal{B}_1$ executes the same algorithm as $\mathsf{Sig}'_{\mathsf{sk}'}(M_i)$ except that $\mathcal{B}_1$ does not execute $\mathsf{Sig}_{\mathsf{sk}}$ but makes a query to the signing oracle. More precisely, $\mathcal{B}_1$ executes the following procedures. $\mathcal{B}_1$ selects $t_i \in \mathbb{Z}_q$ randomly, and sets $C_i = g^{t_i}$. Then $\mathcal{B}_1$ makes the query C_i to its signing oracle, and receives a signature σ_i on C_i as the answer. $\mathcal{B}_1$ computes $m_i = H(M_i || \sigma_i)$, selects $r_i, s_i \in \mathbb{Z}_q$ satisfying $m_i + r_i x + s_i y = t_i \bmod q$, and sets $\sigma'_i = (\sigma_i, r_i, s_i)$. One can easily show that σ'_i is a valid signature on M_i. $\mathcal{B}_1$ finally sends σ'_i to $\mathcal{A}$.

Extraction: Suppose that $\mathcal{A}$ outputs a message M and a valid signature $\sigma' = (\sigma, r, s)$ on M, such that $(M, \sigma') \neq (M_i, \sigma'_i)$ for any i. Recall that $\mathcal{A}$ is a Type 1 adversary. Therefore, $C \neq C_i$ holds for any i. Since $\sigma' = (\sigma, r, s)$ is a valid signature on M, σ is a valid signature on C. Recall that the queries that $\mathcal{B}_1$ has made to the signing oracle are $C_1, \dots, C_{q_S}$. This means that $\mathcal{B}_1$ succeeds in forging the signature σ on the message C, which $\mathcal{B}_1$ has not sent as a query to the signing oracle.

One can easily show that $\mathcal{B}_1$ succeeds in breaking the (t, q_S, ε')-existentially unforgeability of Σ. Here we used the fact that $(2q_S + 2)E \simeq 2q_S E + (\text{lower terms}) \leq q_S S' + (\text{lower terms})$.

Type 2 adversary: Suppose that $\mathcal{A}$ is a Type 2 adversary that breaks the (t, q, ε)-SEU property of Σ'. By using $\mathcal{A}$ as a subroutine, we construct a machine $\mathcal{B}_2$ that can break the (t, ε)-collision resistance of H. $\mathcal{B}_2$ runs $\mathcal{A}$ as follows:

Setup: $\mathcal{B}_2$ firstly executes $\mathsf{Gen}'(1^\kappa)$ and obtains $\mathsf{pk}' = (\mathsf{pk}, g, h_1, h_2)$, and $\mathsf{sk}' = (\mathsf{sk}, x, y)$ as the output of Gen'. Then $\mathcal{B}_2$ provides pk' to $\mathcal{A}$.

Signing Oracle Simulation: By using the secret key sk', $\mathcal{B}_2$ computes a signature on a queried message that $\mathcal{A}$ makes, and sends the signature to $\mathcal{A}$.

Extraction: Suppose that $\mathcal{A}$ outputs a message M and a valid signature $\sigma' = (\sigma, r, s)$ on M, such that $(M, \sigma') \neq (M_i, \sigma'_i)$ for any i. Recall that $\mathcal{A}$ is a Type 2 adversary. Therefore, we can suppose that $(C, r, s) = (C_i, r_i, s_i)$ for some i. Since the data $C = g^{H(M||\sigma)}h_1{}^r h_2{}^s$ equals to $C_i = g^{H(M_i||\sigma_i)}h_1{}^{r_i}h_2{}^{s_i} = g^{H(M_i||\sigma_i)}h_1{}^r h_2{}^s$, the hash value $H(M||\sigma)$ equals $H(M_i||\sigma_i)$. Recall that $(M, \sigma') \neq (M_i, \sigma'_i)$ holds. This means that $(M||\sigma, M_i||\sigma_i)$ is a collision pair of H. Therefore, $\mathcal{B}_2$ outputs $(M||\sigma, M_i||\sigma_i)$ and stops.

One can easily show that $\mathcal{B}_2$ succeeds in breaking (t, ε')-collision resistance of $\{H\}$.

Type 3A adversary: Suppose that $\mathcal{A}$ is a Type 3A adversary that breaks the (t, q, ε)-SEU property of Σ'. By using $\mathcal{A}$ as a subroutine, we construct a machine $\mathcal{B}_{3A}$ that can break the (t, ε)-discrete logarithm problem in $\mathcal{G}$. Let $(h_*, g_*) \in \mathcal{G}^2$ be an instance of the discrete logarithm problem in $\mathcal{G}$. The aim of $\mathcal{B}_3$ is to obtain $z_* \in \mathbb{Z}_q$ satisfying $g_* = h_*{}^{z_*}$. $\mathcal{B}_{3A}$ runs $\mathcal{A}$ as follows:

Setup: $\mathcal{B}_{3A}$ firstly executes the same procedure as Gen$'$, except that $\mathcal{B}_{3A}$ sets (g, h_1) to (g_*, h_*). More precisely, $\mathcal{B}_{3A}$ executes Gen(1^κ), to obtain (pk, sk), selects $y \in \mathbb{Z}_q$ randomly, sets $(g, h_1) = (g_*, h_*)$, $h_2 = g^y$, pk$' = (\text{pk}, g, h_1, h_2)$. Then $\mathcal{B}_{3A}$ provides pk$'$ to $\mathcal{A}$.

Signing Oracle Simulation: Let M_i be the i-th queried message of $\mathcal{A}$. $\mathcal{B}_{3A}$ selects $t'_i, r_i \in \mathbb{Z}_q$ randomly, and sets $C_i = g^{t'_i}h_1{}^{r_i}$. By using the secret key sk, $\mathcal{B}_{3A}$ computes $\sigma_i = \text{Sig}_{\text{sk}}(C_i)$. Then $\mathcal{B}_{3A}$ computes $m_i = H(M_i||\sigma_i)$, selects $s_i \in \mathbb{Z}_q$ satisfying $m_i + s_i y = t'_i \bmod q$, and sets $\sigma'_i = (\sigma_i, r_i, s_i)$. One can easily show that σ'_i is a valid signature on M_i. $\mathcal{B}_{3A}$ finally sends σ'_i to $\mathcal{A}$.

Extraction: Suppose that $\mathcal{A}$ outputs a message M and a valid signature $\sigma' = (\sigma, r, s)$ on M, such that $(M, \sigma') \neq (M_i, \sigma'_i)$ for any i. Recall that $\mathcal{A}$ is a Type 3A adversary. Therefore, $C = C_i$ and $r \neq r_i$ hold for some i. Let m and m_i be $H(M||\sigma)$ and $H(M_i||\sigma_i)$. Since the data $C = g^m h_1{}^r h_2{}^s = g_*{}^{m+sy}h_*{}^r$ is equal to $C_i = g^m h_1{}^{r_i}h_2{}^{s_i} = g_*{}^{m_i+s_i y}h_*{}^{r_i}$, the equation $h_*{}^{r-r_i} = g_*{}^{m_i+s_i y-m-sy}$ holds. Recall that $r \neq r_i$ holds. Therefore, $z_* = (m_i + s_i y - m - sy)/(r - r_i) \bmod q$ is the discrete logarithm of g_* based on h_*. Therefore, $\mathcal{B}_{3A}$ outputs z_* and stops.

One can easily show that $\mathcal{B}_{3A}$ succeeds in breaking the (t, ε')-discrete logarithm assumption in $\mathcal{G}$.

Type 3B adversary: The proof for this type adversary is quite similar to the proof for the Type 3A adversary, although $\mathcal{B}_{3B}$ embeds (g_*, h_*) not to (g, h_1) but to (g, h_2). Therefore, we omit the details. $\square$

One can easily show the following corollary:

Corollary 4. *Suppose that the existential unforgeability of Σ, the discrete logarithm problem in $\mathcal{G}$ and the collision resistance problem of H have difficulty T_1, T_2, and T_3 respectively. Then the SEU property of the proposed scheme Σ' has difficulty T'. Here*

$$T' \geq (2 + 2(E/S)) \min\{T_1, T_2, T_3\} + \text{(lower terms)}.$$

	[BSW06]	Conv. of Sec. 3	Conv. of Sec. 4
Condition on Σ	Partitioned	Nothing	Nothing
Model	Standard	Random Oracle	Standard
Reduction	Tight	Tight	Tight
Precomputation before Signing	0	$S + E$	$S + E$
Signing using Precomp. data	$S + E$	0	0
Total Signing Cost	$S + E$	$S + E$	$S + E$
Verification Cost	$V + E$	$V + E$	$V + E$
Signature Length	$|\sigma| + |q|$	$|\sigma| + |q|$	$|\sigma| + 2|q|$

Fig. 3. Comparison

5 Comparison

Fig. 3 compares the schemes transformed by the conversion of [BSW06], of Section 3, and of Section 4. In this figure, S and V represents the computational cost of the signing and verifying algorithms of the original signature scheme Σ respectively. The value E represents the exponentiation cost on $\mathcal{G}$, $|\sigma|$ represents the bit length of a signature of Σ, and $|q|$ represents the bit length of q.

We assume that one computes $g^m h^r$ by using the simultaneous exponentiation technique [MOV96]. That is, we assume that the computational cost to compute $g^m h^r$ is equal to E. We also assume that the computational cost of a multiplication on $\mathcal{G}$ and a hashing are very small.

The conversion of [BSW06] is applicable for a signature scheme that satisfies the partitioned property [BSW06]. However, the authors do not give any example of the partitioned signature scheme other than the Waters scheme. In contrast, our two conversions are applicable to any signature scheme.

The signing costs of all of three schemes are equal. However, in the case of schemes transformed by our conversions, signers can precompute almost all operations on the signing before they are given messages.

The signature length of the scheme transformed by our second conversion is longer than that of the scheme transformed by our first conversion. But the security of the former scheme is proved without assuming the random oracle, although that of the latter scheme is proved only when one assumes the random oracle.

6 Conclusion

For the first time, we proposed two conversions, which are the first that can transform any secure signature scheme to a SEU signature scheme, and we also introduce the notion difficulty, and estimate the difficulties of our converted schemes. There is trade off between the two conversions. The first conversion requires the random oracle, but the signature scheme transformed by the first conversion has a shorter signature length than the scheme transformed by the second conversion. The second conversion does not require the random oracle. Therefore, if the original signature scheme is of the standard model, the SEU property of the converted signature scheme is proved also in the standard model.

The proposed two conversions ensure the tight security reduction to the underlying security assumptions. Moreover, signers of the converted schemes can precompute almost all operations on the signing before they are given a message. Therefore, the signer can generate signatures quite efficiently.

References

[ADR02]   Jee Hea An, Yevgeniy Dodis, Tal Rabin. On the Security of Joint Signature and Encryption. In Eurocrypt 2002, pp.83-107.

[BR93]   Mihir Bellare, Phillip Rogaway. Random Oracles are Practical: A Paradigm for Designing Efficient Protocols. ACM Conference on Computer and Communications Security 1993. pp.62-73

[BSS99]   Ian F. Blake, Gadiel Seroussi, Nigel P. Smart. Elliptic Curve in Cryptography. Cambridge University Press, 1999.

[BBS04]   Dan Boneh, Xavier Boyen, Hovav Shacham. Short Group Signatures. In Crypto 2004, pp. 41-55.

[BSW06]   Dan Boneh, Emily Shen, and Brent Waters. Strongly Unforgeable Signatures Based on Computational Diffie-Hellman. In PKC 2006, pp. 229-240.

[CHK04]   Ran Canetti, Shai Halevi, and Jonathan Katz. Chosen-Ciphertext Security from Identity-Based Encryption In Eurocrypt 2004, pp. 229-235.

[CG04]   Jan Camenisch, Jens Groth. Group Signatures: Better Efficiency and New Theoretical Aspects. In SCN 2004, pp. 120-133.

[CLS06]   Scott Contini, Arjen K. Lenstra, Ron Steinfeld. VSH, an Efficient and Provable Collision-Resistant Hash Function. Eurocrypt 2006, pp. 165-182.

[DDN00]   Danny Dolev, Cynthia Dwork, Moni Naor. Non-malleable Cryptography. SIAM J. of Computing, 30(2), pp.391-437, 2000.

[GMR88]   Shafi Goldwasser, Silvio Micali, Ronald L. Rivest. A Digital Signature Scheme Secure Against Adaptive Chosen-Message Attacks. SIAM J. Comput. 17(2), pp. 281-308 (1988)

[KR97]   Hugo Krawczyk, Tal Rabin. Chameleon Hashing and Signatures. 1997. http://ibm.com/security/chameleon.ps, http://iacr.org/1998/010.ps.gz

[KR00]   Hugo Krawczyk, Tal Rabin. Chameleon Signatures. In NDSS 2000, pp. 143–154.

[MOV96]   Alfred J. Menezes, Paul C. van Oorschot and Scott A. Vanstone. HANDBOOK of APPLIED CRYPTOGRAPHY, CRC Press, 1996.

[ST01]   Adi Shamir, Yael Tauman. Improved Online/Offline Signature Schemes. Crypto 2001. pp.355-367

[S97]   Victor Shoup. Lower Bound for Discrete Logarithms and Related Problems. In Eurocrypt'97. pp.256-266.

[S97]   Ron Steinfeld, Josef Pieprzyk, and Huaxiong Wang. How to Strengthen any Weakly Unforgeable Signature into a Strongly Unforgeable Signature. CT-RSA 2007.

[W05]   Brent Waters. Efficient identity-based encryption without random oracles. In Eurocrypt 2005, pp. 114-127.

Conditionally Verifiable Signature

(Extended Abstract)

Ian F. Blake[1] and Aldar C-F. Chan[2]

[1] Department of Electrical and Computer Engineering, University of Toronto
Toronto, Ontario M5S 2G5, Canada
[2] INRIA Rhône-Alpes, Inovallée, 38330 Montbonnot Saint Ismier, France

Abstract. We introduce a new digital signature model, called conditionally verifiable signature (CVS), which allows a signer to specify and convince a recipient under what conditions his signature would become valid and verifiable; the resulting signature is not publicly verifiable immediately but can be converted back into an ordinary one (verifiable by anyone) after the recipient has obtained proofs, in the form of signatures/endorsements from a number of third party witnesses, that all the specified conditions have been fulfilled. A fairly wide set of conditions could be specified in CVS. The only job of the witnesses is to certify the fulfillment of a condition and none of them need to be actively involved in the actual signature conversion, thus protecting user privacy. It is guaranteed that the recipient cannot cheat as long as at least one of the specified witnesses does not collude. We formalize the concept of CVS and give a generic CVS construction based on any CPA-secure identity based encryption (IBE) scheme. Theoretically, we show that the existence of IBE with indistinguishability under a chosen plaintext attack (a weaker notion than the standard one) is necessary and sufficient for the construction of a secure CVS.[1]

1 Introduction

Balancing between the accountability and privacy of a signer is an important but largely unanswered issue of digital signatures. A digital signature scheme usually consists of two parties, a signer and a recipient, with the former giving his signature on a message/document to the latter as his commitment or endorsement on the message. To ensure that the signer is held accountable for his commitment, his signature needs to be publicly verifiable. However, public verifiability of a digital signature would put the signer's privacy at risk as a digital signature could be replicated and spread so easily, compared to its handwritten counterpart. More importantly, if the message presents valuable information about the signer, then the signed message itself is a certified piece of that information. Hence, the interests of the signer and the recipient are in conflict.

Of course, ensuring signer privacy and accountability simultaneously seems to be impossible. However, we observe that, in most of the real world scenarios, this conflict could be solved if the signer can ensure non-verifiability of his signature before certain

---

[1] Due to page limit, some proofs are omitted here but could be found in the full version [7].

R. Barua and T. Lange (Eds.): INDOCRYPT 2006, LNCS 4329, pp. 206–220, 2006.

conditions are fulfilled but still be able to convince the recipient that he will be obligated to exercise his commitment; in other words, he needs to give the recipient some guarantee that his commitment or his signature will become effective or publicly verifiable once all the conditions are fulfilled.

To provide a flexible solution to this problem of controllably passing signatures from one party to another without actively involving a trusted third party (i.e. the third party does not have to see or know the signer's message), we introduce a new signature concept called conditionally verifiable signature (CVS). In the CVS model, a signer is allowed to embed a set of verifiability conditions C into his ordinary signature σ to create a partial signature δ that is solely verifiable by the recipient, who cannot immediately convince others of the validity of δ (as δ is no more convincing than any random number and hence nobody can link it to its alleged signer) but can convert it back to the universally verifiable one σ (i.e. verifiable by everyone) after obtaining from a number of witnesses (appointed by the signer) the proofs that all the specified verifiability conditions have been fulfilled.[2] These proofs are in the form of signatures on condition statements, signed by the witnesses, about how the specified conditions are considered as fulfilled. In order to convince the recipient to accept a given partial signature δ on a message M (whose validity could not be verified), the signer runs a proof/confirmation protocol, which could be interactive or non-interactive, with the recipient to convince the latter that δ is indeed his partial signature on M, from which the corresponding ordinary signature could be recovered using the specified witnesses' signatures on the specified verifiability condition statements in C.

Given that $\mathcal{W}$ is the set of all possible witnesses, an instance set of verifiability conditions C is of the form $\{(c_i, W_i) : c_i \in \{0,1\}^*, W_i \in \mathcal{W}\}$ where each condition statement c_i is a string of arbitrary length describing a condition to be fulfilled. Examples of c_i include "A reservation has been made for Alice on flight CX829, 5 Sept 2006.", "A parcel of XXX has been received for delivery to Bob." and so on. The recipient needs to request each one of the specified witnesses, say W_i, to verify whether the condition stated in c_i is fulfilled and in case it is, to sign on c_i to give him a witness signature σ_i. These witness signatures σ_i's would allow the recipient to recover the publicly verifiable, ordinary signature σ from the partial signature δ. It is not necessary for a recipient to present δ or the message M to the witnesses in order to get these σ_i's. The only trust we place on the witnesses is that they only give out their signatures on a condition statement when the specified conditions are indeed fulfilled. In fact, it is not difficult to imagine that the existence of such witnesses is abundant in any business transaction. In addition, we could achieve a fairly high level of privacy in that the witnesses are unaware of the message or the partial signature when verifying the fulfillment of a given condition, namely, he does not learn the deal between the signer and the recipient.

We could view the partial signature as a blinded version of the ordinary signature, that is, nobody could verify its validity. We formulate this non-verifiability property by the notion of *simulatability* in this paper, that is, anyone could use just public information of the signer to simulate a given partial signature while others cannot judge whether

---

[2] Throughout the rest of this paper, we will denote the ordinary (universally verifiable) signature and the CVS partial signature by σ and δ respectively, unless otherwise specified.

it is genuine. In other words, nobody could distinguish between a genuine partial signature and a simulated one.

1.1   Related Work

Related work on controlling the verifiability of a digital signature includes designated verifier signatures [23,30], undeniable signatures [5,8,11,16,14,24,25], designated confirmer signatures [9,6,13,21,28], fair exchanges [1], and timed release of signatures [15]. Despite the considerable amount of work in limiting the verifiability of a digital signature, the conditions that could be incorporated into a digital signature scheme are still very restrictive; the existing protocols merely ensure that only a designated recipient can verify but cannot convince anybody else of the validity of a signature (in designated verifier signatures) and/or collaboration of the signer (in undeniable signatures) or a third party designated by the signer (in designated confirmer signatures, fair exchange) is needed in verifying the signature. More importantly, in these schemes, if a third party is involved to enforce certain verifiability conditions, he needs to know the signer's message, thus violating the privacy of the signer and perhaps the recipient as well. Roughly speaking, CVS can be considered as a generalization of these schemes.

1.2   Our Contributions

The main contribution of this paper is two-fold: First, a new signature model with controllable verifiability, particularly useful in electronic commerce, is introduced. Second, the equivalence between CVS and CPA-secure IBE in terms of existence is shown.

Through the new model of CVS, a signer can incorporate a wide range of verifiability conditions into an ordinary signature scheme to control its verifiability and validity while minimizing the requirement or trust on third-parties. We could possibly view CVS as a more general, unified concept incorporating the ideas of existing work (including undeniable signatures, designated confirmer signature, fair exchange and timed release of signatures), but provides more effective and flexible solutions to the scenarios these existing schemes could not solve satisfactorily, particularly those in electronic commerce. A typical example of these would be the deadlock scenario that may happen in an online purchase between mistrusting parties; using the post office as a witness, CVS would reasonably solve this problem.

We demonstrate the feasibility of CVS by giving a theoretical construction based on any existential-unforgeable signature scheme and any semantic-secure identity based encryption (IBE) scheme [3]. We also show that a secure CVS scheme is equivalent to an IBE scheme which is indistinguishable under a chosen plaintext attack (IND-ID-CPA), a weaker notion than the commonly accepted security notion against an adaptive chosen ciphertext attack (IND-ID-CCA) in IBE. Hence, we believe that CVS could be constructed based on a weaker assumption than IBE.

The rest of this paper is organized as follows. We give the definition of a conditionally verifiable signature scheme and its notions of security in the next section. In Section 3, we give a generic CVS construction and show the equivalent between CVS and IBE. Finally, our conclusions are given in Section 4.

2   Definitions and Security Notions

The players in a CVS scheme include a signer S, a recipient or verifier V, and a number of witnesses $\{W_i\} \subseteq \mathcal{W}$ (let $|\{W_i\}| = L$). Given a security parameter λ, a CVS scheme consists of the following algorithms and a confirmation protocol.

Key Generation (CVKGS, CVKGW). Let $\mathsf{CVKGS}(1^\lambda) \to (PK_S, sk_S)$ and CVKGW $(1^\lambda) \to (PK_W, sk_W)$ be two probabilistic algorithms. Then, (PK_S, sk_S) is the public/private key pair for a signer S and (PK_W, sk_W) is the public/private key pair for a witness W.

Signing and Verification (Ordinary Signatures) (SigS, VerS)/(SigW, VerW). $\mathsf{SigS}(m, sk_S) \to \sigma_S$ is an algorithm generating an ordinary (universally verifiable) signature σ_S of the signer S for a message $m \in \mathcal{M}$. $\mathsf{VerS}(m, \sigma_S, PK_S) \to \{0,1\}$ is the corresponding signature verification algorithm, which outputs 1 if σ_S is a true signature of S on the message m and outputs 0 otherwise. Similarly, $\mathsf{SigW}(m, sk_W) \to \sigma_W$ and $\mathsf{VerW}(m, \sigma_W, PK_W) \to \{0,1\}$ are the signature generation and verification algorithms of the witness W.[3]

Partial Signature Generation (CVSig). Given a set of verifiability conditions $C \subseteq \mathcal{C} \times \mathcal{W}$ and the corresponding set of witness public keys PK_C, the probabilistic algorithm $\mathsf{CVSig}(m, C, sk_S, PK_S, PK_C) \to \delta$ generates the partial signature δ on message $m \in \mathcal{M}$ under the set of verifiability conditions C. *Note that δ is not universally verifiable.*

Ordinary Signature Extraction (CVExtract). $\mathsf{CVExtract}(m, C, \delta, PK_S, \sigma_C) \to \sigma/\perp$ is an algorithm which extracts the corresponding ordinary signature σ from a partial signature δ for a message m under the verifiability condition specified by C and a signing public key PK_S when given the set of witness signatures or endorsements σ_C. The extracted signature σ is a universally verifiable one. In case the extraction fails, it outputs $\perp$. *Note that $\sigma_C = \{\mathsf{SigW}(sk_{W_i}, c_i) : (c_i, W_i) \in C\}$.*

CVS Confirmation/Verification. $\mathsf{CVCon}_{(S,V)} = \langle \mathsf{CVConS}, \mathsf{CVConV} \rangle$ is the signature confirmation protocol between the signer and recipient, which could be interactive or non-interactive:

$$\mathsf{CVCon}_{(S,V)}(m, C, \delta) = \langle \mathsf{CVConS}(\sigma, sk_S, r), \mathsf{CVConV}() \rangle (m, C, \delta, PK_S, PK_C)$$

The common input consists of the message m, the set of verifiability conditions C, the partial signature δ, and the public keys of the signer PK_S and the involved witnesses public keys PK_C. The private input of the signer S is σ, sk_S, and r where σ is the corresponding ordinary signature on m embedded in δ, and r represents all the random coins used. The output is either 1 ("true") or 0 ("false"). In essence, this protocol allows S to prove to V that δ is his partial signature on m, which can be converted back into a publicly verifiable signature σ (i.e. $\mathsf{VerS}(m, \sigma, PK_S) = 1$), once V has obtained all the witness signatures/endorsements on the condition statements as specified in C. Ideally, this protocol should be zero-knowledge, and the interactive version is considered in this paper.

In general, a CVS scheme should satisfy both completeness and perfect convertibility property. Completeness ensures that a valid ordinary signature can be retrieved from a

---

[3] We may write SigW as CVEndW to reflect the fact that it is actually an endorsement of W.

valid partial signature. A CVS scheme is perfectly convertible if nobody can distinguish whether a given ordinary signature is extracted from a partial signature or generated directly. A secure CVS scheme should also satisfy unforgeability, simulatability, cheat-immunity, and have a zero knowledge confirmation protocol.

Oracle Queries — Allowed Adversary Interaction. In our security model, two types of adversary interaction are allowed:

1. **Signing Oracle** $O_S(m, C)$. For fixed keys PK_S, sk_S, $\{PK_{W_i}\}$, $\{sk_{W_i}\}$, on input a signing query $\langle m, C \rangle$ (where $m \in \mathcal{M}$ and $C = \{(c_i, W_i) : c_i \in \mathcal{C}, W_i \in \mathcal{W}\}$ is a set of verifiability conditions), O_S responds by running CVSig to generate the corresponding partial signature δ. After sending δ to the querying party, O_S runs the confirmation protocol $CVCon_{(S,V)}$ with the querying party to confirm the validity of δ. *Note that a malicious verifier is allowed to put in any random number in place of δ when running the confirmation protocol.*
2. **Endorsement Oracle** $O_E(c, W)$. For fixed keys $\{PK_{W_i}\}$, $\{sk_{W_i}\}$, on input an endorsement query $\langle c, W \rangle$, Q_E responds by retrieving the needed witness private key sk_W and then running the witness endorsement/signing algorithm SigW (or CVEndW) to create a witness signature $\sigma_W(c)$ on the condition statement c.

These oracle queries may be asked adaptively, that is, each query may depend on the replies of the previous queries.

2.1 Unforgeability

Unforgeability ensures that there is a negligible probability to forge an ordinary signature even though all the witnesses collude and are given access to other ordinary and partial signatures of their choice.

Definition 1. *A CVS scheme is unforgeable if the probability of winning the game below, $p_{\mathcal{A}}^{UF}$, is negligible in the security parameter λ for all PPT (Probabilistic Polynomial Time) adversaries $\mathcal{A}$.*

In the setup, the challenger takes a security parameter λ, runs the key generation algorithms for the signer and all witnesses, that is, $(PK_S, sk_S) \leftarrow \{CVKGS(1^\lambda)\}$ and $(PK_{W_i}, sk_{W_i}) \leftarrow \{CVKGW(1^\lambda)\}$. The challenger gives the adversary all the public keys, PK_S and $\{PK_{W_i}\}$ and all the witness private keys $\{sk_{W_i}\}$. The challenger keeps the signer's private key sk_S. Then, the adversary is allowed to make queries to O_S to request a partial signature δ_j for $\langle m_j, C_j \rangle$. Note that the adversary has the witness private keys so no O_E query is necessary. Finally, the adversary has to output a message-signature pair (m, σ) where $m \neq m_j$ for all j. The adversary $\mathcal{A}$ is said to win this game if $VerS_S(m, \sigma) = 1$.

2.2 Simulatability

In order to ensure the protection of signer privacy, the partial signature should be (computationally) indistinguishable from the output of a certain **public** PPT simulator: $Fake(m, C, PK_S, PK_C) \rightarrow \delta'$. As can be seen, the simulator only uses public information of the signer; hence, a partial signature is not linkable to its alleged signer, thus protecting his privacy. The notion about the indistinguishability between a genuine partial signature and a simulator output is best described by the following game between a challenger and an adversary:

In the setup, the challenger takes a security parameter λ, runs the key generation algorithms for the signer and all witnesses, that is, $(PK_S, sk_S) \leftarrow \{\text{CVKGS}(1^\lambda)\}$ and $(PK_{W_i}, sk_{W_i}) \leftarrow \{\text{CVKGW}(1^\lambda)\}$. The challenger gives the adversary all the public keys, PK_S and $\{PK_{W_i}\}$. The challenger keeps the private keys $\{sk_{W_i}\}$. We consider the strongest security model in this paper — the signer's private key sk_S is also given to the adversary.[4] Then, the adversary is allowed to make queries to obtain the signer's partial signatures and witness signatures of messages of his choice until it is ready to receive a challenged partial signature. It can make two types of oracle queries: (1) Signing Query $\langle m_j, C_j \rangle$ to O_S; (2) Endorsement Query $\langle c_j, W_j \rangle$ to O_E. As the simulator Fake is publicly known, the adversary could also freely get a simulator output for any message and condition of his choice. Once the adversary decides it is ready for a challenge, it outputs a message $m \in \mathcal{M}$ and a set of conditions $C \subset \mathcal{C} \times \mathcal{W}$ on which it wishes to be challenged. Let C_E^1 denote the set of all endorsement queries sent to O_E previously. The only constraint is that $C \backslash C_E^1 \neq \phi$ (the empty set). The challenger flips a coin $b \in \{0, 1\}$ and outputs the following challenge to the adversary:

$$\delta_b = \begin{cases} \text{CVSig}(m, C, sk_S, PK_S, PK_C), & b = 0 \\ \text{Fake}(m, C, PK_S, PK_C), & b = 1 \end{cases}$$

The adversary is allowed to run until it outputs a guess. Let C_E^2 be the set of queries that have been made to O_E so far after the challenge is issued. The adversary can issue more (but polynomially many) queries, both signing and endorsement, but for any endorsement query (c_j, W_j), the following must hold: $C \backslash (C_E^1 \cup C_E^2 \cup \{(c_j, W_j)\}) \neq \phi$. Finally, the adversary halts and outputs a guess b' for the hidden coin b. The adversary is said to win this game if $b' = b$. The advantage of the adversary $\mathcal{A}$ is defined as: $Adv_{\mathcal{A}}^{Sim}(\lambda) = \left| Pr[b' = b] - \frac{1}{2} \right|$.

Definition 2. *If there exists a PPT simulator Fake such that the advantage of winning the above game is negligible in the security parameter λ for all PPT adversaries, then the given CVS scheme is simulatable (with respect to Fake).*

2.3 Zero Knowledge Confirmation Protocol and Non-transferability

We use the notion of simulatability of the communication transcript as a formulation for the zero knowledge property of the confirmation protocol. In details, any communication transcript recorded in carrying out the confirmation protocol could be simulated by a PPT simulator SimT (using only public information) whose output is indistinguishable from a genuine transcript.

The definition of simulatability of CVS ensures that nobody could associate a partial signature to its signer or tell its validity given just the partial signature. If given also the

---

[4] In addition to O_S queries, the adversary can generate partial signatures of arbitrary messages and conditions on its own. But even on identical input, these signatures may not be the same as those from the challenger since the random coins used are likely to be different.

communication transcript of the confirmation protocol for the partial signature, nobody could still tell its validity, then the CVS scheme is said to be **non-transferable**. The formulation of **non-transferability** is very similar to that of simulatability described previously except that it includes an additional simulator for the transcript of the confirmation protocol, and in the challenge phase, the adversary receives either a genuine partial signature and its confirmation protocol transcript or a fake (simulated) partial signature and its simulated transcript. Note that while the confirmation protocol is carried out in all oracle queries, no confirmation protocol would be carried out in the challenge phase. It can be shown that a CVS scheme is non-transferable if it is simulatable and its confirmation protocol is zero knowledge, which is summarized by the following theorem (We leave out the proof in the full version of the paper.).

Theorem 1. *Given that a CVS scheme is simulatable with respect to a PPT partial signature simulator* Fake, *if its confirmation protocol* $CVCon_{(S,V)}$ *is zero knowledge with respect to a PPT transcript simulator* SimT, *then it is non-transferable in the same attack model with adaptive queries as in the simulatability definition and* SimT *can be used as the transcript simulator* FakeT *for the output of* Fake. *The following two distributions are indistinguishable for all* S, m, C *with adaptive endorsement queries:*

$$\{CVSig_S(m, C),\ \pi_{S,V}^{CVCon}(m, C, CVSig_S(m, C))\},$$
$$\{Fake_S(m, C),\ \pi^{FakeT}(m, C, Fake_S(m, C))\}$$

where $\pi_{S,V}^{CVCon}(\cdot)$ *and* $\pi^{FakeT}(\cdot)$ *are transcript outputs of a real confirmation protocol run and* FakeT *respectively.*

2.4 Cheat-Immunity

Cheat-immunity guarantees that the recipient of a partial signature cannot recover the ordinary signature without collecting all the needed witness signatures. Details are described by the following game:

In the setup, the challenger takes a security parameter λ, runs the key generation algorithms for the signer and all witnesses, that is, $(PK_S, sk_S) \leftarrow \{CVKGS(1^\lambda)\}$ and $(PK_{W_l}, sk_{W_l}) \leftarrow \{CVKGW(1^\lambda)\}$. The challenger gives the adversary all the public keys, PK_S and $\{PK_{W_l}\}$. The challenger keeps all the private keys sk_S and $\{sk_{W_l}\}$. The adversary makes queries to obtain the signer's partial signatures and witness signatures of messages of his choice until it is ready to receive a challenge partial signature. It can make two types of queries: (1) Signing Query $\langle m_j, C_j \rangle$ to O_S; (2) Endorsement Query $\langle c_j, W_j \rangle$ to O_E. *With these two types of queries, the adversary can obtain the ordinary signature of the signer on any message of his choice.* Once the adversary decides it is ready for a challenge, it outputs a message $m \in \mathcal{M}$ not queried before and a set of verifiability conditions $C = \{(c_i, W_i)\} \subset \mathcal{C} \times \mathcal{W}$ on which it wishes to be challenged. Let C_E^1 denote the set of all the endorsement queries made to O_E before the challenge. The only constraint is that $C \backslash C_E^1 \neq \phi$. The challenger uses CVSig to generate a partial signature δ on a message m under the conditions in C. It sends δ as the challenge to the adversary and runs the confirmation protocol $CVCon_{(S,V)}$ with it. Let C_E^2 denote the set of all the endorsement queries made to O_E so far after the challenge is issued.

The adversary can issue more queries, both signing and endorsement, but for any endorsement query (c_j, W_j), the following must hold: $C \backslash (C_E^1 \cup C_E^2 \cup \{(c_j, W_j)\}) \neq \phi$, and for any signing query, the queried message is not the challenged message. Finally, the adversary halts and outputs an ordinary signature σ for message m. The adversary $\mathcal{A}$ is said to win this game if $\mathsf{VerS}_S(m, \sigma) = 1.$[5]

Definition 3. *A CVS scheme is cheat-immune if the probability of winning the above game is negligible in the security parameter λ for all PPT adversaries.*

It can be shown that unforgeability and simulatability imply cheat immunity if the confirmation protocol is zero knowledge as summarized by the following theorem.

Theorem 2. *An unforgeable and simulatable CVS scheme is also cheat-immune given its confirmation protocol is zero knowledge. (Proof in the full version.)*

Hence, proving that a CVS scheme is secure reduces to showing that it is unforgeable and simulatable and its confirmation protocol is zero knowledge. Theorem 2 allows one to ignore the cheat-immunity requirement when designing a CVS scheme.

3   The Existence of a Secure CVS Scheme

In this section, we give a generic CVS construction from IBE and show the equivalence between CVS and IBE.

3.1   A Generic Construction of CVS from IBE

We show how to construct a secure CVS scheme based on the following components (*Please refer to the cited references for details and security definitions of these primitives*): (1) A secure signature scheme $SIG = \langle SKG(1^\lambda) \rightarrow (PK_S, sk_S), Sig(m, sk_S)$ $\rightarrow \sigma, Ver(m, \sigma, PK_S) \rightarrow \{0, 1\}\rangle$ (where m and σ are the message and signature respectively) which is existentially unforgeable against an adaptive chosen message attack [20]; (2) An IBE scheme $IBE = \langle Setup(1^\lambda) \rightarrow (PK_G, sk_G), Extract(ID, sk_G) \rightarrow$ $d_{ID}, Enc(PK_G, ID, M) \rightarrow C, Dec(PK_G, C, d_{ID}) \rightarrow M\rangle$ (where M, ID and C are the plaintext, identity and ciphertext respectively, and PK_G and sk_G are the public and private keys of the private key generator) with semantic security, that is, IND-ID-CPA [2]; (3) A computationally hiding commitment scheme $COM = \langle Com(s, m) \rightarrow c\rangle$ [26,12] where m and c are the secret and commitment respectively; (4) A pseudorandom generator (PRG) [17,22]. Let the plaintext and ciphertext spaces of IBE be $\mathcal{P}_{IBE}$ and $\mathcal{C}_{IBE}$ respectively.

Let the message and signature spaces of SIG be $\mathcal{M}$ (same as the message space of CVS) and $\mathcal{S}_\sigma$ (same as the ordinary signature space of CVS) respectively. Let $h :$ $\{0, 1\}^{l_p} \rightarrow \{0, 1\}^{l_s}$ be a PRG where l_p and l_s are the length of an IBE plaintext and a SIG signature respectively. Let $\mathcal{C}_{COM}$ be the output space of the commitment scheme

---

[5] This model is reasonable as the only restriction in practice is the signer should not give to the same party multiple partial signatures on the same message but with different verifiability conditions. First, the event in question is rare; otherwise, the restriction can be easily achieved by adding a serial number if the same message is signed multiple times.

COM and $Com : \mathcal{P}_{IBE} \times \mathcal{S}_\sigma \to \mathcal{C}_{COM}$ be its committing function. Depending on the number of witnesses, the IBE scheme is used multiple times with each witness W_i being a private key generator (PKG) for its IBE scheme. Assume there are N witnesses and the partial signature is: $\delta \in \mathcal{S}_\sigma \times \mathcal{C}_{IBE}^N \times \mathcal{C}_{COM}$. The generic CVS construction is as follows.

Key Generation. CVKGS $\overset{\text{def}}{=} SKG$ for generating (PK_S, sk_S) for the signer S. CVKGW $\overset{\text{def}}{=} Setup$ for generating (PK_{W_i}, sk_{W_i}) for the witnesses W_i.

Partial Signature Generation. Given an input message $m \in \mathcal{M}$, a condition set $C = \{(c_i, W_i) : 1 \le i \le N\}$, a signing key sk_S, a signer's public key PK_S and the set of witness public keys $PK_C = \{PK_{W_i} : 1 \le i \le N\}$,

1. Generate an ordinary signature using the signing algorithm of SIG: $\sigma = Sig(m, sk_S)$
2. For each $(c_i, W_i) \in C$, pick a random $a_i \in \mathcal{P}_{IBE}, 1 \le i \le N$ and the CVS partial signature is:

$$\delta = \left\langle \sigma \oplus h\left(\bigoplus_i^N a_i\right), \{Enc(PK_{W_i}, c_i, a_i) : 1 \le i \le N\}, Com\left(\sigma, h\left(\bigoplus_i^N a_i\right)\right) \right\rangle$$

where $Enc(PK_{W_i}, c_i, a_i)$ is the IBE ciphertext on message a_i using W_i (witness) as the PKG and c_i (condition statement) as the identity.[6]

Witness Signature Generation. SigW$(c, sk_W) \overset{\text{def}}{=} Extract(c, sk_W)$. Taking the condition statement c as an identity, the witness W could extract the private key d_c^W corresponding to c. d_c^W could be considered as a kind of signature on c as in [4].

Signature Extraction. Given a partial signature $\delta = \langle \alpha, \{\beta_i : 1 \le i \le N\}, \gamma \rangle$ and $\sigma_i = d_{c_i}^{W_i}, 1 \le i \le N$,

1. For $1 \le i \le N$, get $a_i' = Dec(PK_{W_i}, \beta_i, \sigma_i)$.
2. Recover $\sigma' = \alpha \oplus h(\bigoplus_i^N a_i')$. Check if $Com(\sigma', h(\bigoplus_i^N a_i')) \overset{?}{=} \gamma$. If not, output "fail", otherwise, σ' is the ordinary signature.

Signature Verification. VerS $\overset{\text{def}}{=} Ver$ (the signature verification algorithm of SIG).

Confirmation Protocol. Using general interactive zero-knowledge proofs [18] or concurrent zero-knowledge proofs [10], the signer with private input $a_1, \ldots, a_i, \ldots, a_N$ and σ and all the random coins used to generate β_i could convince the verifier that there exists $(\sigma, a_1, \ldots, a_i, \ldots, a_N)$ satisfying the following equations: $\delta = \langle \alpha, \{\beta_1, \beta_2, \ldots, \beta_i, \ldots, \beta_N\}, \gamma \rangle$; $\alpha = \sigma \oplus h\left(\bigoplus_i^N a_i\right)$; $\beta_i = Enc(PK_{W_i}, c_i, a_i)$, $1 \le i \le N$; $\gamma = Com\left(\sigma, h\left(\bigoplus_i^N a_i\right)\right)$; $Ver(m, \sigma, PK_S) = 1$. The common input to the confirmation protocol is PK_S, PK_{W_i} ($1 \le i \le N$), $m, C = \{(c_i, W_i) : 1 \le i \le N\}$ and δ. Since verifying whether a given tuple $(\sigma, a_1, a_2, \ldots, a_i, \ldots a_N)$ satisfies the above equations is a poly-time predicate, a general zero-knowledge proof for it should exist.

Fake Signature Simulator — Fake$(C) : C = \{(c_i, W_i) : 1 \le i \le N\}$

1. Randomly (uniformly) pick $\sigma_f \in \mathcal{S}_\sigma$.

---

[6] For short, we may denote $Enc(PK_{W_i}, c_i, a_i)$ as $Enc_{W_i}(c_i, a_i)$ in the following discussion.

2. Randomly pick $b_i \in \mathcal{P}_{IBE}$, for $1 \leq i \leq N$ and output the fake partial signature:

$$\delta_f =$$

$$\left\langle \sigma_f \oplus h\left(\bigoplus_i^N b_i\right), \{Enc(PK_{W_i}, c_i, b_i) : 1 \leq i \leq N\}, Com\left(\sigma_f, h\left(\bigoplus_i^N b_i\right)\right)\right\rangle$$

Obviously, this simulator is PPT. The generic CVS construction from IBE is slight over-designed: The commitment scheme is generally not needed; it is mainly used to allow detection of failure in ordinary signature extraction which may occur when invalid witness signatures are used in ordinary signature extraction .

Security of the Generic CVS Construction. The completeness of the above CVS construction is guaranteed by the correctness of the underlying IBE scheme. Besides, it is also perfectly convertible. The security of this CVS construction is best summarized with the following lemmas which lead to Theorem 3.

Lemma 1. *If SIG is existentially unforgeable under an adaptive chosen message attack, then the generic CVS construction is unforgeable. (Proof Sketch in Appendix.)*

Lemma 2. *If IBE is IND-ID-CPA secure, COM is a computationally hiding commitment scheme, and h is a PRG, then the generic CVS construction is simulatable with respect to the simulator Fake. (Proof Sketch in Appendix.)*

Theorem 3. *Given any semantically secure IBE scheme (under a chosen plaintext attack) and any existentially unforgeable signature scheme, together with a PRG and a computationally hiding commitment scheme, a secure CVS scheme can be constructed.*

3.2   A Generic Construction of IBE from CVS

We show how to construct a 1-bit IBE scheme with semantic security (i.e. IND-ID-CPA) using a CVS scheme. We assume the CVS scheme is simulatable with respect to a fake partial signature simulator Fake. Our construction is similar to that in the seminal work of probabilistic encryption by Goldwasser and Micali [19]. While they used the indistinguishability between the quadratic residues and non-residues in $\mathbb{Z}_n^*$ for some composite n (Quadratic Residuosity Problem) to encrypt a single bit, we leverage the indistinguishability between a true and a simulated (fake) partial signature of CVS to create a ciphertext. By repeating the operation of the 1-bit scheme k times as in [19], we could construct an IBE scheme for k-bit long messages. We consider a CVS scheme with a single witness $G \in \mathcal{W}$ which is used as the PKG for the IBE scheme. Suppose Fake is a PPT simulator for the CVS scheme. The IBE scheme works as follows.

Key Setup. The public and private keys of the witness G in the CVS scheme are used as the public and private keys of the PRG in the IBE scheme. We set $Setup \overset{\text{def}}{=}$ CVKGW to generate the public private keys of the PRG: CVKGW$(1^\lambda) \rightarrow (PK_G, sk_G)$.

Private Key Extraction. The identity ID_i of any user in IBE could be treated as a condition statement in CVS as they are both a bit string of arbitrary length. We set $Extract \overset{\text{def}}{=}$ SigW/CVEndW, then extracting the private key d_i for ID_i is the same as requesting a signature on the statement ID_i: SigW$(ID_i, sk_G) \rightarrow d_i$.

Encryption. The identity of a user i is ID_i (treated as a condition statement in the underlying CVS scheme) and its private key is the witness signature d_i obtained from G. We consider a 1-bit plaintext $b \in \{0, 1\}$. To encrypt, randomly pick a message $m \in \mathcal{M}$, run $\mathsf{CVKGS}(1^\lambda)$ to generate the public/private key pair (PK_S, sk_S) of the signer, the encryption function is: $Enc(PK_G, ID_i, b) \to (m, \delta_b, PK_S)$, where

$$\delta_b = \begin{cases} \mathsf{CVSig}(m, ID_i, sk_S, PK_S, PK_G), & b = 0 \\ \mathsf{Fake}(m, ID_i, PK_S, PK_G), & b = 1 \end{cases}$$

i.e. When $b = 0$, δ_b is a valid partial signature on m; when $b = 1$, δ_b is a fake one.

Decryption. Given an identity ID_i, a PKG public key PK_G and the user private key d_i, to decrypt a given ciphertext $C = (m', \delta', PK'_S)$, the decryption function $Dec(PK_G, C, d_i) \to b$ is implemented as follows: extract the ordinary signature from δ' using $\mathsf{CVExtract}(m', ID_i, \delta', PK'_S, d_i) \to \sigma'$, and the plaintext b' is given by the following: $b' = 0$ if $\mathsf{VerS}(m', \sigma', PK'_S) = 1$ and 1 otherwise.

Correctness of the CVS-based IBE. The completeness of the CVS scheme guarantees the correctness of decryption in the above IBE scheme. The completeness property of the CVS scheme ensures that, if $\delta = \mathsf{CVSig}(m, ID_i, sk_S, PK_S, PK_G)$ and $d_i = \mathsf{CVEndW}(ID_i, sk_G)$, then the verification must return 1. The CVS scheme also guarantees that with negligible probability a valid ordinary signature on message m could be extracted form $\mathsf{Fake}(m, ID_i, PK_S, PK_G)$, otherwise, the CVS scheme would be forgeable. These together ensure that $Dec(PK_G, Enc(PK_G, ID_i, b), d_i) = b$ with probability almost 1. The security of above IBE is contained in the following theorem.

Theorem 4. (Security of the CVS-based IBE) *The above IBE construction from CVS is semantically secure against a chosen plaintext attack (IND-ID-CPA). (Proof Sketch in Appendix.)*

3.3   The Equivalence Between CVS and IBE

A secure CVS scheme is equivalent to a secure IBE scheme in terms of existence.

Theorem 5. *A secure CVS scheme (unforgeable, simulatable, with zero knowledge confirmation protocol) exists if and only if an IND-ID-CPA-secure IBE scheme exists.*

Proof. The **only if part** follows directly from the CVS-based IBE construction given above. For the **if part**, we assume the existence of a IND-ID-CPA secure IBE. Then a one-way function exists (We could use *Setup* of the IBE scheme to construct a one-way function.), which implies the existence of an ordinary signature scheme existentially unforgeable under an adaptive chosen message attack[29,27]. Besides, this also implies that a PRG exists [22], which in turn implies the existence of a computationally hiding multi-bit commitment function[26]. Finally, the existence of a one-way function also implies the existence of zero-knowledge proofs. By Theorem 3, we could use the generic construction to build a secure CVS scheme. Hence, the existence of a secure IBE scheme implies the existence of a secure CVS scheme.

As can be seen from Theorem 5 above, a weaker notion of IBE, namely, one with IND-ID-CPA security, is necessary and sufficient for the construction of a secure CVS scheme. It is thus fair to say CVS could be constructed based on a weaker assumption than IBE with the standard IND-ID-CCA security [2].

4   Conclusions

In this paper, we introduce a new signature concept called CVS which could provide effective solutions in many digital business scenarios, in particular, those involving mutually distrusting parties. We demonstrate its feasibility by giving a generic construction using IBE and show that it is equivalent to CPA-secure IBE. The equivalence result could imply that CVS can be constructed based on weaker computational assumptions compared with IBE which should usually be CCA-secure. One open problem is whether CVS can be constructed from primitives other than IBE.

References

1. N. Asokan, V. Shoup, and M. Waidner. Optimistic fair exchange of digital signatures. *IEEE Journal on Selected Areas in Communication*, 18(4):591–610, April 2000.
2. D. Boneh and M. Franklin. Identity-based encryption from the Weil pairing. In *Advances in Cryptology — CRYPTO 2001, Springer-Verlag LNCS vol. 2139*, pages 213–229, 2001.
3. D. Boneh and M. Franklin. Identity-based encryption from the Weil pairing. *SIAM Journal on Computing*, 32(3):586–615, 2003.
4. D. Boneh, B. Lynn, and H. Shacham. Short signatures from Weil pairing. In *Advances in Cryptology — Asiacrypt 2001, Springer-Verlag LNCS vol. 2248*, pages 514–532, 2001.
5. J. Boyar, D. Chaum, I. Damgård, and T. Pedersen. Convertible undeniable signatures. In *Advances in Cryptology — CRYPTO 1990, Springer-Verlag LNCS vol. 537*, pages 189–205, 1991.
6. J. Camenisch and M. Michels. Confirmer signature schemes secure against adaptive adversaries. In *Advances in Cryptology — EUROCRYPT 2000, Springer-Verlag LNCS vol. 1870*, pages 243–258, 2000.
7. Aldar C-F. Chan and Ian F. Blake. Conditionally verifiable signatures. Cryptology ePrint Archive, Report 2005/149, 2005. http://eprint.iacr.org/.
8. D. Chaum. Zero-knowledge undeniable signatures. In *Advances in Cryptology — EURO-CRYPT 90, Springer-Verlag LNCS vol. 473*, pages 458–464, 1990.
9. D. Chaum. Designated confirmer signatures. In *Advances in Cryptology — EUROCRYPT 1994, Springer-Verlag LNCS vol. 950*, pages 86–91, 1995.
10. I. Damgård. Efficient concurrent zero-knowledge in the auxiliary string model. In *Advances in Cryptology — EUROCRYPT 2000, Springer-Verlag LNCS vol. 1807*, pages 418–430, 2000.
11. I. Damgård and T. Pedersen. New convertible undeniable signature schemes. In *Advances in Cryptology — EUROCRYPT 1996, Springer-Verlag LNCS vol. 1070*, pages 372–386, 1996.
12. I. Damgård, B. Piftzmann, and T. Pedersen. Statistical secrecy and multi-bit commitments. *IEEE Transaction on Information Theory*, 44:1143–1151, 1998.
13. S. Galbraith and W. Mao. Invisibility and anonymity of undeniable and confirmer signatures. In *Cryptographers' Track RSA Conference (CT-RSA 2003), Springer-Verlag LNCS vol. 2612*, pages 80–97, 2003.
14. S. D. Galbraith, W. Mao, and K. G. Paterson. RSA-based undeniable signatures for general moduli. In *Cryptographers' Track RSA Conference (CT-RSA 2002), Springer-Verlag LNCS vol. 2271*, pages 200–217, 2002.
15. J. Garay and C. Pomerance. Timed fair exchange of standard signatures. In *Financial Cryptography (FC 2003), Springer-Verlag LNCS vol. 2742*, pages 190–203, 2003.
16. R. Gennaro, H. Krawczyk, and T. Rabin. RSA-based undeniable signatures. In *Advances in Cryptology — CRYPTO 1997, Springer-Verlag LNCS vol. 1294*, pages 397–416, 1997.

17. O. Goldreich, S. Goldwasser, and S. Micali. How to construct random functions. *Journal of ACM*, 33(4):792–807, 1986.

18. O. Goldreich, S. Micali, and A. Wigderson.   How to prove all NP-statements in zero-knowledge, and a methodolgy of cryptographic protocol design. In *Advances in Cryptology — CRYPTO 1986, Springer-Verlag LNCS vol. 263*, pages 171–185, 1986.

19. S. Goldwasser and S. Micali. Probabilistic encryption. *Journal of Computer and System Sciences*, 28(2):270–299, 1984.

20. S. Goldwasser, S. Micali, and R. Rivest. A secure signature scheme secure against adaptive chosen-message attacks. *SIAM Journal on Computing*, 17(2):281–308, 1988.

21. S. Goldwasser and E. Waisbard. Transformation of digital signature schemes into designated confirmer signature schemes. In *Theory of Cryptography Conference (TCC 2004), Springer-Verlag LNCS vol. 2951*, pages 77–100, 2004.

22. I. Impagliazzo, L. Levin, and M. Luby. Pseudo-random generation from one-way functions. In *ACM Symposium on Theory of Computing (STOC 1989)*, 1989.

23. M. Jakobsson, K. Sako, and R. Impagliazzo. Designated verifier proofs and their applications. In *Advances in Cryptology — EUROCRYPT 1996, Springer-Verlag LNCS vol. 1070*, pages 143–154, 1996.

24. B. Libert and J. J. Quisquater. Identity based undeniable signatures. In *Cryptographers' Track RSA Conference (CT-RSA 2004), Springer-Verlag LNCS vol. 2964*, pages 112–125, 2004.

25. M. Michels and M. Stadler. Efficient convertible undeniable signature schemes. In *Proceedings International Workshop on Selected Area of Cryptography (SAC'97)*, pages 231–244, 1997.

26. M. Naor. Bit commitment using pseudo-randomness. In *Advances in Cryptology — CRYPTO 1989, Springer-Verlag LNCS vol. 435*, pages 128–136, 1990.

27. M. Naor and M. Yung. Universal one-way hash functions and their cryptographic applications. In *ACM Symposium on Theory of Computing (STOC 1989)*, pages 33–43, 1989.

28. T. Okamoto. Designated confirmer signatures and public-key encryption are equivalent. In *Advances in Cryptology — CRYPTO 1994, Springer-Verlag LNCS vol. 839*, pages 61–74, 1994.

29. J. Rompel. One-way functions are necessary and sufficient for secure signature. In *Proceedings 22ndACM Symposium on Theory of Computing (STOC 1990)*, pages 387–394, 1990.

30. W. Susilo, F. Zhang, and Y. Mu. Identity-based strong designated verifier signature schemes. In *ACISP2004, Springer-Verlag LNCS vol. 3108*, pages 313–324, 2004.

Appendix

Proof Sketch of Lemma 1

We prove the unforgeability property by contradiction. Assume SIG is existentially unforgeable under chosen message attacks. Suppose there is a PPT forging algorithm $\mathcal{F}$ which can forge a CVS partial signature with probability of success $p_{\mathcal{F}}^{CVS}$. We show how to construct another forging algorithm $\mathcal{F}'$ from $\mathcal{F}$ to forge a signature of SIG. $\mathcal{F}'$ runs as follows.

Algorithm $\mathcal{F}'$

In the **Setup** phase: ask its challenger for the signer public key PK_S; run $Setup$ to get all the witness public/private key pairs (PK_{W_i}, sk_{W_i}), $1 \leq i \leq N$; run $\mathcal{F}$ on PK_S and (PK_{W_i}, sk_{W_i}).

In the **Query** phase: when $\mathcal{F}$ issues a O_S query for $\langle m_j, C_j \rangle$ where $C_j = \{(c_{ji}, W_{ji}) : 1 \leq i \leq N\}$, ask its signing orale for an ordinary signature $\sigma_j = Sig(sk_s, m_j)$; randomly choose a_{ji} $(1 \leq i \leq N)$ to create a partial signature:

$$\delta_j = \left\langle \sigma_j \oplus h \left(\bigoplus_i^N a_{ji} \right), \{Enc(PK_{W_{ji}}, c_{ji}, a_{ji})\}, Com \left(\sigma_j, h \left(\bigoplus_i^N a_{ji} \right) \right) \right\rangle;$$

with a_{ji}'s, σ_j, and all random coins used, run the confirmation protocol with $\mathcal{F}$.
Finally, when $\mathcal{F}$ outputs a guess (m, σ), output (m, σ) as the guess of $\mathcal{F}'$.
END.

Obviously, if F is PPT, then F' is also PPT. Note that $\mathcal{F}$ should output $m \neq m_j$, $\forall j$ The probability of success of $\mathcal{F}'$ is: $p_{\mathcal{F}'}^{SIG} = Pr[Ver(m, \sigma, PK_S) = 1] = p_{\mathcal{F}}^{CVS}$. If the CVS scheme is forgeable, that is, $p_{\mathcal{F}}^{CVS}$ is non-negligible, then $p_{\mathcal{F}'}^{SIG}$ is also non-negligible (a contradiction).

Proof Sketch of Lemma 2

It can be shown that if a given CVS scheme is secure for one witness, then the version with finitely many witnesses is also secure. Hence, we consider a single witness case and leave the multiple-witness case in the full version.

Assume IBE is IND-ID-CPA secure, h is a pseudorandom generator, and COM is computationally hiding. Suppose $\mathcal{D}$ is a PPT distinguisher which has non-negligible advantage $Adv_{\mathcal{D}}^{Sim}$ in winning the simulatability game in Definition 2. We can base on $\mathcal{D}$ to construct another distinguisher $\mathcal{D}'$ to break the semantic security of IBE as follows. *To avoid confusion, we should clarify that in the following discussion, we denote the challenge ciphertext of the IBE game by $C_b, b \in \{0, 1\}$ and the queried verifiability condition set by C_j.*

Algorithm $\mathcal{D}'(C_b), b \in \{0, 1\}$
In the **Setup** phase: ask its challenger for the public key PK_G of the PKG; use it as the witness public key for W; run **CVKGS** to generate the signer public/private key pair (PK_S, sk_S); run $\mathcal{D}$ on PK_G and (PK_S, sk_S).
To answer **Signing Query** (O_S) on $\langle m_j, C_j \rangle$ where $C_j = (c_j, W)$: generate $\sigma_j = Sig(m_j, sk_S)$; randomly pick a_j and encrypts itself to generate the partial signature: $\delta_j = \langle \sigma_j \oplus h(a_j), Enc(PK_G, c_j, a_j), Com(\sigma_j, h(a_j)) \rangle$; based on all the random coins used, run the confirmation protocol with $\mathcal{D}$.
To answer **Endorsement Query** (O_E) on (c_j, W): pass all endorsement queries (c_j, W) from $\mathcal{D}$ as extraction queries on c_j to its oracle to get d_j; d_j is equivalent to $\sigma_W(c_j)$.
In the **Challenge** phase: $\mathcal{D}$ outputs m and (c, W) to ask for a challenge; create a signature σ_t on a message m using Sig; randomly pick $\sigma_f \in S_\sigma$; randomly pick $a_t, a_f \in \mathcal{P}_{IBE}$; output a_t and a_f to ask for a challenge C_b where

$$C_b = \begin{cases} Enc(PK_G, c, a_t), & b = 0 \\ Enc(PK_G, c, a_f), & b = 1; \end{cases}$$

flip a coin $e \in \{0, 1\}$ and send the following challenge to $\mathcal{D}$:

$$\delta_e = \begin{cases} \langle \sigma_t \oplus h(a_t), C_b, Com(\sigma_t, h(a_t)) \rangle, & e = 0 \\ \langle \sigma_f \oplus h(a_f), C_b, Com(\sigma_f, h(a_f)) \rangle, & e = 1. \end{cases}$$

Finally, when $\mathcal{D}$ outputs a guess b', $\mathcal{D}'$ outputs b' as a guess for b.
END.

Note: $\langle \sigma_t \oplus h(a_t), Enc(PK_G, c, a_t), Com(\sigma_t, h(a_t)) \rangle$ *is equivalent to* **CVSig**$_S(m, C)$ *and* $\langle \sigma_f \oplus h(a_f), Enc(PK_G, c, a_f), Com(\sigma_f, h(a_f)) \rangle$ *is equivalent to* **Fake**(C).

Obviously, if $\mathcal{D}$ is PPT, so is $\mathcal{D}'$. Let ϵ_h and ϵ_{COM} be the indistinguishability coefficients of the pseudorandom generator and the commitment scheme. Recall that ϵ_h denotes the advantage of the best PPT distinguisher in distinguishing between the output distribution of a pseudorandom generator $h : \{0,1\}^{l_p} \to \{0,1\}^{l_s}$ and a uniform distribution over the output space of h. Whereas, ϵ_{COM} denotes the advantage of the best PPT distinguisher in distinguishing between the output distributions of the commitments of two different input values, say σ_f and σ_t, that is, between $\{r \leftarrow \{0,1\}^* : Com(\sigma_f, r)\}$ and $\{r \leftarrow \{0,1\}^* : Com(\sigma_t, r)\}$. Using hybrid arguments, it can be shown that the advantages of $\mathcal{D}$ and $\mathcal{D}'$ are related as follows:

$$Adv_{\mathcal{D}}^{Sim} < 2Adv_{\mathcal{D}'}^{IBE} + \epsilon_h + \tfrac{1}{2}\epsilon_{COM}.$$

If we assume COM is computationally hiding and h is a pseudorandom generator, then both ϵ_h and ϵ_{COM} should be negligible. Consequently, if $Adv_{\mathcal{D}}^{Sim}$ is non-negligible, the only possibility is either $Adv_{\mathcal{D}'}^{IBE}$ is non-negligible, meaning $\mathcal{D}'$ could break the semantic security of the IBE scheme (a contradiction). In other words, the semantic security of the IBE scheme implies the simulatability of the CVS construction with respect to the given construction of **Fake**. Since **Fake** is PPT, we could conclude that the given generic CVS construction is simulatable.

Proof Sketch of Theorem 4

Assume the CVS scheme is simulatable with respect to **Fake**. Suppose the constructed IBE scheme is not **IND-ID-CPA** secure, that is, there exists an adversary $\mathcal{D}$ which can win the **IND-ID-CPA** game with a non-negligible advantage $Adv_{\mathcal{D}}^{IBE}$. In other words, given a ciphertext (m, δ_b, PK_S) where δ_b is a valid/fake partial signature when $b = 0/1$, $\mathcal{D}$ could tell whether the plaintext bit $b = 0$ or $b = 1$ with a non-negligible advantage. We can construct $\mathcal{D}'$ from $\mathcal{D}$ to tell whether a given partial signature δ_b originates from **CVSig** or **Fake** as follows.

Algorithm $\mathcal{D}'(\delta_b)$

In the **Setup** phase: get the public key PK_G of the witness from its challenger; run $\mathcal{D}$ on PK_G; get the signer's public/private key pair (PK_S, sk_S).

To answer **Extraction Query** $\langle ID_j \rangle$: pass all extraction queries from $\mathcal{D}$ to its endorsement oracle.

In the **Challenge** phase: when $\mathcal{D}$ outputs ID to be challenged (Note the plaintext could only be 0 or 1), randomly select a message $m \in \mathcal{M}$; pass m, ID to its challenger and receive the challenge δ_b; pass $C_b = (m, \delta_b, PK_S)$ as a challenged ciphertext to $\mathcal{D}$. Finally, when $\mathcal{D}$ outputs a guess b', $\mathcal{D}'$ outputs b' as a guess for b.

END.

It obvious that the advantage of $\mathcal{D}'$ with respect to CVS simulatability is the same as the advantage of $\mathcal{D}$ on breaking the semantic security of the IBE scheme. Hence, if the latter is non-negligible, so is the former, a contradiction as we assume the given CVS scheme is simulatable with respect to **Fake**. In conclusion, the constructed IBE scheme is semantically secure as long as the CVS scheme is simulatable.

Constant Phase Bit Optimal Protocols for Perfectly Reliable and Secure Message Transmission

Arpita Patra[1], Ashish Choudhary[1,*], K. Srinathan[2], and C. Pandu Rangan[1,**]

[1] Dept of Computer Science and Engineering
IIT Madras, Chennai India 600036
arpita@cse.iitm.ernet.in, ashishc@cse.iitm.ernet.in, rangan@iitm.ernet.in
[2] International Institute of Information Technology
Hyderabad India 500032
srinathan@iiit.ac.in

Abstract. In this paper, we study the problem of *perfectly reliable message transmission*(PRMT) and *perfectly secure message transmission*(PSMT) between a sender **S** and a receiver **R** in a synchronous network, where **S** and **R** are connected by n vertex disjoint paths called wires, each of which facilitates bidirectional communication. We assume that atmost t of these wires are under the control of adversary. We present two-phase-*bit optimal* PRMT protocol considering Byzantine adversary as well as mixed adversary. We also present a three phase PRMT protocol which reliably sends a message containing l field elements by overall communicating $O(l)$ field elements. This is a significant improvement over the PRMT protocol proposed in [10] to achieve the same task which takes $log(t)$ phases. We also present a three-phase-*bit-optimal* PSMT protocol which securely sends a message consisting of t field elements by communicating $O(t^2)$ field elements.

Keywords: Reliable and Secure Communication, Information Theoretic Security, Communication Efficiency.

1 Introduction

In the problem of *perfectly reliable message transmission* (PRMT), a sender **S** is connected to **R** in an unreliable network by n vertex disjoint paths called wires; **S** wishes to send a message m chosen from a finite field $\mathbb{F}$ reliably to **R**, in a guaranteed manner, inspite of the presence of several kinds of faults in the network. The problem of *perfectly secure message transmission*(PSMT) has an additional constraint that the adversary should get no information about m. The faults in

---

* Work Supported by Project No. CSE/05-06/076/DITX/CPAN on Protocols for Secure Communication and Computation Sponsored by Department of Information Technology, Government of India.
** Work Supported by Microsoft Project No CSE0506075MICOCPAN on Foundation Research in Cryptography.

R. Barua and T. Lange (Eds.): INDOCRYPT 2006, LNCS 4329, pp. 221–235, 2006.

the network is modeled by an *adversary* who controls the actions of nodes in the network in a variety of ways. There are various network settings, fault models and computational models in which PRMT and PSMT problem has been studied extensively [3,2,4,13,12,5,7,11]. The PRMT and PSMT problems are very important primitives in various reliable and secure distributed protocols.

In this paper, we focus on undirected synchronous networks, where the adversary is an adaptive threshold Byzantine adversary having infinite computing power. In the past, PRMT and PSMT had been studied extensively in this setting [3,13,15,10,1]. The problem of PRMT and PSMT in this setting was first posed in [3]. In [3], it is proved that for t Byzantine faults, a two phase or three phase protocol exists iff there exists atleast $2t + 1$ vertex disjoint paths between **S** and **R**. However, the protocols of [3] involve lot of communication overhead. These protocols were improved significantly in [13]. However, the protocols of both [3] and [13] consider the problem of sending only one field element reliably and securely. So, in order to send a message consisting more than one field elements, we have to parallely execute these protocols for individual field elements, which will result in a huge communication overhead. This problem was first addressed in [15], where the authors attempted to give optimal PRMT and PSMT protocols to send messages containing more than one field element. In [15], the authors proved a lower bound of $\Omega(\frac{nl}{n-2t})$ field elements to be communicated to send a message containing l field elements reliably(securely) by using any two phase PRMT(PSMT) protocol. In view of this lower bound, any two phase PRMT(PSMT)protocol, which achieves this bound to send a message containing l field elements, reliably(securely) is called bit optimal two phase PRMT(PSMT) protocol. In [15], the authors claimed a two phase secure and reliable protocol to achieve this bound. However, in [1], the protocol of [15] has been proved to be unreliable. In [1], a two phase optimal PSMT(also PRMT) protocol has been proposed to send a message of size $O(t)$ securely and reliably by communicating overall $O(t^2)$ field elements. However, the protocol performs local computation(computation by **S** and **R**) which is not polynomial in n. Thus there doesnot exist any two phase polynomial time bit optimal PRMT protocol. In this paper, we propose a two phase bit optimal PRMT protocol, which performs local computation which is polynomial in n. We also show how to extend this protocol to incorporate additional faults of type omission and failstop.

In [10], a $log(t)$ phase protocol has been proposed to reliably send a message m of sufficiently large length l ($l = \Omega(nlog^2n)$, by communicating overall $O(l)$ field elements. In this paper, we significantly improve the above result by proposing a three phase protocol that reliably sends a message of size $l(l = \Omega(t^2))$ by communicating $O(l)$ field elements. Thus we can achieve reliability for free! We also propose a three phase PSMT protocol, that securely sends a message of size $O(t)$ by communicating overall $O(t^2)$ field elements. Though, the same task can be done by using the two phase protocol of [1], as mentioned earlier, their protocol involve huge amount of local computation, which is not polynomial in n. However, the three phase protocol proposed in this paper involves only polynomial time computation.

The rest of the paper is organised as follows: In the next section, we mention the network settings. We also recall some of the techniques used in [15], which we have used as blackbox in our protocols. In section 3, we give our two phase bit optimal PRMT protocol, considering only Byzantine adversary. We also show, how to extend this protocol to incorporate fail-stop and omission faults. This is followed by our interesting three phase PRMT protocol for sending a message of l field elements by communicating overall $O(l)$ field elements in section 4. In section 5, we propose a three phase PSMT protocol which securely sends a message of size $O(t)$ field elements by communicating $O(t^2)$ field elements. The paper ends with a brief conclusion and directions for further research.

2   Preliminaries

Here we recall the network settings and some of the algorithms which we have used as blackbox in our protocols from [15]. The underlying network is a synchronous network represented by an undirected graph $\mathcal{N}(\mathcal{P}, \mathcal{E})$, where $\mathcal{P} = \{P_1, P_2, \ldots, P_N\} \cup \{\mathbf{S}, \mathbf{R}\}$ denotes the set of players (nodes) in the network that are connected by 2-way communication links as defined by $\mathcal{E} \subset \mathcal{P} \times \mathcal{P}$. Also, $\mathcal{N}$ is atleast $(2t + 1)$-$(\mathbf{S}, \mathbf{R})$ connected [1]. Following the approach of [3], we abstract away the network entirely and concentrate on solving PRMT and PSMT problem for a single pair of synchronized processors, the *sender* $\mathbf{S}$ and the *receiver* $\mathbf{R}$, connected by n wires $w_1, w_2, \ldots, w_n$, where $n \geq 2t + 1$. We may think of these wires as a collection of vertex-disjoint paths between $\mathbf{S}$ and $\mathbf{R}$ in the underlying network [2]. The adversary is a static [3] adversary that can corrupt upto t of the wires connecting $\mathbf{S}$ and $\mathbf{R}$ and has unbounded computing power. Throughout this paper, we use m to denote the message that $\mathbf{S}$ wishes to send to $\mathbf{R}$ reliably. The message is assumed to be a sequence of ℓ elements from the finite field $\mathbb{F}$. The only constraint on $\mathbb{F}$ is that its size must be no less than the number of wires n. Since we measure the size of the message in terms of the number of field elements, we must also measure the communication complexity in units of field elements. We say that a wire is *faulty* if it is controlled by the adversary; all other wires are called *honest*. A faulty wire is *corrupted* in a specific phase if the value sent along that wire is changed.

2.1   Efficient Single Phase Reliable Communication

In [15], the authors have shown how to convert a t-error correcting code into a protocol **REL-SEND** for single phase reliable communication. In their protocol,

---

[1] We say that a network $\mathcal{N}$ is $n - (P_i, P_j)$-connected if the deletion of no $(n - 1)$ or less nodes from $\mathcal{N}$ disconnects P_i and P_j.

[2] The approach of abstracting the network as a collection of n wires is justifying using Menger's theorem [9] which states that a graph is $c - (\mathbf{S}, \mathbf{R})$-connected iff $\mathbf{S}$ and $\mathbf{R}$ are connected by atleast c vertex disjoint paths.

[3] By static adversary, we mean an adversary that decides on the set of players to corrupt before the start of the protocol.

the authors have used Reed-Solomon codes as t-error correcting code. The Reed-Solomon codes are defined as follows:

Definition 1. *Let* $\mathbb{F}$ *be a finite field and* $\alpha_1, \alpha_2, \ldots \alpha_n$ *be a collection of distinct elements of* $\mathbb{F}$. *Given* $k \leq n \leq |\mathbb{F}|$, *and a block* $\mathbf{B} = [m_0 \ m_1 \ \ldots \ m_{k-1}]$ *the encoding function for the Reed-Solomon code* $RS(n, k)$ *is defined as* $[p_\mathbf{B}(\alpha_1) \ p_\mathbf{B}(\alpha_2) \ \ldots \ p_\mathbf{B}(\alpha_n)]$ *where* $p_\mathbf{B}(x)$ *is the polynomial* $\sum_{i=0}^{k-1} m_i x^i$.

The efficiency (maximum amount of information that can be sent reliably even in the presence of faults) of an error correcting code is subject to **Singleton bound** [8], which states that $k \leq n - d + 1$, where k is the length of the message block, n is the length of the codeword and d is distance of the code. It is well known in coding theory that for a t-error correcting code, the distance d is at least $2t + 1$. Thus, for any t-error correcting block code, we have $k \leq n - 2t$. Since Reed-Solomon codes are also error correcting codes, whose efficiency is bounded by Singleton bound [8], for a t-error correcting RS code, we have $k \leq n - 2t$. We now recall the **REL-SEND**(m, k) protocol [15], which is a single phase protocol which reliably sends a message m by dividing m into blocks of length of k field elements where $k \leq n - 2t$.

Protocol REL-SEND(m, k): optimal single phase reliable message transmission of m. Without loss of generality, we assume that length l of the message m is a multiple of k.

- **S** breaks up m into blocks of length k field elements.
- For each block $\mathbf{B} = [m_0 \ m_1 \ \ldots \ m_{k-1}]$:
  - **S** computes $RS(n, k)$ to obtain $[p_\mathbf{B}(\alpha_1) \ p_\mathbf{B}(\alpha_2) \ \ldots \ p_\mathbf{B}(\alpha_n)]$.
  - **S** sends $p_\mathbf{B}(\alpha_i)$ along the wire w_i.
  - **R** receives the (possibly corrupted) $p_\mathbf{B}(\alpha_i)$'s and applies the Reed-Solomon decoding algorithm and constructs $\mathbf{B}$.
- **R** concatenates the $\mathbf{B}$'s to recover the message m.

Lemma 1 ([15]). *Suppose that the receiver* **R** *knows* f *faults among the* n *wires, and* t' *be the number of faulty wires apart from these* f *wires. Then* **REL-SEND**(m, k) *works correctly even for slightly larger* k. *Specifically,* **REL-SEND**(m, k) *works correctly for all* k, $k \leq n - 2t' - f$.

2.2 Extracting Randomness

In [15], the authors have proposed an algorithm for the following problem: Suppose **S** and **R** by some means agree on a sequence of n numbers $\mathbf{x} = [x_1 x_2 \ldots x_n] \in \mathbb{F}^n$ such that the adversary knows $n - f$ components of $\mathbf{x}$, but the adversary has no information about the other f components of $\mathbf{x}$, however, **S** and **R** do not necessarily know which values are known to the adversary. The goal is for **S** and **R** to agree on a sequence of f numbers $y_1 y_2 \ldots y_f \in \mathbb{F}$ such that the adversary has no information about $y_1 y_2 \ldots y_f$. This is achieved by the following algorithm [15]:

Algorithm EXTRAND$_{n,f}(\mathbf{x})$. Let $\mathbf{V}$ be a $n \times f$ Vandermonde matrix with members in $\mathbb{F}$. This matrix is published as a part of the protocol specification. $\mathbf{S}$ and $\mathbf{R}$ both locally compute the product $[y_1 \ y_2 \ \cdots \ y_f] = [x_1 \ x_2 \ \cdots \ x_n]\mathbf{V}$.

Lemma 2 ([15]). *The adversary has no information about $[y_1 \ y_2 \ \cdots \ y_f]$ computed in algorithm EXTRAND.*

2.3 Communicating Conflict Graph

In the two phase and three phase PSMT protocols proposed in [3,13], we come across the following situation: In the first phase, player $\mathbf{A}$ (which is $\mathbf{S}$ for three phase protocol and $\mathbf{R}$ for two phase protocol) selects at random $n = 2t + 1$ polynomials p_i, $1 \leq i \leq n$ over $\mathbb{F}$, each of degree t. Next through each wire w_i, $\mathbf{A}$ sends to the other player $\mathbf{B}$ (which is $\mathbf{R}$ for three phase protocol and $\mathbf{S}$ for two phase protocol), the following: the polynomials p_i.[4] and for each j, $1 \leq j \leq n$, the value of $p_j(\alpha_i)$ (which we denote by r_{ij} where α_i's are arbitrary distinct publicly specified members of $\mathbb{F}$. Assume that $\mathbf{B}$ receives the polynomials p_i' and the values r_{ij}' along the wire w_i. In the next phase, $\mathbf{B}$ tries to find as many faults as he can find that occurred in the previous phase and communicate all his findings reliably back to $\mathbf{A}$. Towards this, $\mathbf{B}$ first constructs what is known as conflict graph $H = (\mathcal{W}, E)$, where $\mathcal{W} = \{w_1, w_2, \ldots, w_n\}$ and $(w_i, w_j) \in E$ if $r_{ij}' \neq p_j'(\alpha_i)$ or $r_{ji}' \neq p_i'(\alpha_j)$. A naive and straightforward way of reliably sending the conflict graph to $\mathbf{A}$ is to broadcast the entire graph over all the n wires. This approach of broadcasting the set of contradictions is used in the PSMT protocols of [3,13]. Since in the worst case there can be $\Theta(n^2)$ edges in the conflict graph, broadcasting them requires communicating $\Theta(n^3)$ field elements.

In [15], the authors have given a method to communicate the set of $\Theta(n^2)$ edges by communicating $\Theta(n^2)$ field elements. We call this method as Matching technique, which we use as a black box in this paper. We briefly describe the method: Suppose $\mathbf{B}$ receives the polynomials p_i' and the values r_{ij}' along the wire w_i, $1 \leq i \leq n$.

B's computation and communication

- $\mathbf{B}$ initializes his *fault-list*, denoted by L_{fault}, to $\emptyset$. $\mathbf{B}$ then constructs an undirected graph $H = (\mathcal{W}, E)$ where the edge $(w_i, w_j) \in E$ if $r_{ij}' \neq p_j'(\alpha_i)$ or $r_{ji}' \neq p_i'(\alpha_j)$.
- For each i, $1 \leq i \leq n$, such that the degree of node w_i in the graph H constructed above is greater than t (i.e., $degree(w_i) \geq t + 1$), $\mathbf{B}$ adds w_i to L_{fault}.

---

[4] We assume that a polynomial is sent by sending a $(t + 1)$-tuple of field elements.

- Let $H' = (\mathcal{W}', E')$ be the induced subgraph of H on the vertex set $\mathcal{W}' = (\mathcal{W} \setminus L_{fault})$. Next, **B** finds a maximum matching[5] $M \subseteq E'$ of the graph H'.
- For each edge (w_i, w_j) in H that does not belong to M, **B** associates the six-tuple $\{\alpha_i, \alpha_j, r'_{ij}, r'_{ji}, p'_j(\alpha_i), p'_i(\alpha_j)\}$. Let $\{a_1, a_2, \ldots, a_N\}$ be the edges in H that are not in M. Replacing each edge with its associated 6-tuple, **R** gets a set of $6N$ field elements, $X = \{X_1, X_2, \ldots, X_{6N}\}$.
- **B** then sends the following to **A** through all the wires: the set L_{fault} and for each edge $(w_i, w_j) \in M$, the following six field elements: $\{\alpha_i, \alpha_j, r'_{ij}, r'_{ji}, p'_i(\alpha_j)$ and $p'_j(\alpha_i)\}$. **B** also sends the set X as **REL-SEND**$(X, |M| + |L_{fault}| + 1)$ to **A**.

The following lemma and theorems taken from [15] shows that the above method reliably sends the list of $\Theta(n^2)$ contradictions by overall communicating $O(n^2)$ field elements.

Lemma 3 ([15]). **A** *is guaranteed to receive the set X correctly.*

Theorem 1 ([15]). *Given an undirected graph $H = (V, E)$, with a maximum degree of t, and a maximum matching M, the number of edges $|E|$ is less than or equal to $(2|M|^2 + |M|t)$.*

Theorem 2 ([15]). *The overall communication complexity involved in the protocol* **REL-SEND***$(X, M + |L_{fault}| + 1)$ is $O(t^2)$.*

Thus, the entire conflict graph is send in two parts; first a matching M is broadcasted in $O(t^2)$ communication complexity and then the rest of the edges of the conflict graph are send by communicating $O(t^2)$ field elements by using **REL-SEND** protocol. We are now ready to describe our protocols.

3   Two Phase Bit Optimal Perfectly Reliable Message Transmission Protocols

Here we propose a two phase bit optimal perfectly reliable message transmission protocol, which reliably sends a message consisting of $O(t)$ field elements by communicating $O(t^2)$ field elements. As mentioned earlier, our protocol performs polynomial time computation in comparison to the two phase PRMT protocol of [1] which performs huge amount of local computation which is not polynomial in n. We also show how our protocol can incorporate omission and fail-stop errors in addition to Byzantine errors.

3.1   Two Phase PRMT Protocol Considering Only Byzantine Errors

Let $n = 2t + 1$ and $m = [m_0 m_1 \ldots m_t]$ be the message block of size $t + 1$, where each $m_i \in \mathbb{F}$. In our protocol, the first phase is from **R** to **S** and the second phase is from **S** to **R**.

---

[5] A subset M of the edges of H, is called a *matching* in H if no two of the edges in M are adjacent. A matching M is called *maximum* if H has no matching M' with a greater number of edges than M has.

Phase I (R to S)
The receiver **R** selects at random n polynomials p_i, $1 \le i \le n$ over $\mathbb{F}$, each of degree t. Next, through each wire w_i, **R** sends the following to **S**: the polynomials p_i and for each j, $1 \le j \le n$, the value of $p_j(\alpha_i)$ (which we denote by r_{ij}) where α_i's are arbitrary distinct publicly specified members of $\mathbb{F}$.

Phase II (S to R)
Let **S** receive the polynomials p_i' and the values r_{ij}' along the wire w_i.

S's computation and communication

- **S** initializes his *fault-list*, denoted by L_{fault}, to $\emptyset$. **S** then constructs an undirected graph $H = (\mathcal{W}, E)$ where $\mathcal{W} = \{w_1, w_2, \ldots, w_n\}$ and edge $(w_i, w_j) \in E$ if $r_{ij}' \ne p_j'(\alpha_i)$ or $r_{ji}' \ne p_i'(\alpha_j)$. Here H is called the conflict graph.
- For each i, $1 \le i \le n$, such that the degree of node w_i in the graph H constructed above is greater than t (i.e., $degree(w_i) \ge t + 1$), **S** adds w_i to L_{fault}.
- **S** constructs a $3t+1$ degree, message carrying polynomial $s(x) = \sum_{i=0}^{3t+1} k_i x^i$ over $\mathbb{F}$ as follows:

$$\text{Assign } k_i = \begin{cases} m_i & \text{if } 0 \le i \le t. \\ 0 & \text{if } w_{i-t} \in L_{fault}. \\ p_{i-t}'(0) & \text{otherwise.} \end{cases}$$

- **S** constructs the set $Z = \{s(\alpha_1), s(\alpha_2), \ldots, s(\alpha_{2t+1})\}$.
- **S** sends the conflict graph H to **R** by using the matching technique as mentioned in section 2.3.
- **S** sends the set Z and the list L_{fault} along each of the n wires.

Message recovery by R

1. **R** reliably receives L_{fault} and knows that the wires in this set are faulty. He initializes $L_{fault}^{\mathbf{R}} = L_{fault}$.
2. For each edge $(w_i, w_j) \in M$, **R** reliably receives the six tuple $\{\alpha_i, \alpha_j, r_{ij}', r_{ji}',$ $p_i'(\alpha_j)$ and $p_j'(\alpha_i)\}$. **R** locally verifies: $r_{ij}' \overset{?}{=} r_{ij}$ and $p_j'(\alpha_i) \overset{?}{=} p_j(\alpha_i)$.
   (a) If the first test fails, it implies that the value of the polynomial $p_j(x)$ at α_i had been change by the adversary and hence **R** adds wire i to $L_{fault}^{\mathbf{R}}$.
   (b) If the second test fails, **R** concludes that the polynomial $p_j(x)$ had been changed by the adversary and hence adds wire j to $L_{fault}^{\mathbf{R}}$.
   (c) If both the test fails, then **R** knows that both the polynomial $p_j(x)$ as well its value at α_i had been changed and hence adds wire i and j to $L_{fault}^{\mathbf{R}}$.
3. After step 2, at least $|M|$ new faults are caught by **R**.
4. From the correctness of the matching technique, it is clear that **R** receives the conflict graph H reliably. Again **R** locally verifies for each edge's (say (w_i, w_j)) 6-tuple: $r_{ij}' \overset{?}{=} r_{ij}$ and $p_j'(\alpha_i) \overset{?}{=} p_j(\alpha_i)$.

    (a) If the former check fails (that is the values are unequal), it implies that the value of the polynomial $p_j(x)$ at α_i had been changed by the adversary and hence **R** adds w_i to $L^{\mathbf{R}}_{fault}$.

    (b) If the latter check fails, then **R** concludes that the polynomial $p_j(x)$ had been changed by the adversary and hence adds wire j to $L^{\mathbf{R}}_{fault}$.

    (c) If both the check fails, then **R** knows that both the polynomial $p_j(x)$ as well its value at α_i had been changed by the adversary and adds both wire i and j to $L^{\mathbf{R}}_{fault}$.

5. At the end of this step, *all* the faults that occurred during transmission in **Phase I** are guaranteed to have been identified (see Theorem 3 below).

6. **R** will try to reconstruct $s(x)$ as follows:

   – The first $t+1$ coefficients correspond to the the the $t+1$ elements of m and hence are unknown to **R**. So he puts $t+1$ unknown variables at these positions.

   – If wire $w_i \in L_{fault}$, then **R** puts 0 at position $t+1+i$; i.e., if wire w_i was detected as faulty by **S** after phase 1, then **R** knows that **S** had put zero as the coefficient at position $t+1+i$ in $s(x)$.

   – If wire $w_i \notin L_{fault}$, but present in the set $L^{\mathbf{R}}_{fault}$, then **R** puts an unknown variable at position $t+1+i$; i.e., if wire w_i is detected as faulty by **R** after receiving the conflict graph, then **R** knows that **S** has used $p'_i(0)$ as the coefficient at position $t+1+i$ in $s(x)$, such that $p'_i(0) \neq p_i(0)$. Since **R** does not know what $p'_i(0)$ has been used by **S**, **R** puts an unknown variable corresponding to the position $t+1+i$ in $s(x)$.

   – For all other positions k, **R** puts $p_k(0)$

7. There will be maximum $2t+1$ unknowns for **R** corresponding to $s(x)$ ($t+1$ corresponding to m and atmost t corresponding to the faulty polynomials that **S** had received after phase 1), but he is receiving the value of $s(x)$ at $2t+1$ points. So he forms $2t+1$ equations in atmost $2t+1$ unknown variables, solves these equation, gets the values of the unknown variables and correctly reconstruct $s(x)$ and hence correctly retrieve the message m.

Theorem 3. *The above protocol reliably communicates $t+1$ field elements from* **S** *to* **R** *by overall communicating $O(n^2) = O(t^2)$ field elements.*

Proof (sketch). Suppose wire w_i was corrupted in **Phase I**; i.e., either $p_i(x) \neq p'_i(x)$ or $p'_j(\alpha_i) \neq r'_{ij}$. Consider the first case; i.e., $p_i(x) \neq p'_i(x)$. Then the two polynomials can intersect in at most t points, since both are of degree t. Since there are atleast $t+1$ honest wires, $p_i = p'_i$ in atmost t of the α_i's corresponding to these honest wires, so there is atleast one honest wires w_j which will contradict w_i and so the edge (w_i, w_j) will be present in the conflict graph. Corresponding to the edge (w_i, w_j), **R** reliably receives the values $p'_i(\alpha_j)$ and r'_{ji}. In step 2(b) or 4(b) of the protocol, **R** will come to know that $p_i(\alpha_j) \neq p'_i(\alpha_j)$ and will add wire i to $L^{\mathbf{R}}_{fault}$. Suppose, on the other hand that the adversary had changed the value of a correct polynomial $p_j(x)$ at α_i. Then $p'_j(\alpha_i) \neq r'_{ij}$. So the edge (w_i, w_j) will be present in the conflict graph. Corresponding to the edge (w_i, w_j), **R** reliably receives the values r'_{ij} and $p'_j(\alpha_i)$. In step 2(a) or 4(a) of the protocol, $\mathbf{R}$ will come to know that $r_{ij} \neq r'_{ij}$ and adds wire i to $L^{\mathbf{R}}_{fault}$. Since the correct values corresponding to every contradiction have been received, $\mathbf{R}$ eventually knows all the corruptions; i.e., $\mathbf{R}$ will know all the wires w_i over which the either the polynomial p_i had been changed or the value of a correct polynomial $p_j(x)$ at α_i had been changed after **Phase I**. The reliability of the protocol follows from the working of the protocol. In **Phase I** $O(n^2)$ field elements are communicated. In **Phase II**, to send the conflict graph, $O(n^2)$ communication complexity is involved. Broadcasting the set Z also involves $O(n^2)$ communication complexity. Hence the overall communication complexity is $O(n^2)$. □

From [15], the lower bound on communication complexity for reliably sending t field elements, where $n \geq 2t + 1$, using any two phase PRMT protocol is $O(t^2)$. Since our protocol achieves this bound, our protocol has optimal communication complexity.

3.2 Two Phase PRMT Protocol Considering Mixed Adversary

Here, we show how the PRMT protocol of previous section can be extended to incorporate omission and failstop errors in addition to Byzantine errors. For the definition of failstop and omission errors see [6]. Note that unlike Byzantine adversary, which can force P to behave maliciously, omission or fail-stop adversary cannot do so. In the mixed adversary model, the threshold adversary is represented by the 3-tuple (t_b, t_o, t_f), where t_b, t_o and t_f are the number of wires that are under the control of the adversary in Byzantine, omission and fail-stop fashion. The following theorem is taken from [14].

Theorem 4 ([14]). *Perfectly reliable message transmission between* $\mathbf{S}$ *and* $\mathbf{R}$ *in a network* $\mathcal{N}$ *is possible under the presence of a threshold adversary characterized by the 3-tuple* (t_b, t_o, t_f) *iff there exists atleast* $(2t_b + t_o + t_f + 1)$ *wires between* $\mathbf{S}$ *and* $\mathbf{R}$ *in* $\mathcal{N}$.

The following theorem taken from [14] gives the lower bound on the communication complexity of any two phase PRMT protocol to reliably send a message consisting l field elements under the presence of threshold adaptive adversary (t_b, t_o, t_f).

Theorem 5 ([14]). *Any two phase PRMT protocol, which reliably communicates* l *field elements from* $\mathbf{S}$ *to* $\mathbf{R}$, *which are connected by* $n \geq 2t_b + t_o + t_f + 1$ *wires, where* t_b, t_o *and* t_f *are the number of wires under the control of a threshold adaptive adversary in Byzantine, omission and fail-stop fashion, requires communicating atleast* $\Omega\left(\frac{nl}{n-(2t_b+t_o+t_f)}\right)$ *field elements.*

Putting $l = (t_b + t_o + t_f + 1) = O(n)$ and $n = 2t_b + t_o + t_f + 1$, we get a lower bound of $\Omega(n^2)$ field elements to be communicated to reliably send a message of size $O(n)$ from $\mathbf{S}$ to $\mathbf{R}$. We informally describe, how our two phase PRMT protocol of the previous section, can be extended in the mixed adversary model to achieve two phase bit optimal PRMT protocol.

Informal Description of the Protocol

In the first phase, **R** will select $n = 2t_b + t_o + t_f + 1$ polynomials p_i over $\mathbb{F}$ of degree t_b and send over wire w_i, the polynomial p_i and the values $r_{ij} = p_j(\alpha_i)$. In the second phase, **S** will receive the polynomial p_i' and the values r_{ij}' over w_i. **S** will form the conflict graph H and construct a list L_{fault}, where $w_i \in L_{fault}$ iff $degree(w_i) \geq t_b + 1$. Also, **S** will create another list $L_{omitted}$, where $w_i \in L_{omitted}$ if **S** does not receive anything over w_i. It is clear that $|L_{omitted}| \leq t_o + t_f$. **S** will broadcast the list $L_{omitted}$ and L_{fault} to **R** and sends H to **R** using matching technique. **S** also form a secret carrying polynomial $s(x) = \sum_{i=0}^{i=n+t_b+t_o+t_f} k_i x^i$ as follows: let $m = [m_0 m_1 \ldots m_{t_b+t_o+t_f}]$ be the message block.

$$\text{Assign } k_i = \begin{cases} m_i & \text{if } 0 \leq i \leq t_b + t_o + t_f. \\ 0 & \text{if } w_{i-(t_b+t_o+t_f+1)} \in L_{fault} \cup L_{omitted}. \\ p_{i-(t_b+t_o+t_f+1)}'(0) & \text{otherwise.} \end{cases}$$

S will initialize a list $Y = (s(\alpha_1), s(\alpha_2), \ldots, s(\alpha_n))$ and broadcast Y to **R**. After receiving the conflict graph and the list $L_{omitted}$, **R** will find out all the corruptions that had taken place after **Phase I** and constructs the list of faulty wires $L_{fault}^{\mathbf{R}}$. The message recovery by **R** is done in the same way as in the previous protocol, except that here the degree of $s(x)$ is $3t_b + 2t_o + 2t_f + 1$. It is clear that for **R**, there will be atmost $2t_b + t_o + t_f + 1$ unknown variables corresponding to $s(x)$, but he is receiving the value of $s(x)$ at $n = 2t_b + t_o + t_f + 1$ points. Hence, he can easily form a system of equations and reconstruct $s(x)$ and retrieve m. It is easy to see that the overall communication complexity in each phase is $O(n^2)$, which is matching the lower bound for bit optimal two phase PRMT protocol involving mixed adversary. Hence, our protocol is bit optimal.

4   A Three Phase PRMT Protocol Considering Byzantine Adversary

Here we propose a three phase PRMT protocol to reliably send a message consisting n^2 field elements by communicating overall $O(n^2)$ field elements, where $n = 2t + 1$. Thus we get reliability for "free". In [10], a $log(t)$ phase protocol has been proposed to do the same. Thus our protocol is a significant improvement over the protocol of [10]. In [14], it is shown that any r-phase PRMT protocol for $r \geq 1$, which reliably sends a message of size l from **S** to **R** needs to communicate atleast $\Omega(l)$ field elements. Since, our protocol achieves this bound, our protocol is bit optimal.

In our protocol, the first phase is a send from **S** to **R**, the second phase is from **R** to **S** and the third and final phase is again from **S** to **R**. Let the sequence of (n^2) field elements that **S** wishes to transmit be denoted by m_{ij}, $0 \leq i, j \leq (n-1)$.

Phase I (S to R)

Using the m_{ij} values, **S** defines a bivariate polynomial $q(x, y)$ of degree $(n-1)$ in each variable:

$$q(x, y) = \sum_{i=0, j=0}^{\substack{i=n-1 \\ j=n-1}} m_{ij} x^i y^j$$

The sender **S** then defines n polynomials $p_i(x)$, $1 \leq i \leq n$ over $\mathbb{F}$, each of degree $(n-1)$ as follows:

$$p_i(x) = q(x, \alpha_i)$$

where α_i's are arbitrary distinct publicly specified members of $\mathbb{F}$. **S** then writes each polynomial $p_i(x)$ which is of degree $n - 1 = 2t$, in the following form

$$p_i(x) = (\theta_i(x) * \beta_i(x)) + \gamma_i(x)$$

where $\theta_i(x)$ and $\beta_i(x)$ are polynomials of degree t and $\gamma_i(x)$ is a polynomial of degree atmost $t - 1$. Next, through each wire w_i, **S** sends the following to **R**: the polynomial θ_i, β_i and γ_i and for each j, $1 \leq j \leq n$, the value of $\theta_j(\alpha_i), \beta_j(\alpha_i)$ and $\gamma_j(\alpha_i)$, which we denote by θ_{ij}, β_{ij} and γ_{ij} respectively.

Phase II (R to S)

Phase II is similar to **Phase II** of the PRMT protocol of section 3, except that here **R** performs the computation instead of **S**. Also, here **R** will receive three polynomials and $3n$ values over each wire. **R** will construct three conflict graphs, H_1, H_2 and H_3 corresponding to the polynomials θ_i, β_i and γ_i respectively. **R** also constructs the list L_{fault}, where $w_i \in L_{fault}$ if $degree(w_i) \geq t + 1$ in any of the conflict graphs H_1, H_2 or H_3. **R** then sends the list L_{fault} to **S** by broadcasting it over all the n wires. **R** also sends the conflict graphs H_1, H_2 and H_3 to **S** by using matching technique.

Phase III (S to R)

In **Phase III**, **S** will reliably receive the conflict graphs H_1, H_2 and H_3 and the list L_{fault}. As in two phase PRMT protocol of section 3, **S** will locally check each contradiction present in the conflict graphs H_1, H_2 and H_3 and find out all additional faulty wires w_i, over which atleast one of the polynomials θ_i, β_i or γ_i has been changed after **Phase I**. **S** will now have the list of all faulty wires $L_{fault}^{\mathbf{S}}$. Note that $|L_{fault}^{\mathbf{S}}| \leq t$. **S** knows that for all other wires $w_i \notin L_{fault}^{\mathbf{S}}$, **R** has received the polynomials θ_i, β_i and γ_i correctly. **S** will now send the correct polynomials θ_i, β_i and γ_i corresponding to each faulty wire $w_i \in L_{fault}^{\mathbf{S}}$ to **R** as follows: **S** sends the set $L_{fault}^{\mathbf{S}}$ to **R** through all the wires. Let Z be the list of coefficients of all the polynomials $\theta_i(x), \beta_i(x)$ and $\gamma_i(x)$ for all $w_i \in L_{fault}^{\mathbf{S}}$. Thus $|Z| \leq 3(t+1) * |L_{fault}^{\mathbf{s}}|$. **S** will send Z reliably using protocol **REL-SEND** as **REL-SEND**$(Z, |L_{fault}^{\mathbf{s}}|)$.

Message Recovery by R

R reliably receives the list $L_{fault}^{\mathbf{S}}$ and list Z. From the list $L_{fault}^{\mathbf{S}}$, **R** will come to know all the wires w_i, over which he had received atleast one faulty polynomial

θ_i, β_i or γ_i after **Phase I**. From the list Z, **R** will receive the correct θ_i, β_i and γ_i corresponding to each wire $w_i \in L^S_{fault}$. **R** will now have correct θ_i, β_i and γ_i for all the wires $w_i, 1 \le i \le n$. Thus he reconstruct all the $p_i(x) = q(x, \alpha_i)$ as

$$p_i(x) = (\theta_i(x) * \beta_i(x)) + \gamma_i(x)$$

R reconstructs the polynomial $q(x, y)$ using n correct $q(x, \alpha_i)$ and hence retrieve all $m_{ij}, 0 \le i, j, \le (n-1)$.

Theorem 6. *The above protocol reliably sends a message consisting of n^2 field elements by communicating $O(n^2)$ field elements.*

Proof (Sketch). To reconstruct the message, **R** should be able to reconstruct the polynomial $Q(x, y)$. If **R** knows n correct $Q(x, i)$'s, then he can easily reconstruct $Q(x, y)$, because $Q(x, i)$ is a polynomial in x of degree $n - 1$. In **Phase III**, **S** will find out all the wires over which atleast one of the three polynomials has been changed. **S** will reliably send all the three polynomials for all such identified faulty wires by **REL-SEND**$(Z, |L^S_{fault}|)$ to **R**. From lemma 1, the REL-SEND$(\cdot, k)$ protocol succeeds provided that $n - f \ge k + 2(t - f)$, where f is the number of faults that **R** already knows; here, $k = |L^S_{faults}|$ and $n = 2t+1$. Therefore, REL-SEND succeeds if $(2t+1) - f - (|L^S_{faults}|) \ge 2t - 2f$, or if, $f \ge |L^S_{faults}|$. Since, in **Phase III**, **S** broadcasts the list L^S_{fault}, **R** is guaranteed to have identified at least $|L^S_{faults}|$ faulty wires. Thus **R** will receive the list Z correctly. The reliability of the protocol follows from the working of the protocol. The overall communication complexity in **Phase I** is $O(n^2)$. In **Phase II**, **R** needs to send three conflict graphs using matching technique, which involve an overall communication complexity of $O(n^2)$. The communication complexity involved in sending Z through **REL-SEND** is $O\left(\frac{|Z|*n}{|L^S_{fault}|}\right) = O(n^2)$ because $|Z| = O(t * |L^S_{fault}|) = O(n * |L^S_{fault}|)$ as $t = O(n)$. Thus the overall communication of the protocol is $O(n^2)$. Hence the theorem.   □

5 A Three Phase PSMT Protocol

Here we propose a three phase PSMT protocol which securely sends a message m consisting of $t + 1$ field elements by communicating $O(n^2)$ field elements, where $n = 2t + 1$. In [1], a two phase PSMT protocol to do the same task is proposed. However, the protocol in [1] performs a huge amount of local computation by **R**, which is not polynomial in n. However, our protocol performs polynomial time local computation.

Phase I (S to R)

- **S** selects n polynomial $p_1(x), p_2(x), \ldots, p_n(x)$ over $\mathbb{F}$, each of degree t. Over each wire w_i, **S** sends the polynomial p_i and the values $p_j(\alpha_i)$, denoted by r_{ij}, for $1 \le j \le n$, where α_i are elements in $\mathbb{F}$.

Phase II (R to S)

- **R** receives over each wire the polynomial p'_i and the values r'_{ij} and constructs the conflict graph H based on the values p'_i and r'_{ij}. **R** also prepares a list L_{fault}, where $w_i \in L_{fault}$ if $degree(w_i) \geq t+1$ in H. **R** sends the conflict graph H using matching technique and broadcasts the list L_{fault} over $2t+1$ wires.

Phase III (S to R)

- **S** reliably receives the list L_{fault} and the conflict graph H. **S** will perform local checking for each contradiction (w_i, w_j) present in the conflict graph H. After this **S** will have the list L^S_{fault} which contains all the wires w_i over which the polynomial p_i has been changed after **Phase I**. Note that $|L^S_{fault}| \leq t$.
- **S** will form a $n - |L^S_{fault}|$ tuple $x = [x_{i_1} x_{i_2} \ldots x_{i_{n-|L^S_{fault}|}}]$, where $x_{i_j} = p_{i_j}(0)$, for $1 \leq j \leq n - |L^S_{fault}|$; i.e., the vector x is the collection of the constant terms of all the polynomials p_i, such that $w_i \notin L^S_{fault}$. **S** will execute **EXTRAND**$_{n-|L^S_{fault}|, t+1}(x)$ algorithm of section 2.2 to get a sequence of $t+1$ field elements $y = [y_1 y_2 \ldots y_{t+1}]$
- **S** computes $c = [c_1 c_2 \ldots c_{t+1}] = y \oplus m$, where $c_i = m_i \oplus y_i$, for $1 \leq i \leq t+1$. Here $m = [m_1 m_2 \ldots m_{t+1}]$ denotes the message block. **S** then broadcasts the list c and L^S_{fault} over all the n wires.

The overall communication complexity of the protocol is $O(n^2)$. The message recovery by **R** is done as follows: **R** will reliably receive the list L^S_{fault} and c. From L^S_{fault}, **R** will know the identity of all the faulty wires w_i, **R** will ignore all the polynomials p'_i and consider the constant term of the remaining $n - |L^S_{fault}|$ polynomials and reconstruct the vector x. **R** will then execute **EXTRAND**$_{n-|L^S_{fault}|, t+1}(x)$ algorithm of section 2.2 to recover y. **R** will then recover the message m as $m = c \oplus y$, where $m_i = c_i \oplus y_i$.

Theorem 7. *The above protocol securely transmits a message m of size $t+1$ field elements by communicating $O(n^2)$ field elements.*

Proof (sketch). Since in each of the three phases, the overall communication complexity is $O(n^2)$, the overall communication complexity of the above protocol is $O(n^2)$. The security of the above protocol is argued as follows: there can be maximum t wires under the control of the adversary. So even if the adversary passively listen to the contents of these wires, he will come to know only about the constant terms of these t polynomials. However, he will have no idea about the constant terms of the remaining $t+1$ polynomials corresponding to $t+1$ honest wires, because each of these polynomials if of degree t and to interpolate a polynomial of degree t, we require $t+1$ points on that polynomial. However, the adversary is getting only t points corresponding to an honest polynomial and hence the constant term of such an honest polynomial is information theoretically secure. Since there are atleast $t+1$ honest wires and hence $t+1$ honest polynomials, the adversary has no information at all about the constant terms of these $t+1$ honest polynomials. In **Phase III**, **S** knows that for all $i \in L^{\mathbf{S}}_{fault}$, the polynomial $p_i(x)$ had been changed by the adversary and the rest of the $n - L^{\mathbf{S}}_{fault}$ polynomials had reached correctly to **R** after **Phase I**. However, among these $n - L^{\mathbf{S}}_{fault}$ polynomials, the adversary knows atmost $t - L^{\mathbf{S}}_{fault}$ polynomials by passively listening the contents of the corresponding wires. Since **S** doesnot knows which $t - L^{\mathbf{S}}_{fault}$ wires are under the control of the adversary, **S** executes $\mathbf{EXTRAND}_{n-|L^{\mathbf{S}}_{fault}|, t+1}(x)$ to obtain the vector y. By the correctness of the $\mathbf{EXTRAND}_{n-|L^{\mathbf{S}}_{fault}|, t+1}(x)$, the adversary will have no idea about the vector y and hence the message m. Since **R** will also be able to reconstruct y, he will be able to reconstruct the message. Hence the theorem. □

In [15], it is shown that any three phase PSMT protocol which sends l field elements securely and reliably from **S** to **R**, needs to communicate atleast $\Omega\left(\frac{nl}{n-2t}\right)$ field elements. Thus our three phase PSMT protocol is bit optimal.

6   Conclusion

In this paper, we have proposed bit optimal two phase PRMT protocol for threshold adversary. Our protocol performs polynomial time computation in comparison to two phase bit optimal PRMT protocol of [1], which performs almost exponential amount of local computation. We have also shown how this protocol can be extended in the mixed adversary model. We have also given a three phase PRMT protocol for reliably sending l field elements by communicating $O(l)$ field elements. Comparing this protocol with the $log(t)$ phase PRMT protocol of [10], we find that our protocol has significantly reduced the number of phases. We have also proposed a three phase PSMT protocol to securely send a message containing t field elements by overall communicating $O(t^2)$ field elements. Again comparing this protocol with the two phase PSMT protocol of [1] which performs almost exponential amount of computation, we find that by increasing the number of phases by 1 (but keeping the connectivity of the network same), we can bring down the amount of computation to be polynomial in the number of wires. However as stated in [1], to do the same task in two phases, involving polynomial computation still remains an open problem. Extending the protocol of section 4 and section 5 to mixed adversary model is another interesting and challenging problem.

References

1. S. Agarwal, R. Cramer, and R. de Haan. Asymptotically optimal two-round perfectly secure message transmission. In C. Dwork, editor, *Proc. of Advances in Cryptology: CRYPTO 2006*, LNCS 4117, pages 394–408. Springer-Verlag, 2006.
2. Y. Desmedt and Y. Wang. Perfectly secure message transmission revisited. In *Proc. of Advances in Cryptology: Eurocrypt 2002*, LNCS 2332, pages 502–517. Springer-Verlag, 2003.

3. D. Dolev, C. Dwork, O. Waarts, and M. Yung. Perfectly secure message transmission. *JACM*, 40(1):17–47, 1993.
4. M. Franklin and R. Wright. Secure communication in minimal connectivity models. *Journal of Cryptology*, 13(1):9–30, 2000.
5. J. A. Garay and K. J. Perry. A continuum of failure models for distributed computing. In *Proc. of 6th WDAG*, pages 153–165, 1992.
6. R. Guerraoui and L. Rodrigues. *Introduction to Reliable Distributed Programming*. Springer Verlag, 2006.
7. M. V. N. A. Kumar, P. R. Goundan, K. Srinathan, and C. Pandu Rangan. On perfectly secure communication over arbitrary networks. In *Proc. of 21st PODC*, pages 193–202. ACM Press, 2002.
8. F. J. MacWilliams and N. J. A. Sloane. *The Theory of Error Correcting Codes*. North-Holland Publishing Company, 1978.
9. K. Menger. Zur allgemeinen kurventheorie. *Fundamenta Mathematicae*, 10:96–115, 1927.
10. A. Narayanan, K. Srinathan, and C. Pandu Rangan. Perfectly reliable message transmission. *Information Processing Letters*, 11(46):1–6, 2006.
11. R. Ostrovsky and M. Yung. How to withstand mobile virus attacks. In *Proc. of 10th PODC*, pages 51–61. ACM Press, 1991.
12. H. Sayeed and H. Abu-Amara. Perfectly secure message transmission in asynchronous networks. In *Proc. of Seventh IEEE Symposium on Parallel and Distributed Processing*, 1995.
13. H. Sayeed and H. Abu-Amara. Efficient perfectly secure message transmission in synchronous networks. *Information and Computation*, 126(1):53–61, 1996.
14. K. Srinathan. Secure distributed communication. PhD Thesis, IIT Madras, 2006.
15. K. Srinathan, A. Narayanan, and C. Pandu Rangan. Optimal perfectly secure message transmission. In *Proc. of Advances in Cryptology: CRYPTO 2004*, LNCS 3152, pages 545–561. Springer-Verlag, 2004.

Using Wiedemann's Algorithm to Compute the Immunity Against Algebraic and Fast Algebraic Attacks

Frédéric Didier

Projet CODES, INRIA Rocquencourt, Domaine de Voluceau,
78153 Le Chesnay cedex
frederic.didier@inria.fr

Abstract. We show in this paper how to apply well known methods from sparse linear algebra to the problem of computing the immunity of a Boolean function against algebraic or fast algebraic attacks. For an n-variable Boolean function, this approach gives an algorithm that works for both attacks in $O(n2^n D)$ complexity and $O(n2^n)$ memory. Here $D = \binom{n}{d}$ and d corresponds to the degree of the algebraic system to be solved in the last step of the attacks. For algebraic attacks, our algorithm needs significantly less memory than the algorithm in [ACG$^+$06] with roughly the same time complexity (and it is precisely the memory usage which is the real bottleneck of the last algorithm). For fast algebraic attacks, it does not only improve the memory complexity, it is also the algorithm with the best time complexity known so far for most values of the degree constraints.

Keywords: algebraic attacks, algebraic immunity, fast algebraic attacks, Wiedemann's algorithm.

1 Introduction

Algebraic attacks and fast algebraic attacks have proved to be a powerful class of attacks against stream ciphers [CM03, Cou03, Arm04]. The idea is to set up an algebraic system of equations satisfied by the key bits and to try to solve it. For instance, this kind of approach can be quite effective [CM03] on stream ciphers which consist of a linear pseudo-random generator hidden with non-linear combining functions acting on the outputs of the generator to produce the final output.

For such an attack to work, it is crucial that the combining functions satisfy low degree relations. The reason for this is that it ensures that the algebraic system of equations verified by the secret key is also of small degree, which is in general essential for being able to solve it. This raises the fundamental issue of determining whether or not a given function admits non-trivial low degree relations [MPC04, Car04, DGM04, BP05, DMS05].

For algebraic attacks, in order to find relations of degree at most d satisfied by an n-variable combining Boolean function f, only two algorithmic approaches

R. Barua and T. Lange (Eds.): INDOCRYPT 2006, LNCS 4329, pp. 236–250, 2006.
© Springer-Verlag Berlin Heidelberg 2006

are known for the time being. The first one relies on Gröbner bases [FA03] and consists of finding minimal degree elements in the polynomial ideal spanned by the ANF of f and the field equations. The second strategy relies on linear algebra, more precisely we can associate to f and d a matrix M such that the elements in the kernel of M give us low degree relations. Building M is in general easy so the issue here is to find non-trivial elements in the kernel of a given matrix or to show that the kernel is trivial.

The linear approach has lead to the best algorithm known so far [ACG+06] that works in $O(D^2)$ where $D = \binom{n}{d}$. There is also the algorithm of [DT06] which performs well in practice and which is more efficient when d is small. Actually, when d is fixed and $n \to \infty$, this last algorithm will be able to prove the nonexistence of low degree relation in $O(D)$ for almost all Boolean functions. Note however that if the ANF of f is simple or has a lot of structure, it is possible that the Gröbner basis approach outperforms these algorithms, especially if the number of variables is large (more than 30).

For fast algebraic attacks, only the linear algebra approach has been used. There are now two degree constraints $d < e$ and the best algorithms are the one of [ACG+06] working in $O(ED^2)$ where $E = \binom{n}{e}$ and the one of [BLP06] working in $O(ED^2 + E^2)$.

All these algorithms relying on the linear algebra approach use some refinements of Gaussian elimination in order to find the kernel of a matrix M. Efficiency is achieved using the special structure of M. We will use here a different approach. The idea is that the peculiar structure of M allows for a fast matrix vector product that will lead to efficient methods to compute its kernel. This comes from the following facts:

- There are algorithms for solving linear systems of equations which perform only matrix vector products. Over the finite fields, there is an adaptation of the conjugate gradient and Lanczos algorithm [COS86, Odl84] or the Wiedemann algorithm [Wie86]. These algorithms were developed at the origin for solving large sparse systems of linear equations where one can compute a matrix vector product efficiently. A lot of work has been done on the subject because of important applications in public key cryptography. Actually, these algorithms are crucial in the last step of the best factorization or discrete logarithm algorithms. Notice as well that these algorithms were also used in the context of algebraic attacks against HFE cryptosystems (see [FJ03]).
- Computing a matrix vector product of the matrix involved in the algebraic immunity computation can be done using only the binary Moebius transform. It is an involution which transforms the two main representations of a Boolean function into each other (namely the list of its ANF coefficients and the list of its images).
- The Moebius transform of an n-variable Boolean function can be computed efficiently in $O(n2^n)$ complexity and $O(2^n)$ memory. We will call the corresponding algorithm the fast Moebius transform by analogy with the fast Fourier transform (note that they both rely on the same principle).

We will focus here on the Wiedemann algorithm and derive an algorithm for computing algebraic attacks or fast algebraic attacks relations in $O(n2^n D)$. Wiedemann's algorithm is probabilistic and so is our algorithm, however we can get a failure probability as small as we want with the same asymptotic complexity. When d or e are close to $n/2$ the asymptotic complexity is very good for fast algebraic attacks but is a little bit less efficient for algebraic attacks than the one presented in [ACG$^+$06]. However this algorithm presents another advantage, its memory usage is very efficient, $O(n2^n)$ to be compared with $O(D^2)$. This may not seem really important, but in fact the memory is actually the bottleneck of the other algorithms.

The outline of this paper is as follows. We first recall in Section 1 some basic facts about Boolean functions, algebraic and fast algebraic attacks, and the linear algebra approach used by almost all the known algorithms. Then, we present in Section 2 the Wiedemann algorithm and how we can apply it to our problem. We present in Section 3 some benchmark results of our implementation of this algorithm. We finally conclude in Section 4.

2   Preliminary

In this section we recall basic facts about Boolean functions, algebraic attacks and fast algebraic attacks. We also present the linear algebra approach used by almost all the actual algorithm to compute relations for these attacks.

2.1   Boolean Functions

In all this paper, we consider the binary vector space of n-variable Boolean functions, that is the space of functions from $\{0,1\}^n$ to $\{0,1\}$. It will be convenient to view $\{0,1\}$ as the field over two elements, what we denote by $\mathbf{F}_2$. It is well known that such a function f can be written in an unique way as an n-variable polynomial over $\mathbf{F}_2$ where the degree in each variable is at most 1 using the Algebraic Normal Form (ANF) :

$$f(x_1,\ldots,x_n) = \sum_{u \in \mathbf{F}_2^n} f_u \left(\prod_{i=1}^{n} x_i^{u_i} \right) \quad f_u \in \mathbf{F}_2, u = (u_1,\ldots,u_n) \in \mathbf{F}_2^n \quad (1)$$

By *monomial*, we mean in what follows, a polynomial of the form $\prod_{i=1}^{n} x_i^{u_i}$. We will heavily make use in what follows of the notation f_u, which denotes for a point u in $\mathbf{F}_2^n$ and an n-variable Boolean function f, the coefficient of the monomial associated to u in the ANF (1) of f. Each monomial associated to a u in $\mathbf{F}_2^n$ can be seen as a function having only this monomial as its ANF. Such function is only equal to 1 on points x such that $u \subseteq x$ where

$$\text{for } u, x \in \mathbf{F}_2^n \quad u \subseteq x \quad \text{iff} \quad \{i, u_i \neq 0\} \subseteq \{i, x_i \neq 0\} \quad (2)$$

The *degree* of f is the maximum weight of the u's for which $f_u \neq 0$. By listing the images of a Boolean function f over all possible values of the variables, we can also view it as a binary word of length 2^n. For that, we will order the points of $\mathbf{F}_2^n$ in lexicographic order

$$(0,\ldots,0,0)(0,\ldots,0,1)(0,\ldots,1,0)\ldots(1,\ldots,1,1) \tag{3}$$

The *weight* of a Boolean function f is denoted by $|f|$ and is equal to $\sum_{x \in \mathbf{F}_2^n} f(x)$ (the sum being performed over the integers). We also denote in the same way the (Hamming) weight of a binary 2^n-tuple or the cardinal of a set. A *balanced* Boolean function is a function with weight equal to half its length, that is 2^{n-1}.

There is an important involutive (meaning its own inverse) transformation linking the two representations of a Boolean function f, namely its image list $(f(x))_{x \in \mathbf{F}_2^n}$ and its ANF coefficient list $(f_u)_{u \in \mathbf{F}_2^n}$. This transformation is known as the *binary Moebius transform* and is given by

$$f(x) = \sum_{u \subseteq x} f_u \quad \text{and} \quad f_u = \sum_{x \subseteq u} f(x) \tag{4}$$

Here u and x both lie in $\mathbf{F}_2^n$ and we use the notation introduced in (1) for the ANF coefficients of f.

Dealing with algebraic attacks, we will be interested in the subspace of all Boolean functions of degree at most d. Note that the set of monomials of degree at most d forms a basis of this subspace. By counting the number of such monomials we obtain that its dimension is given by $D \stackrel{\text{def}}{=} \sum_{i=0}^{d} \binom{n}{i}$. In the following, a Boolean function g of degree at most d will be represented by its ANF coefficients $(g_i)_{|i| \leq d}$. Notice as well that we will need another degree constraint e and that we will write E for the dimension of the subspace of Boolean functions with degree at most e.

2.2 Algebraic and Fast Algebraic Attacks

We will briefly describe here how algebraic and fast algebraic attacks work on a filtered LFSR. In the following, L is an LFSR on n bits with initial state $(x_1,\ldots,x_n)$ and a filtering function f. The idea behind algebraic attacks is just to recover the initial state given the keystream bits $(z_i)_{i \geq 0}$ by solving the algebraic system given by the equations $f(L^i(x_1,\ldots,x_n)) = z_i$. However the algebraic degree of f is usually too high, so one has to perform further work.

In the original algebraic attacks [CM03], the first step is to find *annihilators* of f. This means functions g of low degree such that $fg = 0$ where fg stands for the function defined by

$$\forall x \in \mathbf{F}_2^n, \quad fg(x) = f(x)g(x) \tag{5}$$

in particular $fg = 0$ if and only if

$$\forall x \in \mathbf{F}_2^n, \quad f(x) = 1 \quad \Longrightarrow \quad g(x) = 0 \tag{6}$$

So we obtain a new system involving equations of the form

$$g(L^i(x_1,\ldots,x_n)) = 0 \quad \text{for } i \geq 0 \text{ and } z_i = 1 \tag{7}$$

that can be solved if the degree d of g is low enough. Remark that for the i's such that $z_i = 0$, we can use the same technique with the annihilators of $1 + f$ instead.

In order to quantify the resistance of a function f to algebraic attacks, the notion of algebraic immunity was introduced in [MPC04]. By definition, the algebraic immunity of f is the smallest degree d such that f or $1 + f$ admits a non-trivial annihilator of degree d. It has been shown in [CM03] that for an n-variable Boolean function a non-trivial annihilator of degree at most $\lceil n/2 \rceil$ always exists.

Sometimes, annihilators of low degree do not exist, but another relation involving f can be exploited. That is what is done in fast algebraic attacks introduced in [Cou03] and further confirmed and improved in [Arm04, HR04]. The aim is to find a function g of low degree d and a function h of larger degree e such that $fg = h$. We now get equations of the form

$$z_i\, g(L^i(x_1, \ldots, x_n)) = h(L^i(x_1, \ldots, x_n)) \quad \text{for} \quad i \geq 0 \tag{8}$$

In the second step of fast algebraic attacks, one has to find a linear relation between successive equations [Cou03] in order to get rid of the terms with degree greater than d. Remark that these terms come only from h and so such a relation does not involve the keystream bits z_i. More precisely, we are looking for an integer l and binary coefficients c_i such that all the terms of degree greater than d cancel out in the sum

$$\sum_{i=0}^{i<l} c_i\, h(L^i(x_1, \ldots, x_n)) \tag{9}$$

One can search this relation offline and apply it not only from time 0 but also shifted at every time $i \geq 0$. In the end, we get an algebraic system of degree d that we have to solve in the last step of the attack.

In this paper, we will focus on the first step of these attacks. We are given a function f and we will discuss algorithms to compute efficiently its immunity against algebraic and fast algebraic attacks.

2.3   Linear Algebra Approach

We will formulate here the problem of finding low-degree relations for a given function f in terms of linear algebra. In the following, all the lists of points in $\mathbf{F}_2^n$ are always ordered using the order defined in Subsection 2.1.

Let us start by the case of the classical algebraic attacks. Recall that a function g of degree at most d is an annihilator of f if and only if

$$\forall x \in \mathbf{F}_2^n, \quad f(x) = 1 \quad \Longrightarrow \quad g(x) = 0 \tag{10}$$

So, for each x such that $f(x) = 1$ we get from (4) a linear equation in the D coefficients $(g_u)_{|u| \leq d}$ of g, namely

$$\sum_{u \subseteq x} g_u = 0 \quad u \in \mathbf{F}_2^n,\ |u| \leq d \tag{11}$$

This gives rise to a linear system that we can write $M_1((g_u)_{|u|\le d})^t = 0$ where M_1 is an $|f| \times D$ binary matrix and the t indicates transposition. Each row of M_1 corresponds to an x such that $f(x) = 1$. Actually, we see that the matrix vector product of M_1 with $((g_u)_{|u|\le d})^t$ is just an evaluation of a function g with ANF coefficients $(g_u)_{|u|\le d}$ on all the points x such that $f(x) = 1$. We will encounter again such type of matrices and we will introduce a special notation. Let $\mathcal{A}$ and $\mathcal{B}$ be two subsets of $\mathbf{F}_2^n$, we will write

$$V_{\mathcal{B}}^{\mathcal{A}} = (v_{i,j})_{i=1...|\mathcal{B}|, j=1...|\mathcal{A}|} \tag{12}$$

for the matrix corresponding to an evaluation over all the points in $\mathcal{B}$ of a Boolean function with non-zero ANF coefficients in $\mathcal{A}$. The V stands for evaluation, and we have $v_{i,j} = 1$ if and only if the j-th point of $\mathcal{B}$ is included (notation $\subseteq$ over $\mathbf{F}_2^n$) in the i-th point of $\mathcal{A}$. With this notation, we get

$$M_1 = V_{\{x,f(x)=1\}}^{\{u,|u|\le d\}} = V_{\{x,f(x)=1\}}^d \tag{13}$$

The exponent d is a shortcut for $\{u, |u| \le d\}$, in particular an exponent n means all the points in $\mathbf{F}_2^n$. It is important to understand this notion of evaluation because if we look a little ahead, we see that performing a matrix vector product for such a matrix is nothing but performing a binary Moebius transform.

Now, a function g with ANF $(g_u)_{|u|\le d}$ is an annihilator of f if and only if $((g_u)_{|u|\le d})^t \in \ker(M_1)$. A non-trivial annihilator of degree smaller than or equal to d exists if and only if this matrix is not of full rank.

For the fast algebraic attacks, we obtain the same description with a different linear system. Functions g of degree at most d and h of degree at most e such that $fg = h$ exist if and only if

$$\forall x \in \mathbf{F}_2^n \quad h(x) + f(x)g(x) = 0 \tag{14}$$

Here the unknowns are the D coefficients $(g_u)_{|u|\le d}$ of g and the E coefficients $(h_u)_{|u|\le e}$ of h. So, for each point x we derive by using (4) the following equation on these coefficients

$$\sum_{u\subseteq x, |u|\le e} h_u \quad + \quad f(x) \sum_{u\subseteq x, |u|\le d} g_u = 0 \tag{15}$$

And we obtain a system that we can write $M_2((h_u)_{|u|\le e}, (g_u)_{|u|\le d})^t = 0$ where M_2 is an $N \times (E + D)$ binary matrix given by

$$M_2 = \left(V_n^e \mid \mathrm{Diag}((f(x))_{x\in\mathbf{F}_2^n})V_n^d \right) \tag{16}$$

The multiplication by f in (15) corresponds here with the product by the diagonal matrix. With this new matrix, each kernel element corresponds to functions g and h such that $fg = h$.

There is a way to create a smaller linear system for the fast algebraic attacks. This follows the idea in [BLP06]. Actually, as pointed out in [DM06], the matrix V_e^e is an involutive $E \times E$ matrix. The idea is then to start by computing the

$(h_u)_{|u| \leq e}$ using the values that h has to take on the points x with $|x| \leq e$. That is, taking for the unknowns the D coefficients $(g_u)_{|u| \leq d}$ of g, the values $(h(x))_{|x| \leq e}$ of h are given by

$$((h(x))_{|x| \leq e})^t = \text{Diag}((f(x))_{|x| \leq e})V_e^d((g_u)_{|u| \leq d})^t \qquad (17)$$

We can then find the ANF coefficients $(h_u)_{|u| \leq e}$ of h by applying V_e^e on the left because this matrix is involutive. We can then evaluate h over all $\mathbf{F}_2^n$ by multiplying on the left by V_n^e. In the end, we obtain a new linear system $M_3((g_u)_{|u| \leq d})^t = 0$ where M_3 is the following $N \times D$ matrix

$$M_3 = \text{Diag}((f(x))_{x \in \mathbf{F}_2^n})V_n^d + V_n^e V_e^e \text{Diag}((f(x))_{|x| \leq e})V_e^d \qquad (18)$$

Here $\text{Diag}((f(x))_{x \in \mathbf{F}_2^n})V_n^d$ corresponds to the evaluation of fg on all the points in $\mathbf{F}_2^n$ and the other part to the evaluation of h on the same points. Remark that the rows corresponding to $|x| \leq e$ are null by construction, so M_3 can be reduced to an $(N - E) \times D$ matrix.

Up to now, all the known algorithms relying on the linear algebra approach ([MPC04, DT06, BLP06, ACG$^+$06]) worked by computing the kernel of these matrices using some refinements of Gaussian elimination. Efficiency was achieved using the very special structure involved. We will use here a different approach. The idea is that the special structure behind these linear systems allows a fast matrix vector product. We will actually be able to compute $M_1((g_i)_{|i| \leq d})^t$ or $M_3((g_i)_{|i| \leq d})^t$ in $O(n2^n)$. This will lead to an algorithm in $O(n2^n D)$ complexity and $O(n2^n)$ memory.

3   Using Wiedemann's Algorithm

In this section we describe how the Wiedemann algorithm [Wie86] can be used efficiently on our problem. We focused on this algorithm (instead of Lanczos' or conjugate gradient algorithm) because it is easier to analyze and it does not need any assumption on the matrix.

3.1   Fast Evaluation

The first ingredient for Wiedemann's algorithm to be efficient on a given matrix, is that we can compute the matrix vector product for this particular matrix efficiently. This is for example the case for a sparse matrix and will also be the case for the matrices M_1, M_2 or M_3 involved in the algebraic immunity computation. In the following we still use the order defined in Subsection 2.1 for all the lists of points in $\mathbf{F}_2^n$.

So we want to compute efficiently a matrix vector product of M_1, M_2 or M_3. For that, looking at the definition of these matrices (see (13), (16) and (18)) it is enough to be able to compute efficiently a matrix vector product of diagonal matrices and of the matrices $V_\mathcal{B}^\mathcal{A}$ ($\mathcal{A}$ and $\mathcal{B}$ being two subsets of $\mathbf{F}_2^n$). Then, we will just compute this kind of product for all the matrices appearing in the previous definitions to get the final product.

Computing a product between a diagonal matrix and a vector is easy, it can be computed in $O(2^n)$ using a binary AND between the vector and the list of the diagonal elements. Regarding the matrices $V_{\mathcal{B}}^{\mathcal{A}}$, performing the product is almost the same as doing a Moebius transform as we have seen in Subsection 2.3. The details are explained in the following algorithm.

Algorithm 1 (Matrix vector product of $V_{\mathcal{B}}^{\mathcal{A}}$). Given n, two subsets ($\mathcal{A}$ and $\mathcal{B}$) of $\mathbf{F}_2^n$ and a vector $v = (v_i)_{i=1\ldots|\mathcal{A}|}$, this algorithm computes the matrix vector product of $V_{\mathcal{B}}^{\mathcal{A}}$ and v.

1. [pack] Initialize a vector $s = (s_u)_{u \in \mathbf{F}_2^n}$ as follows If u is the i-th point in $\mathcal{A}$ then set $s_u = v_i$. Otherwise (that is $u \notin \mathcal{A}$) set $s_u = 0$.
2. [Moebius] Compute the fast binary Moebius transform of s in place.
3. [Extract] The result is given by the (s_u) with $u \in \mathcal{B}$.

So, the key point in a fast matrix vector product here is that we can compute the binary Moebius transform efficiently. The following algorithm called the fast Moebius transform works in $O(n2^n)$ and uses the same idea as the fast Fourier transform algorithm. In the end, we are able to perform a matrix vector product of M_1, M_2 or M_3 with the same complexity. Remark as well that for all these algorithms, the memory usage is in $O(2^n)$.

Algorithm 2 (Fast binary Moebius transform). Given an n-variable Boolean function f in the form of a list of ANF coefficients $(f_u)_{u \in \mathbf{F}_2^n}$, this algorithm computes its image list $(f(x))_{x \in \mathbf{F}_2^n}$ recursively. In both cases the list must be ordered using the order described in section 2.1. The algorithm can work in place (meaning the result overwrites the $(f_u)_{u \in \mathbf{F}_2^n}$ without modifications).

1. [stop] If $n = 0$ then $f(0) = f_0$. Exit the function.
2. [left recursion] Perform the Moebius transform for a $n - 1$ variable function $f^{(0)}$ whose coefficients are given by the first half of the coefficient list of f, that is the f_u's with $u = (u_1, \ldots, u_n)$ and $u_1 = 0$.
3. [right recursion] Perform the Moebius transform for a $n - 1$ variable function $f^{(1)}$ whose coefficients are given by the second half of the coefficient list of f, that is the f_u with $u = (u_1, \ldots, u_n)$ and $u_1 = 1$.
4. [combine] We have $f(x_1, \ldots, x_n) = f^{(0)}(x_2, \ldots, x_n) + x_1 f^{(1)}(x_2, \ldots, x_n)$.

The complexity in $O(n2^n)$ comes from the fact that at each call, we apply the algorithm over two problems of half the size of the original one. The correctness is easy to prove provided that the equality at step 4 is correct. But using the definition of the Moebius transform, we have

$$f(x) = \sum_{u \subseteq x} f_u(x) = \sum_{u \subseteq x, u_1 = 0} f_u(x) + \sum_{u \subseteq x, u_1 = 1} f_u(x) \qquad (19)$$

The second sum is zero if $x_1 = 0$, so we can write

$$f(x) = \sum_{(u_2,...,u_m) \subseteq (x_2,...,x_n)} f_{(0,u_2,...,u_n)}(x) \quad + \quad x_1 \sum_{(u_2,...,u_n) \subseteq (x_2,...,x_n)} f_{(1,u_2,...,u_n)}(x) \quad (20)$$

and we retrieve the equality at step 4.

3.2   The Wiedemann Algorithm

We present here the Wiedemann algorithm for an $n \times n$ square matrix A. We will deal with the non-square case in the next subsection.

The approach used by Wiedemann's algorithm (and more generally blackbox algorithms) is to start from a vector b and to compute the so called Krylov sequence

$$b, Ab, A^2b, \ldots, A^nb, \ldots \quad (21)$$

This sequence is linearly generated and admits a minimal polynomial $P_b \in F_2[X]$ such that $P_b(A)b = 0$. Moreover, P_b divides the minimal polynomial of the matrix A and is of maximum degree n.

The idea of Wiedemann's algorithm is to find P_b using the Berlekamp-Massey algorithm. For that, we take a random vector u^t in $\mathbf{F}_2^n$ and compute the inner products

$$u.b, u.Ab, u.A^2b, \ldots, u.A^{2n}b \quad (22)$$

The complexity of this step is in $2n$ evaluations of the matrix A. This sequence is still linearly generated, and we can find its minimal polynomial $P_{u,b}$ in $O(n^2)$ using the Berlekamp-Massey Algorithm ([Mas69]). Moreover, $P_{u,b}$ divides P_b and they are equal with probability bounded away from 0 by $1/(6 \log n)$ (see [Wie86]). Notice that if X divides $P_{u,b}$, then A is singular since 0 is then one of its eigenvalues.

Now, let us assume that we have computed P_b and that $P_b(x) = c_0 + xQ(x)$ with $Q \in \mathbf{F}_2[X]$. If $c_0 \neq 0$ (and therefore $c_0 = 1$) then $AQ(A)b = b$ and $Q(A)b$ is a solution x of the system $Ax = b$. If $c_0 = 0$ then $AQ(A)b = 0$ and $Q(A)b$ is a non-trivial (by minimality of P_b) element of $\ker(A)$. So, we can either find a solution of $Ax = b$ or a non-trivial element in $\ker(A)$ with complexity the number of steps of computing $Q(A)$ that is n evaluations of A.

Remark that in both cases we can verify the coherence of the result (even when $P_{u,b} \neq P_b$) with only one evaluation of A. Moreover, when A is singular, we are sure to find a non-trivial kernel element if b does not lie in $\mathrm{Im}(A)$. This happens with probability always greater than $1/2$ over $\mathbf{F}_2$.

If we are only interested in knowing if a matrix is singular then, as already pointed out, we just need that X divides $P_{u,b}$ and we have:

Theorem 1. *If an $n \times n$ square matrix A over F_2 is singular, then applying Wiedemann's algorithm with a random choice of b and u will prove that the matrix is singular with probability greater than $1/4$ and $O(n)$ evaluations of A.*

Proof. Let us decompose $E = \mathbf{F}_2^n$ into the characteristic subspaces of A. In particular we have $E = E_0 \oplus E_1$ where E_0 is the subspace associated with the eigenvalue 0 and A restricted to E_1 is invertible. Using this decomposition, let us write $b = b_0 + b_1$. Let P_0 and P_1 be the minimal polynomial associated to the sequences $(u.b_0, u.Ab_0, \dots)$ and $(u.b_1, u.Ab_1, \dots)$. We know that P_0 is just a power of X and that the LFSR generating the second sequence is non degenerate. So the minimal polynomial associated to the sum is equal to $P_0 P_1$. To conclude, we see that $X/P_{u,b}$ if $u.b_0 \neq 0$ and over $\mathbf{F}_2$ this happens for a random choice of u and b with a probability greater than $1/4$.

In the end, the algorithm consists in trying different values of b and u until we have a large enough probability that A is singular or not. When b is fixed, we just described a Monte-Carlo algorithm here but there is also a Las-Vegas version (Algorithm 1 of [Wie86]) that works better in this case. It gives P_b (so at the end a kernel element or a solution to $Ax = b$) in $O(n \log n)$ matrix evaluations on average.

Remark that when X divides $P_{u,b}$ we are sure that A is singular. So the three possible outputs of the algorithm are the following :
- either A is singular and we know it for sure,
- or we know that A is non-singular with a very high probability,
- finally when $P_{u,b}$ is of maximum degree (that is n) then we know the minimal polynomial of A. So if it is not divisible by X then we are sure that the matrix is of full rank.

3.3   Non Square Case

The square case could be applied directly when we try to show the maximum algebraic immunity of a balanced Boolean function with an odd number of variables (because in this case $d = (n-1)/2$ and $|f| = 2^{n-1}$ is equal to D). However, in the general case we do not have a square matrix.

One method to extend this could be to select randomly a subsquare matrix until we find an invertible one or until we have done so many choices that we are pretty sure that the initial matrix is not of full rank. This method is however quite inefficient when the matrix is far from being a square matrix and in this case there is a better way to perform this task.

Let us consider an $n \times k$ matrix A with $k < n$, the idea is to generate a $k \times n$ random sparse matrix Q such that with high probability QA will be of rank k if A is non-singular. From [Wie86] we have the following result.

Theorem 2. *If A is a non-singular $n \times k$ matrix with $k < n$, let us construct a $k \times n$ matrix Q as follows. A bit of the row i from 1 to k is set to 1 with probability $w_i = \min(1/2, 2\log n/(k+1-i))$. Then with probability at least $1/8$ the following statements hold*
- *The $k \times k$ matrix QA is non-singular.*
- *The total Hamming weight of the rows in Q is at most $2n(2 + \log n)^2$.*

Notice that in [Wie86], another generating method is given to generate Q such that QA is non-singular with a probability bounded away from 0 and such that the total Hamming weight of Q is in $O(n \log n)$. This is better asymptotically but less applicable in practice since the probability is smaller than the $1/8$ of Theorem 2.

We are now back to the square case with a little overhead because we have to compute Q times a vector at each step of the Wiedemann algorithm. This is why we have generated Q as sparse as possible to minimize this overhead. In particular, when Q has a total Hamming weight in $O(n \log n)$ then we can perform the matrix vector product in $O(n \log n)$. Notice as well that we need $O(n \log n)$ extra memory in order to store the matrix Q.

In order to know if A is singular or not, the algorithm is the following. We generate a matrix Q and test the non singularity of QA with Wiedemann's Algorithm. If this matrix is non-singular, then we know that A is non-singular as well (with the failure probability of Wiedemann's algorithm). Otherwise, we can go on a few times (say i) with different matrices Q and if all the products are singular then A is singular with probability greater than $1 - (7/8)^i$.

Remark that with negligible complexity overhead, we can compute a non-trivial kernel element x of QA when this matrix is singular. And if A is singular, with a probability greater than $1/8$ we will also have $Ax = 0$. So we can run the algorithm until we are sure that A is singular (when we get a non-trivial kernel element) or until we have a very high probability that A is non-singular.

4   Implementation Results

We have implemented all the algorithms described in this article and we give their performances in this section. All the experiments were done on a Pentium 4 running at 3.2Ghz with 2Gb of memory.

First of all, let us summarize the complexity of our algorithms. Both for algebraic and fast algebraic attacks, we will have to perform Wiedemann's algorithm on a $D \times D$ matrix. This requires $O(D)$ matrix vector products of this matrix plus $O(D^2)$ operations for the Berlekamp-Massey algorithm. We have seen that we can perform the product in $O(n2^n)$ operations, so we get in all cases a final complexity in $O(n2^n D)$. Notice that in order to get a small failure probability, we will have to perform this task a few times. This is especially true for the non-square case since we will have to check different matrices Q. However, only a constant number of times is needed to get a given probability, so the asymptotic complexity is still the same.

Regarding the memory, the matrix evaluation needs $O(2^n)$ memory for the square case and an extra $O(n2^n)$ memory when we have to store a matrix Q for the non-square case. All the other operations need only an $O(D)$ memory.

The running time of the algorithms is almost independent of the function f involved except for the Berlekamp-Massey part. However, this part is clearly not the most time consuming. In particular, it is a good idea in Wiedemann's algorithm to perform the computation with more than one random u per vector b. If we perform 4 Berlekamp-Massey steps per random vector b, then we will be able to detect a singularity with a probability almost one half and a very small

Table 1. Running time for the square case: $n = 2d + 1$, $D = 2^{n-1}$ and f is a random balanced n-variable Boolean function. Optimized implementation using the SSE2 instructions set.

d,n	4,9	5,11	6,13	7,15	8,17	9,19	10,21	11,23	12,25
time	0s	0s	0.01s	0.3s	5s	102s	30m	11h	20d

Table 2. Time and memory for computing the immunity against normal algebraic attacks using Wiedemann's algorithm

d,n	2,22	2,23	3,19	3,20	3,21	4,19	5,19
D	254	277	1160	1351	1562	5036	16664
time	113s	264s	100s	252s	630s	640s	2706s
memory	656Mb	1397Mb	118Mb	255Mb	547Mb	160Mb	194Mb

overhead. Remark as well that all the experiments where done for random balanced functions but the running time will be roughly the same for real functions used in stream cipher.

When we compute the immunity against normal algebraic attacks for a square matrix, we can implement the code in a very efficient way. In particular, using the transposition of the Moebius transform, we can merge step 1 and 3 of Algorithm 1 between two consecutive evaluations. This is because this transposition will map back all the positions in the set $\mathcal{B}$ into the set $\mathcal{A}$. Moreover, if M_1 is square and invertible, applying $(V^d_{\{x, f(x)=1\}})^t$ on the left will result in another invertible matrix. In this way, we obtain a fully parallelizable algorithm because we can perform a fast Moebius transform (or its transpose) dealing with 32 bits at a time (even more with SSE2). In the end we get a very efficient implementation with the running time for a random choice of b and four random choices of u given in Table 1. Moreover, the memory usage is negligible in this case since there is no matrix Q to store.

In this case, since $n = 2D + 1$, the complexity is in $O(D^2 \log D)$ and the memory usage in $O(D)$. So the asymptotic complexity is a little worse than the one of [ACG+06] (in $O(D^2)$) but the memory usage is a lot better (to be compared with $O(D^2)$). We see that two consecutive sets of parameters differ by a factor of 16 in the computation time as could be inferred from the asymptotic complexity. This factor increase a little with the number of variables but we were still able do deal with as many as 25 variables. Notice that this implementation also breaks the previous record which was 20 variables in the papers [DT06] and [ACG+06].

For the non-square case however, the results are less impressive. The reason is that there is no simple way to do the multiplication by Q in a parallel way. Hence we loose a factor 32 in the process. To overcome this difficulty a block version of Wiedemann's algorithm might be use (see [Cop94]), but we did not have time to implement it. Another issue is that the memory to store the sparse matrix Q may become two large. Moreover, the code used for the following experiments is not as optimized as the one for the square case. We might divide the time and memory by a factor 2 roughly with a careful implementation.

Table 3. Time and memory for computing the immunity against fast algebraic attacks using Wiedemann's algorithm. Here we chose $n = 2e + 1$.

d/e,n	2/8,17	3/8,17	3/9,19	3/10,21	4/8,17	5/8,17	6/8,17
D	154	834	1160	1562	3214	9402	21778
time	1s	15s	101s	614s	82s	297s	801s
memory	14Mb	25Mb	118Mb	547Mb	33Mb	40Mb	45Mb

Table 4. Dependence of fast algebraic attacks immunity computation in the parameter e. In all cases $D = 1160$.

d/e,n	3/7,19	3/8,19	3/9,19	3/10,19	3/11,19
time	154s	130s	101s	70s	43s
memory	192Mb	159Mb	118Mb	77Mb	43Mb

We give in Table 2 the time to compute the immunity against normal algebraic attacks. Wiedemann's algorithm is executed for one random matrix Q a random b and four random u's. In order to obtain a small enough probability of error, one will have to execute this a few times (16 gives an error probability of 0.1 and 32 of 0.01).

What is interesting is that the time for computing immunity against fast algebraic attacks is almost the same as the one for normal algebraic attacks (see Table 3). There is only little influence of the degree e (see Table 4) on the performance because the size of Q depends on it. But the time and memory will always stay within a factor two compared to the case where e is equal to $n/2$.

5   Conclusion

In this paper, we devised a new algorithm to compute the immunity of a Boolean function against both algebraic and fast algebraic attacks. This algorithm presents a few advantages:

- It is easy to understand since it is based on a well known sparse linear algebra algorithm.
- Its complexity is quite good, especially for the fast algebraic attacks.
- It uses little memory compared to the other known algorithms which make it able to deal with more variables.
- And it is quite general since it can work for both attacks with little modification. In particular, if in the future one is interested in other kind of relations defined point by point, then the same approach can be used.

Acknoweldegement

I want to thank Jean-Pierre Tillich for his great help in writing this paper and Yann Laigle-Chapuy for his efficient implementation of the fast binary Moebius transform in SSE2.

References

[ACG+06] Frederik Armknetcht, Claude Carlet, Philippe Gaborit, Simon Künzli, Willi Meier, and Olivier Ruatta. Efficient computation of algebraic immunity for algebraic and fast algebraic attacks. *EUROCRYPT 2006*, 2006.

[Arm04] Frederick Armknetch. Improving fast algebraic attacks. In *Fast Software Encryption, FSE*, volume 3017 of *LNCS*, pages 65–82. Springer Verlag, 2004. http://eprint.iacr.org/2004/185/.

[BLP06] An Braeken, Joseph Lano, and Bart Preneel. Evaluating the resistance of stream ciphers with linear feedback against fast algebraic attacks. *To appears in ACISP 06*, 2006.

[BP05] An Braeken and Bart Preneel. On the algebraic immunity of symmetric Boolean functions. 2005. http://eprint.iacr.org/2005/245/.

[Car04] Claude Carlet. Improving the algebraic immunity of resilient and nonlinear functions and constructing bent functions. 2004. http://eprint.iacr.org/2004/276/.

[CM03] Nicolas Courtois and Willi Meier. Algebraic attacks on stream ciphers with linear feedback. *Advances in Cryptology – EUROCRYPT 2003*, LNCS 2656:346–359, 2003.

[Cop94] Don Coppersmith. Solving linear equations over GF(2) via block Wiedemann algorithm. *Math. Comp.*, 62(205):333–350, January 1994.

[COS86] D. Coppersmith, A. Odlyzko, and R. Schroeppel. Discrete logarithms in GF(p). *Algorithmitica*, 1:1–15, 1986.

[Cou03] Nicolas Courtois. Fast algebraic attacks on stream ciphers with linear feedback. In *Advances in Cryptology-CRYPTO 2003*, volume 2729 of *LNCS*, pages 176–194. Springer Verlag, 2003.

[DGM04] Deepak Kumar Dalai, Kishan Chand Gupta, and Subhamoy Maitra. Results on algebraic immunity for cryptographically significant Boolean functions. In *INDOCRYPT*, volume 3348 of *LNCS*, pages 92–106. Springer, 2004.

[DM06] Deepak Kumar Dalai and Subhamoy Maitra. Reducing the number of homogeneous linear equations in finding annihilators. *to appears in SETA 2006*, 2006.

[DMS05] Deepak Kumar Dalai, Subhamoy Maitra, and Sumanta Sarkar. Basic theory in construction of Boolean functions with maximum possible annihilator immunity. 2005. http://eprint.iacr.org/2005/229/.

[DT06] Frédéric Didier and Jean-Pierre Tillich. Computing the algebraic immunity efficiently. *Fast Software Encryption, FSE*, 2006.

[FA03] J.-C. Faugère and G. Ars. An algebraic cryptanalysis of nonlinear filter generator using Gröbner bases. *Rapport de Recherche INRIA*, 4739, 2003.

[FJ03] J.-C. Faugère and Antoine Joux. Algebraic cryptanalysis of Hidden Field Equation (HFE) cryptosystems using Gröbner bases. In *Advances in Cryptology-CRYPTO 2003*, volume 2729 of *LNCS*, pages 44–60. Springer Verlag, 2003.

[HR04] P. Hawkes and G. C. Rose. Rewriting variables: The complexity of fast algebraic attacks on stream ciphers. In *Advances in Cryptology-CRYPTO 2004*, volume 3152 of *LNCS*, pages 390–406. Springer Verlag, 2004.

[Mas69] J. L. Massey. Shift-register synthesis and BCH decoding. *IEEE Trans. Inf. Theory*, IT-15:122–127, 1969.

[MPC04]   Willi Meier, Enes Pasalic, and Claude Carlet. Algebraic attacks and decomposition of Boolean functions. *LNCS*, 3027:474–491, April 2004.

[Odl84]   Andrew M. Odlyzko. Discrete logarithms in finite fields and their cryptographic significance. In *Theory and Application of Cryptographic Techniques*, pages 224–314, 1984.

[Wie86]   Douglas H. Wiedemann. Solving sparse linear equations over finite fields. *IEEE Trans. Inf. Theory*, IT-32:54–62, January 1986.

Enciphering with Arbitrary Small Finite Domains

Valery Pryamikov

Harper Security Consulting, Vestre Rosten 81, 7075 Trondheim, Norway
valery@harper.no

Abstract. In this paper we present a new block cipher over a small finite domain $\mathcal{T}$ where $|\mathcal{T}| = k$ is either 2^{16} or 2^{32} . After that we suggest a use of this cipher for enciphering members of arbitrary small finite domains $\mathcal{M}$ where $\mathcal{M} \subseteq \mathcal{T}$. With cost of an extra mapping, this method could be further extended for enciphering in arbitrary domain $\mathcal{M}'$ where $|\mathcal{M}'| = k' \leq k$. At last, in a discussion section we suggest a few interesting usage scenarios for such a cipher as an argument that enciphering with arbitrary small finite domains is a very useful primitive on its own rights, as well as for designing of a higher level protocols.

Keywords: Block Ciphers, Symmetric Encryption, Pseudorandom Permutations, Modes of Operations.

1 Introduction

Our motivation for this research was ignited by an obvious dissonance. Pseudo-random permutations are a very useful tool for many tasks, starting from the shuffling of a card deck to generation of bankcard numbers, pin codes and one time passwords, collecting samples from real-time data and many more. There is a tool that conveniently defines families of pseudo-random permutations - block ciphers. The modern block ciphers have very strong pseudo-random properties, compactly index selected permutation by an encryption key, and operate with block size that is very well suited for secure encryption of large amount of data. However, the large domains of the modern block ciphers make the size of their pseudo-random permutation too big for being practical for the tasks outlined above. The task of enciphering in arbitrary finite domain was previously considered by creators of Hasty Pudding Cipher, Schroeppel and Orman [13][1]. Later a rigorous treatment of the iterative encryption method, together with two other methods of encryption messages in arbitrary finite domains, were presented in a paper of Black and Rogaway [1]. Another method of construction of variable input length ciphers was also presented earlier in [2].

However, in cases when the size of required permutation is small, the previously suggested techniques either don't work [2] or become quite inefficient [13, 1], especially when considering small 8-bits microcontrollers.

---

[1] Schroeppel believes that the idea of iterative application of the encryption until proper domain point is reached, dates back to the rotor machines used in the early twentieth century.

R. Barua and T. Lange (Eds.): INDOCRYPT 2006, LNCS 4329, pp. 251–265, 2006.

To solve this task we have designed a cipher with a small block size of either 16 or 32 bits that we named TinyPRP. The TinyPRP cipher is key-alternating block cipher [5] that uses wide trail design strategy [3, 5] – the methodology of AES [4, 5] cipher. Additionally, TinyPRP cipher reuses some of the elements of the AES cipher, such as AES's S-box with optimal linear and differential resistance properties. In order to fit the small block size, we have designed the linear step of the round transformation function of our cipher in a different way than AES, however it is also based on Maximum Distance Separable codes and provide the maximal branch number that we could have achieved with such a small block size. Analogously to AES, TinyPRP cipher could be efficiently implemented in software and hardware, including but not limited to small 8-bits microcontrollers.

Due to a small block size, the security goals of the design of TinyPRP cipher follows definitions from [1] and could be outlined as: even when an adversary having access to the encryption oracle, has collected encryption of all but the two last points of the domain, the adversary should not be able to distinguish encryptions of the remaining two points significantly better than a random guess.

TinyPRP is very well suited for the tasks such as non-expanding encryption of small fields of database, for example numeric fields, that are usually smaller than the block size of standard symmetric encryption algorithms such as AES. Note, that we are not encouraging the use of TinyPRP for encryption of large messages – where the standard ciphers such as AES is a superior choice.

When used together with iterative encryption until the proper domain point is reached of [13, 1], TinyPRP cipher provides efficient method of encrypting messages in arbitrary small domain, that could be used for tasks such as shuffling a card deck, generation and verification of pin codes and onetime passwords, generation and verification of onetime credit card numbers, and many others.

In this paper we present design of the TinyPRP cipher. Reference implementation of the key elements of the algorithm is presented in appendixes. A complete reference implementation of the cipher could be downloaded from the authors website [14].

2   TinyPRP Specification

TinyPRP is an iterated block cipher with a variable block length and a variable key length. The cipher supports the block length of 16 or 32 bits and the key length of 96 or 128 bits that could be independently specified[2] . The cipher iteratively transforms the intermediate state that has the size of block. The state is initialized with a block of plain text before application of the algorithm. The cipher starts and ends with addition of round key[3] which we denote as function

---

[2] The algorithm allows selecting key size as any multiple of 32 bits, however the key sizes below 96 bits are considered as insecure and should be used for experimenting purposes only.

[3] Any transformation of the state before the first addition of the round key or after the last addition of the round key is known and therefore could be easily factored out by cryptanalyst.

$\sigma(k_r)$. The cipher applies variable number of rounds to the intermediate state. Number of rounds depends on the block size and the size of encryption key. Round transformation consists of:

- γ – A local non-linear transformation that consists of parallel application of S-boxes to each byte of the state. For TinyPRP cipher we used AES S-boxes.
- λ – A linear mixing transformation that guarantees high diffusion over multiple rounds.
- σ – A round key addition, that is a simple XOR of the round key to the intermediate state.

Figures 1- 2 show schematic graphical representation of the TinyPRP encryption and decryption with the block-sizes 32 and 16 bits. Since γ and σ are essentially the same as in AES[4], following description of the cipher will concentrate on λ linear mixing transformation step.

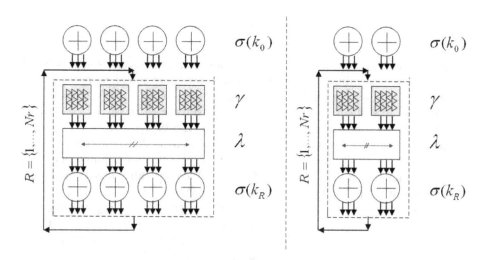

Fig. 1. Encryption

2.1   Linear Mixing Transformation λ

λ transformation of all rounds but the last consists of the two operations Shift-Cells, and MixCells. For the symmetry of the decryption operation, MixCells is missing from the λ transformation of the last round[5].

Due to limitations imposed by a small block size, ShiftCells and MixCells treats structure of the state in a different ways:

---

[4] We use AES S-boxes for γ and σ is XOR of the round key with the intermediate state.

[5] Difference in the last round transformation is customary for many ciphers including AES and it could be easily shown that the absence of the last MixCells doesn't improve or reduce the security of the cipher.

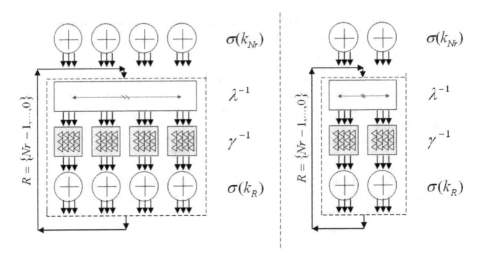

Fig. 2. Decryption

- ShiftCells treats the state as a table consisting of 4 columns and either 2 or 4 rows depending on the block size. Rows correspond to bytes of the intermediate state with $byte_0 \rightarrow row_0$, $byte_1 \rightarrow row_1$ ... . Columns correspond to bits of corresponding byte of the state with $bits_{\{0,1\}} \rightarrow column_0$, $bits_{\{2,3\}} \rightarrow column_1$, $bits_{\{4,5\}} \rightarrow column_2$ and $bits_{\{6,7\}} \rightarrow column_3$. Mapping of a byte to the table row used by ShiftCells is shown on Figure 3.
- MixCells treats the state as either one polynomial or a combination of two polynomials with coefficients in $GF(2^4)$ which we call state-polynomials. Coefficients of polynomial are obtained following way:
  - (32 bits block) two polynomials $b_{0,3}x^3 + b_{0,2}x^2 + b_{0,1}x + b_{0,0}$ and $b_{1,3}x^3 + b_{1,2}x^2 + b_{1,1}x + b_{1,0}$ with coefficients $b_{\{0,x\}} = bits_{\{0,1,2,3\}}$ of $byte_x$ where $x \in \{0,1,2,3\}$, $b_{\{1,x\}} = bits_{\{4,5,6,7\}}$ of $byte_x$ where $x \in \{0,1,2,3\}$.
  - (16 bits block) one polynomial $b_3x^3 + b_2x^2 + b_1x + b_0$ where $b_0 = bits_{\{0,1,2,3\}}$ of $byte_0$; $b_1 = bits_{\{0,1,2,3\}}$ of $byte_1$; $b_2 = bits_{\{4,5,6,7\}}$ of $byte_0$; $b_3 = bits_{\{4,5,6,7\}}$ of $byte_1$.

ShiftCells uses cyclic bit rotation operation for two types of transformation of the state matrix:

- cyclic rotation of the rows of the intermediate state with different offsets, where offsets were chosen to ensure that nearby bits spreads over the state providing protection against truncated differentials (*see left part of Figure 4*). Rotation offsets are given in Table 1;
- cyclic rotation of the state (*2 or 4 bytes depending on the block size*) for two cells (*4 bits*) to ensure good diffusion when combined with MixCells transformation (*see right part of Figure 4*);

MixCells is multiplication of the state polynomials with constant polynomial $e(x) \mod x^4 + 1$ and coefficients in $GF(2^4)$. Multiplication in $GF(2^4)$ is done

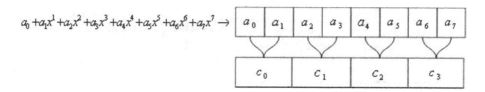

Fig. 3. Mapping of a byte to the table row used by ShiftCells operation

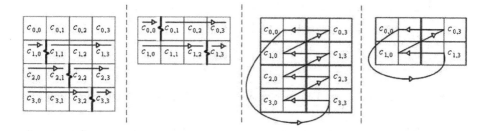

Fig. 4. ShiftCells Operation

Table 1. Rotation offsets (in cells) of ShiftCells operation

row index;	32 bits;	16 bits;
0	0	1
1	1	3
2	2	n/a
3	3	n/a

mod primitive polynomial $x^4 + x + 1$. For TinyPRP we chose the same constant polynomial $e(x)$ as in AES: $03x^3 + 01x^2 + 01x + 02$, which is relatively prime with $x^4 + 1$ and coefficients in $GF(2^4)$ and ensures the maximal branch number for the MixCells. Figure 5 shows how the state polynomials are constructed from the intermediate state.

Polynomial $d(x)$ used by InvMixCells is also the same as in AES: $0Bx^3 + 0Dx^2 + 09x + 0E$, which is inverse polynomial to $e(x)$ mod $x^4 + 1$ and coefficients in $GF(2^4)$.

2.2   Key Schedule

Key schedule was designed to satisfy following criteria:

- to ensure that generated round keys depend on all bits of the key;
- to introduce a sufficient amount non-linearity for increased resistance to key related attacks or partially known key attacks;
- to eliminate symmetry by using round constants;

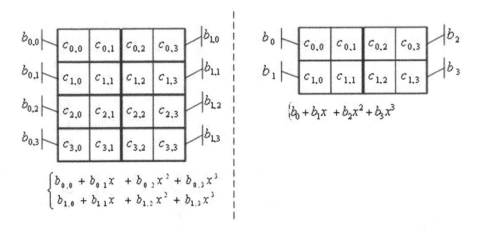

Fig. 5. Mapping of Coefficients of State Polynomials for MixCells Operation

- to minimize memory requirements - it should be possible to execute key schedule using a small amount of working memory and we only need to keep a number of sequential round keys with total size equal to the size of the key to be able to recalculate missing round keys;
- to compensate for small entropy of the round key - sequential round keys with the total size equal to the size of the key are required for restoring the key or the other round keys;

The key schedule generates $N_R + 1$ round keys. The number of rounds N_R for different key sizes and block sizes of TinyPRP cipher is given in Table 2. [6]

Table 2. Number of Rounds

	32 bits	16 bits
96 bits key	8	15
128 bits key	9	17

We have defined ten 16 bits round constants where the first round constant is 0000 and all following round constants are $\{x^{r-1} \bmod x^8 + x^4 + x^3 + x + 1, x^{r-1} \bmod x^8 + x^4 + x^3 + x + 1\}$ for $1 \leq r \leq 12$. There are six 32 bits round constants that are defined by concatenation of two sequential 16 bits round constants and treating result as little endian 32 bits integers.

- For 16 bits block size we defined $\mathrm{rcon}_{16} = \{0000, 0101, 0202, 0404, 0808, 1010, 2020, 4040, 8080, 1B1B\}$.
- For 32 bits block size we defined $\mathrm{rcon}_{32} = \{01010000, 04040202, 10100808, 40402020, 1B1B8080, 6c6c3636\}$.

The algorithm of the key schedule is shown on Figure 6.

---

[6] Twofold increase of rounds for 16 bits block size is explained by necessity to compensate for small entropy of the round keys.

$$N_{R_0} \leftarrow \frac{\text{sizeof(key)}}{\text{sizeof(block)}} \; ;$$

$$N_R \leftarrow \text{countof} \left(\text{rcon}_{\text{sizeof(block)}} \right) + N_{R_0} \; ;$$

for $(0 \leq k < N_{R_0})$

    $R_k \leftarrow \text{getBlock}^{(k)} \, (\text{key}) \; ;$

for $(N_{R_0} \leq k < N_R)$

    $R_k \leftarrow \text{rcon}_{\text{sizeof(block)}}^{(k)} \; ;$

      for $(0 \leq i < N_R)$

         $R_k \leftarrow R_k \oplus R_{k-i-1} \; ;$

         $R_k \leftarrow \text{SubBytes} \, (R_k) \; ;$

         $R_k \leftarrow \text{ShiftCells} \, (R_k) \; ;$

         $R_k \leftarrow \text{MixCells} \, (R_k) \; ;$

Fig. 6. Key Schedule Algorithm

2.3   TinyPRP Algorithm

Description of TinyPRP algorithm is presented on Figure 7.

Appendix A contains reference implementation of the key elements of the TinyPRP algorithm. A complete reference implementation of the cipher could be downloaded from the authors website [14]. Source code of reference implementation of the TinyPRP algorithm uses straightforward implementation of polynomial multiplication   (mod $x^4 + 1$) with coefficients in $\text{GF}(2^4)$. Performance measures of non-optimized reference implementation could be found in section 2.5. However we would like to note that table-lookup implementation of the linear mixing step would provide much better performance. Also note that most of other optimizations of Rijndael cipher could be applied to TinyPRP cipher as well.

Encryption	Decryption
$\text{RKey} \leftarrow \text{ExpandKey} \, (\text{key}) \; ;$	$\text{RKey} \leftarrow \text{ExpandKey} \, (\text{key}) \; ;$
$N_R \leftarrow \text{countof} \, (\text{RKey}) \; ;$	$N_R \leftarrow \text{countof} \, (\text{RKey}) \; ;$
$\text{State} \leftarrow \text{State} \oplus \text{RKey}^{(0)} \; ;$	$\text{State} \leftarrow \text{State} \oplus \text{RKey}^{(N_R-1)} \; ;$
for $(1 \leq k < N_R - 1)$	for $(1 \leq k < N_R - 1)$
$\text{State} \leftarrow \text{SubBytes} \, (\text{State}) \; ;$	$\text{State} \leftarrow \text{ShiftCells}^{-1} \, (\text{State}) \; ;$
$\text{State} \leftarrow \text{ShiftCells} \, (\text{State}) \; ;$	$\text{State} \leftarrow \text{SubBytes}^{-1} \, (\text{State}) \; ;$
$\text{State} \leftarrow \text{MixCells} \, (\text{State}) \; ;$	$\text{State} \leftarrow \text{State} \oplus \text{RKey}^{(N_R-1-k)} \; ;$
$\text{State} \leftarrow \text{State} \oplus \text{RKey}^{(k)} \; ;$	$\text{State} \leftarrow \text{MixCells}^{-1} \, (\text{State}) \; ;$
$\text{State} \leftarrow \text{SubBytes} \, (\text{State}) \; ;$	$\text{State} \leftarrow \text{ShiftCells}^{-1} \, (\text{State}) \; ;$
$\text{State} \leftarrow \text{ShiftCells} \, (\text{State}) \; ;$	$\text{State} \leftarrow \text{SubBytes}^{-1} \, (\text{State}) \; ;$
$\text{State} \leftarrow \text{State} \oplus \text{RKey}^{(N_R-1)} \; ;$	$\text{State} \leftarrow \text{State} \oplus \text{RKey}^{(0)} \; ;$

Fig. 7. TinyPRP Algorithm

2.4   Security Goals and Considered Attacks

It is clear that the small block size of TinyPRP cipher makes it easy to collect all correspondent pairs of plain–cipher texts. Therefore, the security goals of the cipher' design follows definitions from [1] and could be outlined as:

```
Even when an adversary having access to the encryption oracle, has
collected encryption of all but the last two points of the domain,
the adversary should not be able to distinguish encryptions of the
remaining points significantly better than a random guess.
```

A new challenge of designing a cipher with a small block size appears to be an interesting shift of the attack vector. Traditional attacks, such as linear [11] and differential [7] cryptanalysis and truncated differentials [12] appears to be far less dangerous for such a cipher, because of a small block size, which significantly reduces amount of possible patterns available to adversary. For example, 16 bits block size even allows to enumerate all possible difference patterns as well as check correlation contributions of all possible linear trails (that could make such a cipher an interesting tools for studying these classical attacks). The same concerns to saturation attack [10, 9] (also known as Square attack [9]) which is also applicable, but far less dangerous for TinyPRP than for any other Square/Rijndael based ciphers, again because of extremely small block size of TinyPRP cipher.

What appears to be far more dangerous for such a cipher is a small entropy of round keys as well as a small total size of the round keys. For example, 17 rounds of 16 bits cipher gives only 288 bits of round-key material, from these – 128 bits are the key and only 160 bits are derived round keys. Therefore, a special care should be taken of the key schedule to eliminate weak keys and rounds symmetry (by using round constants) and to protect against slide attack [8] and related keys attack [6]. Key schedule must ensure that the recovery of the round key(s) should not contribute to the recovery of the encryption key more than the size of recovered round key(s). For example, with 16 bits blocks, the recovery of two round keys must not help adversary to recover 128 bits encryption key faster than 2^{95}. In order to provide better suitability of TinyPRP cipher for small 8-bits processors, additional goal of our key schedule was a requirement that the knowledge of sequential round keys of the same size as the size of the key must be sufficient to restore the rest of round keys and the key itself.

Wide trail design strategy [3, 5], combination of optimal linear and differential resistance properties of AES S-boxes with high diffusion properties of linear mixing transformation designed for TinyPRP cipher ensures that after two rounds of TinyPRP, all output bits of second round depends on every input bit of the first round. The key schedule, that uses the same round transformation, generates round keys with total size that is equal to the size of the key plus two round keys (to ensure that the last round key depends on all previous round keys). That also gives raise to amount of rounds that ensures that TinyPRP cipher is immune to linear and differential analysis, truncated differentials and saturation attacks. An extended security analysis of TinyPRP cipher will be provided in extended paper available from the authors website shortly after publication of this article.

2.5   Performance

Tables 3 and 4 show performance of TinyPRP cipher on Intel Pentium M processor with Intel Centrino technology in cycles per block. Note that performance is given for non-optimized reference implementation of the cipher that could be downloaded from the authors website [14].

Table 3. Performance of non-optimized TinyPRP cipher with 128-bit key on Intel Pentium M Centrino processor (cycles/per-block)

	key-setup;	encryption;	decryption;
16-bits blocks	10280	2400	4320
32-bits blocks	6100	1780	3420

Table 4. Performance of non-optimized TinyPRP cipher with 96-bit key on Intel Pentium M Centrino processor (cycles/per-block)

	key-setup;	encryption;	decryption;
16-bits blocks	7900	2140	3800
32-bits blocks	4940	1690	2990

3   Enciphering with Arbitrary Small Finite Domains

The method of enciphering in smaller domain by using iterative encryption until proper point of domain is reached was first suggested by creators of Hasty Pudding Cipher, Schroeppel and Orman [13]. Schroeppel believes that the idea dates back to the rotor machines used in the early twentieth century. A rigorous treatment of the iterative encryption method was given in a paper of Black and Rogaway [1], who called this method Cycle-Walking Cipher (*see Figure 8*). However, we believe that would be more correct to name it as cycle-walking mode of operation or the cycle-walking cipher mode.

The idea is based on the fact that any permutation is uniquely identified by product of permutation cycles (also called orbits). Block cipher encryption with a fixed key defines a permutation of the domain points and encryption of a domain point is a closed operation which guarantees that after enough encryptions we will eventually get back to the original point. Or in other words: when we start our encryption from the point of a smaller domain $\mathcal{M}$, then we have a guarantee that after enough encryptions we will find at least one point belonging to the domain $\mathcal{M}$, even so it might be the original point. In this context, the encryption

Encryption	Decryption
if $(p \notin \mathcal{M})$	if $(p \notin \mathcal{M})$
stop ;	stop ;
do	do
$p \leftarrow E_K(p)$;	$p \leftarrow E_K^{-1}(p)$;
while $(p \notin \mathcal{M})$;	while $(p \notin \mathcal{M})$;
return(p) ;	return(p) ;

Fig. 8. Cycle-Walking Cipher Mode

operation could be thought as walking forward the orbit, while as decryption is walking backward the same orbit. Since neighboring elements of a fixed orbit are always the same, means that iterative application of encryption until the first proper point of the domain is reached also forms a permutation.

A natural question arises: "how much security do we lose in deriving such permutation?". One of the results of the [1] was a proof that there is no loss of security when the cipher is used with cycle-walking mode of operations. Here we added the statement of the theorem asserting security of cycle-walking cipher mode from [1]:

Theorem 1 (Security of Cycle-Walking Cipher Mode). *Fix $k \geq 1$ and let $\mathcal{M} = [0, k-1]$. Let $E_K(\cdot)$ be an ideal block cipher on the set T where $\mathcal{M} \subseteq T$. Choose a key K uniformly at random and then construct $C_{Y_K}(\cdot)$ using $E_K(\cdot)$. Then $C_{Y_K}(\cdot)$ is a uniform random permutation on $\mathcal{M}$.*

We refer to [1] for the proof of this theorem.

The main problem of using cycle-walking cipher mode for enciphering members of small domains with conventional ciphers is that it may require unacceptably high number of iterations for reaching the proper point of the domain. For example in case of 128 bits cipher such as AES, it might be computationally unfeasible to reach the required point of the domain which might require up to the order of 2^{128} encryptions. Even 64 bits block ciphers, such as DES, have unacceptably high computational cost for using them with cycle-walking cipher mode when the density of the points of sub-domain $\mathcal{M}$ inside the domain T is relatively low. In other words – the density of the points of the sub-domain $\mathcal{M}$ inside the domain T must be high enough for the cycle-walking cipher mode being practical. In case if we don't want to limit lower bound for the size of sub-domain $\mathcal{M}$, then the only viable solution is reducing the size of the domain T. The TinyPRP cipher, which compared to other existing block ciphers operates on unusually small domain, is very practical for use with cycle-walking cipher mode over sub-domains of very small size. For example, if used for generation (and validation) of 4 digits one time passwords, expected amount of encryptions with 16 bits TinyPRP is just about half a dozen, and even the worst case scenario, which could only occurs with negligible probability, is upperbounded by $2^{16} - 10^4$ encryptions. In other words, amount of TinyPRP 16-bits encryptions required for enciphering with cycle-walking cipher mode in sub-domain of 4 decimal digits numbers is acceptable even for small 8-bits processors, even for the worst case scenarios.

4   Discussion

This naturally leads to a question about applicability and usage scenarios where TinyPRP cipher is suited better than existing tools such as pseudorandom bits generators and stream ciphers. We envision several interesting usage scenarios, starting from generation of non-colliding pseudorandom numbers in domains different than $(0...2^N - 1)$ (for example, credit card number generation [1], non-colliding indexes for distributed sampling, or even simple mixing of a card deck); to non-expanding type-aware encryption of strongly typed data fields and generation of one-time passwords on embedded devices and 8 bits microcontrollers.

Probably, the one of the most typical usage scenarios for TinyPRP cipher would be non-expanding type-aware encryption of strongly typed data fields. Software industry is now experiencing change of security attitude with many existing applications being upgraded to support a better level of privacy protection. Usual measures include encryption of privacy-sensitive information, such as social security numbers, addresses, birthdays and others. And the most usual challenge is often related to necessity of protecting strongly typed fields of enterprise databases. Changes of the datatype or the storage size often introduces nontrivial challenges related to tracking down and correcting the code that may rely on implicit assumptions. With enterprise class systems, that often means error-prone, highly expensive process that may lead to severe disruption of normal business operation and other consequences.

TinyPRP together with cycle-walking mode of operations provides means to relax some of these problems by providing non-expanding datatype aware encryption. For example, for protection of the birthday field with TinyPRP cipher in cycle-walking mode of operation, the birthday could be encrypted to a valid date that additionally may be restricted to a specific age interval (as for example between 20 and 67 years old). The encryption and decryption is a straightforward modification of cycle-walking algorithm with priory mapping of date to an integer less than 10^8 and domain check that tries to map result of encryption back to date datatype and in case of success, checks for required age interval.

When it concerns to existing tools: Using pseudorandom bits generators with domain that is different than exact multiple of bits may easily produce undesirable results. For example, naive using of a modulo N operator may introduce bias because numbers in interval $\{N, ..., 2^{\lceil \log(N) \rceil}\}$ will contribute to the interval $\{0, ..., 2^{\lceil \log(N) \rceil} - N\}$. Additionally, the possibility of collisions dictated by the birthday paradox significantly reduces amount of non-colliding messages that could be randomly generated, and that could be highly undesirable for an already small message domain. Using pseudorandom bits generator for one-time passwords generation devices encumbered by patents that covers any reasonable way of accounting the clock drift and synchronization of the clock. Necessity of frequent reseeding of pseudorandom generator on such devices depletes batteries.

Stream ciphers are effective tool for encrypting messages of arbitrary length $\{0, ..., 2^L - 1\}$, but their predictable behavior to a simple bit-flipping may be unacceptable when source message has strongly defined type or a structure, such as for example a counter or a birth date. That makes stream cipher unusable for applications such as bankcard number generation [1] or non-expanding encryption of fields in database. Problems also arise using stream cipher for encrypting messages in arbitrary domain that isn't exact multiple of bits. Cycle-walking mode of operations or its modifications aren't compatible with the stream ciphers. Any two sequential encryption with the same stream state negates each other. Attempts of reusing keystream between encryptions result in expected collisions in $\approx \sqrt{2^{\lceil \log(N) \rceil}}$ because blocks of the keystream of the size $\lceil \log(N) \rceil$ are better modeled as pseudo random function rather than pseudo random permutation. As result of collisions – the decryption will not work.

We believe that TinyPRP combined with cycle-walking cipher mode provides a sound solution for problems mentioned above and provides high security combined with low processing cost and suitability for embedded devices and 8 bits microcontrollers.

Acknowledments

Special thanks to Professor Daniel J. Bernstein who made many useful comments and suggestions on an earlier draft.

References

[1] Black, J., Rogaway, P.:, Ciphers with Arbitrary Finite Domains, In Proceedings of the Cryptographer's Track at the RSA Conference 2002

[2] Bellare, M., and Rogaway, P.: On the construction of variable-input-length ciphers. In Fast Software Encryption (1999), vol. 1636 of Lecture Notes in Computer Science, SpringerVerlag

[3] Daemen, J.: Cipher and Hash Function Design, Strategies Based on Linear and Differential Cryptanalysis. Doctoral Dissertation, Katolische Universiteit Leuven, Belgium, March 1995

[4] National Institute of Standards and Technology: Advanced Encryption Standard (AES), FIPS Publication 197, 26 Nov. 2001.

[5] Daemen, J., Rijmen, V.: The Design of Rijndael: AES. The Advanced Encryption Standard. Springer, 2002 ISBN 3540425802

[6] Biham, E.: New Types of Cryptanalytic Attacks Using Related Keys. Journal of Cryptology, Vol.7, Springer-Verlag (1994)

[7] Biham, E., Shamir, A.: Differential Cryptanalysis of DES-like Cryptosystems. Journal of Cryptology, vol. 4, No.1, 1991, SpringerVerlag

[8] Biryukov, A., Wagner, D.: Slide Attacks. Fast Software Encryption 1999, vol. 1636 of Lecture Notes in Computer Science, SpringerVerlag

[9] Daemen, J., Knudsen,L.R., Rijmen, V.: The Block Cipher SQUARE. 4th Fast Software Encryption Workshop 1997, Vol. 1267. of Lecture Notes in Computer Science, SpringerVerlag

[10] Lucks, S.: The saturation attack - a bait for twofish. Fast Software Encryption 2001, vol. 2355 of Lecture Notes in Computer Science, SpringerVerlag

[11] Matsui, M.: Linear Cryptanalysis Method for DES Chiper. Advances in Cryptology, Proc. Eurocrypt 1993, vol. 765 of Lecture Notes in Computer Science, SpringerVerlag

[12] Knudsen, L.R.: Truncated and High Order Differentials. Fast Software Encryption 1994, vol. 1008 of Lecture Notes in Computer Science, SpringerVerlag

[13] Schroeppel, R., Orman, H.: Specification for the Hasty Pudding Cipher. In Proceedings of the First Advanced Encryption Standard Candidate Conference, National Institute of Standards and Technology, Aug. 1998.

[14] Pryamikov, V.: TinyPRP-reference implementation. Aug. 2006.

A   Appendix - Elements of TinyPRP Algorithm

```
inline void MixCells(unsigned char c[4]) {
unsigned char tm, ta = c[0],
 t = c[0] ^ c[1] ^ c[2] ^ c[3];
tm = c[0] ^ c[1]; tm = xtime(tm); c[0] ^= tm ^ t;
tm = c[1] ^ c[2]; tm = xtime(tm); c[1] ^= tm ^ t;
tm = c[2] ^ c[3]; tm = xtime(tm); c[2] ^= tm ^ t;
tm = c[3] ^ ta ; tm = xtime(tm); c[3] ^= tm ^ t;
}
inline void InvMixCells(unsigned char c[4]) {
unsigned char t[4][3], ta[4];
 t[0][0] = xtime(c[0]); t[0][1]=xtime(t[0][0]); t[0][2] = xtime(t[0][1]);
 t[1][0] = xtime(c[1]); t[1][1]=xtime(t[1][0]); t[1][2] = xtime(t[1][1]);
 t[2][0] = xtime(c[2]); t[2][1]=xtime(t[2][0]); t[2][2] = xtime(t[2][1]);
 t[3][0] = xtime(c[3]); t[3][1]=xtime(t[3][0]); t[3][2] = xtime(t[3][1]);
 ta[0] = t[0][2] ^ t[0][1] ^ t[0][0] ^ t[1][2] ^ t[1][0] ^ c[1] ^
         t[2][2] ^ t[2][1] ^ c[2] ^ t[3][2] ^ c[3];
 ta[1] = t[1][2] ^ t[1][1] ^ t[1][0] ^ t[2][2] ^ t[2][0] ^ c[2] ^
         t[3][2] ^ t[3][1] ^ c[3] ^ t[0][2] ^ c[0];
 ta[2] = t[2][2] ^ t[2][1] ^ t[2][0] ^ t[3][2] ^ t[3][0] ^ c[3] ^
         t[0][2] ^ t[0][1] ^ c[0] ^ t[1][2] ^ c[1];
 ta[3] = t[3][2] ^ t[3][1] ^ t[3][0] ^ t[0][2] ^ t[0][0] ^ c[0] ^
         t[1][2] ^ t[1][1] ^ c[1] ^ t[2][2] ^ c[2];
 *((unsigned int *)c) = *((unsigned int*)ta);
}
inline void MixCells_i(unsigned char a[4]) {
 unsigned char c[2][4] = {
  {(unsigned char) (a[0] >> 4 ), (unsigned char) (a[1] >> 4 ),
   (unsigned char) (a[2] >> 4 ), (unsigned char) (a[3] >> 4 )},
  {(unsigned char) (a[0] & 0xF), (unsigned char) (a[1] & 0xF),
   (unsigned char) (a[2] & 0xF), (unsigned char) (a[3] & 0xF)}};
 MixCells(c[0]);
 MixCells(c[1]);
```

```
 a[0] = (c[0][0] << 4) | (c[1][0] & 0xF);
 a[1] = (c[0][1] << 4) | (c[1][1] & 0xF);
 a[2] = (c[0][2] << 4) | (c[1][2] & 0xF);
 a[3] = (c[0][3] << 4) | (c[1][3] & 0xF);
}
inline void ShiftCells_i(unsigned char r[4]) {
 r[1] = r[1] << 2 | r[1] >> 6;
 r[2] = r[2] << 4 | r[2] >> 4;
 r[3] = r[3] << 6 | r[3] >> 2;
 unsigned int &ri = *((unsigned int*)r);
 ri = ri << 4 | ri >> 28;
}
inline void InvMixCells_i(unsigned char a[4]) {
 unsigned char c[2][4] = {
  {(unsigned char) (a[0] >> 4 ), (unsigned char) (a[1] >> 4 ),
   (unsigned char) (a[2] >> 4 ), (unsigned char) (a[3] >> 4 )},
  {(unsigned char) (a[0] & 0xF), (unsigned char) (a[1] & 0xF),
   (unsigned char) (a[2] & 0xF), (unsigned char) (a[3] & 0xF)}};
 InvMixCells(c[0]);
 InvMixCells(c[1]);
 a[0] = (c[0][0] << 4) | (c[1][0] & 0xF);
 a[1] = (c[0][1] << 4) | (c[1][1] & 0xF);
 a[2] = (c[0][2] << 4) | (c[1][2] & 0xF);
 a[3] = (c[0][3] << 4) | (c[1][3] & 0xF);
}
inline void ExpandKey(unsigned int key[],
        ExpandedKey_i::KeySize size, ExpandedKey_i *expandedKey) {
 expandedKey->keySize = size;
 int nSize = size+1, nTotalSize=ExpandedKey_i::round0+nSize;
 for (int i = 0; i < nTotalSize; i++) {
  if (i < nSize)
    expandedKey->rkeys[i] = key[i];
  else {
   expandedKey->rkeys[i] = rcon_i[i-size-1];
   for (int k = 0; k < nSize; k++) {
    expandedKey->rkeys[i] ^= expandedKey->rkeys[i-k-1];
    SubBytes_i((unsigned char*)(expandedKey->rkeys+i));
    ShiftCells_i((unsigned char*)(expandedKey->rkeys+i));
    MixCells_i((unsigned char*)(expandedKey->rkeys+i));
}}}}

void encrypt(ExpandedKey_i *expandedKey, unsigned char a[4]) {
 *((unsigned int*)a) ^= expandedKey->rkeys[0];
 int i = 1;
 for (; i < expandedKey->keySize + ExpandedKey_i::round0 - 1; i++) {
  SubBytes_i(a);
  ShiftCells_i(a);
  MixCells_i(a);
  *((unsigned int*)a) ^= expandedKey->rkeys[i];
 }
```

```
 SubBytes_i(a);
 ShiftCells_i(a);
 *((unsigned int*)a) ^= expandedKey->rkeys[i];
}
void decrypt(ExpandedKey_i *expandedKey, unsigned char a[4]) {
 *((unsigned int*)a) ^= expandedKey->rkeys[expandedKey->keySize +
                        ExpandedKey_i::round0 - 1];
 int i = expandedKey->keySize + ExpandedKey_i::round0 - 2;
 for (; i > 0; i--) {
  InvShiftCells_i(a);
  InvSubBytes_i(a);
  *((unsigned int*)a) ^= expandedKey->rkeys[i];
  InvMixCells_i(a);
 }
 InvShiftCells_i(a);
 InvSubBytes_i(a);
 *((unsigned int*)a) ^= expandedKey->rkeys[0];
}
```

Enumeration of 9-Variable Rotation Symmetric Boolean Functions Having Nonlinearity > 240

Selçuk Kavut[1], Subhamoy Maitra[2], Sumanta Sarkar[2], and Melek D. Yücel[1]

[1] Department of Electrical Engineering and Institute of Applied Mathematics,
Middle East Technical University(METU – ODTÜ), 06531 Ankara, Türkiye
selcukkavut@gmail.com, yucel@eee.metu.edu.tr
[2] Applied Statistics Unit, Indian Statistical Institute, 203, B.T. Road,
Kolkata 700 108, India
{subho, sumanta_r}@isical.ac.in

Abstract. The existence of 9-variable Boolean functions having nonlinearity strictly greater than 240 has been shown very recently (May 2006) by Kavut, Maitra and Yücel; a few functions with nonlinearity 241 have been identified by a heuristic search in the class of Rotation Symmetric Boolean Functions (RSBFs). In this paper, using combinatorial results related to the Walsh spectra of RSBFs, we efficiently perform the exhaustive search to enumerate the 9-variable RSBFs having nonlinearity > 240 and found that there are 8×189 many functions with nonlinearity 241 and there is no RSBF having nonlinearity > 241. We further prove that among these functions, there are only two which are different up to the affine equivalence. This is found by utilizing the binary nonsingular circulant matrices and their variants. Finally we explain the coding theoretic significance of these functions. This is the first time orphan cosets of $R(1, n)$ having minimum weight 241 are demonstrated for $n = 9$. Further they provide odd weight orphans for $n = 9$; earlier these were known for certain $n \geq 11$.

Keywords: Boolean Functions, Covering Radius, Reed-Muller Code, Idempotents, Nonlinearity, Rotational Symmetry, Walsh Transform.

1 Introduction

Nonlinearity is one of the most important cryptographic properties of a Boolean function to be used as a primitive in any crypto system. High nonlinearity resists Best Affine Approximation (BAA) attacks [8] in case of stream ciphers and Linear cryptanalysis [18] in case of block ciphers. One may like to access the references in this paper and the references there in to study the extremely rich literature on Boolean functions having high nonlinearity with other cryptographic properties. Nonlinearity is important in coding theoretic aspects too.

The class of Rotation Symmetric Boolean functions has received a lot of attention in terms of their cryptographic and combinatorial properties [4,5,9,10, 11,19,20,23,26,27,6,14,7]. The nonlinearity and correlation immunity of such functions have been studied in detail in [4,11,19,20,26,27,14]. It is now clear that

R. Barua and T. Lange (Eds.): INDOCRYPT 2006, LNCS 4329, pp. 266–279, 2006.

the RSBF class is extremely rich in terms of these properties. As an important support of that, very recently 9-variable Boolean functions having nonlinearity 241 have been discovered [15] in the RSBF class, which had been open for almost three decades. One should note that the space of the RSBF class is much smaller ($\approx 2^{\frac{2^n}{n}}$) than the total space of Boolean functions (2^{2^n}) on n variables.

The Boolean functions attaining maximum nonlinearity are called bent [25] which occurs only for even number of input variables n and the nonlinearity is $2^{n-1} - 2^{\frac{n}{2}-1}$. For odd number of variables n, the maximum nonlinearity (upper bound) can be at most $2\lfloor 2^{n-2} - 2^{\frac{n}{2}-2} \rfloor$ [13]. Before [15], the following results related to maximum nonlinearity (actually attained) of Boolean functions have been known. In 1972 [1], it was shown that the maximum nonlinearity of 5-variable Boolean functions is 12 and in 1980 [21] it was proved that the maximum nonlinearity of 7-variable Boolean functions is 56. Thus for odd $n \leq 7$, the maximum nonlinearity of n-variable functions is $2^{n-1} - 2^{\frac{n-1}{2}}$. In 1983 [22], Boolean functions on 15 variables having nonlinearity 16276 were demonstrated and using this result one can show that for odd $n \geq 15$, it is possible to get Boolean functions having nonlinearity $2^{n-1} - 2^{\frac{n-1}{2}} + 20 \cdot 2^{\frac{n-15}{2}}$. There was a gap for $n = 9, 11, 13$ and the maximum nonlinearity known for these cases prior to [15] was $2^{n-1} - 2^{\frac{n-1}{2}}$. Very recently [15] 9-variable Boolean functions having nonlinearity 241 have been discovered which belong to the class of Rotation Symmetric Boolean functions. The technique used to find such functions is a suitably modified steepest-descent based iterative heuristic [14, 15].

As the functions could be found by heuristic search only [15], there is a theoretical need to study the complete RSBF class of 9-variables for nonlinearity > 240. Given the nice combinatorial structure of the Walsh spectra for RSBFs on odd number of variables [19], such a search becomes feasible with considerable computational effort. The complete details of the exhaustive search strategy is explained in Section 2 of this paper. The search shows that the maximum nonlinearity of 9-variable RSBFs is 241. We exploit certain results related to binary nonsingular circulant matrices and their variants to show that there are actually two different 9-variable nonlinearity 241 functions in the 9-variable RSBF class up to the affine equivalence. This is described in Section 3. As the maximum nonlinearity issue of Boolean functions is related to the covering radius of first order Reed-Muller code, we briefly outline the coding theoretic implications of our results in Section 4.

Let us consider any Boolean function as a mapping from $GF(2^n) \rightarrow GF(2)$. Then the functions for which $f(\alpha^2) = f(\alpha)$, for any $\alpha \in GF(2^n)$ are referred as idempotents [9, 10] as it follows from $f^2 = f$ in multiplicative algebra. In [9, 10] the idempotents were studied for $n = 9$ with the motivation that the Patterson-Wiedemann functions [22] for $n = 15$ were idempotents. However, in [9, 10] the search was not exhaustive and that is why the functions with nonlinearity 241 could not be discovered. In fact, the idempotents can be seen as RSBFs [9, 10]. Interestingly if one looks at an RSBF, the nice structure [19] in the Walsh spectrum can be exploited to execute an efficient search which is not immediate if one looks at the functions as idempotents.

1.1   Preliminaries

A Boolean function on n variables may be viewed as a mapping from $V_n = \{0,1\}^n$ into $\{0,1\}$. The *truth table* of a Boolean function $f(x_1,\ldots,x_n)$ is a binary string of length 2^n, $f = [f(0,0,\cdots,0), f(1,0,\cdots,0), f(0,1,\cdots,0), \ldots, f(1,1,\cdots,1)]$. The *Hamming weight* of a binary string S is the number of 1's in S denoted by $wt(S)$. An n-variable function f is said to be *balanced* if its truth table contains an equal number of 0's and 1's, i.e., $wt(f) = 2^{n-1}$. Also, the *Hamming distance* between equidimensional binary strings S_1 and S_2 is defined by $d(S_1, S_2) = wt(S_1 \oplus S_2)$, where $\oplus$ denotes the addition over $GF(2)$, i.e., XOR.

An n-variable Boolean function $f(x_1,\ldots,x_n)$ can be considered to be a multivariate polynomial over $GF(2)$. This polynomial can be expressed as a sum of products representation of all distinct k-th order products $(0 \le k \le n)$ of the variables. More precisely, $f(x_1,\ldots,x_n)$ can be written as $a_0 \oplus \bigoplus_{1 \le i \le n} a_i x_i \oplus \bigoplus_{1 \le i < j \le n} a_{ij} x_i x_j \oplus \ldots \oplus a_{12\ldots n} x_1 x_2 \ldots x_n$, where $a_0, a_{ij}, \ldots, a_{12\ldots n} \in \{0,1\}$. This representation of f is called the *algebraic normal form* (ANF) of f. The number of variables in the highest order product term with nonzero coefficient is called the *algebraic degree*, or simply the degree of f and denoted by $deg(f)$. Functions of degree at most one are called *affine* functions. An affine function with constant term equal to zero is called a *linear* function. The set of all n-variable affine (respectively linear) functions is denoted by $A(n)$ (respectively $L(n)$). The nonlinearity of an n-variable function f is $nl(f) = min_{g \in A(n)}(d(f,g))$, i.e., the minimum distance from the set of all n-variable affine functions.

Let $x = (x_1,\ldots,x_n)$ and $\omega = (\omega_1,\ldots,\omega_n)$ both belonging to $\{0,1\}^n$ and $x \cdot \omega = x_1\omega_1 \oplus \ldots \oplus x_n\omega_n$. Let $f(x)$ be a Boolean function on n variables. Then the *Walsh transform* of $f(x)$ is a real valued function over $\{0,1\}^n$ which is defined as $W_f(\omega) = \sum_{x \in \{0,1\}^n} (-1)^{f(x) \oplus x \cdot \omega}$. In terms of Walsh spectrum, the nonlinearity of f is given by $nl(f) = 2^{n-1} - \frac{1}{2} max_{\omega \in \{0,1\}^n} |W_f(\omega)|$.

Towards cryptographic applications, one needs to consider the autocorrelation spectrum [24,28] of a Boolean function. Let $\alpha \in \{0,1\}^n$ and f be an n-variable Boolean function. The autocorrelation value of the Boolean function f with respect to the vector α is $\Delta_f(\alpha) = \sum_{x \in \{0,1\}^n} (-1)^{f(x) \oplus f(x \oplus \alpha)}$. Further $\Delta_f = max_{\alpha \in \{0,1\}^n, \alpha \ne (0,\ldots,0)} |\Delta_f(\alpha)|$ is called the absolute indicator. A function is said to satisfy PC(k), if $\Delta_f(\alpha) = 0$ *for* $1 \le wt(\alpha) \le k$.

1.2   Rotation Symmetric Boolean Functions

The study of rotation symmetric functions for good cryptographic properties was initiated in [9] and they were called idempotents in that paper. Let $x_i \in \{0,1\}$ for $1 \le i \le n$. For some integer $k \ge 0$ we define $\rho_n^k(x_i)$ as $\rho_n^k(x_i) = x_{i+k \bmod n}$, with the exception that when $i + k \equiv 0 \bmod n$, then we will assign $i + k \bmod n$ by n instead of 0. This is to cope up with the input variable indices $1,\ldots,n$ for $x_1,\ldots,x_n$. Let $(x_1, x_2, \ldots, x_{n-1}, x_n) \in V_n$. Then we extend the definition as $\rho_n^k(x_1, x_2, \ldots, x_{n-1}, x_n) = (\rho_n^k(x_1), \rho_n^k(x_2), \ldots, \rho_n^k(x_{n-1}), \rho_n^k(x_n))$. Hence, ρ_n^k acts as k-cyclic rotation on an n-bit vector. A Boolean function f is called *rotation symmetric (RSBF)* if for each input $(x_1,\ldots,x_n) \in \{0,1\}^n$,

$f(\rho_n^k(x_1,\ldots,x_n)) = f(x_1,\ldots,x_n)$ for $1 \leq k \leq n-1$. That is, the rotation symmetric Boolean functions are invariant under cyclic rotation of inputs. The inputs of a rotation symmetric Boolean function can be divided into *orbits* so that each orbit consists of all cyclic shifts of one input. An orbit is generated by $G_n(x_1, x_2, \ldots, x_n) = \{\rho_n^k(x_1, x_2, \ldots, x_n) | 1 \leq k \leq n\}$ and the number of such orbits is denoted by g_n. Thus the number of n-variable RSBFs is 2^{g_n}. Let ϕ be Euler's *phi*-function, then it can be shown by Burnside's lemma that (see also [26]) $g_n = \frac{1}{n} \sum_{k|n} \phi(k) 2^{\frac{n}{k}}$.

An *orbit* is completely determined by its *representative element* $\Lambda_{n,i}$, which is the lexicographically first element belonging to the orbit [27]. These representative elements are again arranged lexicographically as $\Lambda_{n,0}, \ldots, \Lambda_{n,g_n-1}$. In [27] it was shown that the Walsh transform takes the same value for all elements belonging to the same orbit, i.e., $W_f(u) = W_f(v)$ if $u \in G_n(v)$. In analyzing the Walsh spectrum of an RSBF, the $_n\mathcal{A}$ matrix has been introduced [27]. The matrix $_n\mathcal{A} = (_n\mathcal{A}_{i,j})_{g_n \times g_n}$ is defined as $_n\mathcal{A}_{i,j} = \sum_{x \in G_n(\Lambda_{n,i})}(-1)^{x \cdot \Lambda_{n,j}}$, for an n-variable RSBF. Using this $g_n \times g_n$ matrix, the Walsh spectrum for an RSBF can be calculated as $W_f(\Lambda_{n,j}) = \sum_{i=0}^{g_n-1}(-1)^{f(\Lambda_{n,i})} {}_n\mathcal{A}_{i,j}$.

2 Search Algorithm

In this section we present the search algorithm that exhausts the 9-variable RSBFs having nonlinearity > 240. To understand the search method, we first need to study the structure of $_n\mathcal{A}$ under some permutation of the orbit leaders as explained in [19].

2.1 Structure of $_n\mathcal{A}$ for n Odd

The structure of $_n\mathcal{A}$ has been studied in detail for odd n in [19]. Instead of ordering the representative elements in lexicographical manner, the ordering was considered in a different way to get better combinatorial structures. Define $\hat{\Lambda}_{n,i}$ as the representative element of $G_n(x_1, x_2, \ldots, x_n)$ that contains complement of $\Lambda_{n,i}$. For odd n, there is a one-to-one correspondence between the classes of even weight $\Lambda_{n,i}$'s and the classes of odd weight $\Lambda_{n,i}$'s by $\Lambda_{n,i} \rightarrow \hat{\Lambda}_{n,i}$. Hence, the set of orbits can be divided into two parts (of same cardinality) containing representative elements of even weights and odd weights respectively.

Let us consider the ordering of the representative elements in the following way. First the representative elements of even weights are arranged in lexicographical order and they are termed as $\Lambda_{n,i}$, for $i = 0$ to $\frac{g_n}{2} - 1$. Then the next $\frac{g_n}{2}$ representative elements correspond to the complements of the even weight ones, i.e., these are of odd weights. These are recognized as $\Lambda_{n,i} = \hat{\Lambda}_{n,i-\frac{g_n}{2}}$ for $i = \frac{g_n}{2}$ to $g_n - 1$. Thus following [19], the matrix $_n\mathcal{A}$ needs to be reorganized. The resulting matrix is denoted by $_n\mathcal{A}^\pi$, which has the form [19]

$$_n\mathcal{A}^\pi = \left(\begin{array}{c|c} _n\mathcal{H} & _n\mathcal{H} \\ \hline _n\mathcal{H} & -_n\mathcal{H} \end{array} \right),$$

where $_n\mathcal{H}$ is a sub matrix (of order $\frac{g_n}{2} \times \frac{g_n}{2}$) of $_n\mathcal{A}^\pi$. Using this matrix $_n\mathcal{A}^\pi$, the authors of [19] showed that Walsh spectrum calculation could be reduced by almost half of the amount compared to [27].

Given the new ordering of $\Lambda_{n,i}$'s, we represent two strings $\mu_f = ((-1)^{f(\Lambda_{n,0})},$ $\ldots, (-1)^{f(\Lambda_{n,\frac{g_n}{2}-1})})$ and $\nu_f = ((-1)^{f(\Lambda_{n,\frac{g_n}{2}})}, \ldots, (-1)^{f(\Lambda_{n,g_n-1})})$ corresponding to an n-variable RSBF f. Note that μ_f, ν_f are vectors of dimension $\frac{g_n}{2}$.

Let us now consider the vectors $u_f = \mu_f \,_n\mathcal{H}, v_f = \nu_f \,_n\mathcal{H}$. Then the Walsh spectrum values of f will be $(u_f[i] + v_f[i])$ for the first $\frac{g_n}{2}$ many representative elements (which are of even weights) and $(u_f[i] - v_f[i])$ for the next $\frac{g_n}{2}$ many representative elements (which are of odd weights).

2.2   Walsh Spectra of 9-Variable RSBFs Having Nonlinearity > 240

Let us start with a technical result which is easy to prove.

Proposition 1. *Let a, b and M be three integers with $M > 0$. Then $|a+b| \leq M$, $|a - b| \leq M$ iff $|a| + |b| \leq M$.*

The matrix $_9\mathcal{A}^\pi$ for 9-variable RSBFs is a 60×60 matrix, as the number of distinct orbits is 60. The matrix $_9\mathcal{H}$ is a 30×30 matrix.

For an RSBF f on 9 variables, which has nonlinearity strictly greater than 240, the values in the Walsh spectrum are in the range $[-30, 30]$. Thus for a pattern $\mu_f || \nu_f$, one must get $|u_f[i] + v_f[i]| \leq 30$ and $|u_f[i] - v_f[i]| \leq 30$; using Proposition 1, these two conditions are equivalent to $|u_f[i]| + |v_f[i]| \leq 30$ for $0 \leq i \leq \frac{g_9}{2} - 1 = 29$.

Thus one needs to find a 9-variable RSBF f (represented by a 60-bit vector $\mu_f || \nu_f$) such that $|u_f[i]| + |v_f[i]| \leq 30$ for $0 \leq i \leq 29$. By a naive method this requires to exhaust the search space of 2^{60}, i.e., generating all the $\mu_f || \nu_f$ patterns and then checking whether the condition $|u_f[i]| + |v_f[i]| \leq 30$ is satisfied for $0 \leq i \leq 29$ for each of such patterns.

Next we present an efficient method for this. Note that the conditions $|u_f[i] + v_f[i]| \leq 30$ and $|u_f[i] - v_f[i]| \leq 30$ lead to $|u_f[i]| \leq 30$ and $|v_f[i]| \leq 30$; i.e., each of the $\mu_f || \nu_f$ patterns must satisfy the necessary conditions $|u_f[i]| \leq 30$ and $|v_f[i]| \leq 30$ respectively for $0 \leq i \leq 29$. Thus we first search for all the patterns μ_f's such that $|u_f[i]| \leq 30$ for $0 \leq i \leq 29$. Let us denote the set of such patterns by S. This search requires checking for 2^{29} such patterns by fixing $\mu_f[0] = (-1)^0 = 1$. The reason why we fix $u_f[0]$ is presented in Proposition 2 and the discussion after it. In a computer with the specification 3.6 Ghz Intel Xeon and 4 GB RAM, it took little less than half an hour to generate the file containing all these patterns and it contains 24037027 many records, i.e., $|S| = 24037027$. Note that $2^{24} < 24037027 < 2^{25}$.

Clearly the search for all the patterns ν_f's such that $|v_f[i]| \leq 30$ for $0 \leq i \leq 29$ will produce the same set S. Hence the search for $\mu_f || \nu_f$ with the property $|u_f[i]| + |v_f[i]| \leq 30$ for $0 \leq i \leq 29$ requires choosing any two patterns μ_f, ν_f from S and checking them. To explain how we select two patterns, we first need to present the following technical result.

Proposition 2. *Consider a 9-variable RSBF f which is represented as $\mu_f \| \nu_f$ such that $|u_f[i]| + |v_f[i]| \leq 30$ for $0 \leq i \leq 29$. Consider the functions g such that any of the following holds:*

1. *$\mu_g = \mu_f$, $\nu_g = \nu_f^c$, i.e., $g(x_1 \ldots x_9) = f(x_1, \ldots, x_9) \oplus l_9$,*
2. *$\mu_g = \mu_f^c$, $\nu_g = \nu_f$, i.e., $g(x_1 \ldots x_9) = f(x_1, \ldots, x_9) \oplus l_9 \oplus 1$,*
3. *$\mu_g = \mu_f^c$, $\nu_g = \nu_f^c$, i.e., $g(x_1 \ldots x_9) = f(x_1, \ldots, x_9) \oplus 1$,*
4. *$\mu_g = \nu_f$, $\nu_g = \mu_f$, i.e., $g(x_1 \ldots x_9) = f(1 \oplus x_1, \ldots, 1 \oplus x_9)$,*
5. *$\mu_g = \nu_f$, $\nu_g = \mu_f^c$, i.e., $g(x_1 \ldots x_9) = f(1 \oplus x_1, \ldots, 1 \oplus x_9) \oplus l_9$,*
6. *$\mu_g = \nu_f^c$, $\nu_g = \mu_f$, i.e., $g(x_1 \ldots x_9) = f(1 \oplus x_1, \ldots, 1 \oplus x_9) \oplus l_9 \oplus 1$,*
7. *$\mu_g = \nu_f^c$, $\nu_g = \mu_f^c$, i.e., $g(x_1 \ldots x_9) = f(1 \oplus x_1, \ldots, 1 \oplus x_9) \oplus 1$,*

where $l_9 = x_1 \oplus x_2 \ldots \oplus x_8 \oplus x_9$, the rotation symmetric linear function containing all the variables. Then $|u_g[i]| + |v_g[i]| \leq 30$ for $0 \leq i \leq 29$.

Thus from a single 9-variable RSBF f one can get 8 many (including f) RSBFs having the same nonlinearity. This is the reason we fix $\mu_f[0] = 1$. We initially check that repeating a pattern from S twice (i.e., $\mu_f \| \nu_f$, when $\nu_f = \mu_f$) one can not satisfy the condition $|u_f[i]| + |v_f[i]| \leq 30$ for $0 \leq i \leq 29$. Thus one requires $\binom{24037027}{2} = 288889321480851$ many pairs to check. Note that $2^{48} < 288889321480851 < 2^{49}$.

We first apply a sieving method to reduce the size of S. The idea is to fix some t, $0 \leq t \leq 29$ and list all the μ_f patterns from S such that $|u_f[t]| = 30$ and store them in the set $S_{30,t}$. Similarly, we form the set $S_{0,t}$ consisting of ν_f patterns from the same set S such that $|v_f[t]| = 0$. Then we choose each of the μ_f patterns from $S_{30,t}$ and each of the ν_f patterns from $S_{0,t}$. If for some μ_f and ν_f, the condition $|u_f[i]| + |v_f[i]| \leq 30$ for all i such that $0 \leq i \leq 29$ holds, then $\mu_f \| \nu_f$ is a 9-variable RSBF having nonlinearity 241. We store these RSBFs with nonlinearity 241 and update S by $S \setminus S_{30,t}$ as the elements of $S_{30,t}$ when considered as μ_f, can not be attached with any ν_f of S except the elements of $S_{0,t}$ to generate an RSBF having nonlinearity > 240.

We do this by fixing t taking all integers in $[0, 29]$. The result found is presented in the following table. Before running the algorithm we like to note the following two observations.

1. For $t = 28$, in the set S, there is no μ_f such that $|u_f[28]| \leq 2$. Thus we initially remove all the μ_f patterns such that $28 \leq |u_f[28]| \leq 30$. This reduces the number of patterns in S from 24037027 to 18999780.
2. For $t = 0$, in the set S, there is no μ_f such that $|u_f[0]| = 30$. Thus we do not consider this.

In Table 1, we try to fix t such that more number of rows can be removed by lesser search, however this is done only by observation and no specific strategy is involved here. That is the reason the indices in the table are not in order. We find $7 \times 27 = 189$ many RSBFs by this method and hence following Proposition 2, we get 8×189 many 9-variable RSBFs having nonlinearity 241. Thus after this experiment the set S is reduced to 9540580 elements which is less than half of

Table 1. Initial search result for 9-variable RSBFs having nonlinearity> 241

t	$|S_{30,t}|$	$|S_{0,t}|$	# of μ_f $\|\nu_f\|$ such that $nl(f) = 241$
29	747073	37584	0
15	552651	77328	27
1	687215	37584	0
27	613686	37584	0
26	542078	37584	0
24	597941	37584	0
16	531456	37584	0
4	545152	37584	0
2	514474	37584	0
19	495350	37584	0
12	464475	37584	0
5	408014	37584	0
14	385125	37584	0
13	364029	37584	0
8	338321	37584	0
23	320685	37584	0
20	272767	37584	0
6	255915	37584	0
10	237525	37584	0
17	222237	37584	0
9	206952	37584	0
21	192113	37584	0
3	132406	77328	27
7	126821	77328	27
11	121290	77328	27
18	115705	77328	27
25	110174	77328	27
22	104643	77328	27

its initial size 24037027. The experiment requires little more than a day in a PC having 3.6 Ghz Intel Xeon and 4 GB RAM.

Then we go for exhaustive search by taking any two patterns in $\binom{9540580}{2}$ ways. Note that $2^{45} < \binom{9540580}{2} < 2^{46}$. We use 20 computers in parallel that work for 30 hours to check this and we do not find any other function having nonlinearity > 240. The specification of computers are 2.8 GHz Pentium IV with 256 MB RAM.

Thus we have the following result.

Theorem 1. *There are 8×189 many 9-variable RSBFs having nonlinearity 241 and this is the highest nonlinearity for the 9-variable RSBF class.*

Now let us present the distribution of Walsh spectra of the 189 functions available from Table 1 and interestingly all of them are same.

Table 2. Distribution of Walsh spectra of the functions found in Table 1

$W_f(\omega)$	-30	-22	-14	-6	2	10	18	26
# of ω's	127	27	36	18	55	39	54	156

We found two classes of functions out of them (63 functions in one class and rest in another class) having different distribution of autocorrelation spectra as follows.

Table 3. Distribution of autocorrelation spectra of the functions found in Table 1

$\Delta_f(\omega)$		-52	-44	-36	-20	-12	-4	4	12	28
# of nonzero ω's		9	9	9	18	81	85	198	81	21

$\Delta_f(\omega)$	-76	-36	-28	-20	-12	-4	4	12	20	28
# of nonzero ω's	1	9	18	36	81	135	108	54	48	21

Thus it is expected that the 189 functions found in Table 1 are linear transformations of two different functions up to affine equivalence and we justify this in the next section.

3   Affine Equivalence of RSBFs Having Nonlinearity 241

Given two Boolean functions f, g on n variables, we call them affinely equivalent if there exist a binary nonsingular $n \times n$ matrix A, two n-bit binary vectors b, d and a binary constant c such that $g(x) = f(xA \oplus b) \oplus d \cdot x \oplus c$. Thus it is clear that given the function f in Proposition 2, all the other seven functions are affinely equivalent to f. In this section we will try to find out affine equivalence among the 189 functions available from Table 1.

For $(a_1, \ldots, a_n) \in \{0, 1\}^n$, the $n \times n$ circulant matrix generated by $(a_1, \ldots, a_n)$ is given by

$$C(a_1, a_2, \ldots, a_n) = \begin{bmatrix} a_1 & a_2 & a_3 & \ldots & a_n \\ a_n & a_1 & a_2 & \ldots & a_{n-1} \\ a_{n-1} & a_n & a_1 & \ldots & a_{n-2} \\ \vdots & & & & \vdots \\ a_2 & a_3 & a_4 & \ldots & a_1 \end{bmatrix}.$$

The determinant of the matrix $C(a_1, a_2, \ldots, a_n)$ is

$$det[C(a_1, a_2, \ldots, a_n)] = \prod_{i=0}^{n-1} (a_1 + a_2\omega_i + a_3\omega_i^2 + \ldots + a_n\omega_i^{n-1}),$$

where ω_i's $(0 \leq i \leq n-1)$ are the n-th roots of unity. In particular we denote $\omega_0 = 1$. We are interested in the binary circulant matrices which are nonsingular.

Proposition 3. *Let $\alpha, \beta \in \{0,1\}^n$ such that $\alpha \in G_n(\beta)$ and A be an $n \times n$ nonsingular binary circulant matrix. Then $\alpha A \in G_n(\beta A)$.*

Proof. As $\alpha \in G(\beta)$, we have $\alpha = \rho^k(\beta)$, for some k such that $0 \le k < n$. It is also clear that the columns $A_1, A_2, \ldots, A_n$ of the matrix $A = C(a_1, a_2, \ldots, a_n)$ are cyclic shift of each other, precisely, $A_j = \rho^{j-1}(A_1)$. Now,

$$
\begin{aligned}
\beta A &= (\beta A_1, \qquad \beta A_2, \qquad \beta A_3, \ldots, \qquad \beta A_n) \\
&= (\beta A_1, \quad \beta \rho^1(A_1), \quad \beta \rho^2(A_1), \ldots, \beta \rho^{n-1}(A_1)) \\
&= (\beta A_1, \rho^{n-1}(\beta) A_1, \rho^{n-2}(\beta) A_1, \ldots, \quad \rho^1(\beta) A_1)
\end{aligned}
$$

Again,

$$
\begin{aligned}
\alpha A &= (\alpha A_1, \alpha A_2, \alpha A_3, \ldots, \alpha A_k, \alpha A_{k+1}, \alpha A_{k+2}, \ldots, \alpha A_n) \\
&= (\alpha A_1, \rho^{n-1}(\alpha) A_1, \rho^{n-2}(\alpha) A_1, \ldots, \rho^{n-k+1}(\alpha) A_1, \rho^{n-k}(\alpha) A_1, \\
&\quad \rho^{n-k-1}(\alpha) A_1, \ldots, \rho^1(\alpha) A_1) \\
&= (\rho^k(\beta) A_1, \rho^{n-1}(\rho^k(\beta)) A_1, \rho^{n-2}(\rho^k(\beta)) A_1, \ldots, \rho^{n-k+1}(\rho^k(\beta)) A_1, \\
&\quad \rho^{n-k}(\rho^k(\beta)) A_1, \rho^{n-k-1}(\rho^k(\beta)) A_1, \ldots, \rho^1(\rho^k(\beta)) A_1) \\
&= (\rho^k(\beta) A_1, \rho^{n-1+k}(\beta) A_1, \rho^{n-2+k}(\beta) A_1, \ldots, \rho^{n-k+1+k}(\beta) A_1, \\
&\quad \rho^{n-k+k}(\beta) A_1, \rho^{n-k-1+k}(\beta) A_1, \ldots, \rho^{1+k}(\beta) A_1) \\
&= (\rho^k(\beta) A_1, \rho^{k-1}(\beta) A_1, \rho^{k-2}(\beta) A_1, \ldots, \rho^1(\beta) A_1, \beta A_1, \rho^{n-1}(\beta) A_1, \\
&\quad \ldots, \rho^{k+1}(\beta) A_1)
\end{aligned}
$$

This shows $\alpha A \in G_n(\beta A)$.

Proposition 4. *Let $f(x)$ be an n-variable RSBF and A be an $n \times n$ nonsingular binary circulant matrix. Then $f(xA)$ is also an RSBF.*

Proof. Let $g(x) = f(xA)$. Consider $x_1, x_2 \in G_n(\Lambda)$. Now $g(x_1) = f(x_1 A)$ and $g(x_2) = f(x_2 A)$. As $x_1 A, x_2 A \in G_n(\Lambda A)$ (from Proposition 3) and f is an RSBF, $g(x_1) = f(x_1 A) = f(x_2 A) = g(x_2)$. Thus g is also an RSBF.

We have enumerated all the 21 distinct nonsingular binary circulant 9×9 matrices up to equivalence corresponding to the row permutations. Based on Proposition 4 we first try to identify whether one of the 189 functions found in Table 1 are affinely equivalent to another function using any of these 21 matrices. We find that this is indeed true and the 189 functions can be divided into 9 classes each containing 21 functions. One example matrix used for this purpose is

$$
A = C(0, 0, 0, 1, 0, 1, 1, 1, 1) =
\begin{bmatrix}
0 & 0 & 0 & 1 & 0 & 1 & 1 & 1 & 1 \\
1 & 0 & 0 & 0 & 1 & 0 & 1 & 1 & 1 \\
1 & 1 & 0 & 0 & 0 & 1 & 0 & 1 & 1 \\
1 & 1 & 1 & 0 & 0 & 0 & 1 & 0 & 1 \\
1 & 1 & 1 & 1 & 0 & 0 & 0 & 1 & 0 \\
0 & 1 & 1 & 1 & 1 & 0 & 0 & 0 & 1 \\
1 & 0 & 1 & 1 & 1 & 1 & 0 & 0 & 0 \\
0 & 1 & 0 & 1 & 1 & 1 & 1 & 0 & 0 \\
0 & 0 & 1 & 0 & 1 & 1 & 1 & 1 & 0
\end{bmatrix}.
$$

Given one function $f(x)$, the other functions are generated as $f(xA), f(xA^2), \ldots,$ $f(xA^{20})$ (in each class containing 21 functions). There are 9 such classes containing 21 functions each and the functions in each class are affinely equivalent. Now let us take one function from each of the 9 classes. Out of these nine functions, three functions follow the distribution of autocorrelation spectrum presented in the top sub-table of Table 3 and six functions follow the distribution of the autocorrelation spectrum presented in the bottom one of Table 3. However, using these 21 matrices no further affine equivalence could be achieved.

Thus we need to concentrate on some larger class of nonsingular matrices than the binary circulant matrices. We study the matrices whose rows are certain kinds of permutation of the rows of binary circulant matrices. Note that if a circulant matrix is nonsingular, then by making the permutation of rows the nonsingularity will not be disturbed. In a circulant matrix we start with a row and then rotate the row one place (we use the right rotation in this paper) to generate the next row. Instead, given the first row, we may go for i-rotation where i, n are coprime.

Define $C^i(a_1, a_2, \ldots, a_n)$ as the matrix obtained by taking $(a_1, a_2, \ldots, a_n)$ as the first row and each of the other rows of the matrix is the i-rotations of its previous row, i.e., $C^i(a_1, a_2, \ldots, a_n) =$

$$
\begin{bmatrix}
a_1 & a_2 & a_3 & \cdots & a_n \\
a_{n+1-i} & a_{n+2-i} & a_{n+3-i} & \cdots & a_{n+n-i} \\
a_{2n+1-2i} & a_{2n+2-2i} & a_{2n+3-2i} & \cdots & a_{2n+n-2i} \\
\vdots & & & & \vdots \\
a_{(n-1)n+1-(n-1)i} & a_{(n-1)n+2-(n-1)i} & a_{(n-1)n+1-(n-1)i} & \cdots & a_{(n-1)n+1-(n-1)i}
\end{bmatrix}.
$$

Proposition 5. *Let $\alpha, \beta \in \{0,1\}^n$ such that $\alpha \in G_n(\beta)$. Let B be a nonsingular matrix, $B = C^i(a_1, a_2, \ldots, a_n)$, where n and i are coprime and $(a_1, a_2, \ldots, a_n) \in \{0,1\}^n$. Then $\alpha B \in G_n(\beta B)$.*

Proof. As $\alpha \in G(\beta)$, then $\alpha = \rho^k(\beta)$, for some k such that $1 \leq k < n$. It is also clear that each of the $B_1, B_2, \ldots, B_n$ columns of the matrix $B = C^i(a_1, a_2, \ldots, a_n)$ is i-cyclic shift of the previous column, i.e., $B_j = \rho^{(j-1)i} B_1$. Now,

$$
\begin{aligned}
\beta B &= (\beta B_1, \quad \beta B_2, \quad \beta B_3, \ldots, \quad \beta B_n) \\
&= (\beta B_1, \quad \beta \rho^i(B_1), \quad \beta \rho^{2i}(B_1), \ldots, \beta \rho^{(n-1)i}(B_1)) \\
&= (\beta B_1, \quad \rho^{n-i}(\beta) B_1, \quad \rho^{n-2i}(\beta) B_1, \ldots, \quad \rho^i(\beta) B_1)
\end{aligned}
$$

Again, $\alpha B = (\alpha B_1, \alpha B_2, \alpha B_3, \ldots, \alpha B_n)$

$$
\begin{aligned}
&= (\alpha B_1, \rho^{n-i}(\alpha) B_1, \rho^{n-2i}(\alpha) B_1, \ldots, \rho^i(\alpha) B_1) \\
&= (\rho^k(\beta) B_1, \rho^{n-i}(\rho^k(\beta)) B_1, \rho^{n-2i}(\rho^k(\beta)) B_1, \ldots, \rho^i(\rho^k(\beta)) B_1) \\
&= (\rho^k(\beta) B_1, \rho^{n-i+k}(\beta) B_1, \rho^{n-2i+k}(\beta) B_1, \ldots, \rho^{i+k}(\beta) B_1).
\end{aligned}
$$

Since i and n are coprime, for some integer γ we have, $\gamma i \equiv 1 \bmod n$, i.e., $\gamma k i \equiv k \bmod n$, i.e., $ri \equiv k \bmod n$, as $\gamma k \equiv r \bmod n$, for some r, $0 \leq r < n$. Therefore, in the expression of αB, we have, $\rho^{(n-ri+k)}(\beta) B_1 = \beta B_1$, $\rho^{(n-(r+1)i+k)}(\beta) B_1 =$

$\rho^{(n-i)}(\beta)B_1$ and in this way all the elements of $\{\beta B_1,\, \rho^{n-i}(\beta)B_1,\, \rho^{n-2i}(\beta)B_1, \ldots,$ $\rho^i(\beta)B_1\}$ will appear in αB in the same sequence in which they occur in βB. If τ be the term of βB, which occurs as the n-th term of αB, then all the remaining terms of βB after τ will appear in the same sequence starting from the 1st position up to the $(r-2)$-th position in αB. Therefore $\alpha B \in G_n(\beta B)$. Hence the proof.

Similar to the Proposition 4, using Proposition 5 we get the following.

Proposition 6. *Let $f(x)$ be an n-variable RSBF and B be an $n \times n$ nonsingular binary matrix as explained in Proposition 5. Then $f(xB)$ is also an RSBF.*

In our case, $n = 9$ and we choose $i = 2$. As for example, one may consider the matrix

$$B = \begin{bmatrix} 0&0&0&0&0&0&0&0&1 \\ 0&1&0&0&0&0&0&0&0 \\ 0&0&0&1&0&0&0&0&0 \\ 0&0&0&0&0&1&0&0&0 \\ 0&0&0&0&0&0&0&1&0 \\ 1&0&0&0&0&0&0&0&0 \\ 0&0&1&0&0&0&0&0&0 \\ 0&0&0&0&1&0&0&0&0 \\ 0&0&0&0&0&0&1&0&0 \end{bmatrix}.$$

Using this matrix we find that the nine RSBFs can be represented by two distinct functions up to affine equivalence. Note that these two functions are not affinely equivalent as their autocorrelation spectra are different as given in Table 3. Below we present these two functions, the first one with maximum absolute value in the autocorrelation spectrum 52 and the second one with 76. The two functions are as follows.

05777A7A6ED82E887CFCE3C549E994947AE4FBA5B91FE46674C3AC8386609671
3FCCAC20EE9B9966CAD357AAE921286D7A20A55A8DF0910BC03C3C51866D2B16

04757A727ED96F087EFCE2C768EB04947AECFBA5B91DE42E7CC1AC8B1060D671
2FCCEDB0EE8B8926CAD357A2E92148ED3AB4A1128DF0918B46143C51A66D2B16

4   Coding Theoretic Implications

As the question of maximum nonlinearity for Boolean functions is related to the covering radius of First order Reed-Muller code $R(1, n)$, we explain the coding theoretic implications of the 9-variable functions having nonlinearity 241. We like to refer to the papers [2, 3, 10, 12, 13, 16] for relevant coding theoretic discussions.

We present the basic definitions following [16] Let us consider a binary code C of length N. Here we consider $R(1, n)$, i.e., C consists of the 2^{n+1} many truth tables (of length $N = 2^n$) of the affine functions on n variables. Now consider any coset D of the code C, i.e., the elements of the coset D are $f \oplus l$, where $l \in R(1, n)$ and f is a nonlinear Boolean function. The weight of the minimum weight vector in D is considered as the weight of D. Let the minimum weight be w. Then all the vectors having weight w constitute the set of the leaders in D, denoted as $L(D)$. One can define a partial ordering on F_2^N by $S \leq T$ between two binary vectors S, T of length N if $S_i \leq T_i$ for $0 \leq i \leq N - 1$. A partial ordering on the space of cosets of C can be defined as follows. Given two cosets D, D' of C, $D \leq D'$ means there exist $S \in L(D)$ and $S' \in L(D')$ such that $S \leq S'$. We define a coset D as an orphan or urcoset of C if D is a maximal coset in this partial ordering. This concept was first presented in [12] as urcoset and then in [2,3] as orphan coset. One can check [16] that a coset D is an orphan or urcoset when $\cup_{g \in L(D)} supp(g) = \{0, 1\}^N$.

We have checked by running computer program that given any of the two functions described in the previous function (say f, g), each of the cosets $f \oplus R(1, n)$ and $g \oplus R(1, n)$ is an orphan or urcoset. It is clear from Table 2 that the weight of each of the leaders is 241 and there are 127 leaders in each coset. Since each coset is an odd weight orphan, according to [16, Proposition 7], one coordinate position (out of the 512 positions numbered as 0 to 511) must be covered by all the 127 leaders (i.e., the leaders will have the value 1 at that position). In both of the cosets, the 0th position, the output of the 9-variable function corresponding to input $(0, 0, \ldots, 0, 0)$, is covered by all the leaders.

In [10], orphan cosets having minimum weight of 240 have been reported. This is the first time orphan cosets having minimum weight 241 are demonstrated. Further it is reported in [2, Page 401] that X.-d. Hou has constructed odd weight orphans of $R(1, n)$ for certain $n \geq 11$. Our result shows that this is true for $n = 9$ also.

Let $\rho(C)$ be the covering radius [17, 22] of C, the weight of the coset of C having largest weight. We like to point out a conjecture in this direction presented in [3]. The conjecture says that the covering radius of $R(1, n)$ is even. For $n = 9$ we found that the covering radius is at least 241, and searching the space of 9-variable RSBFs we could not get higher nonlinearity. In fact some heuristic attempts to increase the nonlinearity did not work yet. It will be an interesting open question to settle the covering radius of $R(1, 9)$. The bound presented in [13] for $R(1, 9)$ gives the value 244.

5 Conclusion

In this paper we present an efficient exhaustive search strategy to enumerate the 9-variable RSBFs having nonlinearity > 240. We find 8×189 many functions with nonlinearity 241 and it is found that there is no function having more nonlinearity in the 9-variable RSBF class. Using binary nonsingular circulant matrices and some variants of them, we could show that there are only two different 9-variable functions having nonlinearity 241 in the RSBF class up to affine equivalence. Towards the end we present the coding theoretic implications of these functions. The most important open question is to study outside the RSBF class to see if there is any function having nonlinearity > 241.

278   S. Kavut et al.

Acknowledgment. We like to thank Prof. Philippe Langevin for pointing out the coding theoretic issues presented in Section 4.

References

1. E. R. Berlekamp and L. R. Welch. Weight distributions of the cosets of the $(32,6)$ Reed-Muller code. *IEEE Transactions on Information Theory*, IT-18(1):203–207, January 1972.
2. R. A. Brualdi and V. S. Pless. Orphans of the first order Reed-Muller codes. *IEEE Transactions on Information Theory*, 36(2):399–401, 1990.
3. R. A. Brualdi, N. Cai and V. Pless. Orphan structure of the first order Reed-Muller codes. *Discrete Mathematics*, No. 102, pages 239–247, 1992.
4. J. Clark, J. Jacob, S. Maitra and P. Stănică. Almost Boolean Functions: The Design of Boolean Functions by Spectral Inversion. *Computational Intelligence*, Pages 450–462, Volume 20, Number 3, 2004.
5. T. W. Cusick and P. Stănică. Fast Evaluation, Weights and Nonlinearity of Rotation-Symmetric Functions. *Discrete Mathematics* **258**, 289–301, 2002.
6. D. K. Dalai, K. C. Gupta and S. Maitra. Results on Algebraic Immunity for Cryptographically Significant Boolean Functions. In *INDOCRYPT 2004*, number 3348 in Lecture Notes in Computer Science, Page 92–106, Springer Verlag, December 2004.
7. D. K. Dalai, S. Maitra and S. Sarkar. Results on rotation symmetric Bent functions. In *Second International Workshop on Boolean Functions: Cryptography and Applications, BFCA'06*, March 2006.
8. C. Ding, G. Xiao, and W. Shan. *The Stability Theory of Stream Ciphers*. Number 561 in Lecture Notes in Computer Science. Springer-Verlag, 1991.
9. E. Filiol and C. Fontaine. Highly nonlinear balanced Boolean functions with a good correlation-immunity. In *Advances in Cryptology - EUROCRYPT'98*, Springer-Verlag, 1998.
10. C. Fontaine. On some cosets of the First-Order Reed-Muller code with high minimum weight. *IEEE Transactions on Information Theory*, 45(4):1237–1243, 1999.
11. M. Hell, A. Maximov and S. Maitra. On efficient implementation of search strategy for rotation symmetric Boolean functions. In *Ninth International Workshop on Algebraic and Combinatoral Coding Theory, ACCT 2004*, June 19–25, 2004, Black Sea Coast, Bulgaria.
12. T. Helleseth, T. Kløve and J. Mykkeltveit. On the covering radius of binary codes. *IEEE Transactions on Information Theory*, volume IT-24, pages 627–628, September 1978.
13. X. -d. Hou. On the norm and covering radius of the first order Reed-Muller codes. *IEEE Transactions on Information Theory*, 43(3):1025–1027, 1997.
14. S. Kavut, S. Maitra, M. D. Yucel. Autocorrelation spectra of balanced boolean functions on odd number input variables with maximum absolute value $< 2^{\frac{n+1}{2}}$. In *Second International Workshop on Boolean Functions: Cryptography and Applications, BFCA 06*, March 13–15, 2006, LIFAR, University of Rouen, France.
15. S. Kavut, S. Maitra and M. D. Yücel. There exist Boolean functions on n (odd) variables having nonlinearity $> 2^{n-1} - 2^{\frac{n-1}{2}}$ if and only if $n > 7$. http://eprint.iacr.org/2006/181.
16. P. Langevin. On the orphans and covering radius of the Reed-Muller codes. In *9th International Symposium on Applied Algebra, Algebraic Algorithms and Error-Correcting Codes, AAECC 1991*, LNCS 539, Springer Verlag, 234–240, 1991.

17. F. J. MacWilliams and N. J. A. Sloane. *The Theory of Error Correcting Codes.* North Holland, 1977.

18. M. Matsui. Linear cryptanalysis method for DES cipher. In *Advances in Cryptology - EUROCRYPT'93*, Lecture Notes in Computer Science, pages 386–397. Springer-Verlag, 1994.

19. A. Maximov, M. Hell and S. Maitra. Plateaued Rotation Symmetric Boolean Functions on Odd Number of Variables. In *First Workshop on Boolean Functions: Cryptography and Applications, BFCA 05,* March 7–9, 2005, LIFAR, University of Rouen, France.

20. A. Maximov. Classes of Plateaued Rotation Symmetric Boolean functions under Transformation of Walsh Spectra. In WCC 2005, Pages 325–334. See also IACR eprint server, no. 2004/354.

21. J. J. Mykkeltveit. The covering radius of the $(128, 8)$ Reed-Muller code is 56. *IEEE Transactions on Information Theory,* IT-26(3):359–362, 1980.

22. N. J. Patterson and D. H. Wiedemann. The covering radius of the $(2^{15}, 16)$ Reed-Muller code is at least 16276. *IEEE Transactions on Information Theory,* IT-29(3):354–356, 1983. See also the correction in *IEEE Transactions on Information Theory,* IT-36(2):443, 1990.

23. J. Pieprzyk and C. X. Qu. Fast Hashing and Rotation-Symmetric Functions. *Journal of Universal Computer Science* **5**, 20–31, 1999.

24. B. Preneel, W. Van Leekwijck, L. Van Linden, R. Govaerts, and J. Vandewalle. Propagation characteristics of Boolean functions. In *Advances in Cryptology - EUROCRYPT'90*, Lecture Notes in Computer Science, pages 161–173. Springer-Verlag, 1991.

25. O. S. Rothaus. On bent functions. *Journal of Combinatorial Theory, Series A,* pages 300–305, vol 20, 1976.

26. P. Stănică and S. Maitra. Rotation Symmetric Boolean Functions – Count and Cryptographic Properties. In *R. C. Bose Centenary Symposium on Discrete Mathematics and Applications,* December 2002. Electronic Notes in Discrete Mathematics, Elsevier, Vol 15.

27. P. Stănică, S. Maitra and J. Clark. Results on Rotation Symmetric Bent and Correlation Immune Boolean Functions. *Fast Software Encryption Workshop (FSE 2004),* New Delhi, INDIA, LNCS 3017, Springer Verlag, 161–177, 2004.

28. X. M. Zhang and Y. Zheng. GAC - the criterion for global avalanche characteristics of cryptographic functions. *Journal of Universal Computer Science,* 1(5):316–333, 1995.

Symmetric Nonce Respecting Security Model and the MEM Mode of Operation

Peng Wang[1], Dengguo Feng[1,2], and Wenling Wu[2]

[1] State Key Laboratory of Information Security
Graduate School of Chinese Academy of Sciences, Beijing 100049, China
wp@is.ac.cn
[2] State Key Laboratory of Information Security
Institution of Software of Chinese Academy of Sciences, Beijing 100080, China
{feng, wwl}@is.iscas.ac.cn

Abstract. The MEM mode is a nonce-based encryption mode of operation proposed by Chakraborty and Sarkar, which was claimed to be secure against symmetric nonce respecting adversaries. We first compare this security model with two similar models and then show that MEM is not secure under symmetric respecting attacks. One attack needs one decryption and one encryption queries, and the other only needs one encryption query.

Keywords: Blockcipher, tweakable blockcipher, modes of operation, nonce-based encryption, security model.

1 Introduction

A *mode of operation*, or mode, for short, is a scheme that specifies how to use a *blockcipher* to provide some cryptographic services, such as privacy, authenticity or both. Recently, Chakraborty and Sarkar [2] proposed a new security model in which the adversary is *symmetric nonce respecting* and the MEM (Mask Encrypt Mask) mode, a *nonce-based* encryption mode of operation, which was claimed to be secure in this model.

Suppose the underlying blockcipher is

$$E : \mathcal{K} \times \{0,1\}^n \to \{0,1\}^n$$

where $\mathcal{K}$ is a key space, then the MEM mode is

$$\text{MEM}[E] : \mathcal{K} \times \mathcal{N} \times (\{0,1\}^n)^+ \to (\{0,1\}^n)^+$$

where $\mathcal{N} = \{0,1\}^n$ is a nonce space and the key space $\mathcal{K}$ is same as that of the underlying blockcipher E. Let $\mathbf{E}_K(\cdot, \cdot)$ and $\mathbf{D}_K(\cdot, \cdot)$ be the encryption and decryption algorithms in an encryption scheme respectively, which is just MEM in section 3. Let D_K be the inverse of E_K.

R. Barua and T. Lange (Eds.): INDOCRYPT 2006, LNCS 4329, pp. 280–286, 2006.

1.1   Symmetric Nonce Respecting Adversaries

The *nonce-based* symmetric encryption [10] is a syntax for an encryption scheme where the encryption process is a deterministic algorithm, which surfaces an initial vector. The initial vector which is supplied by the user and not by the encryption algorithm is usually a *nonce* — a value, like a counter, used at most once within a session. This syntax was advocated by Rogaway, Bellare, *et al.* [12,11], and first used in the OCB mode [12,10]. In a nonce-based encryption scheme, the encryption algorithm is $\mathbf{E} : \mathcal{K} \times \mathcal{N} \times \mathcal{M} \to \mathcal{M}$, where $\mathcal{K}$ is a key space, $\mathcal{N}$ is a nonce space and $\mathcal{M}$ is a message space. We often write $\mathbf{E}(K, N, M)$ as $\mathbf{E}_K(N, M)$ or $\mathbf{E}_K^N(M)$.

IND\$-SNR. The security model of MEM assumes that the adversary be symmetric nonce respecting, i.e., the adversary can not repeat nonce in either encryption or decryption query. Note that an adversary is allowed to choose the same nonce for both the encryption and the decryption queries. Without loss of generality, we also assume that the adversary does not make *pointless* query, such as as a decryption query of (N, C) after getting it as an answer to an encryption query, etc. IND\$-SNR is a reasonable model in certain scenarios [2].

If any symmetric nonce respecting adversary cannot distinguish $\mathbf{E}_K(\cdot, \cdot)$ and $\mathbf{D}_K(\cdot, \cdot)$ from that of a *random tweakable permutation* and its inverse, we say that $\mathbf{E}$ is secure against symmetric nonce respecting attacks. Note that all the adversaries in this paper can only use reasonable resources, such as polynomial time and queries. Indistinguishability means that the advantage of the adversary is negligible. Or equivalently [2], this kind of adversary cannot distinguish $\mathbf{E}_K(\cdot, \cdot)$ and $\mathbf{D}_K(\cdot, \cdot)$ from \$$(\cdot, \cdot)$ and \$$(\cdot, \cdot)$, where \$$(N, P)$ returns a random string of length $|P|$. If $\approx$ denotes indistinguishability, we can write it as

$$\mathbf{E}_K(\cdot, \cdot), \mathbf{D}_K(\cdot, \cdot) \approx \$(\cdot, \cdot), \$(\cdot, \cdot).$$

We denote this security model as IND\$-SNR.

IND\$-SSTB. Without the symmetric nonce respecting restriction, the IND\$-SNR security model is exactly that of strong secure *tweakable blockcipher* [7]. More specifically [3,4], $\mathbf{E}$ is an strong security tweakable blockcipher, if for any adversary making no pointless queries

$$\mathbf{E}_K(\cdot, \cdot), \mathbf{D}_K(\cdot, \cdot) \approx \$(\cdot, \cdot), \$(\cdot, \cdot).$$

Dedicated strong secure tweakable blockcipher constructions, such as CMC [3], EME [4], HCTR [13] etc. are of course secure against symmetric nonce respecting adversaries. We denote this security model as IND\$-SSTB.

IND\$-CCA. The other similar security model is IND\$-CCA [11], i.e. the chosen ciphertext security model for nonce based encryption scheme. In this model, the adversary is nonce respecting when makes an encryption query and

$$\mathbf{E}_K(\cdot, \cdot), \mathbf{D}_K(\cdot, \cdot) \approx \$(\cdot, \cdot), \mathbf{D}_K(\cdot, \cdot)$$

where $ returns a $|\mathbf{E}_K(N, P)|$ bits random string to each query (N, P). This model is adopted by the OCB mode [12,10], the GCM mode [8], the CWC mode [6], the EAX mode [1], etc.

1.2 Security Claimed by [2]

Chakraborty, *et al.* claimed that MEM was secure under the IND$-SNR model. They "proved" that:

$$\mathbf{E}_K(\cdot, \cdot), \mathbf{D}_K(\cdot, \cdot) \approx \$(\cdot, \cdot), \$(\cdot, \cdot).$$

Unfortunately, it was wrong. We will give two very simple attacks in section 3.

The security proof is always a subtle thing, especial a long one. For example, the EMD mode [9] proposed by Rogaway also had a detailed proof, but was soon broken by Joux [5].

1.3 Our Contributions

We first analyze the relations among IND$-SNR IND$-SSTB and IND$-CCA. The results show that IND$-SSTB is the strongest model and IND$-SNR and IND$-CCA are incomparable, i.e. there exists an encryption scheme which is IND$-SNR secure but not IND$-CCA secure and there exists an encryption scheme which is IND$-CCA secure but not IND$-SNR secure.

We then show that EME is not secure against symmetric nonce respecting adversaries at all. The first attack makes one decryption and one encryption queries. The second attack makes only one encryption query.

2   Relations Among Three Security Models

We sum up the query restrictions of adversaries in IND$-SNR, IND$-SSTB and IND$-CCA, respectively, in following table.

Table 1. The restrictions of adversaries in three models

	When query	and get	then these queries are disallowed:
IND$-SNR	$\mathbf{E}_K(N, P)$	C	$\mathbf{E}_K(N, \cdot), \mathbf{D}_K(N, C)$
	$\mathbf{D}_K(N, C)$	P	$\mathbf{D}_K(N, \cdot), \mathbf{E}_K(N, P)$
IND$-SSTB	$\mathbf{E}_K(N, P)$	C	$\mathbf{E}_K(N, P), \mathbf{D}_K(N, C)$
	$\mathbf{D}_K(N, C)$	P	$\mathbf{D}_K(N, C), \mathbf{E}_K(N, P)$
IND$-CCA	$\mathbf{E}_K(N, P)$	C	$\mathbf{E}_K(N, \cdot), \mathbf{D}_K(N, C)$
	$\mathbf{D}_K(N, C)$	P	$\mathbf{E}_K(N, P)$

We write MA$\longrightarrow$MB to denote that for any scheme secure under model MA is also secure under model MB; and write MA$\longrightarrow\!\!\!\!/\,$MB to denote that there exists an scheme which is secure under model MA but is not secure under model MB. The relations among IND\$-SNR, IND\$-SSTB and IND\$-CCA are represented in Figure 1.

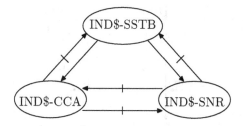

Fig. 1. Relations among IND\$-SNR, IND\$-SSTB and IND\$-CCA

According to the ability of the adversary, it is obviously that IND\$-SSTB$\longrightarrow$ IND\$-SNR, IND\$-SSTB$\longrightarrow$IND\$-CCA, IND\$-SNR $\not\longrightarrow$IND\$-SSTB and IND\$-CCA$\not\longrightarrow$IND\$-SSTB. For the sake of simplicity, in the following encryption schemes we only describe the encryption algorithms.

IND\$-SNR$\not\longrightarrow$IND\$-CCA. First notice that the encryption scheme whose encryption algorithm is $\mathbf{E}'_{K_1, K_2}(M) = K_1 \cdot M \oplus K_2$ is IND\$-SNR secure if the adversary only makes one encryption query and one decryption query. We construct $\mathbf{E}^N_K(M) = E_K(N||0) \cdot M \oplus E_K(N||1)$, where E is an secure blockcipher and N is a nonce. In the IND\$-SNR model, each nonce can only appear in one encryption query and one decryption query. When N is new, $E_K(N||0)$ and $E_K(N||1)$ generate two independent pseudorandom strings. So this scheme is IND\$-SNR secure. Obviously it is not IND\$-CCA secure.

IND\$-CCA$\not\longrightarrow$IND\$-SNR. For example, most of IND\$-CCA secure modes, such as OCB etc., return $\perp$ denoting the invalidity of the ciphertext in a decryption query most of time. These modes are all not IND\$-SNR secure.

3   Cryptanalysis of MEM

3.1   Specifications of MEM

An n-bit string is viewed as an element in the finite field $GF(2)[x]/\tau(x)$, where $\tau(x)$ is a fixed irreducible polynomial of degree n.

MEM makes use of the polynomials $p_i(x)$ which are defined as following. For $0 < i < m$ and $(n + 1) \nmid i$, define $p_i(x) = x^{j-1} + x^j$, where $j = (i - 1)$ mod $(n + 1) + 1$; for $0 < i < m$ and $(n + 1) \mid i$, define $p_i(x) = x^n + 1$; for $i = m$, define $p_i(x) = x^{j-1} + 1$, where $j = (i - 1)$ mod $(n + 1) + 1$.

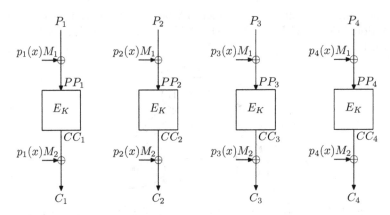

Fig. 2. Encryption of a four-block message $P_1P_2P_3P_4$ under MEM. Set $M_1 = E_K(P_1 \oplus P_2 \oplus P_3 \oplus P_4 \oplus E_K(N))$ and $M_2 = E_K(CC_1 \oplus CC_2 \oplus CC_3 \oplus CC_4 \oplus E_K(xE_K(N)))$.

The encryption and decryption algorithms of EME are listed in the figure 3, which consists of three cases: $m = 1$, $m = 2$ and $m > 2$. In our attacks we only make use of the case $m > 2$. Figure 2 shows the encryption of a four-block message.

3.2 Distinguishers Against MEM

We can distinguish $\mathbf{E}_K(\cdot, \cdot), \mathbf{D}_K(\cdot, \cdot)$ from $\$(\cdot, \cdot), \$(\cdot, \cdot)$ with overwhelming advantage of $1 - 1/2^n$. The first distinguisher makes one decryption and one encryption queries. The second distinguisher makes only one encryption.

Two-Query Distinguisher. This distinguisher is similar to the one used in [2] to show that MEM is not secure against nonce repeating adversary. The difference is that the one in [2] made two encryption queries with the same nonce and we make one decryption and one encryption queries with the same nonce.

The distinguisher is as following:

1. Make a decryption query $(N^s, C_1^s, C_2^s, C_3^s, C_4^s)$ and get $(P_1^s, P_2^s, P_3^s, P_4^s)$;
2. Make an encryption query $(N^t, P_1^t, P_2^t, P_3^t, P_4^t)$ and get $(C_1^t, C_2^t, C_3^t, C_4^t)$, where $N^s = N^t$, $P_1^s = P_1^t$, $P_2^s = P_2^t$, $P_3^s \neq P_3^t$, and $P_3^s \oplus P_4^s = P_3^t \oplus P_4^t$.
3. Calculate $X_1 = p_1(x)^{-1}(C_1^s \oplus C_1^t)$ and $X_2 = p_2(x)^{-1}(C_2^s \oplus C_2^t)$.
4. If $X_1 = X_2$, then return 1, else return 0.

When the distinguisher queries $\mathbf{E}_K(\cdot, \cdot), \mathbf{D}_K(\cdot, \cdot)$,

$$M_2^s = p_1(x)^{-1}(CC_1^s \oplus C_1^s) = p_2(x)^{-1}(CC_2^s \oplus C_2^s)$$

and

$$M_2^t = p_1(x)^{-1}(CC_1^t \oplus C_1^t) = p_2(x)^{-1}(CC_2^t \oplus C_2^t).$$

Algorithm $E_K^N(P_1, P_2, \cdots, P_m)$	Algorithm $D_K^N(C_1, C_2, \cdots, C_m)$
$EN \leftarrow E_K(N)$; $EEN \leftarrow E_K(xEN)$;	$EN \leftarrow E_K(N)$; $EEN \leftarrow E_K(xEN)$;
$MP \leftarrow P_1 \oplus P_2 \cdots \oplus P_m$;	$MC \leftarrow C_1 \oplus C_2 \cdots \oplus C_m$;
if $m = 1$ **then**	**if** $m = 1$ **then**
$\quad M_1 \leftarrow E_K(MP \oplus EN)$;	$\quad M_2 \leftarrow D_K(MC \oplus xEEN)$;
$\quad C_1 \leftarrow M_1 \oplus xEEN$;	$\quad P_1 \leftarrow M_2 \oplus EEN$;
$\quad$ **return** C_1	$\quad$ **return** P_1
if $m = 2$ **then**	**if** $m = 2$ **then**
$\quad M_1 \leftarrow E_K(MP \oplus EN)$;	$\quad M_2 \leftarrow E_K(MC \oplus EN \oplus EEN)$;
$\quad PP_1 \leftarrow M_1 \oplus P_1$; $PP_2 \leftarrow M_1 \oplus EEN \oplus P_2$;	$\quad CC_1 \leftarrow M_2 \oplus C_1$; $CC_2 \leftarrow M_2 \oplus EN \oplus C_2$;
$\quad CC_1 \leftarrow E_K(PP_1)$; $CC_2 \leftarrow E_K(PP_2)$;	$\quad PP_1 \leftarrow D_K(CC_1)$; $PP_2 \leftarrow D_K(CC_2)$;
$\quad M_2 \leftarrow E_K(CC_1 \oplus CC_2 \oplus EEN)$;	$\quad M_1 \leftarrow E_K(PP_1 \oplus PP_2 \oplus EEN \oplus EN)$;
$\quad C_1 \leftarrow M_2 \oplus CC_1$; $C_2 \leftarrow EN \oplus CC_2$;	$\quad P_1 \leftarrow M_1 \oplus PP_1$; $P_2 \leftarrow M_1 \oplus EEN \oplus PP_2$;
$\quad$ **return** C_1, C_2	$\quad$ **return** P_1, P_2
if $m > 2$ **then**	**if** $m > 2$ **then**
$\quad M_1 \leftarrow E_K(MP \oplus EN)$; $MC \leftarrow 0^n$;	$\quad M_2 \leftarrow E_K(MC \oplus EEN)$; $MP \leftarrow 0^n$;
$\quad$ **for** $i = 1$ **to** m	$\quad$ **for** $i = 1$ **to** m
$\quad\quad$ **if** $(i - 1 > 0 \land i - 1 \mod (n+1) = 0)$	$\quad\quad$ **if** $(i - 1 > 0 \land i - 1 \mod (n+1) = 0)$
$\quad\quad\quad M_1 \leftarrow E_K(M_1)$;	$\quad\quad\quad M_2 \leftarrow E_K(M_2)$;
$\quad\quad PP_i \leftarrow P_i \oplus p_i(x)M_1$;	$\quad\quad CC_i \leftarrow C_i \oplus p_i(x)M_2$;
$\quad\quad CC_i \leftarrow E_K(PP_i)$; $MC \leftarrow MC \oplus CC_i$;	$\quad\quad PP_i \leftarrow D_K(CC_i)$; $MP \leftarrow MP \oplus PP_i$;
$\quad$ **for** $i = 1$ **to** m	$\quad$ **for** $i = 1$ **to** m
$\quad\quad$ **if** $(i - 1 > 0 \land i - 1 \mod (n+1) = 0)$	$\quad\quad$ **if** $(i - 1 > 0 \land i - 1 \mod (n+1) = 0)$
$\quad\quad\quad M_2 \leftarrow E_K(M_2)$;	$\quad\quad\quad M_1 \leftarrow E_K(M_1)$;
$\quad\quad C_i \leftarrow CC_i \oplus p_i(x)M_2$;	$\quad\quad P_i \leftarrow PP_i \oplus p_i(x)M_1$;
$\quad$ **return** $C_1, C_2, \cdots, C_m$	$\quad$ **return** $P_1, P_2, \cdots, P_m$

Fig. 3. The MEM mode

Notice that $CC_1^s = CC_1^t$ and $CC_2^s = CC_2^t$, we get that

$$M_2^s \oplus M_2^t = p_1(x)^{-1}(C_1^s \oplus C_1^t) = p_2(x)^{-1}(C_2^s \oplus C_2^t).$$

So the probability of $X_1 = X_2$ is 1.

When the distinguisher queries $\$(\cdot, \cdot), \$(\cdot, \cdot)$, then C_1^t and C_2^t are two independently random strings. So the probability of $X_1 = X_2$ is $1/2^n$.

From the above analysis, the advantage of the distinguisher is $1 - 1/2^n$.

One-Query Distinguisher. This distinguisher only makes one encryption query. Notice that when the message length is $m = n + 3$ blocks, $p^{n+2}(x) = p^{n+3}(x) = 1 + x$. We make an encryption query of $(N, P_1, P_2, \cdots, P_{m+3})$, where $P_1 = P_2 = ... = P_{n+3} = 0^n$, and get $(C_1, C_2, \cdots, C_{n+3})$. If $C_{n+2} = C_{n+3}$ then return 1, else return 0.

When the distinguisher queries $E(\cdot, \cdot)$, we always have $C_{n+2} = C_{n+3}$. When the distinguisher queries $\$(\cdot, \cdot)$, the probability of $C_{n+2} = C_{n+3}$ is $1/2^n$.

From the above analysis, the advantage of the distinguisher is also $1 - 1/2^n$.

Acknowledgment

We thank Debrup Chakraborty for providing their paper and the anonymous referees for their many helpful comments. This research is supported by the National Natural Science Foundation Of China (No. 60673083, 60373047 and 90604036); the National Grand Fundamental Research 973 Program of China (No. 2004CB318004).

References

1. M. Bellare, P. Rogaway, and D. Wagner. The EAX mode of operation. In B. Roy and W. Meier, editors, *Fast Software Encryption 2004*, volume 3017 of *Lecture Notes in Computer Science*, pages 389–407. Springer-Verlag, 2004.
2. D. Chakraborty and P. Sarkar. A new mode of encryption secure against symmetric nonce respecting adversaries, extended pre-proceedings version of FSE'06 paper. Cryptology ePrint Archive, Report 2006/062, 2006. http://eprint.iacr.org/.
3. S. Halevi and P. Rogaway. A tweakable enciphering mode. In D. Boneh, editor, *Advances in Cryptology – CRYPTO 2003*, volume 2729 of *Lecture Notes in Computer Science*, pages 482–499. Springer-Verlag, 2003.
4. S. Halevi and P. Rogaway. A parallelizable enciphering mode. In T. Okamoto, editor, *The Cryptographers' Track at RSA Conference – CT-RSA 2004*, volume 2964 of *Lecture Notes in Computer Science*. Springer-Verlag, 2004.
5. A. Joux. Cryptanalysis of the EMD mode of operation. In L. R. Knudsen, editor, *Advances in Cryptology – EUROCRYPT 2003*, volume 2656 of *Lecture Notes in Computer Science*, pages 1–16. Springer-Verlag, 2003.
6. T. Kohno, J. Viega, and D. Whiting. CWC: A high-performance conventional authenticated encryption mode. In W. M. Bimal Roy, editor, *Fast Software Encryption 2004*, volume 3017 of *Lecture Notes in Computer Science*, pages 408–426. Springer-Verlag, 2004.
7. M. Liskov, R. L. Rivest, and D. Wagner. Tweakable block ciphers. In M. Yung, editor, *Advances in Cryptology – CRYPTO 2002*, volume 2442 of *Lecture Notes in Computer Science*, pages 31–46. Springer-Verlag, 2002.
8. D. A. McGrew and J. Viega. The security and performance of the galois/counter mode (GCM) of operation. In A. Canteaut and K. Viswanathan, editors, *Advances in Cryptology – INDOCRYPT 2004*, volume 3348 of *Lecture Notes in Computer Science*, pages 343–355. Springer-Verlag, 2002.
9. P. Rogaway. The EMD mode of operation (tweaked, wide-blocksize, strong PRP), 2002. http://eprint.iacr.org/2002/148.pdf
10. P. Rogaway. Efficient instantiations of tweakable blockciphers and refinements to modes OCB and PMAC. In P. J. Lee, editor, *Advances in Cryptology – ASIACRYPT 2004*, volume 3329 of *Lecture Notes in Computer Science*, pages 16–31. Springer-Verlag, 2004.
11. P. Rogaway. Nonce-based symmetric encryption. In B. Roy and W. Meier, editors, *Fast Software Encryption 2004*, volume 3017 of *Lecture Notes in Computer Science*, pages 348–359. Springer-Verlag, 2004.
12. P. Rogaway, M. Bellare, J. Black, and T. Krovetz. OCB: a block-cipher mode of operation for efficient authenticated encryptiona. In *Proceedings of the 8th ACM Conference on Computer and Communications Security*, pages 196–205, 2001.
13. P. Wang, D. Feng, and W. Wu. HCTR: A variable-input-length enciphering mode. In D. Feng, D. Lin, and M. Yung, editors, *SKLOIS Conference on Information Security and Cryptology, CISC 2005*, volume 3822 of *Lecture Notes in Computer Science*, pages 175–188. Springer-Verlag, 2005.

HCH: A New Tweakable Enciphering Scheme Using the Hash-Encrypt-Hash Approach

Debrup Chakraborty[1] and Palash Sarkar[2]

[1] Computer Science Department
CINVESTAV-IPN
Mexico, D.F., 07360, Mexico
debrup@cs.cinvestav.mx
[2] Applied Statistics Unit
Indian Statistical Institute
Kolkata 700108, India
palash@isical.ac.in

Abstract. The notion and the first construction of a tweakable enciphering scheme, called CMC, was presented by Halevi-Rogaway at Crypto 2003. In this paper, we present HCH, which is a new construction of such a scheme. The construction uses the hash-encrypt-hash approach introduced by Naor-Reingold. This approach has recently been used in the constructions of tweakable enciphering schemes HCTR and PEP. HCH has several advantages over the previous schemes CMC, EME, EME*, HCTR, and PEP. CMC, EME, and EME* use two block-cipher invocations per message block, while HCTR, PEP, and HCH use only one. PEP uses four multiplications per block, while HCTR and HCH use only two. In HCTR, the security bound is cubic, while in HCH security bound is quadratic.[1]

Keywords: modes of operations, tweakable encryption, strong pseudorandom permutation.

1 Introduction

A block cipher is one of the basic primitives used in cryptography. Depending upon the application goals, there are many uses of a block cipher. A particular method of using a block cipher is called a mode of operation. The literature describes different modes of operations of a block cipher achieving goals such as confidentiality, authentication, authenticated encryption, etcetera. For several years, NIST of USA [1] has been running an open domain process to standardize modes of operations for achieving various functionalities. Currently, there are around twenty different modes of operations proposals for different tasks.

One particular interesting functionality is a tweakable enciphering scheme [4]. (We note that this functionality is currently not covered by NIST's standardization efforts.) This is based on the notion of tweakable block ciphers introduced

---

[1] The last three sentences of the abstract are due to a reviewer who suggested that these accurately capture the contribution of the paper.

R. Barua and T. Lange (Eds.): INDOCRYPT 2006, LNCS 4329, pp. 287–302, 2006.
© Springer-Verlag Berlin Heidelberg 2006

in [6]. A tweakable enciphering scheme is a length preserving encryption protocol which can encrypt messages of varying lengths. The security goal is to satisfy the notion of the tweakable strong pseudo-random permutation (SPRP). As pointed out in [4], one of the most important applications of a tweakable enciphering scheme is disk encryption. Currently, there are several proposals CMC [4], EME [5], EME* [3], HCTR [9] and PEP [2] for tweakable enciphering schemes.

Our Contribution: In this paper, we present HCH, which is a construction of a new tweakable enciphering scheme. HCH uses a single key, can encrypt arbitrary length messages and has a quadratic security bound. Our construction is based on HCTR. It uses a counter mode of encryption sandwiched between two polynomial hashes. HCTR uses two keys and has a cubic security bound. To avoid these problems, we use certain ideas (and analysis) from PEP. In particular, the idea of appropriately encrypting the tweak and the message length is adopted from PEP. In addition, we initialize the counter mode by the output of a block cipher encryption; a feature not present in HCTR. The combination of all these features leads us to the desired goal.

HCH is based on the hash-encrypt-hash approach to the construction of strong pseudo-random permutation. The hash is a Wegman-Carter [10] type of polynomial hash. This approach was originally suggested by Naor-Reingold [8,7], though they did not consider tweaks, a notion which appeared later in the literature. The constructions HCTR and PEP are also based on the hash-encrypt-hash approach. On the other hand, CMC [4] introduced the encrypt-mask-encrypt approach, i.e., to have two layers of encryption with a masking layer in-between. CMC used CBC for the encryption layer, while the later works EME and EME* used ECB for the encryption layers.

In terms of efficiency, HCH and HCTR have roughly the same efficiency; HCH performs a few extra block cipher invocations, while HCTR performs a few extra $GF(2^n)$ multiplications. In a sequential mode, HCH is faster than PEP; though in a parallel mode all three of HCTR, PEP and HCH have roughly the same efficiency. The comparison to CMC (and EME*) depends on the relative efficiency of a block cipher invocation and a $GF(2^n)$ multiplication. It is currently believed, that a single AES-128 invocation is faster than a $GF(2^n)$ multiplication. Hence, used with AES-128, CMC will be faster than HCH (or HCTR, PEP). On the other hand, if one invocation of the underlying block cipher is slower than one $GF(2^n)$ multiplication, then HCH (and HCTR, PEP) will be faster than CMC. Thus, the comparison is really between the two approaches rather than individual constructions. We believe both approaches are interesting and can be pursued further.

2   Specification of HCH

We construct the tweakable enciphering scheme HCH from a block cipher $E :$ $\mathcal{K} \times \{0,1\}^n \to \{0,1\}^n$ and call it HCH[E]. The key space of HCH[E] is same as that of the underlying block cipher E and the tweak space is $\mathcal{T} = \{0,1\}^n$. The message space consists of all binary strings of length greater than n.

An n-bit string can be viewed as an element of $GF(2^n)$. We will consider each n-bit string in the specification of HCH as a polynomial over $GF(2)$ of degree less than n, and multiplication will be done modulo a fixed irreducible polynomial $\tau(x)$ of degree n. Thus, if A and B are n-bit strings, then by AB we will mean the n-bit string representing the product $A(x)B(x) \bmod \tau(x)$. Also, the notation xQ denotes the n-bit string representing $xQ(x) \bmod \tau(x)$. The operation $\oplus$ denotes addition over $GF(2^n)$.

For an n-bit string X, by $\mathsf{pad}_t(X)$ we denote the string $X\|0^t$ and by $\mathsf{drop}_t(X)$ we denote the prefix of X obtained by dropping the last t bits of X. For $0 \leq i \leq 2^t - 1$, by $\mathsf{bin}_t(i)$ we denote the t-bit binary representation of the integer i.

Let $R, A_1, \ldots, A_m$ be n-bit strings. We define

$$
\left.
\begin{aligned}
G_R(A_1,\ldots,A_m) &= A_1 R \oplus \cdots \oplus A_m R^m. \\
H_{R,Q}(A_1,\ldots,A_m) &= Q \oplus A_1 \oplus G_R(A_2,\ldots,A_m) \\
&= Q \oplus A_1 \oplus A_2 R \oplus \cdots \oplus A_m R^{m-1}.
\end{aligned}
\right\} \quad (1)
$$

The above operations are over $GF(2^n)$, i.e., $\oplus$ denotes addition over $GF(2^n)$ and terms of the form $A_i R^{i-1}$ denote the product $A_i(x)R^{i-1}(x) \bmod \tau(x)$. The final value of $H_{R,Q}(A_1,\ldots,A_m)$ is an element of $GF(2^n)$ given by its n-bit string representation with respect to $\tau(x)$. From the definition of $H_{R,Q}()$, we have the following simple property which is required for proper decryption.

$$\text{If } B_1 = H_{R,Q}(A_1, A_2,\ldots,A_m), \text{ then } A_1 = H_{R,Q}(B_1, A_2,\ldots,A_m). \quad (2)$$

HCH requires a counter mode of operation. We define the counter mode based on [9] but in a more general form. Let $f_1,\ldots,f_m$ be a sequence of bijective functions $f_i : \{0,1\}^n \rightarrow \{0,1\}^n$ such that for each n-bit string S, and for $i \neq j$, we have $f_i(S) \neq f_j(S)$. In other words, for each S, the sequence $f_1(S), f_2(S),\ldots,f_m(S)$ is a sequence of distinct n-bit strings. One simple way of defining f_i is to set $f_i(S) = S \oplus \mathsf{bin}_n(i)$ as has been done in [9]. On the other hand for nonzero S, we can also have $f_i(S) = L_i$, where L_i is the ith state of a maximal length LFSR initialized by S. Given an n-bit string S and a key K, we define the counter mode in the following manner.

$$\mathsf{Ctr}_{K,S}(A_1,\ldots,A_m) = (A_1 \oplus E_K(f_1(S)),\ldots,A_m \oplus E_K(f_m(S))). \quad (3)$$

Details about message parsing are the following.

1. The message length is l bits, where $n < l < 2^n - 1$.
   We write $l = n(m-1) + r$, with $1 \leq r \leq n$.
2. The message consists of blocks $P_1,\ldots,P_m$,
   with $|P_1| = \cdots = |P_{m-1}| = n$ and $1 \leq |P_m| = r \leq n$.
3. The ciphertext is of the same length as the message, i.e.,
   the ciphertext blocks are $C_1,\ldots,C_m$, with $|C_i| = |P_i|$ for $1 \leq i \leq m$.

The maximum length l of a message can be $2^n - 1$. Since, for secure block ciphers, we have $n \geq 128$, the maximum value of l is sufficient for all practical purposes.

The complete encryption and decryption algorithm of HCH is given in Figure 1. A schematic diagram of encryption is given in Figure 2.

Algorithm $\mathbf{E}_K^T(P_1, \ldots, P_m)$	Algorithm $\mathbf{D}_K^T(C_1, \ldots, C_m)$
1. $R = E_K(T)$; $Q = E_K(R \oplus \mathsf{bin}_n(l))$;	1. $R = E_K(T)$; $Q = E_K(R \oplus \mathsf{bin}_n(l))$;
2. $M_m = \mathsf{pad}_{n-r}(P_m)$;	2. $U_m = \mathsf{pad}_{n-r}(C_m)$;
3. $M_1 = H_{R,Q}(P_1, \ldots, P_{m-1}, M_m)$;	3. $U_1 = H_{R,xQ}(C_1, \ldots, C_{m-1}, U_m)$;
4. $U_1 = E_K(M_1)$; $I = M_1 \oplus U_1$;	4. $M_1 = E_K^{-1}(U_1)$; $I = M_1 \oplus U_1$;
5. $S = E_K(I)$;	5. $S = E_K(I)$;
6. $(C_2, \ldots, C_{m-1}, D_m)$	6. $(P_2, \ldots, P_{m-1}, V_m)$
$= \mathsf{Ctr}_{K,S}(P_2, \ldots, P_{m-1}, M_m)$;	$= \mathsf{Ctr}_{K,S}(C_2, \ldots, C_{m-1}, U_m)$;
7. $C_m = \mathsf{drop}_{n-r}(D_m)$;	7. $P_m = \mathsf{drop}_{n-r}(V_m)$;
8. $U_m = \mathsf{pad}_{n-r}(C_m)$;	8. $M_m = \mathsf{pad}_{n-r}(P_m)$;
9. $C_1 = H_{R,xQ}(U_1, C_2, \ldots, C_{m-1}, U_m)$;	9. $P_1 = H_{R,Q}(M_1, P_2, \ldots, P_{m-1}, M_m)$;
10. return $(C_1, \ldots, C_m)$.	10. return $(C_1, \ldots, C_m)$.

Fig. 1. Encryption and Decryption using HCH

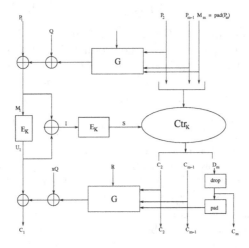

Fig. 2. Encryption using HCH. Here $R = E_K(T)$ and $Q = E_K(R \oplus \mathsf{bin}_n(l))$.

2.1   HCH$^+$

HCH can handle strings of length l greater than n. We have excluded the case of $l = n$ from the specification. This is because if $l = n$, then there is only a single block message and the counter part becomes vacuous. As a result, the block cipher call to produce S is no longer required. (The quantity Q is still required.) Thus, this case needs to be tackled separately. We define HCH$^+$ to be the mode of operation defined in the following manner. If $l > n$, then use HCH to encrypt, and if $l = n$, then there is a single message block P_1 whose corresponding ciphertext block C_1 is defined to be $C_1 = xQ \oplus E_K(P_1 \oplus Q)$. This requires a total of 3 block cipher calls (2 to produce Q and one extra).

The security of HCH$^+$ cannot be generically derived from that of HCH. We need to have a separate proof for HCH$^+$. On the other hand, this proof will be very similar to that of HCH, with the only difference that we will have to take care of the possibilities of domain and range collisions due to single block adversarial queries. We do not actually present the proof for HCH$^+$. Instead, we present the complete proof for HCH, from which the reader should be able to easily obtain a proof for HCH$^+$.

The case $l < n$ is not tackled by any modes of operations and hence also not by HCH. Tackling such values of l is a difficult problem as discussed by Halevi [3].

3   Discussion and Comparison

The structure of HCH is based on HCTR, though there are several important differences. We mention the similarities and differences below.

- The basic structure of the two polynomial hashes, separate treatment for the first message block and the counter mode are taken from HCTR.
- The actual definition of the polynomial hashes are different from that of HCTR. In HCTR, the tweak is provided as input to the polynomial hash, whereas in HCH the tweak is encrypted to obtain R, which is XORed with $\mathrm{bin}_n(l)$, which is again encrypted to obtain Q. This increases the number of block cipher invocations in HCH, whereas HCTR requires more $GF(2^n)$ multiplications. Additionally, for computing the polynomial hashes, HCTR requires an extra key different from the block cipher key.
- In HCH, the counter mode is initialized by S, which is the output of a block cipher encryption. On the other hand, in HCTR, the counter is initialized directly by the XOR of the input and output of the first block cipher invocation. Using the additional invocation to produce S is important in obtaining a quadratic security bound for HCH compared to the cubic security bound for HCTR.
- The definition of counter mode used in HCH is somewhat more general than the counter mode used in HCTR.

In Table 1, we present a comparison of HCH with the previous algorithms. (The construction given in [7] is not tweakable and hence we do not include it in our comparison.) EME tackles only strings of very specific lengths and hence we do not consider it any further here. From the table, we see that EME*, HCTR and HCH are the only algorithms which can tackle arbitrary length strings (HCH can be easily modified to also tackle strings of length n, see Section 2.1.) Among these, EME* uses three keys and has a quadratic security bound while HCTR uses two keys and has a cubic security bound. On the other hand, HCH uses a single key and has a quadratic security bound. Thus, HCH is the only algorithm to-date which tackles arbitrary length messages, uses a single key and has a quadratic security bound.

The algorithms CMC and EME* are based on the encrypt-mask-encrypt approach, while HCTR, PEP and HCH are based on the hash-encrypt-hash approach. The first approach requires more block cipher calls while the second approach requires more finite field multiplications. Let us first compare HCTR,

Table 1. Comparison of SPRPs using an n-bit block cipher, an n-bit tweak and for m message blocks. Note that HCH can be modified to also tackle strings of length n (see Section 2.1). For PEP we assume $m \geq 3$ and for HCH we have $m \geq 2$. [BC]: one block cipher invocation; [M]: one $GF(2^n)$ multiplication.

Mode	sec. bnd.	computation cost	keys	msg. len.	passes	enc. layers	parallel?
CMC	$\sigma_n^2/2^n$	$(2m+1)$[BC]	1	$mn, m \geq 1$	2	2	No
EME	$\sigma_n^2/2^n$	$(2m+2)$[BC]	1	$mn, 1 \leq m \leq n$	2	2	Yes
EME*	$\sigma_n^2/2^n$	$(2m+\frac{m}{n}+1)$[BC]	3	$\geq n$	2	2	Yes
HCTR	$\sigma_n^3/2^n$	m[BC] $+2(m+1)$[M]	2	$\geq n$	3	1	partial
PEP	$\sigma_n^2/2^n$	$(m+5)$[BC] $+(4m-6)$[M]	1	$mn, m \geq 1$	3	1	Yes
HCH	$\sigma_n^2/2^n$	$(m+3)$[BC] $+2(m-1)$[M]	1	$> n$	3	1	partial

PEP and HCH. In a sequential mode, HCH is clearly better than PEP. Compared to HCTR, it requires three extra block cipher calls but four less finite field multiplications. The net effect is that both HCTR and HCH have roughly the same computation cost. The comparison to CMC and EME* is based on the relative cost of a block cipher call versus a finite field multiplication. A good AES-128 implementation can be faster than a good $GF(2^{128})$ multiplication and hence used with AES-128, a sequential version of CMC and EME* will be faster than HCH. (Of course, as mentioned above, HCH has other features which CMC and EME* do not have.) On the other hand, a mode of operation is not intended to be used with only one block cipher. It is conceivable that there are (possibly proprietary) block ciphers for which a single invocation is costlier than a finite field multiplication. Used with such block ciphers, HCH will be faster than CMC and EME*.

For hardware implementation, it is of interest to have parallel implementation of the different algorithms. CMC is a strictly sequential algorithm. EME* is parallel, though it requires a re-keying after every n blocks. HCTR and HCH are partially parallel in the sense that the counter part is fully parallel but the hash function computations are not. The hash function computations can be made parallel, but this roughly doubles the total number of multiplications of both HCTR and HCH and make them comparable to PEP. Of course, in a parallel implementation, the total number of multiplications is not important; what is important is the total number of multiplication rounds. This issue is discussed in more details in [2].

The number of passes in the hash-encrypt-hash modes is three while this value is two for encrypt-mask-encrypt modes. In both cases, it is more than one, which implies that the encryption cannot be on-line, i.e., the ciphertext cannot be produced without reading (or processing) the entire message. This property is natural to expect from a *strong* pseudorandom permutation, since such a primitive tries to make each bit of the ciphertext depend on the entire message. Since the encryption cannot be on-line, storage space is required for the entire message. Then it does not matter too much whether the number of passes is two or three. The important thing is to make the encryption efficient. We have already discussed the issue of efficiency.

4 Security of HCH

4.1 Definitions and Notation

An n-bit block cipher is a function $E : \mathcal{K} \times \{0,1\}^n \to \{0,1\}^n$, where $\mathcal{K} \neq \emptyset$ is the key space and for any $K \in \mathcal{K}$, $E(K,.)$ is a permutation. We write $E_K()$ instead of $E(K,.)$.

An adversary A is a probabilistic algorithm which has access to some oracles and which outputs either 0 or 1. Oracles are written as superscripts. The notation $A^{\mathcal{O}_1,\mathcal{O}_2} \Rightarrow 1$ denotes the event that the adversary A, interacts with the oracles $\mathcal{O}_1, \mathcal{O}_2$, and finally outputs the bit 1. In what follows, by the notation $X \xleftarrow{\$} \mathcal{S}$, we will denote the event of choosing X uniformly at random from the set $\mathcal{S}$.

Let $\mathrm{Perm}(n)$ denote the set of all permutations on $\{0,1\}^n$. We require $E(,)$ to be a strong pseudorandom permutation. The advantage of an adversary in breaking the strong pseudorandomness of $E(,)$ is defined in the following manner.

$$\mathbf{Adv}_E^{\pm\mathrm{prp}}(A) = \left| \Pr\left[K \xleftarrow{\$} \mathcal{K} : A^{E_K(),E_K^{-1}()} \Rightarrow 1\right] - \Pr\left[\pi \xleftarrow{\$} \mathrm{Perm}(n) : A^{\pi(),\pi^{-1}()} \Rightarrow 1\right] \right|.$$

Formally, a tweakable enciphering scheme is a function $\mathbf{E} : \mathcal{K} \times \mathcal{T} \times \mathcal{M} \to \mathcal{M}$, where $\mathcal{K} \neq \emptyset$ and $\mathcal{T} \neq \emptyset$ are the key space and the tweak space respectively. The message and the cipher spaces are $\mathcal{M}$. For HCH we have $\mathcal{M} = \cup_{i>n}\{0,1\}^i$. We shall write $\mathbf{E}_K^T(.)$ instead of $\mathbf{E}(K,T,.)$. The inverse of an enciphering scheme is $\mathbf{D} = \mathbf{E}^{-1}$ where $X = \mathbf{D}_K^T(Y)$ if and only if $\mathbf{E}_K^T(X) = Y$.

Let $\mathrm{Perm}^T(\mathcal{M})$ denote the set of all functions $\pi : \mathcal{T} \times \mathcal{M} \to \mathcal{M}$ where $\pi(\mathcal{T},.)$ is a length preserving permutation. Such a $\pi \in \mathrm{Perm}^T(\mathcal{M})$ is called a tweak indexed permutation. For a tweakable enciphering scheme $\mathbf{E} : \mathcal{K} \times \mathcal{T} \times \mathcal{M} \to \mathcal{M}$, we define the advantage an adversary A has in distinguishing $\mathbf{E}$ and its inverse from a random tweak indexed permutation and its inverse in the following manner.

$$\mathbf{Adv}_{\mathbf{E}}^{\pm\widetilde{\mathrm{prp}}}(A) = \left| \Pr\left[K \xleftarrow{\$} \mathcal{K} : A^{\mathbf{E}_K(.,.),\mathbf{E}_K^{-1}(.,.)} \Rightarrow 1\right] - \Pr\left[\pi \xleftarrow{\$} \mathrm{Perm}^T(\mathcal{M}) : A^{\pi(.,.),\pi^{-1}(.,.)} \Rightarrow 1\right] \right|.$$

Pointless queries: We assume that an adversary never repeats a query, i.e., it does not ask the encryption oracle with a particular value of (T,P) more than once and neither does it ask the decryption oracle with a particular value of (T,C) more than once. Furthermore, an adversary never queries its deciphering oracle with (T, C) if it got C in response to an encipher query (T, P) for some P. Similarly, the adversary never queries its enciphering oracle with (T, P) if it got P as a response to a decipher query of (T, C) for some C. These queries are called *pointless* as the adversary knows what it would get as responses for such queries.

Following [4], we define the query complexity σ_n of an adversary as follows. A string X contributes $\max(|X|/n, 1)$ to the query complexity. A tuple of strings $(X_1, X_2, \ldots)$ contributes the sum of the contributions from all oracle queries plus the contribution from the adversary's output. Suppose an adversary makes q queries where the number of n-bit blocks in the ith query is ℓ_i. Then, $\sigma_n = 1 + \sum_{i=1}^{q}(1 + \ell_i) \geq 2q$. Let ρ be a list of resources used by the adversary A and suppose $\mathbf{Adv}_{\mathcal{E}}^{\pm \mathrm{xxx}}(A)$ has been defined where $\mathcal{E}$ is either a block cipher or a tweakable enciphering scheme. $\mathbf{Adv}_{\mathcal{E}}^{\pm \mathrm{xxx}}(\rho)$ denotes the maximal value of $\mathbf{Adv}_{\mathcal{E}}^{\pm \mathrm{xxx}}(A)$ over all adversaries A using resources at most ρ. Usual resources of interest are the running time t of the adversary, the number of oracle queries q made by the adversary and the query complexity σ_n ($n \geq 1$).

The notation HCH[E] denotes a tweakable enciphering scheme, where the n-bit block cipher E is used in the manner specified by HCH. Our purpose is to show that HCH[E] is secure if E is secure. The notation HCH[Perm(n)] denotes a tweakable enciphering scheme obtained by plugging in a random permutation from Perm(n) into the structure of HCH. For an adversary attacking HCH[Perm(n)], we do not put any bound on the running time of the adversary, though we still put a bound on the query complexity σ_n. We show the information theoretic security of HCH[Perm(n)] by obtaining an upper bound on $\mathbf{Adv}_{\mathrm{HCH[Perm}(n)]}^{\pm \widetilde{\mathrm{prp}}}(\sigma_n)$. The upper bound is obtained in terms of n and σ_n. For a fixed block cipher E, we bound $\mathbf{Adv}_{\mathrm{HCH}[E]}^{\pm \widetilde{\mathrm{prp}}}(\sigma_n, t)$ in terms of $\mathbf{Adv}_{E}^{\pm \mathrm{prp}}(\sigma_n, t')$, where $t' = t + O(\sigma_n)$. We use the notation $\mathbf{E}_\pi$ as a shorthand for HCH[Perm(n)] and $\mathbf{D}_\pi$ will denote the inverse of $\mathbf{E}_\pi$. Thus, the notation $A^{\mathbf{E}_\pi, \mathbf{D}_\pi}$ will denote an adversary interacting with the oracles $\mathbf{E}_\pi$ and $\mathbf{D}_\pi$.

4.2   Statement of Result

The following theorem specifies the security of HCH.

Theorem 1. *Fix n, q and $\sigma_n \geq 2q$ to be positive integers and an n-bit block cipher $E : \mathcal{K} \times \{0, 1\}^n \rightarrow \{0, 1\}^n$. Then*

$$\mathbf{Adv}_{\mathrm{HCH[Perm}(n)]}^{\pm \widetilde{\mathrm{prp}}}(\sigma_n) \leq \frac{7\sigma_n^2}{2^n}. \tag{4}$$

$$\mathbf{Adv}_{\mathrm{HCH}[E]}^{\pm \widetilde{\mathrm{prp}}}(\sigma_n, t) \leq \frac{7\sigma_n^2}{2^n} + \mathbf{Adv}_{E}^{\pm \mathrm{prp}}(\sigma_n, t') \tag{5}$$

where $t' = t + O(\sigma_n)$.

The above result and its proof is similar to previous work (see for example [4,5,2]). As mentioned in [4], Equation (5) embodies a standard way to pass from the information theoretic setting to the complexity theoretic setting. (A brief description of how (4) can be obtained from (5) is given in [2].)

For proving (4), we need to consider an adversary's advantage in distinguishing a tweakable enciphering scheme $\mathbf{E}$ from an oracle which simply returns random bit strings. This advantage is defined in the following manner.

$$\mathbf{Adv}^{\pm\mathrm{rnd}}_{\mathrm{HCH}[\mathrm{Perm}(n)]}(A) = \left| \Pr\left[\pi \xleftarrow{\$} \mathrm{Perm}(n) : A^{\mathbf{E}_\pi, \mathbf{D}_\pi} \Rightarrow 1 \right] - \right.$$
$$\left. \Pr\left[A^{\$(.,.),\$(.,.)} \Rightarrow 1 \right] \right|$$

where $\$(.,.)$ returns random bits of length $|M|$. It can be shown that

$$\mathbf{Adv}^{\pm\widetilde{\mathrm{prp}}}_{\mathrm{HCH}[\mathrm{Perm}(n)]}(A) \le \mathbf{Adv}^{\pm\mathrm{rnd}}_{\mathrm{HCH}[\mathrm{Perm}(n)]}(A) + \binom{q}{2}\frac{1}{2^n} \qquad (6)$$

where q is the number of queries made by the adversary. For more details see [4,2]. The main task of the proof now reduces to obtaining an upper bound on $\mathbf{Adv}^{\pm\mathrm{rnd}}_{\mathrm{HCH}[\mathrm{Perm}(n)]}(\sigma_n)$. This proof is provided in Section 5, where we show (see (17)) that for any adversary having query complexity σ_n, we have

$$\mathbf{Adv}^{\pm\mathrm{rnd}}_{\mathrm{HCH}[\mathrm{Perm}(n)]}(\sigma_n) \le \frac{6\sigma_n^2}{2^n}. \qquad (7)$$

Using this and (6), we obtain

$$\mathbf{Adv}^{\pm\widetilde{\mathrm{prp}}}_{\mathrm{HCH}[\mathrm{Perm}(n)]}(\sigma_n) \le \frac{7\sigma_n^2}{2^n}.$$

5   Proof of Theorem 1

Here we prove the upper bound on $\mathbf{Adv}^{\pm\mathrm{rnd}}_{\mathrm{HCH}[\mathrm{Perm}(n)]}(q, \sigma_n)$. We model the adversary's interaction with the oracles $\mathbf{E}_\pi$ and $\mathbf{D}_\pi$ as a game. In the usual game, which we call HCH1, the adversary submits queries to $\mathbf{E}_\pi$ and $\mathbf{D}_\pi$ and gets appropriate answers. Starting from this game, we modify it in successive steps to obtain games where the adversary is provided random bit strings of appropriate lengths. This results in a sequence of games: HCH1, RAND1, RAND2, RAND3 and NON. For lack of space we do not give the complete description of these games. They can be found in an extended version of this paper at the eprint server maintained by IACR.

By an abuse of notation, we will use A^{HCH1} to denote an adversary A's interaction with the oracles while playing game HCH1. We will use similar notations for the other games.

Game HCH1: We describe the attack scenario of the adversary A through a probabilistic game which we call HCH1. In HCH1, the adversary interacts with $\mathbf{E}_\pi$ and $\mathbf{D}_\pi$ where π is a randomly chosen permutation from $\mathrm{Perm}(n)$. Instead of initially choosing π, we build up π in the following manner.

Initially π is assumed to be undefined everywhere. When $\pi(X)$ is needed, but the value of π is not yet defined at X, then a random value is chosen among the available range values. Similarly when $\pi^{-1}(Y)$ is required and there is no X yet defined for which $\pi(X) = Y$, we choose a random value for $\pi^{-1}(Y)$ from the available domain values. Thus, in this game the calls to π and π^{-1} are answered randomly, but maintaining the fact that π and π^{-1} are permutations.

The game HCH1 accurately represents the attack scenario, and by our choice of notation, we can write

$$\Pr[A^{\mathbf{E}_\pi, \mathbf{D}_\pi} \Rightarrow 1] = \Pr[A^{\mathrm{HCH1}} \Rightarrow 1]. \tag{8}$$

Game RAND1: We modify HCH1 to RAND1. In RAND1 the permutation is not maintained. But if there is any collision in the domain or range set of π then a bad flag is set. Thus, the games HCH1 and RAND1 are identical apart from what happens when the bad flag is set. So,

$$|\Pr[A^{\mathrm{HCH1}} \Rightarrow 1] - \Pr[A^{\mathrm{RAND1}} \Rightarrow 1]| \leq \Pr[A^{\mathrm{RAND1}} \text{ sets bad}] \tag{9}$$

Game RAND2: We make certain changes to the game RAND1 which are invisible to the adversary. In this game, for an encryption query, we choose the ciphertext blocks to be random n-bit strings and return to the adversary. Then we adjust the internal variables so as to ensure that the particular choice of ciphertext blocks is consistent as per the protocol. Similarly, for a decryption query, we choose the plaintext blocks to be random n-bit strings and return to the adversary and then adjust the internal variables. This does not alter the adversary's view of the game since for each such change the adversary obtains a random n-bit string both before and after the change. Thus,

$$\Pr[A^{\mathrm{RAND1}} \Rightarrow 1] = \Pr[A^{\mathrm{RAND2}} \Rightarrow 1] \tag{10}$$

also,

$$\Pr[A^{\mathrm{RAND1}} \text{ sets bad}] = \Pr[A^{\mathrm{RAND2}} \text{ sets bad}] \tag{11}$$

In RAND2 the adversary is supplied with random bits as response to queries to both the encrypt and the decrypt oracles. Hence,

$$\Pr[A^{\mathrm{RAND2}} \Rightarrow 1] = \Pr[A^{\$(\cdot,\cdot),\$(\cdot,\cdot)} \Rightarrow 1] \tag{12}$$

Now, from Equation (8), (9), (10), (11) and (12) we get

$$\mathbf{Adv}_{\mathrm{HCH[Perm}(n)]}^{\pm \mathrm{rnd}}(A) = |\Pr[A^{\mathbf{E}_\pi, \mathbf{D}_\pi} \Rightarrow 1] - \Pr[A^{\$(\cdot,\cdot),\$(\cdot,\cdot)} \Rightarrow 1]| \tag{13}$$

$$\leq \Pr[A^{\mathrm{RAND2}} \text{ sets bad}] \tag{14}$$

Our task is thus to bound $\Pr[A^{\mathrm{RAND2}} \text{ sets bad}]$.

Game RAND3: Here we make two subtle changes to the game RAND2. Here instead of the *Domain* and *Range* sets we use multisets $\mathcal{D}$ and $\mathcal{R}$ respectively. In the game RAND3 on either an encryption or a decryption query by the adversary a random string is given as output. Next, the internal variables are adjusted in the first phase of the finalization step. The bad flag is set at the second phase of the finalization step by checking whether a value occurs in either $\mathcal{R}$ or $\mathcal{D}$ more than once. The game RAND3 is shown in Figure 3.

RAND3 sets bad in exactly the same conditions in which RAND2 sets bad, hence

$$\Pr[A^{\text{RAND2}} \text{ sets bad}] = \Pr[A^{\text{RAND3}} \text{ sets bad}]. \tag{15}$$

Respond to the sth adversary query as follows:
ENCIPHER QUERY $\mathsf{Enc}(T^s; P_1^s, P_2^s, \ldots, P_{m^s}^s)$

$\qquad ty^s = \mathsf{Enc}; \; C_1^s||C_2^s||\ldots||C_{m^s-1}^s||D_{m^s}^s \xleftarrow{\$} \{0,1\}^{nm^s};$
$\qquad C_{m^s}^s \leftarrow \mathsf{drop}_{n-r}(D_{m^s}) \; \textbf{return } C_1^s||C_2^s||\ldots||C_{m^s}^s;$
DECIPHER QUERY $\mathsf{Dec}(T^s; C_1^s, C_2^s, \ldots, C_{m^s}^s)$

$\qquad ty^s = \mathsf{Dec}; \; P_1^s||P_2^s||\ldots||P_{m^s-1}^s||V_{m^s}^s \xleftarrow{\$} \{0,1\}^{nm^s};$
$\qquad P_{m^s}^s \leftarrow \mathsf{drop}_{n-r}(V_{m^s}) \; \textbf{return } P_1^s||P_2^s||\ldots||P_{m^s}^s;$

Finalization:

FIRST PHASE

$\qquad$ **if** $T^s = T^t$ for some $t < s$ **then**
$\qquad\qquad R^s \leftarrow R^t;$
$\qquad\qquad$ **if** $l^s = l^t$ **then**
$\qquad\qquad\qquad Q^s \leftarrow Q^t;$
$\qquad\qquad$ **else**
$\qquad\qquad\qquad Q^s \xleftarrow{\$} \{0,1\}^n;$
$\qquad\qquad\qquad \mathcal{D} \leftarrow \mathcal{D} \cup \{R^s \oplus \mathsf{bin}(l^s)\}; \; \mathcal{R} \leftarrow \mathcal{R} \cup \{Q^s\};$
$\qquad\qquad$ **endif**
$\qquad$ **else**
$\qquad\qquad R^s \xleftarrow{\$} \{0,1\}^n;$
$\qquad\qquad \mathcal{D} \leftarrow \mathcal{D} \cup \{T^s\}; \; \mathcal{R} \leftarrow \mathcal{R} \cup \{R^s\};$
$\qquad\qquad Q^s \xleftarrow{\$} \{0,1\}^n;$
$\qquad\qquad \mathcal{D} \leftarrow \mathcal{D} \cup \{R^s \oplus \mathsf{bin}(l^s)\}; \; \mathcal{R} \leftarrow \mathcal{R} \cup \{Q^s\};$
$\qquad$ **endif**

Case $ty^s = \mathsf{Enc}$:	Case $ty^s = \mathsf{Dec}$:
$M_{m^s}^s \leftarrow \mathsf{pad}_{n-r}(P_{m^s}^s);$	$U_{m^s}^s \leftarrow \mathsf{pad}_{n-r}(C_{m^s}^s);$
$M_1^s \leftarrow H_{R^s,Q^s}(P_1^s, P_2^s, \ldots, P_{m^s-1}^s, M_m^s);$	$U_1^s \leftarrow H_{R^s,xQ^s}(C_1^s, C_2^s, \ldots, C_{m^s-1}^s, U_m^s);$
$S^s \xleftarrow{\$} \{0,1\}^n;$	$S^s \xleftarrow{\$} \{0,1\}^n;$
for $i = 2$ to $m^s - 1$,	**for** $i = 2$ to $m^s - 1$,
$\qquad Y_i^s \leftarrow C_i^s \oplus P_i^s;$	$\qquad Y_i^s \leftarrow C_i^s \oplus P_i^s;$
$\qquad \mathcal{D} \leftarrow \mathcal{D} \cup \{f_{i-1}(S^s)\}; \; \mathcal{R} \leftarrow \mathcal{R} \cup \{Y_i^s\};$	$\qquad \mathcal{D} \leftarrow \mathcal{D} \cup \{f_{i-1}(S^s)\}; \; \mathcal{R} \leftarrow \mathcal{R} \cup \{Y_i^s\};$
end for	**end for**
$Y_{m^s}^s \leftarrow D_{m^s}^s \oplus M_{m^s}^s$	$Y_{m^s}^s \leftarrow V_{m^s}^s \oplus U_{m^s}^s$
$\mathcal{D} \leftarrow \mathcal{D} \cup \{f_{m^s-1}(S^s)\}; \; \mathcal{R} \leftarrow \mathcal{R} \cup \{Y_{m^s}^s\};$	$\mathcal{D} \leftarrow \mathcal{D} \cup \{f_{m^s-1}(S^s)\}; \; \mathcal{R} \leftarrow \mathcal{R} \cup \{Y_{m^s}^s\};$
$U_{m^s}^s \leftarrow \mathsf{pad}_{n-r}(C_{m^s}^s);$	$M_{m^s}^s \leftarrow \mathsf{pad}_{n-r}(P_{m^s}^s);$
$U_1^s \leftarrow H_{R^s,xQ^s}(C_1^s, C_2^s, \ldots, C_{m-1}^s, U_{m^s}^s);$	$M_1^s \leftarrow H_{R^s,Q^s}(P_1^s, P_2^s, \ldots, P_{m^s-1}^s, M_m^s);$
$\mathcal{D} \leftarrow \mathcal{D} \cup \{M_1^s\}; \; \mathcal{R} \leftarrow \mathcal{R} \cup \{U_1^s\};$	$\mathcal{D} \leftarrow \mathcal{D} \cup \{M_1^s\}; \; \mathcal{R} \leftarrow \mathcal{R} \cup \{U_1^s\};$
$\mathcal{D} \leftarrow \mathcal{D} \cup \{M_1^s \oplus U_1^s\}; \; \mathcal{R} \leftarrow \mathcal{R} \cup \{S^s\};$	$\mathcal{D} \leftarrow \mathcal{D} \cup \{M_1^s \oplus U_1^s\}; \; \mathcal{R} \leftarrow \mathcal{R} \cup \{S^s\};$

SECOND PHASE
$\qquad$ **if** (some value occurs more than once in $\mathcal{D}$) **then** bad = true **endif**;
$\qquad$ **if** (some value occurs more than once in $\mathcal{R}$) **then** bad = true **endif**.

Fig. 3. Game RAND3

Game NON: In Game RAND3 consider the variable Y_i^s which is defined as

$$Y_i^s = C_i^s \oplus P_i^s, \text{ when } 2 \leq i \leq m^s - 1$$
$$= D_{m^s}^s \oplus M_{m^s}^s, \text{ when } i = m^s \text{ and } ty = \mathsf{Enc}$$
$$= U_{m^s}^s \oplus V_{m^s}^s, \text{ when } i = m^s \text{ and } ty = \mathsf{Dec}$$

For $2 \leq i \leq m^s$, the variable Y_i^s enters the range set and it is always an n-bit random quantity. As, when $2 \leq i \leq m^s - 1$, then for a encryption (resp. decryption) query C_i^s (resp. P_i^s) is a randomly chosen n-bit string. When $i = m^s$, then for an encryption (resp. decryption) query $D_{m^s}^s$ (resp. $V_{m^s}^s$) is a randomly chosen n-bit string. Thus, for $(s, i) \neq (t, j)$, $\Pr[Y_i^s = Y_j^t] = \frac{1}{2^n}$. The condition $Y_i^s = Y_j^t$ for some $(s, i) \neq (t, j)$ leads to a collision in the range and results in bad being set to true. The total probability of bad being set to true due to collisions of this kind is at most $\binom{|\mathcal{R}|}{2}/2^n \leq \sigma_n^2/2^n$. (Note that $|\mathcal{R}| \leq \sigma_n$, where σ_n is the query complexity.) Let $\mathbf{X}$ be the event that bad is set to true in Game RAND3 due to collisions of this kind. Then we have

$$\begin{aligned}
\Pr[A^{\mathrm{RAND3}} \text{ sets bad}] &= \Pr[(A^{\mathrm{RAND3}} \text{ sets bad}) \wedge (\mathbf{X} \vee \overline{\mathbf{X}})] \\
&= \Pr[(A^{\mathrm{RAND3}} \text{ sets bad}) \wedge \mathbf{X}] + \\
&\quad \Pr[(A^{\mathrm{RAND3}} \text{ sets bad}) \wedge \overline{\mathbf{X}}] \\
&= \Pr[(A^{\mathrm{RAND3}} \text{ sets bad})|\mathbf{X}]\Pr[\mathbf{X}] + \\
&\quad \Pr[(A^{\mathrm{RAND3}} \text{ sets bad})|\overline{\mathbf{X}}]\Pr[\overline{\mathbf{X}}] \\
&\leq \Pr[\mathbf{X}] + \Pr[(A^{\mathrm{RAND3}} \text{ sets bad})|\overline{\mathbf{X}}] \\
&\leq \frac{\sigma_n^2}{2^n} + \Pr[(A^{\mathrm{RAND3}} \text{ sets bad})|\overline{\mathbf{X}}].
\end{aligned}$$

Our next task is to upper bound $\Pr[(A^{\mathrm{RAND3}} \text{ sets bad})|\overline{\mathbf{X}}]$. The condition $\overline{\mathbf{X}}$ translates into the fact that we can assume all the Y_i^s's to be distinct. We consider the adversarial behaviour under this condition.

In the previous games, for an encipher query, the adversary specified the tweak and the plaintext; and for a decipher query, he specified the tweak and the ciphertext. We now consider the stronger condition, whereby the adversary specifies the tweak, the plaintext and the ciphertext in both the encryption and the decryption queries subject to the condition that the Y_i^s's are all distinct. For $2 \leq i \leq m^s - 1$, $Y_i^s = P_i^s \oplus C_i^s$ and hence is determined entirely by the transcript. On the other hand, the last block can be partial and hence in this case $Y_{m^s}^s$ is not entirely determined by $P_{m^s}^s \oplus C_{m_s}^s$. There are two ways to tackle this situation. In the first way, we allow the adversary to specify an additional $(n - r^s)$-bit string, which when appended to $(P_{m^s}^s \oplus C_{m_s}^s)$ forms $Y_{m^s}^s$. In the second way, we can generate this $(n - r^s)$-bit string within the game itself. We prefer the first way, since this is notationally simpler. The effect of both the methods are same since we require that Y_i^s's are distinct for all s and $i = 2, \ldots, m^s$.

We do this by modifying the game RAND3 into a new game NON (non-interactive). NON depends on a fixed transcript $\mathrm{tr} = (\mathbf{ty}, \mathbf{T}, \mathbf{P}, \mathbf{C}, \mathbf{E})$ with $\mathbf{ty} = (ty^1, ty^2, \ldots, ty^q)$, $\mathbf{T} = (\mathsf{T}^1, \mathsf{T}^2, \ldots \mathsf{T}^q)$, $\mathbf{P} = (\mathsf{P}^1, \mathsf{P}^2, \ldots \mathsf{P}^q)$, $\mathbf{C} = (\mathsf{C}^1, \mathsf{C}^2, \ldots \mathsf{C}^q)$,

$\mathbf{E} = (\mathsf{E}^1, \mathsf{E}^2, \dots, \mathsf{E}^q)$ where $ty^s = \{\mathsf{Enc}, \mathsf{Dec}\}$, $\mathsf{T}^s \in \{0,1\}^n$, $\mathsf{P}^s = \mathsf{P}^s_1, \dots, \mathsf{P}^s_{m_s}$, $\mathsf{C}^s = \mathsf{C}^s_1, \mathsf{C}^s_2, \dots, \mathsf{C}^s_{m_s}$ and each E^s is a string of length $(n - r^s)$ such that $\mathsf{Y}^s_{m^s} = (\mathsf{P}^s_{m_s} \oplus \mathsf{C}^s_{m_s}) \| \mathsf{E}^s$. If this fixed transcript does not contain pointless queries and satisfies the condition that the Y^s_i's are all distinct, then the transcript is called *allowed*.

Now fix an allowed transcript tr which maximizes the probability of **bad** being set. This transcript tr is hardwired into the game NON. The syntax of NON is the same as the syntax of RAND3, except that the part before the finalization step is not present in NON. The main difference between NON and RAND3 is in the interpretation of the variables. The tweaks, plaintext and ciphertext blocks in RAND3 are given by the adversary while in NON they are part of the transcript tr which is hardwired into the game. We denote this difference by using the symbols $\mathsf{T}^s, \mathsf{P}^s_i, \mathsf{C}^s_i$ and Y^s_i to denote the tweaks, plaintext, ciphertext and the XOR blocks respectively in game NON. We have

$$\Pr[A^{\mathrm{RAND3}} \text{ sets bad}] \leq \Pr[A^{\mathrm{NON}} \text{ sets bad}] + \frac{\sigma_n^2}{2^n}. \tag{16}$$

5.1   Analysis of NON

In the analysis we consider the sets $\mathcal{D}$ and $\mathcal{R}$ to consist of the formal variables instead of their values. For example, whenever we set $\mathcal{D} \leftarrow \mathcal{D} \cup \{X\}$ for some variable X we think of it as setting $\mathcal{D} \leftarrow \mathcal{D} \cup \{``X"\}$ where $``X"$ is the name of that formal variable. This is the same technique as used in [4]. Our goal is to bound the probability that two formal variables in the sets $\mathcal{D}$ and $\mathcal{R}$ take the same value. The formal variables which enter $\mathcal{D}$ and $\mathcal{R}$ are the following:

Elements in $\mathcal{D}$ $\mathsf{T}^s, R^s \oplus \mathrm{bin}(l^s)$,

$M_1^s = Q^s \oplus \mathsf{P}_1^s \oplus R\mathsf{P}_2^s \oplus \cdots \oplus R^{m^s-2}\mathsf{P}_{m^s-1}^s \oplus R^{m^s-1}\mathsf{M}_{m^s}^s$,

$I^s = M_1^s \oplus U_1^s = (x+1)Q^s \oplus Y_1^s \oplus RY_2^s \oplus \cdots \oplus R^{m^s-1}\mathsf{Y}_{m^s}^s$,

$f_1(S^s), f_2(S^s), \dots, f_{m^s-1}(S^s)$.

Elements in $\mathcal{R}$ R^s, Q^s,

$U_1^s = xQ^s \oplus \mathsf{C}_1^s \oplus R\mathsf{C}_2^s \oplus \cdots \oplus R^{m^s-2}\mathsf{C}_{m^s-1}^s \oplus R^{m^s-1}U_{m^s}^s$,

S^s,

$(\mathsf{P}_2^s \oplus \mathsf{C}_2^s), \dots, (\mathsf{P}_{m^s-1}^s \oplus \mathsf{C}_{m^s-1}^s), (\mathsf{M}_{m^s}^s \oplus U_{m^s}^s)$.

The analysis will require a result, which we state and prove in Section A. Now let us consider the probability of **bad** being set under the particular (allowed) transcript hardwired into NON. The variable **bad** can be set either as a result of a collision in the domain or as a result of a collision in the range. We consider these two separately. The number of blocks in a particular adversarial query is the length of the query. Suppose the distinct lengths of the queries made by the adversary are $l_1, \dots, l_p$ and the adversary makes t_k queries of length l_k. Then $\sum_{k=1}^p l_k t_k = \sum_{s=1}^q m^s$.

There are $\binom{|\mathcal{D}|}{2}$ (unordered) pairs of distinct variables in $\mathcal{D}$. We have to consider the probability that such a pair collides, i.e., both the variables of the pair get the same value. We identify the following two types of pairs of variables and call them special pairs.

1. (M_1^s, M_1^t), such that $s \neq t$ and $(T^s, l^s) = (T^t, l^t)$.
2. (I^s, I^t), such that $s \neq t$ and $(T^s, l^s) = (T^t, l^t)$.

The number of pairs of each kind is at most $\sum_{k=1}^{p} \binom{l_k}{2}$. The total number of special pairs is at most $2 \sum_{k=1}^{p} \binom{l_k}{2}$.

First let us consider the collision probability of the special pairs. The total (over all s and t) probability of the pairs of either the first or the second kind giving rise to a collision is given by Proposition 1 in Section A to be at most $\sigma_n^2/2^n$. Note that since queries are not pointless, we will have $(P_1^s, P_2^s, \ldots, P_{m^s-1}^s, P_{m^s}^s) \neq (P_1^t, P_2^t, \ldots, P_{m^t-1}^t, P_{m^t}^t)$ when $(T^s, l^s) = (T^t, l^t)$. Also, in an allowed transcript, all the Y_i^s's (for $1 \leq i \leq m^s - 1$) are distinct. These ensure that we can apply Proposition 1 to the above two cases.

For any non-special pair, the probability of collision is either 0 or equal to $1/2^n$. The actual proof is a tedious case analysis, but is based on a few observations. Which we do not present here due to lack of space.

The total probability of a domain collision is the sum of the probabilities of collision between special pairs and non-special pairs. The total number of non-special pairs is at most $\binom{|\mathcal{D}|}{2})|$. Thus, the total probability of a domain collision is at most

$$\binom{|\mathcal{D}|}{2} \frac{1}{2^n} + \frac{2\sigma_n^2}{2^n} \leq \frac{|\mathcal{D}|^2}{2^n} + \frac{2\sigma_n^2}{2^n} \leq \frac{3\sigma_n^2}{2^n}.$$

Now consider pairs of elements from $\mathcal{R}$. First, leave out the pairs (U_1^s, U_1^t), such that $s \neq t$ and $(T^s, l^s) = (T^t, l^t)$. These are now the special pairs and there are a total of $\sum_{k=1}^{p} \binom{l_k}{2}$ of such pairs. The total (over all s and t) probability of such pairs giving rise to a collision is given by Proposition 1 to be at most $\sigma_n^2/2^n$. Here we use the fact that queries are not pointless to note that $(C_1^s, C_2^s, \ldots, C_{m^s-1}^s, C_{m^s}^s) \neq (C_1^t, C_2^t, \ldots, C_{m^t-1}^t, C_{m^t}^t)$. This ensures that we can apply Proposition 1.

The probability of any non-special pair of elements from $\mathcal{R}$ colliding is either 0 or is equal to $1/2^n$. This analysis is again similar to that for the domain. The additional thing to note is that since the transcript is allowed, we have $(P_i^s \oplus C_i^s) \neq (P_j^t \oplus C_j^t)$ for $(s, i) \neq (j, t)$. In other words, the elements $(P_i^s \oplus C_i^s)$ are all distinct and so the probability of any two such elements colliding is zero.

There are at most $\binom{|\mathcal{R}|}{2}$ non-special pairs for $\mathcal{R}$. As in the case of domain elements, we can now show that the probability of a pair of elements in $\mathcal{R}$ colliding is at most $2\sigma_n^2/2^n$. (Note that the corresponding value for $\mathcal{D}$ has $3\sigma^2$. We get 2 here because there is only one type of special pairs from $\mathcal{R}$.)

Combining the domain and range collision probabilities, we obtain the probability of bad being set to true in NON to be at most $5\sigma_n^2/2^n$. Combining (16), (15) and (14), we have

$$\mathbf{Adv}_{\mathrm{HCH[Perm}(n)]}^{\pm\mathrm{rnd}}(A) \leq \frac{6\sigma_n^2}{2^n}. \tag{17}$$

This completes the proof of Theorem 1. $\square$

6   Conclusion

In this paper, we have presented HCH, which is a new tweakable enciphering scheme. Our approach to the construction is based on the hash-encrypt-hash approach. The important features of HCH are the use of a single key, ability to encrypt arbitrary length messages and a quadratic security bound. To the best of our knowledge, HCH is the first construction to simultaneously achieve all the above three properties. Compared to currently known schemes, HCH is an attractive alternative to a designer of a practical disk encryption algorithm.

References

1. http://csrc.nist.gov/CryptoToolkit/modes/.
2. Debrup Chakraborty and Palash Sarkar. A new mode of operation providing a tweakable strong pseudorandom permutation. In *Fast Software Encryption*, volume 4047 of *Lecture Notes in Computer Science*, pages 293–309. Springer, 2006.
3. Shai Halevi. EME*: Extending EME to handle arbitrary-length messages with associated data. In Anne Canteaut and Kapalee Viswanathan, editors, *IN-DOCRYPT*, volume 3348 of *Lecture Notes in Computer Science*, pages 315–327. Springer, 2004.
4. Shai Halevi and Phillip Rogaway. A tweakable enciphering mode. In Dan Boneh, editor, *CRYPTO*, volume 2729 of *Lecture Notes in Computer Science*, pages 482–499. Springer, 2003.
5. Shai Halevi and Phillip Rogaway. A parallelizable enciphering mode. In Tatsuaki Okamoto, editor, *CT-RSA*, volume 2964 of *Lecture Notes in Computer Science*, pages 292–304. Springer, 2004.
6. Moses Liskov, Ronald L. Rivest, and David Wagner. Tweakable block ciphers. In Moti Yung, editor, *CRYPTO*, volume 2442 of *Lecture Notes in Computer Science*, pages 31–46. Springer, 2002.
7. M. Naor and O. Reingold. A pseudo-random encryption mode. Manuscript available from www.wisdom.weizmann.ac.il/~naor.
8. Moni Naor and Omer Reingold. On the construction of pseudorandom permutations: Luby-Rackoff revisited. *J. Cryptology*, 12(1):29–66, 1999.
9. Peng Wang, Dengguo Feng, and Wenling Wu. HCTR: A variable-input-length enciphering mode. In Dengguo Feng, Dongdai Lin, and Moti Yung, editors, *CISC*, volume 3822 of *Lecture Notes in Computer Science*, pages 175–188. Springer, 2005.
10. Mark N. Wegman and Larry Carter. New hash functions and their use in authentication and set equality. *J. Comput. Syst. Sci.*, 22(3):265–279, 1981.

A   A Useful Result

Let $\mathcal{L} = (\mathcal{L}^1, \ldots, \mathcal{L}^q)$ be a list of vectors, with $\mathcal{L}^{i_1} \neq \mathcal{L}^{i_2}$ for $i_1 \neq i_2$ and where $\mathcal{L}^i = (L_1^i, \ldots, L_{m^i}^i)$ with each L_j^i is an n-bit string considered to be an element of $GF(2^n)$. Let $\mathcal{S}$ be the set of all i, j such that $m^i = m^j$. In other words, if $i, j \in \mathcal{S}$, then $\mathcal{L}^i$ and $\mathcal{L}^j$ have the same number of components. For $\{i, j\} \in \mathcal{S}$, define a polynomial $P_{i,j}(X) \in GF(2^n)[X]$ as

$$P_{i,j}(X) = (L_1^i \oplus L_1^j) \oplus (L_2^i \oplus L_2^j)X \oplus \cdots \oplus (L_{m^j}^i \oplus L_{m^i}^j)X^{m^i-1}. \quad (18)$$

We do not distinguish between $P_{i,j}(X)$ and $P_{j,i}(X)$. The coefficients of this polynomial are elements of $GF(2^n)$. Let $\mathbf{e}_{i,j}$ be the following event. Choose a random element R from $GF(2^n)$ and evaluate $P_{i,j}(R)$: $\mathbf{e}_{i,j}$ is the event $P_{i,j}(R) = 0$. Define

$$\mathbf{e} = \bigvee_{\{i,j\} \in \mathcal{S}} \mathbf{e}_{i,j}. \tag{19}$$

The probability of $\mathbf{e}$ is given by the following proposition whose proof can be found in the extended version of the paper at the eprint server maintained by IACR.

Proposition 1. $\Pr[\mathbf{e}] \leq \frac{1}{2^n} \left(\sum_{j=1}^{q} m^j \right)^2.$

Efficient Shared-Key Authentication Scheme from Any Weak Pseudorandom Function[*]

Ryo Nojima[1], Kazukuni Kobara[2,3], and Hideki Imai[2,3]

[1] Information Security Research Center,
National Institute of Information and Communications Technology,
4-2-1 Nukui Kitamachi, Koganei-shi, Tokyo 184-8795, Japan
`ryo-no@nict.go.jp`
[2] Research Center for Information Security,
National Institute of Advanced Industrial Science and Technology,
1102 Akihabara Daibiru, 1-18-13 Sotokanda, Chiyoda-ku, Tokyo 101-0021, Japan
`{k-kobara, h-imai}@aist.go.jp`
[3] Faculty of Science and Engineering,
Department of Electrical Electronics and Communication Engineering
Chuo University,
1-13-27 Kasuga, Bunkyo-ku, Tokyo 112-8551, Japan

Abstract. One of the most widely used shared-key authentication schemes today is a challenge-response scheme. In this scheme, a function such as a message authentication code or a symmetric encryption scheme plays an important role. To ensure the security, we need to assume that these functions are included in a certain kind of functions family, e.g., a pseudorandom functions family. For example, functions such as SHA1-HMAC, DES and AES often assumed as the pseudorandom functions. But unfortunately, nobody knows that these functions are really pseudorandom functions and if not, then the security of the challenge-response scheme is not ensured any more. The common way to reduce this kind of fear is to construct the shared-key authentication scheme which can be proven secure with a weaker assumption on these functions. In this paper, we show that a *blind-challenge-response* shared-key authentication scheme which is a simple modified version of the original challenge-response authentication scheme can be constructed from a weaker cryptographic assumption known as *weak pseudorandom functions*.

1 Introduction

The challenge-response scheme is one of the most widely used shared-key authentication schemes among our lives. In this scheme, two parties, say Alice and Bob, share a secret key (shared-key) beforehand, and, when Alice wants to authenticate to Bob, Alice proves that she has a key without disclosing it entirely.

---

[*] The essential part of this paper was done when the authors were in the university of Tokyo.

R. Barua and T. Lange (Eds.): INDOCRYPT 2006, LNCS 4329, pp. 303–316, 2006.

The reason why this scheme deploys widely is that it can be implemented easily with small devices such as RFID tags, mobile phones, or even humans [7,8,9].

Intuitively, we say that a shared-key authentication scheme is secure if the adversary, say Eve, who attempts to impersonate Alice, cannot be identified as Alice by Bob. We can classify the security levels more precisely concerning the ability of Eve. The weakest one is that Eve has ability of eavesdropping the interactions between Alice and Bob before impersonation attempt. We say a shared-key authentication scheme is *secure against passive attacks* if it is secure against this type of an adversary. The stronger one is *secure against active attacks* where Eve can actively play the role of Bob, i.e., Eve can interact with Alice numerous times before the impersonation attempt. Security against active attack has been the goal of the shared-key authentication schemes [7]. In this paper, we concentrate on the shared-key authentication scheme which is secure against active attacks.

The challenge-response scheme is secure against active attacks if the function used inside has a certain property which is similar to the pseudorandom functions (PRFs). Good candidacies for the PRFs are DES, AES, or SHA1-HMAC [3]. But unfortunately, nobody knows that these functions are really PRFs and also design criteria of these functions are different from it. In fact, if these functions are not pseudorandom functions then there is a possibility that these functions embedded challenge-response scheme is not secure anymore. The common choice of reducing this kind of fear in the cryptography is to construct the shared-key authentication scheme which can be proven secure with a weaker assumption on them. For instance, consider the following recent situation. SHA1 has conjectured to be a collision-resistant hash function but it seems not [15]. As a result, security of SHA1-HMAC [3] which security was proven under this conjecture becomes danger.[1] Therefore, constructing the cryptographic schemes with a weaker assumption is important not only from the theoretical viewpoint but also from the practical viewpoint.

In this paper, we show an efficient shared-key authentication scheme which can be proven secure with a weak assumption. The scheme is a simple modified version of the challenge-response scheme, named a *blind-challenge-response* authentication scheme. The good property of this scheme is that we can construct and prove the security from any *weak PRF* (WPRF for short). The WPRF was first defined explicitly in [14] and there are many applications [1,6,10,12,13]. Highly efficient candidacies for WPRFs are described in [5]. The WPRFs are not studied extensively as PRFs, but the notion of a WPRF is substantially weaker than the notion of a PRF. Thus, potentially, there must be a lot of WPRFs compared to the PRFs.

RELATED WORKS: As authors know, the blind-challenge-response authentication scheme was first appeared in [7] with a specific function. This scheme has been proposed to be suitable for small devices, such as RFID tags. The function they employ is based on the Learning Parity with Noise (LPN) problem and

---

[1] Later this was repaired [2].

they prove that the scheme is secure against active attacks. Later in [9], Katz et. al. proved that the protocol is secure even if the adversary concurrently accesses to Alice. But they did not say anything about what kind of functions we can employ in the blind-challenge-response scheme. On the other hand, we are interested in a more generic result, that is "what is the sufficient condition for functions to obtain the secure blind-challenge-response scheme?" Thus we can regard our work as a generic framework for constructing the blind-challenge-response scheme.

ORGANIZATION: In Section 2, we define notations and a functions family we employ in this paper. In Section 3, we first describe a blind-challenge-response scheme, and we show the main theorem which states that the blind-challenge-response scheme which employs a WPRF is secure against active attacks. In this section, we will also introduce one of the candidacies of WPRF which was proposed in [14]. We conclude this paper in Section 5.

2   Preliminaries

Let $s \overset{\$}{\leftarrow} S$ denote the operation of selecting s uniformly at random from the set S. If $\mathcal{D}$ is a probability distribution over S then $s \leftarrow \mathcal{D}$ denotes the operation of selecting s at random according to $\mathcal{D}$. Let $\mathcal{U}_n$ denote the uniform distribution over $\{0,1\}^n$. Let $\mathcal{R}_{L,l}$ be the the random functions with a range $\{0,1\}^l$ and a domain $\{0,1\}^L$. $\mathcal{O}^f$ denotes an oracle which, if invoked, returns $(r, f(r))$, where f is a function and r a uniformly at random input of f. Let $H : \{0,1\}^\kappa \times \{0,1\}^n \to \{0,1\}^l$ be a functions family. We regard κ as a security parameter. A functions family we consider in this paper has two algorithms Gen_H and $Eval_H$. $Gen(1^\kappa)$ is a probabilistic polynomial time (for short PPT) algorithm. It takes (unary) security parameter 1^κ as input and outputs a key $k \in \{0,1\}^\kappa$. $Eval_H$ is a deterministic polynomial-time algorithm which takes $k \in \{0,1\}^\kappa$, $x \in \{0,1\}^n$ as input and outputs $H_k(x) \in \{0,1\}^l$. For simplicity we assume that a key k is chosen uniformly at random from $\{0,1\}^\kappa$, i.e., $k \leftarrow \mathcal{U}_\kappa$. We only consider a function family H which has PPT algorithms Gen_H and $Eval_H$.

Intuitively, a WPRF is a function that cannot be efficiently distinguished from a uniform random function when given a sequence of random inputs and the corresponding outputs. More formally, we can define as follows:

Definition 1. *Let $H : \{0,1\}^\kappa \times \{0,1\}^n \to \{0,1\}^l$ be a functions family. The advantage of an algorithm (distinguisher) D is defined as*

$$\mathrm{Adv}_{H,D}^{\mathrm{WPRF}} \overset{\mathrm{def}}{=} \Pr\left[k \leftarrow \mathcal{U}_\kappa, d \leftarrow D^{\mathcal{O}^{H_k}} \mid d = 1\right]$$
$$- \Pr\left[R \leftarrow \mathcal{R}_{n,l}, d \leftarrow D^{\mathcal{O}^R} \mid d = 1\right].$$

And the corresponding maximal advantage as

$$\mathrm{Adv}_H^{\mathrm{WPRF}}(t, q) \overset{\mathrm{def}}{=} \max_D \{\mathrm{Adv}_{H,D}^{\mathrm{WPRF}}\},$$

where the maximum is taken over all D restricted to q invocations of its oracle and the standard time-complexity t.

Informally, a PRF is a function with a secret key that cannot be distinguished from a uniform random function even when a distinguisher can query arbitrary input and can obtain corresponding output. It is easy to see that the every PRF is also a WPRF, but not every WPRF is a PRF. This comes from a simple argument. Let $H : \{0,1\}^\kappa \times \{0,1\}^n \to \{0,1\}^\kappa$ be a WPRF. From H we can construct a function H' which is a WPRF but not a PRF. Let define the function H'_k as follows: if an input is the form 0^n, that is n one's, then a corresponding output is a key k, and otherwise an output is $H(k,x)$, where x is an input. Explicitly, this is a WPRF since the distinguisher cannot chose 0^κ by him/herself, but this is not a PRF since the distinguisher can chose 0^κ and can obtain the key k. The notion of a PRF is very strong and it is not clear whether functions such as block ciphers or message authentication codes proposed in the literature have this very strong security property. On the other hand, the notion of the WPRF is promising compared to the PRF's since, potentially, there must be a lot of WPRFs compared to PRFs.

3   The Blind-Challenge-Response Scheme from WPRFs

From any PRF, we can construct a challenge-response (CR) scheme which is secure against active attacks (in the sense of the definition similar to the definition in this paper), but we show that a WPRF is sufficient to construct an efficient shared-key authentication scheme which we named a *blind-challenge-response* (for short *BCR*) scheme. We show its construction in this section. Differences between the CR scheme and the BCR scheme are that the BCR scheme is three moves protocol instead of two, the number of keys Alice and Bob must share beforehand becomes two instead of one, and also the assumption on the function becomes weaker.

3.1   Construction

In this scheme, we consider two kinds of functions for a WPRF. We named them a one-bit WPRF and a multiple-bit WPRF, respectively. We call a WPRF H is one-bit WPRF if the length of its range is one-bit, that is $l = 1$. Also we call a WPRF H is a multiple-bit WPRF if the length of its range is multiple-bit. More precisely, for the multiple-bit WPRF H, we consider $p_1(\kappa) \leq l \leq p_2(\kappa)$, where p_1 and p_2 are some positive polynomials. The reason why we consider these two is that the proofs are completely different. Proving the BCR scheme with a multiple-bit WPRF is simpler than that of a one-bit WPRF case.

Let H be a one-bit or a multiple-bit WPRF. In this scheme, Alice and Bob share two shared-keys beforehand. We denote these two keys k_b and k_c. Two entities proceed the authentication phase as follows:

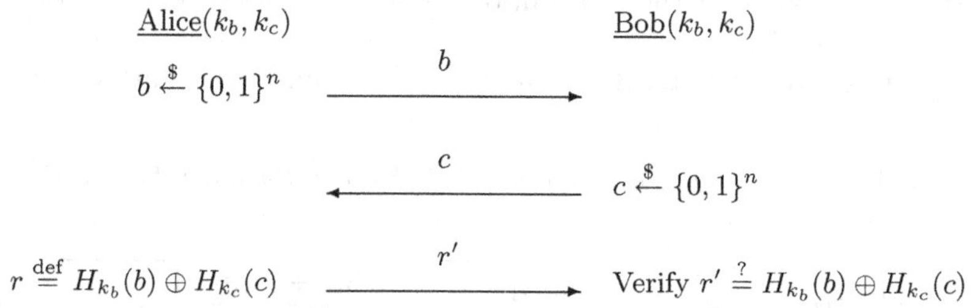

$$\underline{\text{Alice}(k_b, k_c)} \qquad\qquad\qquad\qquad \underline{\text{Bob}(k_b, k_c)}$$

$$b \xleftarrow{\$} \{0,1\}^n \qquad \xrightarrow{\quad b \quad}$$

$$\xleftarrow{\quad c \quad} \qquad c \xleftarrow{\$} \{0,1\}^n$$

$$r \overset{\text{def}}{=} H_{k_b}(b) \oplus H_{k_c}(c) \qquad \xrightarrow{\quad r' \quad} \qquad \text{Verify } r' \overset{?}{=} H_{k_b}(b) \oplus H_{k_c}(c)$$

Fig. 1. The BCR Scheme

- Alice selects a blind $b \in \{0,1\}^n$ randomly and sends it to Bob.
- Bob selects a challenge $c \in \{0,1\}^n$ randomly and sends it to Alice.
- Receiving c, Alice computes a response $r = H_{k_b}(b) \oplus H_{k_c}(c)$ and sends r to Bob.
- Receiving r', Bob verifies whether $r' = H_{k_b}(b) \oplus H_{k_c}(c)$ satisfies or not. If it satisfies then Bob accepts and rejects otherwise.

If H is the one-bit weak pseudorandom function then Alice and Bob proceed the above several times, say m times. As a result the probability of an adversary succeeding impersonation attack will bounded by $\frac{1}{2^m} + \epsilon$, where ϵ is a negligible function with respect to n and κ.

We define the security of the BCR scheme. Let $A_{(k_b, k_c)}$ denote the Alice's algorithm when A holds shared-keys $k_b, k_c \in \{0,1\}^n$, and let $B_{(k_b, k_c)}$ denote the algorithm run by Bob. If Bob accepts with the interaction with Alice then we denote it by $\langle A_{(k_b, k_c)}, B_{(k_b, k_c)} \rangle = 1$ and otherwise $\langle A_{(k_b, k_c)}, B_{(k_b, k_c)} \rangle = 0$.

We consider a two-stage attacker (impersonator) I. Intuitively the first stage of I plays a role of B and tries to obtain the information of k_b and k_c. In the second stage, I tries to impersonate $A_{(k_b, k_c)}$. More formally, we define the success probability of I, denoting $\text{Succ}_{H,I}^{\text{BCR}}$, as follows:

$$\text{Succ}_{H,I}^{\text{BCR}} \overset{\text{def}}{=} \Pr\left[k_b, k_c \leftarrow \mathcal{U}_\kappa, \text{state} \leftarrow I^{A_{(k_b, k_c)}} \mid \langle I(\text{state}), B_{(k_b, k_c)} \rangle = 1 \right].$$

And corresponding maximal probability as

$$\text{Succ}_H^{\text{BCR}}(t, q) \overset{\text{def}}{=} \max_I \{ \text{Succ}_{H,I}^{\text{BCR}} \},$$

where the maximum is taken over all I restricted to q invocation of its oracle and the standard time-complexity t.

We define the advantage of an adversary in both of a one-bit WPRF case and a multiple-bit WPRF case. If H is a multiple-bit WPRF then we define the advantage as

$$\text{Adv}_H^{\text{BCR}}(t, q) \overset{\text{def}}{=} \text{Succ}_H^{\text{BCR}}(t, q).$$

If H is a one-bit WPRF then we define the advantage as

$$\text{Adv}_H^{\text{BCR}}(t, q) \overset{\text{def}}{=} \text{Succ}_H^{\text{BCR}}(t, q) - \frac{1}{2}.$$

Note that the essence of these definitions is identical with the definition (for active attacks) in [7].

We can prove that the BCR scheme is secure against active attacks in both cases.

Theorem 1. *If H is a family of one-bit WPRFs then the advantage of the BCR scheme is*

$$\mathrm{Adv}_H^{\mathrm{BCR}}(t,q) \leq \sqrt{\frac{3}{2}\mathrm{Adv}_H^{\mathrm{WPRF}}(t',q+2) + \frac{3(q+2)(q+1)}{2^{n+2}} + \frac{1}{2^{n+1}}}$$

for any q and t, where $t' \stackrel{\mathrm{def}}{=} 2t + 2t_H + O(l_{\max}(q+1))$.

Theorem 2. *If H is a family of multiple-bit WPRFs then the advantage of the BCR scheme is*

$$\mathrm{Adv}_H^{\mathrm{BCR}}(t,q) \leq \sqrt{3\mathrm{Adv}_H^{\mathrm{WPRF}}(t',q+2) + \frac{3(q+2)(q+1)}{2^{n+1}} + \frac{1}{2^{2(n+1)}} + \frac{1}{2^l} + \frac{1}{2^{n+1}}}$$

for any t and q, where $t' = 2t + 2t_H + O(l_{\max}(q+1))$.

Note that, in Theorem 2, if $l = 1$, then the advantage does not become negligible even when $n, \kappa \to \infty$. This means that the proof does not work if $l = 1$. However, this security proof works when l and n are large. The proofs of Theorem 1 and 2 are appeared in Section 4 and Appendix, respectively.

3.2 Note on WPRFs

Candidacies for one-bit WPRFs are proposed in [5], which states that some "hard to learn" problems, such as learning a DNF formula, can be functions for a WPRF. Also in [14], Naor and Reingold showed an efficient conversion method which transforms from a weak message authentication code (WMAC) into a one-bit WPRF.

Here we give the informal definition of a WMAC. Let $H : \{0,1\}^\kappa \times \{0,1\}^n \to \{0,1\}^l$ be a functions family, and let $\mathrm{Adv}_A^{\mathrm{WMAC}}$ be the advantage of an algorithm A defined as

$$\mathrm{Adv}_A^{\mathrm{WMAC}} \stackrel{\mathrm{def}}{=} \Pr\left[k \leftarrow \mathcal{U}_\kappa, x \leftarrow \mathcal{U}_n, \mathsf{state} \leftarrow A^{\mathcal{O}^{H_k}} \mid A(\mathsf{state}, x) = H_k(x)\right].$$

Definition 2. *We say that H is a WMAC if $\mathrm{Adv}_A^{\mathrm{WMAC}}$ is negligible for every PPT adversary A.*

Note that this definition is a weaker security notion than that of the MAC which satisfies unforgeability against chosen-message attacks (UF-CMA). Generally, UF-CMA is a common goal of security for the MAC, and thus almost all the MAC proposed in literatures are also WMACs.

Intuitively next theorem states that the existence of a WMAC implies the existence of a one-bit WPRF.

Theorem 3 (Theorem 5.1 of [14]). *Let $x, y \in \{0,1\}^l$ be two bit strings, and $x \cdot y$ denotes their inner product.*

Let H be a WMAC and $r \in \{0,1\}^l$ be a random string. Then $H'_r(k, x) \overset{\text{def}}{=} H_k(x) \cdot r$ is a one-bit WPRF, where r is a public information.

Another known result related to the WPRFs is a generic conversion method which transforms a one-bit WPRF to a multiple-bit WPRF [6,11]. Therefore, if we have a one-bit WPRF then we can obtain a multi-bit WPRF as well. But note that using a multiple-bit WPRF which is made from this transformation in the BCR scheme is not recommended since the number of keys each user must hold becomes large.

4   Evaluation of the Security

4.1   Reset and Negative Reset Lemma

The reset lemma was proposed by Bellare and Palacio in [4]. This lemma upper bounds the probability that a cheating impersonator can convince Bob to accept with the probability of a certain experiment. Intuitively, this experiment relates to the behavior of an impersonator who can produce two accepting conversation transcripts. (This lemma was applied only to the public-key identification scheme in [4] but is also applicable to the secret-key setting.)

We modify the original reset lemma of [4] but its essence is completely the same.

Lemma 1 (Reset and Negative Reset Lemma). *Let* state *refer to some state information. Let A be an algorithm who wants to authenticate to B in the BCR scheme, and let α, β the input for A and B, respectively. Let $acc_d(\alpha, \beta)$ be the probability that the B outputs d in its interaction with A, namely the probability that the following experiment returns d:*

Choose random tape R for A; state $\overset{\text{def}}{=} (\alpha, R)$; *$(b, \text{state}) \leftarrow A(\text{state})$*
$c \leftarrow \mathcal{U}_n$; $(r, \text{state}) \leftarrow A(c, \text{state})$; $d' \leftarrow B(b, c, r)$
Return d'

Let $res_d(\alpha, \beta)$ be the probability that the following reset experiment returns d:

Choose random tape R for A; state $\overset{\text{def}}{=} (\alpha, R)$; *$(b, \text{state}) \leftarrow A(\text{state})$*
$c_1 \leftarrow \mathcal{U}_n$; $(r_1, \text{state}_1) \leftarrow A(c_1, \text{state})$; $d_1 \leftarrow B(b, c_1, r_1)$
$c_2 \leftarrow \mathcal{U}_n$; $(r_2, \text{state}_2) \leftarrow A(c_2, \text{state})$; $d_1 \leftarrow B(b, c_2, r_2)$
If $(d_1 = d \wedge d_2 = d \wedge c_1 \neq c_2)$ then return d else return $1 \oplus d$

Then

$$res_d(\alpha, \beta) \geq acc_d(\alpha, \beta)^2 - \frac{acc_d(\alpha, \beta)}{2^n}.$$

Proof Sketch: Let us first consider the case $d = 0$. Let X and Y be random variables. Let X denote the probability, taken over the challenge set $\{0,1\}^n$, that Bob rejects. Let Y denote the probability, taken over the two challenge sets $\{0,1\}^n \times \{0,1\}^n$, that Alice is reset and runs twice, a different challenge is generated each time and Bob rejects each time.

It is obvious that $Y \geq X(X - \frac{1}{2^n})$. From Jensen's inequality,

$$\mathbb{E}[Y] \geq \mathbb{E}\left[X^2 - \frac{X}{2^n}\right] \geq \mathbb{E}[X]^2 - \frac{\mathbb{E}[X]}{2^n},$$

where $\mathbb{E}[Z]$ is an expectation of a random variable Z. Since $\mathbb{E}[X] = acc_0(\alpha, \beta)$ and $\mathbb{E}[Y] = res_0(\alpha, \beta)$,

$$res_0(\alpha, \beta) \geq acc_0(\alpha, \beta)^2 - \frac{acc_0(\alpha, \beta)}{2^n}.$$

The proof for the case $d = 1$ can be done in the similar way.   □

From the reset and negative reset lemma, we can easily derive the following corollary. This corollary becomes a useful tool to prove Theorem 1.

Corollary 1. *Let n, α, β, acc_d, and res_d be described in the above lemma. Then*

$$res_0(\alpha, \beta) + res_1(\alpha, \beta) \geq 2\left(acc_1(\alpha, \beta) - \frac{1}{2}\right)^2 + \frac{1}{2} - \frac{1}{2^n}.$$

4.2   Other Lemmas

To prove the theorems, we need to show two intermediate lemmas, Lemma 2 and Lemma 3. Lemma 2 was proven in [11] and this lemma states that a sequence $b_1, H_{k_b}(b_1), b_2, H_{k_b}(b_2), \cdots, b_q, H_{k_b}(b_q)$ is *pseudorandom* as long as b_i's and k_b are randomly chosen from an appropriate set, respectively.

More formally, let D be an algorithm and let Π_1 denote the following game:

$$b_1, b_2, \cdots b_q \leftarrow \mathcal{U}_n; k_b \leftarrow \mathcal{U}_\kappa;$$
$$d \leftarrow D(b_1, H_{k_b}(b_1), b_2, H_{k_b}(b_2) \cdots, b_q, H_{k_b}(b_q)).$$

Also let Π_0 be the game $z \leftarrow \mathcal{U}_{q(n+l)}; d \leftarrow D(z)$. Let S_i denote the event that $d = 1$ in the game Π_i. The advantage of the distinguisher D is defined as

$$\mathrm{Adv}_{H,D}^{\mathrm{PR}} \stackrel{\mathrm{def}}{=} \Pr[S_1] - \Pr[S_0].$$

And the corresponding maximal advantage as

$$\mathrm{Adv}_H^{\mathrm{PR}}(t, q) \stackrel{\mathrm{def}}{=} \max_D \{\mathrm{Adv}_{H,D}^{\mathrm{PR}}\},$$

where q is the number of a pair of $(b_i, H_{k_b}(b_i))$ and the maximum is taken over all D which time-complexity is up to t.

Lemma 2 (Lemma 1 in [11]). *Let $H : \{0,1\}^\kappa \times \{0,1\}^n \to \{0,1\}^l$ be a WPRF. Then for any t, q,*

$$\mathrm{Adv}_H^{\mathrm{PR}}(t,q) \le \mathrm{Adv}_H^{\mathrm{WPRF}}(t,q) + \frac{q(q-1)}{2^{n+1}}.$$

Intuitively, Lemma 3 states that due to the pseudorandomness of H_{k_b}, distinguishing between H_{k_c} and a random function is still hard for the impersonator I even after I interacts with Alice. More formally, let D be a two-stage algorithm and let consider the following game Λ:

$$k_b, k_c \leftarrow \mathcal{U}_\kappa; \tilde{c}_1, \tilde{c}_2 \leftarrow \mathcal{U}_n; z_1^0, z_2^0 \leftarrow \mathcal{U}_l; \text{state} \leftarrow D^{A_{(k_b,k_c)}};$$

$$z_1^1 \stackrel{\mathrm{def}}{=} H_{k_c}(\tilde{c}_1); z_2^1 \stackrel{\mathrm{def}}{=} H_{k_c}(\tilde{c}_2); d \leftarrow \mathcal{U}_1; d' \leftarrow D(\text{state}, \tilde{c}_1, \tilde{c}_2, z_1^d, z_2^d)$$

The advantage of the distinguisher D is defined as

$$\mathrm{Adv}_{H,D}^\Lambda \stackrel{\mathrm{def}}{=} \Pr[d = d'] - \frac{1}{2}.$$

And the corresponding maximal advantage as

$$\mathrm{Adv}_H^\Lambda(t,q) \stackrel{\mathrm{def}}{=} \max_D \{\mathrm{Adv}_{H,D}^\Lambda\},$$

where the maximum is taken over all D restricted to q invocations of its oracle $A_{(k_b,k_c)}$ and the standard time-complexity t.

Lemma 3. *Let $H : \{0,1\}^\kappa \times \{0,1\}^n \to \{0,1\}^l$ be a WPRF, let t_H the worst case time-complexity of computing H, and $l_{\max} = \max(\kappa, l, n)$.*
Then for any t, q,

$$\mathrm{Adv}_H^\Lambda(t,q) \le \frac{3}{2}\left(\mathrm{Adv}_H^{\mathrm{WPRF}}(t',q+2) + \frac{(q+2)(q+1)}{2^{n+1}}\right),$$

where $t' = t + 2t_H + O(l_{\max} + q l_{\max})$.

Proof: Let assume that an adversary D distinguishes between $(\tilde{c}_1, z_1^1)$, $(\tilde{c}_2, z_2^1)$ and $((\tilde{c}_1, z_1^0), (\tilde{c}_2, z_2^0))$ with the time-complexity being t and the number of queries being q.

We say that D succeeds if and only if $d = d'$. We denote this event by T, and we are ultimately interested in the amount of $\Pr[T]$. We construct a distinguisher D' which distinguishes a random sequence from a sequence of a pair of $(b_i, H_{k_b}(b_i))$ in the sense of Lemma 2. Let such sequence be $(b_1, s_1, b_2, s_2, \cdots, b_q, s_q)$, where each b_i is n-bit and s_i is l-bit. The distinguisher D' works as follows:

1. Generate $k_c \leftarrow \mathcal{U}_\kappa$.
2. On i-th D's oracle invocation, where $1 \le i \le q$:
   Blind Phase: Send b_i to D
   Response Phase: On receiving $c_i \in \{0,1\}^n$, return $s_i \oplus H_{k_c}(c_i)$
3. Receive the state information state from D, and generate $d \stackrel{\$}{\leftarrow} \{0,1\}$

4. Generate $\tilde{c}_1, \tilde{c}_2 \leftarrow \mathcal{U}_n$, and $z_1^0, z_2^0 \leftarrow \mathcal{U}_l$. Also let $z_1^1 \overset{\text{def}}{=} H_{k_c}(\tilde{c}_1), z_2^1 \overset{\text{def}}{=} H_{k_c}(\tilde{c}_2)$
5. Input $(\text{state}, \tilde{c}_1, \tilde{c}_2, z_1^d, z_2^d)$ into D, and obtain d'
6. Output 1 if and only if $d = d'$

Remind that S_0 is the event that the input $b_1, s_1, b_2, s_2, \cdots, b_q, s_q$ is the random sequence, and S_1 is the event that the input is the sequence $b_1, H_{k_b}(b_1)$, $b_2, H_{k_b}(b_2), \cdots, b_q, H_{k_b}(b_q)$. From Lemma 2 and the algorithm D', we know that

$$\Pr[D' = 1 \mid S_1] - \Pr[D' = 1 \mid S_0] \leq \text{Adv}_H^{\text{PR}}(t', q),$$

where $t' \overset{\text{def}}{=} t + 2t_H + O(l_{\max}(1 + q))$.

First, we claim that $\Pr[D' = 1 \mid S_1] = \Pr[T]$. This is because, when S_1 occurs, k_b is used just as this would be in the game Π_1 and D' outputs 1 if and only if $d = d'$. Thus this follows the claim.

To complete the proof, we need to estimate the amount of $\Pr[D' = 1 \mid S_0]$. Since if the event S_0 occurs then what the distinguisher can obtain from its oracle is only a random sequence, D does not obtain any information about the key k_c. Thus, for any t,

$$\Pr[D' = 1 \mid S_0] - \frac{1}{2} \leq \frac{1}{2}\text{Adv}_H^{\text{PR}}(t', 2).$$

Combining these, for any t and q, we have

$$\Pr[T] - \frac{1}{2} = \text{Adv}_H^A(t, q) \leq \text{Adv}_H^{\text{PR}}(t', q) + \frac{1}{2}\text{Adv}_H^{\text{PR}}(t', 2),$$

where $t' = t + 2t_H + O(l_{\max}(1 + q))$. Combining with Lemma 2 concludes the proof as follows:

$$\begin{aligned}
\text{Adv}_H^A(t, q) &\leq \text{Adv}_H^{\text{PR}}(t', q) + \frac{1}{2}\text{Adv}_H^{\text{PR}}(t', 2) \\
&\leq \frac{3}{2}\text{Adv}_H^{\text{PR}}(t', q + 2) \\
&\leq \frac{3}{2}\left(\text{Adv}_H^{\text{WPRF}}(t', q + 2) + \frac{(q + 2)(q + 1)}{2^{n+1}}\right)
\end{aligned}$$

for any t and q. $\square$

4.3   Proof of Theorem 1

Using Lemma 3 with $l = 1$, we can prove Theorem 1.

Assuming an impersonator I attacking the BCR scheme actively. Let assume I accesses to the oracle q times and runs in time t. We denote the event of I succeeding the attack by U. We are interested in the amount of $\Pr[U]$.

We construct a distinguisher D which distinguishes $(\tilde{c}_1, H_{k_c}(\tilde{c}_1)), (\tilde{c}_2, H_{k_c}(\tilde{c}_2))$ and $(\tilde{c}_1, z_1^0), (\tilde{c}_2, z_2^0)$ in the sense of Lemma 3. Note that since the distinguisher D can access to the oracle $A_{(k_b, k_c)}$, there is no difficulty in answering I's oracle queries. We construct the distinguisher D as follows:

1. Invoke the first stage of the impersonator I.
2. Whatever a query received from I, send it to $A_{(k_b,k_c)}$. Also send back its response to I. Repeat this q times
3. Output I's output, i.e., state
   // Second stage of D is invoked with state, $(\tilde{c}_1, z_1^d)$, and $(\tilde{c}_2, z_2^d)$.
   // D needs to decide whether $d = 1$ or $d = 0$.
4. Invoke the second stage of the impersonator I with state
5. On receiving a blind b from I, return a challenge $\tilde{c}_1$, and then receive the corresponding response r_1
6. Rewind I to the step 4
7. On receiving the same blind b from I, return a challenge $\tilde{c}_2$, and then receive its response r_2
8. Output 1 if and only if $r_1 \oplus z_1^d \oplus r_2 \oplus z_2^d = 0$

Let V_0 be the event that $d = 0$ and let V_1 be the event that $d = 1$. We know from Lemma 3 and the algorithm D that

$$\Pr[D = 1 \mid V_1] - \Pr[D = 1 \mid V_0] \leq 2\mathrm{Adv}_H^A(O(l_{\max}(q + 1)) + 2t, q). \quad (1)$$

First, we consider the case that V_0 occurs. We claim that, due to the randomness of z_1^0 and z_2^0, $\Pr[D = 1 \mid V_0] = \frac{1}{2}$. Since z_1^0 and z_2^0 distribute uniformly at random from the viewpoint of I, whatever I outputs, the probability of $r_1 \oplus z_1^0 \oplus r_2 \oplus z_2^0 = 0$ is $\frac{1}{2}$. By this, the claim follows.

Next we concentrate on the case that the event V_1 occurs. That is, we estimate the amount of $\Pr[D = 1 \mid V_1]$. Consider the following two events. The one, say R_1, is that r_1 and r_2 satisfy $r_1 = H_{k_b}(b) \oplus H_{k_c}(\tilde{c}_1)$ and $r_2 = H_{k_b}(b) \oplus H_{k_c}(\tilde{c}_2)$. The other, say R_0, is that r_1 and r_2 satisfy $r_1 = H_{k_b}(b) \oplus H_{k_c}(\tilde{c}_1) \oplus 1$ and $r_2 = H_{k_b}(b) \oplus H_{k_c}(\tilde{c}_2) \oplus 1$. Note that, D outputs 1 if and only if R_0 or R_1 occurs. Thus $\Pr[D = 1 \mid V_1] = \Pr[R_0] + \Pr[R_1]$.

We estimate $\Pr[D = 1 \mid V_1]$. Remind Lemma 1 and let $\alpha = \mathsf{state}$ and $\beta = (k_b, k_c)$. Then from Corollary 1, and also from Jensen's inequality,

$$
\begin{aligned}
\Pr[D = 1 \mid V_1] &= \Pr[R_0] + \Pr[R_1] \\
&= \mathbb{E}\left[res_0(\mathsf{state}, (k_b, k_c))\right] + \mathbb{E}\left[res_1(\mathsf{state}, (k_b, k_c))\right] \\
&\geq 2\mathbb{E}\left[\left(acc_1(\mathsf{state}, (k_b, k_c)) - \frac{1}{2}\right)^2\right] + \frac{1}{2} - \frac{1}{2^n} \\
&\geq 2\left(\mathbb{E}\left[acc_1(\mathsf{state}, (k_b, k_c))\right] - \frac{1}{2}\right)^2 + \frac{1}{2} - \frac{1}{2^n} \\
&= 2\left(\mathrm{Succ}_H^{\mathrm{BCR}}(t, q) - \frac{1}{2}\right)^2 + \frac{1}{2} - \frac{1}{2^n} \\
&= 2\mathrm{Adv}_H^{\mathrm{BCR}}(t, q)^2 + \frac{1}{2} - \frac{1}{2^n},
\end{aligned}
$$

where expectation is taken over the choice of (k_b, k_c) and **state**. Combining the inequality (1), $\Pr[D = 1 \mid V_0] = \frac{1}{2}$ and this, we obtain

$$\mathrm{Adv}_H^{\mathrm{BCR}}(t, q) \le \sqrt{\mathrm{Adv}_H^A(O(l_{\max}(q+1)) + 2t, q) + \frac{1}{2^{n+1}}},$$

for any t and q. With the result of Lemma 3,

$$\mathrm{Adv}_H^{\mathrm{BCR}}(t, q) \le \sqrt{\mathrm{Adv}_H^A(O(l_{\max}(q+1)) + 2t, q) + \frac{1}{2^{n+1}}}$$

$$\le \sqrt{\frac{3}{2}\mathrm{Adv}_H^{\mathrm{WPRF}}(t', q+2) + \frac{3(q+2)(q+1)}{2^{n+2}} + \frac{1}{2^{n+1}}}$$

for any t and q, where $t' = 2t + 2t_H + O(l_{\max}(q+1))$. This concludes the proof.
$\square$

5   Open Problems and Concluding Remarks

In this paper, we showed a shared-key authentication scheme, named the blind-challenge-response scheme, that is secure against active attacks. The good property of our scheme is that the scheme can construct from any WPRF. Thus, we can regard our work as a generic framework for constructing the blind-challenge-response scheme.

As is in [9], we could not treat the man in the middle attack in this paper, but the discussion in [9] states that, for some applications such as RFID-tags, mounting the man-in-the middle attack is not easy for attackers. However, construction of an efficient shared-key authentication scheme secure against man in the middle attack is a still important work for protecting the future communication channel.

References

1. W. Aiello, S. Rajagopalan, R. Venkatesan, *High-Speed Pseudorandom Number Generation with Small Memory*, Fast Software Encryption 1999, LNCS 1636, pp. 290–304, 1999.
2. M. Bellare, *New Proofs for NMAC and HMAC: Security without Collision-Resistance*, 2006. Available from http://www-cse.ucsd.edu/~mihir/papers/hmac-new.html
3. M. Bellare, R. Canetti, H. Krawczyk, *Keying hash functions for message authentication*, CRYPTO 1996, LNCS 1109, pp. 1–15, 1996.
4. M. Bellare, A. Palacio, *GQ and Schnorr Identification Schemes: Proofs of Security against Impersonation under Active and Concurrent Attacks*, CRYPTO 2002, LNCS 2442, pp. 162–177, 2002.
5. A. Blum, M.L. Furst, M.J. Kearns, R.J. Lipton, *Cryptographic Primitives Based on Hard Learning Problems*, CRYPTO 1993, LNCS 773, pp. 278–291, 1993.
6. I. Damgård, J.B. Nielsen, *Expanding Pseudorandom Functions; or: From Known-Plaintext Security to Chosen-Plaintext Security*, CRYPTO 2002, LNCS 2442, pp. 449–464, 2002.

7. A. Juels, S.A. Weis, *Authenticating Pervasive Devices with Human Protocols*, CRYPTO 2005, LNCS 3621, pp. 293–308, 2005.
8. N.J. Hopper, M. Blum, *Secure Human Identification Protocols*, ASIACRYPT 2001, LNCS 2248, pp. 52–66, 2001.
9. J. Katz, J.S. Shin, *Parallel and Concurrent Security of the HB and HB+ Protocols*, EUROCRYPT 2006, LNCS 4004, pp. 73–87, 2006.
10. U.M. Maurer, Y.A. Oswald, K. Pietrzak, J. Sjödin, *Luby-Rackoff Ciphers from Weak Round Functions?*, EUROCRYPT 2006, LNCS 4004, pp. 391–408, 2006.
11. U. Maurer, J. Sjödin, *From Known-Plaintext to Chosen-Ciphertext Security*, Cryptology ePrint Archive: Report 2006/071, 2006.
12. M. Naor, O. Reingold, *Synthesizers and Their Application to the Parallel Construction of Pseudo-Random Functions*, J. Comput. Syst. Sci. 58(2), pp. 336–375, 1999.
13. M. Naor, O. Reingold, *Number-theoretic constructions of efficient pseudo-random functions*, J. ACM 51(2), pp. 231–262, 2004.
14. M. Naor, O. Reingold, *From Unpredictability to Indistinguishability: A Simple Construction of Pseudo-Random Functions from MACs*, CRYPTO 1998, LNCS 1462, pp.267–282, 1998. Available from http://www.wisdom.weizmann.ac.il/~naor/PAPERS/mac_abs.html
15. X. Wang, Y.L. Yin, H. Yu, *Finding Collisions in the Full SHA-1*, CRYPTO 2005, LNCS 3621, pp. 17–36, 2005.

Appendix

Proof of Theorem 2

Main difference with Theorem 1 is $p_1(\kappa) \leq l \leq p_2(\kappa)$, i.e., l is not constant, where p_1 and p_2 are some positive polynomials. Proving the theorem can be done directly from Lemma 3 without using Corollary 1.

Let I be the impersonator which attacks the BCR scheme in time t and in q queries. We construct the distinguisher D in the sense of Lemma 3. The algorithm D is the completely the same with the algorithm described in the proof of Theorem 1.

Let V_0 be the event that $d = 0$ and let V_1 the event that $d = 1$. We know from Lemma 3 and the description of D that

$$\Pr[D = 1 \mid V_1] - \Pr[D = 1 \mid V_0] \leq 2\mathrm{Adv}_H^A(O(l_{\max}(q+1)) + 2t, q). \quad (2)$$

We claim that, due to the randomness of z_1^0 and z_2^0, $\Pr[D = 1 \mid V_0] = \frac{1}{2^l}$. Since z_1^0 and z_2^0 distributes uniformly at random from the viewpoint of I, whatever I outputs, the probability of being $r_1 \oplus z_1^0 \oplus r_2 \oplus z_2^0 = 0$ is $\frac{1}{2^l}$. By this, the claim follows.

To complete the proof, we need to estimate the amount of $\Pr[D = 1 \mid V_1]$. Let W be the event that r_1 and r_2 satisfies $r_1 = H_{k_b}(b) \oplus H_{k_c}(\tilde{c}_1)$ and $r_2 = H_{k_b}(b) \oplus H_{k_c}(\tilde{c}_2)$, respectively. Note that, D outputs 1 if the event W occurs since $r_1 \oplus z_1^d \oplus r_2 \oplus z_2^d = 0$ satisfies with this event. Thus $\Pr[D = 1 \mid V_1] \geq \Pr[W]$.

We estimate the probability. Remind Lemma 1 and let $\alpha = \mathsf{state}$ and $\beta = (k_b, k_c)$. Then

$$
\begin{aligned}
\Pr[D = 1 \mid V_1] &\geq \Pr[W] \\
&= \mathbb{E}\left[res_1(\mathsf{state}, (k_b, k_c))\right] \\
&\geq \mathbb{E}\left[acc_1(\mathsf{state}, (k_b, k_c))^2 - \frac{acc_1(\mathsf{state}, (k_b, k_c))}{2^n}\right] \\
&\geq \mathbb{E}\left[acc_1(\mathsf{state}, (k_b, k_c))\right]^2 - \frac{\mathbb{E}\left[acc_1(\mathsf{state}, (k_b, k_c))\right]}{2^n} \\
&= \mathrm{Adv}_H^{\mathrm{BCR}}(t, q)^2 - \frac{\mathrm{Adv}_H^{\mathrm{BCR}}(t, q)}{2^n} \\
&= \left(\mathrm{Adv}_H^{\mathrm{BCR}}(t, q) - \frac{1}{2^{n+1}}\right)^2 - \frac{1}{2^{2(n+1)}},
\end{aligned}
$$

where expectation is taken over the choice of state and (k_b, k_c). Combining with inequality (2), $\Pr[D = 1 \mid V_0] = \frac{1}{2^l}$ and Lemma 3, we obtain

$$
\begin{aligned}
\mathrm{Adv}_H^{\mathrm{BCR}}(t, q) &\leq \sqrt{2\mathrm{Adv}_H^{\Lambda}(O(l_{\max}(q+1)) + 2t, q) + \frac{1}{2^{2(n+1)}} + \frac{1}{2^l} + \frac{1}{2^{n+1}}} \\
&\leq \sqrt{3\mathrm{Adv}_H^{\mathrm{WPRF}}(t', q+2) + \frac{3(q+2)(q+1)}{2^{n+1}} + \frac{1}{2^{2(n+1)}} + \frac{1}{2^l} + \frac{1}{2^{n+1}}},
\end{aligned}
$$

where $t' = 2t + 2t_H + O(l_{\max} + ql_{\max})$. This concludes the proof. $\square$

A Simple and Unified Method of Proving Indistinguishability
(Extended Abstract)

Mridul Nandi

David R. Cheriton School of Computer Science, University of Waterloo, Canada
m2nandi@cs.uwaterloo.ca

Abstract. Recently Bernstein [4] has provided a simpler proof of indistinguishability of CBC construction [3] which is giving insight of the construction. Indistinguishability of any function intuitively means that the function behaves very closely to a uniform random function. In this paper we make a unifying and simple approach to prove indistinguishability of many existing constructions. We first revisit Bernstein's proof. Using this idea we can show a simpler proof of indistinguishability of a class of DAG based construction [8], XCBC [5], TMAC [9], OMAC [7] and PMAC [6]. We also provide a simpler proof for stronger bound of CBC [1] and a simpler proof of security of on-line Hash-CBC [2]. We note that there is a flaw in the security proof of Hash-CBC given in [2]. This paper will help to understand security analysis of indistinguishability of many constructions in a simpler way.

1 Introduction

This paper deals how one can obtain a simple proof for a bound of distinguishing advantage of two classes of object, mainly two classes of functions. We consider several constructions and show how simply the distinguishing advantage can be obtained. Here we mainly consider distinguishing attack of existing constructions with popularly known *random function* (in this paper, we term it as **uniform random function** [4]). *Indistinguishability* of a construction intuitively means that there is no efficient distinguisher which distinguishes this from the uniform random function. Bernstein has provided a simple proof of indistinguishability of **CBC-MAC** (*Cipher Block Chaining-Message Authentication Code*) [4] which is the main motivation of this paper. We first revisit his proof [4] and show how simply one can extend the proof idea for a class of **DAG** (*Directed Acyclic Graph*) based general construction due to Jutla [8]. This class contains many constructions including CBC and a variant of PMAC [6]. We give a simpler proof of partial result of improved security analysis of CBC-MAC [1]. We also study distinguishing advantage with a different class known as *uniform random on-line function* introduced in Crypto 2001 [2]. We show that same idea of proof is also applicable in this scenario and we obtain a simpler proof of Hash-CBC construction [2]. The idea of all these proofs is based on statistical distribution of the *view of the distinguisher*.

R. Barua and T. Lange (Eds.): INDOCRYPT 2006, LNCS 4329, pp. 317–334, 2006.
© Springer-Verlag Berlin Heidelberg 2006

Thus, it gives information theoretic security and hence the security bound holds for computationally unbounded distinguishers also.

This simple idea can help to understand better about the insight of the construction and can help to come up with very nice constructions and results. For example, we modify slightly the DAG based class due to Jutla [8], so that it will include all known constructions like XCBC [5], TMAC [9], OMAC [7], PMAC [6] etc.

Organization of the paper. In this paper, we first build mathematics for the security bound of the distinguisher in Section 2 which would be used throughout the paper. Then we rewrite the simple proof of security of CBC given by D. J. Bernstein in Section 3 and we show a similar result in case of CBC based on uniform random permutation. In Section 4, we generalize his idea of proof to have a simple proof for a general class proposed by Jutla. We see that security of arbitrary length MAC construction like XCBC, TMAC, OMAC, PMAC etc. can be derived from it. In Section 5 we provide a simpler proof of security of Hash-CBC. We note that in the original paper there is a flaw in the proof. Finally we conclude.

2   Mathematics for Security Proof in Distinguishing Attack

2.1   Different Notion of Distances and Its Cryptographic Significance

(1) Statistical Distance: Let X and Y be two random variables taking values on a finite set S. We define *statistical distance* between two random variables by

$$d_{\text{stat}}(X, Y) := \max_{T \subseteq S} \left| \Pr[X \in T] - \Pr[Y \in T] \right|.$$

Note that, $\Pr[X \in T] - \Pr[Y \in T] = \Pr[Y \notin T] - \Pr[X \notin T]$ and hence $d_{\text{stat}}(X, Y) = \max_{T \subseteq S} \Pr[X \in T] - \Pr[Y \in T]$. It measures the distance between the distribution of the random variables. In fact, it is really a *metric* or *distance function* on the set of all distributions on S. It measures how close their distributions are. For identically distributed random variables X and Y, $d_{\text{stat}}(X, Y) = 0$ and if the random variables are disjoint[1] then the statistical distance is one. In all other cases it lies between zero and one. Now we prove an equivalent definition of statistical distance and study some standard examples. Proof of all lemmas stated in this section are given in Appendix A.

Lemma 1. $d_{\text{stat}}(X, Y) = \Pr[X \in T_0] - \Pr[Y \in T_0] = \frac{1}{2} \times \sum_{a \in S} \left| \Pr[X = a] - \Pr[Y = a] \right|$, *where* $T_0 = \{a \in S : \Pr[X = a] \geq \Pr[Y = a]\}$.

---

[1] X and Y are said to be disjoint if X occurs with some positive probability then Y does occur with probability zero and vice versa. More precisely, there exists a subset T such that $\Pr[X \in T] = 1$ and $\Pr[Y \in T] = 0$.

Example 1. Let X and Y be uniformly distributed on S and $T \subset S$ respectively. Then by Lemma 1, $d_{stat}(X,Y) = \frac{1}{2} \times ((\frac{1}{|T|} - \frac{1}{|S|}) \times |T| + \frac{|S|-|T|}{|S|}) = 1 - \frac{|T|}{|S|}$. Thus, if size of T is very close to S then statistical distance is also very close to zero. On the other hand, if size of T is negligible compare to that of S then statistical distance is close to one.

Example 2. Let $S = \text{Func}(G,G)$ where $\text{Func}(H,G)$ denotes the set of all functions from H to G. Let $T = \text{Func}^{inj}(G,G)$ be the subset containing all injective functions (or permutation since domain and range are same). We say u (or v) is a *uniform random function* (or *uniform random injective function*) if it is uniformly distributed on S (or T respectively). Thus from Example 1 we know that $d_{stat}(u,v) = 1 - \frac{N!}{N^N}$ which is very close to one for large N, where $|G| = N$.

Example 3. Given any distinct $x_1, \cdots, x_k \in G$, let the k-*sampling output* of u be $(u(x_1), \cdots, u(x_k))$ and denoted as $u[k](x_1, \cdots, x_k)$. Let $X = (u(x_1), \cdots, u(x_k))$ and $Y = (v(x_1), \cdots, v(x_k))$. Then we can see that X is uniformly distributed on $S = G^k$ and Y is uniformly distributed on $T = G[k] := \{(y_1, \cdots, y_k) \in G^k : y_i\text{'s are distinct}\}$ and hence (again by Example 1) $d_{stat}(X,Y) = 1 - \frac{N(N-1)\cdots(N-k+1)}{N^k} \approx 1 - \exp^{-k(k-1)/2N}$. Here we note that if $k << \sqrt{N}$ then the statistical distance is very close to zero.

Now, we state two results which will help to give an upper bound of statistical distance of two distributions. If the probability of the event $\{X = a\}$ is not small compare to that of $\{Y = a\}$ for all choices of a (or on a set with high probability) then the statistical distance is also small. More precisely, we have the following two lemmas.

Lemma 2. *Let X and Y be two random variable taking values on S and $\epsilon > 0$. If $\Pr[X = a] \geq (1 - \epsilon) \times \Pr[Y = a]$, $\forall a \in S$ or $\Pr[X = a] \leq (1 + \epsilon) \times \Pr[Y = a]$, $\forall a \in S$ then $d_{stat}(X,Y) \leq \epsilon$.*

Lemma 3. *Let X and Y be two random variables taking values on S. Let for a subset $T \subset S$, $\Pr[X = a] \geq (1 - \epsilon_1) \times \Pr[Y = a]$, $\forall a \in T$ and $\Pr[Y \notin T] \leq \epsilon_2$ then $d_{stat}(X,Y) \leq 2\epsilon_1 + 2\epsilon_2$.*

(2) Computational Distance: The statistical distance is also popularly known as information theoretic distance. In cryptography, there is another notion of distance, known as *computational distance*. Let $\mathcal{A}(\cdot)$ be a *probabilistic algorithm* which runs with an input $a \in S$ and giving output 0 or 1. Define, $\mathcal{A}$-*distance* between X and Y as follows;

$$d^{\mathcal{A}}(X,Y) = |\Pr[\mathcal{A}(X) = 1] - \Pr[\mathcal{A}(Y) = 1]|.$$

Here, $\mathcal{A}(X)$ means the distribution of output of $\mathcal{A}(z)$ where z follows the distribution of X. Similarly for $\mathcal{A}(Y)$. As $\mathcal{A}$ is a probabilistic algorithm it can use a string r chosen from some set $\mathcal{R}$ with a distribution which is *independent* with X and Y. So we consider that $\mathcal{A}$ is having two inputs $r \in \mathcal{R}$ and $z \in S$. We state a fact which shows a relationship between statistical and computational distances. Proof is given in the Appendix A.

Lemma 4. *For any* $\mathcal{A}$, $\mathrm{d}^{\mathcal{A}}(X,Y) \leq \mathrm{d}_{\mathrm{stat}}(X,Y)$. *Conversely, there exists an algorithm* $\mathcal{A}_0$ *(may not be efficient) such that* $\mathrm{d}^{\mathcal{A}_0}(X,Y) = \mathrm{d}_{\mathrm{stat}}(X,Y)$.

In the above proof note that $\mathcal{A}_0$ may not be efficient and does not use any random string. One can consider only deterministic algorithm when it has unbounded computational power. Intuitively, one can make computation for all random choices and choose the random string where it has the best performance. Later, we will show that we can ignore the random string while we distinguish two classes of functions by using unbounded computation.

2.2   Distinguisher of Families of Functions or Random Functions

In this section we describe how a distinguisher can behave. We also show that how the advantage of the distinguisher can be obtained by computing the statistical distance of *view* of the distinguisher.

By random function we mean some distribution on the set $\mathrm{Func}(H,G)$, set of all functions from H to G. In Example 2, we have already defined two random functions, they are uniform random function and uniform random injective function. In cryptography, they are used as ideal candidates. In this paper we will also study another ideal function known as *uniform random on-line injective function*. We will define this in Section 5. Now we follow the notations used in Example 2 and 3. Let f be a random function. For each $\mathbf{x} = (x_1, \cdots, x_k) \in H[k]$, $f[k](\mathbf{x}) = (f(x_1), \cdots, f(x_k))$ follows the distribution induced by the distribution of f. More precisely, for any $\mathbf{y} = (y_1, \cdots, y_k) \in G^k$,

$$\Pr[f[k](\mathbf{x}) = \mathbf{y}] = \sum_{f_0 \in I} \Pr[f = f_0], \quad \text{where} \quad I := \{f \in \mathrm{Func}(H,G) : f[k](\mathbf{x}) = \mathbf{y}\}.$$

Let f and g be two random functions and a distinguisher $\mathcal{D}$ has a function oracle which can be either chosen from f or from g. Distinguisher is behaving as follows.
- First it chooses a random string r from $\mathcal{R}$.
- Based on r it makes query $x_1 := x_1(r) \in H$ and obtains $y_1 \in G$.
- Then it makes queries $x_2 = x_2(r, y_1) \in H$ and obtains $y_2 \in G$ and so on.

Even if x_2 can depend on x_1, it is a function of r and y_1 since x_1 is a function of r only. Thus, x_i is a function of $(r, y_1, \cdots, y_{i-1})$. We say these functions $x_1, x_2, \cdots$ are *query functions* (or $\mathbf{x} = (x_1, \cdots, x_k)$ is k-query function) and the tuple $(y_1, \cdots, y_k) \in G^k$ is the *conditional view* of the distinguisher (condition on the random string r) where k is the number of queries. Note that the output of $\mathcal{D}$ is completely determined by the chosen random string r and the conditional view $(y_1, \cdots, y_k)$. We define the distinguishing advantage of $\mathcal{D}^O$ to distinguish between f and g as $\mathrm{Adv}_{f,g}(\mathcal{D}) = |\Pr[\mathcal{D}^f = 1] - \Pr[\mathcal{D}^g = 1]|$. Define $\mathrm{d}_{f,g}(k) = \max_{\mathcal{D}} \mathrm{Adv}_{f,g}(\mathcal{D})$, where maximum is taken over all oracle algorithms $\mathcal{D}$ which make at most k queries. This denotes the maximum distinguishing advantage for two random functions f and g where the attacker is making at most k queries. Note that there is no restriction on the computational resources of $\mathcal{D}$. We can think $\mathcal{D}$ as a tuple of function $(x_1, \cdots, x_k, \mathcal{A})$ where

x_i's are query functions and $\mathcal{A}$ is the final output function which takes input as $(r, y_1, \cdots, y_k)$. Denote this view without the random string $(y_1, \cdots, y_k)$ by $f[k]_{r,x_1,\cdots,x_k}$ or $g[k]_{r,x_1,\cdots,x_k}$ (in short, $f[k]_{r,\mathbf{x}}$ or $g[k]_{r,\mathbf{x}}$) for the random function f and g respectively. Here, $\mathcal{A}$ is distinguishing two families of random variable $\{f[k]_{r,x_1,\cdots,x_k}\}_{r \in \mathcal{R}}$ and $\{g[k]_{r,x_1,\cdots,x_k}\}_{r \in \mathcal{R}}$. Thus,

$$\mathrm{Adv}_{f,g}(\mathcal{D}) = \Big| \sum_{r \in \mathcal{R}} \Pr[\mathcal{A}(r, f[k]_{r,\mathbf{x}}) = 1] \times \Pr[r] - \sum_{r \in \mathcal{R}} \Pr[\mathcal{A}(r, g[k]_{r,\mathbf{x}}) = 1] \times \Pr[r] \Big|$$

$$= \sum_{r \in \mathcal{R}} \Pr[r] \times \mathrm{d}^{\mathcal{A}}(f[k]_{r,\mathbf{x}}, g[k]_{r,\mathbf{x}})$$

$$\leq \sum_{r \in \mathcal{R}} \Pr[r] \times \mathrm{d}_{\mathrm{stat}}(f[k]_{r,\mathbf{x}}, g[k]_{r,\mathbf{x}})$$

So, given any probabilistic distinguisher $\mathcal{D} = (x_1, \cdots, x_k, \mathcal{A})$ one can define a deterministic distinguisher $\mathcal{D}_0 = (x_1, \cdots, x_k, \mathcal{A}_0)$ such that $\mathrm{Adv}_{f,g}(\mathcal{D}) \leq \mathrm{Adv}_{f,g}(\mathcal{D}_0)$. Here, $\mathcal{D}_0$ chooses a random string r_0 with probability one (i.e., a deterministic algorithm) such that $\mathrm{d}_{\mathrm{stat}}(f[k]_{r,\mathbf{x}}, g[k]_{r,\mathbf{x}}) = \max_{r \in \mathcal{R}} \mathrm{d}_{\mathrm{stat}}(f[k]_{r,\mathbf{x}}, g[k]_{r,\mathbf{x}})$ and $\mathcal{A}_0$ behaves as in Lemma 4. Now we will make following assumptions in this paper.

Assumption 1 (Distinguishers are deterministic): We assume that all distinguishing algorithms are deterministic. Thus, x_1 is a constat and x_i is a function of $(y_1, \cdots, y_{i-1})$.

Assumption 2 (Query functions are distinct): To avoid complicity of notations we use the same notation x_i to denote the function as well as the output of the function. We will assume that all outputs of x_i's (or x_i as a functional value) are distinct (otherwise one can restrict on the set of distinct values of x_i).

Now we use the notation $f[k]_{x_1,\cdots,x_k}$ instead of $f[k]_{r,x_1,\cdots,x_k}$ to denote the view of the distinguisher. We can write that $d_{f,g}(k) = \max_{\mathbf{x}} \mathrm{d}_{\mathrm{stat}}(f[k]_{\mathbf{x}}, g[k]_{\mathbf{x}})$, where maximum is taken over all k-query functions $\mathbf{x} = (x_1, \cdots, x_k)$. Thus, to obtain an upper bound of $d_{f,g}(k)$, it would be enough to bound $\mathrm{d}_{\mathrm{stat}}(f[k]_{\mathbf{x}}, g[k]_{\mathbf{x}})$ for each k-query functions $\mathbf{x}$. The following theorem says how one can obtain this. This theorem has been stated and proved By D. J. Bernstein [4] (a proof is given in Appendix A).

Theorem 1. *If* $\Pr[f[k](\mathbf{a}) = \mathbf{y}] \geq (1 - \epsilon) \times \Pr[g[k](\mathbf{a}) = \mathbf{y}]$ *for each* $\mathbf{a} \in H[k]$ *and* $\mathbf{y} \in G^k$, *then for any* k-*query function* $\mathbf{x} = (x_1, \cdots, x_k)$, $\mathrm{d}_{\mathrm{stat}}(f[k]_{\mathbf{x}}, g[k]_{\mathbf{x}})$ $\leq \epsilon$ *and hence* $d_{f,g}(k) \leq \epsilon$.

3   A Short Proof of the Indistinguishability of CBC Due to D.J. Bernstein [4]

Here, we rewrite the security proof of CBC based on uniform random function given by Bernstein [4]. We also show that the similar result can be obtained for uniform random injective function.

Let f be a function on a group $(G, +)$ (i.e, from $(G, +)$ to $(G, +)$) where $|G| = N$. For $m \geq 1$, define the iterated functions recursively as follow :

$$f^+(g_1, \cdots, g_m) := f_m^+(g_1, \cdots, g_m) = f(f_{m-1}^+(g_1, \cdots, g_{m-1}) + g_m),$$

where $g_i \in G$, $f_0^+() = f_0^+(\lambda) = 0$ and λ is the empty string. Let $\mathbf{x} = (x_1, \cdots, x_k) \in (G^m)^k$ and $(y_1, \cdots, y_k) \in G^k$ where $x_1, \cdots, x_k$ are distinct elements of G^m. We define $\mathbb{P} := \mathbb{P}(\mathbf{x}) \subset G \cup \cdots \cup G^m$, by the set of all non-empty prefixes of x_i's. Note that $|\mathbb{P}(\mathbf{x})| \leq mk$ for any $\mathbf{x} \in (G^m)^k$. Let $\mathbb{P}_1 := \mathbb{P}_1(\mathbf{x}) = \mathbb{P}(\mathbf{x}) \setminus \{x_1, \cdots, x_k\}$.

Example 4. Let $G = \mathbb{Z}_{100}$ and $\mathbf{x} = ((1, 2, 2), (1, 2, 3), (2, 2, 2))$ then $\mathbb{P}(\mathbf{x}) = \{1, 2, (1, 2), (2, 2), (1, 2, 2), (1, 2, 3), (2, 2, 2)\}$ and $\mathbb{P}_1(\mathbf{x}) = \{1, 2, (1, 2), (2, 2)\}$.

We fix any $\mathbf{x}$. Given any f, define the *intermediate induced output function* (or simply induced output function) $op_f : \mathbb{P}_1(\mathbf{x}) \to G$ as $op_f(p) = f^+(p)$. Any function from $\mathbb{P}_1(\mathbf{x})$ to G is called as output function. Note that all output functions may not be an induced output function. We characterize the output functions which are induced output functions. Given op define a *corresponding input function* $ip : \mathbb{P} \to G$ such that

$$\left. \begin{array}{ll} ip(p) = op(\text{chop}(p)) + \text{last}(p) & \text{if } p \notin G \\ = p & \text{if } p \in G \end{array} \right\} \tag{1}$$

where if $p = (q, g') \in G^i$, $\text{chop}(p) := q \in G^{i-1}$, $\text{last}(p) := g' \in G, i \geq 2$.

Lemma 5. *Let op be an output function and ip be its corresponding input function. An output function op is an induced output function if and only if $op(p_1) = op(p_2)$ whenever $ip(p_1) = ip(p_2)$. In particular, op is an induced output function if corresponding input function is injective (the above condition is vacuously true).*

Proof. Given any f, $op_f(p) = f^+(p) = f(ip(p))$ where ip is the corresponding input function of op_f. Thus, the converse of the statement is also true. Now we prove the forward implication of the Lemma. Given any op and its corresponding input function ip, we define

$$\left. \begin{array}{ll} f(x) = op(p) & \text{if } ip(p) = x \\ = * & \text{otherwise} \end{array} \right\} \tag{2}$$

Here, $*$ means that we can choose any arbitrary element from G. This is well defined as $ip(p_1) = ip(p_2) = x$ implies $op(p_1) = op(p_2)$. Recursively, one can check that $f^+(p) = op(p)$ and hence $op = op_f$. ∎

Example 4. (contd.) Let $op(1) = op(1, 2) = 99$, $op(2) = 1$ and $op(2, 2) = 0$. Note that it satisfies the condition of above Lemma. For example, $op(1) = op(1, 2)$ where, $ip((1, 2)) = ip(1) = 1$. Thus for any f such that $f(1) = 99, f(2) = 1$ and $f(3) = 0$, $op_f = op$. Here note that $ip((1, 2, 2)) = 1$, $ip((1, 2, 3)) = 2$ and $ip((2, 2, 2)) = 2$. So, for this output function and for any f such that $op_f = op$, we have $f^+((1, 2, 2)) = 99$, $f^+((1, 2, 3)) = 1$ and $f^+((2, 2, 2)) = 1$.

Following lemma count the number of functions which induce a given induced output function.

Lemma 6. *Let op be an induced output function such that* $|ip(\mathbb{P}_1)| = q$ *where ip is the corresponding input function and* $ip(\mathbb{P}_1) = \{ip(p) : p \in \mathbb{P}_1\}$ *is the range of it. Then there are exactly* N^{N-q} *many f such that* $op = op_f$.

Proof. This is immediate from the construction of f in Equation 2. ∎

Corollary 1. *If op is an output function such that corresponding input function ip is injective then there are* $N^{N-|\mathbb{P}_1|}$ *many f's such that* $op_f = op$ *and there are* $N^{N-|\mathbb{P}_1|-k}$ *many f's such that* $op_f = op$ *and* $f^+[k](\mathbf{x}) = \mathbf{y}$.

Example 4. (contd.) In this example, $ip(\mathbb{P}_1) = \{1, 2, 3\}$ and hence we have 100^{97} many f's such that $op_f = op$. More precisely, all functions f such that $f(1) = 99, f(2) = 1$ and $f(3) = 0$ hold.

Now we give a lower bound of the number of output functions such that corresponding input function is injective. For each $p_1 \neq p_2 \in \mathbb{P}$, let C_{p_1,p_2} be the set of all output functions such that the corresponding input function has same value on p_1 and p_2. Let C be the set of all output functions such that the induced input function is not injective. Thus, $C = \bigcup_{p_1 \neq p_2 \in \mathbb{P}} C_{p_1,p_2}$. Now for each $p_1 \neq p_2$ with $p_1 = (q_1, g_1)$ and (q_2, g_2) where $g_i \in G$,

$$C_{p_1,p_2} = \{op ; \quad op(q_1) - op(q_2) = g_2 - g_1\} \text{ if } q_1 \neq q_2 \atop = \emptyset \qquad\qquad\qquad\qquad\qquad\qquad \text{ if } q_1 = q_2 \} \tag{3}$$

Here, we define $op() = 0$. So we obtain that $|C_{p_1,p_2}| \leq N^{|\mathbb{P}_1|-1}$ and hence $|\neg C| \geq N^{|\mathbb{P}_1|}(1 - \frac{|\mathbb{P}|(|\mathbb{P}|-1)}{2N})$ (note that the total number of output functions is $N^{|\mathbb{P}_1|}$). Let $E = \{f \in \text{Func}(G, G) ; \quad f_m^+[k](\mathbf{x}) = \mathbf{y}\}$ then by Corollary 1

$$|E| \geq |\neg C| \times N^{N-|\mathbb{P}_1|-k} \geq N^{N-k}(1 - \frac{|\mathbb{P}|(|\mathbb{P}|-1)}{2N}).$$

Thus,

$$\Pr[u^+[k](x_1, \cdots, x_k) = (y_1, \cdots, y_k)] \geq \frac{(1 - \epsilon)}{N^k},$$

where u is a uniform random function and $\epsilon = \frac{mk(mk-1)}{2N}$ since we have $|\mathbb{P}_1| \leq mk$. By Theorem 1 we have the following main Theorem of this section.

Theorem 2. *For any* $\mathbf{x} = (x_1, \cdots, x_k) \in G[k]$ *and* $\mathbf{y} = (y_1, \cdots, y_k) \in G^k$ *we have* $\Pr[u^+[k](\mathbf{x}) = \mathbf{y}] \geq \frac{(1-\epsilon)}{N^k}$, *where* $\epsilon = \frac{mk(mk-1)}{2N}$. *We also have,* $d_{\text{stat}}(u_m^+[k]_{\mathbf{x}}, u^{(m)}[k]_{\mathbf{x}}) \leq \frac{mk(mk-1)}{2N}$ *and hence* $d_{u_m^+,u^{(m)}}(k) \leq \frac{mk(mk-1)}{2N}$ *where* $u^{(m)}$ *is the uniform random function on* $\text{Func}(G^m, G)$ *and* $\mathbf{x}$ *is any k-query function.*

3.1   CBC Based on Uniform Random Injective Function

In the original CBC security is provided based on uniform random injective function or uniform random permutation. Here we prove a similar result for uniform random injective function v. The proof is exactly same except in the place of counting the set $\{v : v^+[k](\mathbf{x}) = \mathbf{y}\}$, where y_i's are distinct. So we fix any $\mathbf{y} \in G[k]$. Let for each $p_1 \neq p_2 \in \mathbb{P}$, $C^1_{p_1,p_2}$ be the set of all output functions op such that $op(p_1) = op(p_2)$ and $C^1 = \bigcup_{p_1 \neq p_2} C^1_{p_1,p_2}$. We define $C^* = C \cup C^1$. Thus, $op \notin C^*$ means that both input and output functions are injective. It is easy to check that $|C^1| \leq N^{|\mathbb{P}_1|-1} \times \frac{(|\mathbb{P}_1|)(|\mathbb{P}_1|-1)}{2}$ and hence we have

$$- |\neg C^*| \geq N^{|\mathbb{P}_1|} \times (1 - \frac{(mk-k)(mk-k-1)}{2N} - \frac{mk(mk-1)}{2N}) \geq N^{|\mathbb{P}_1|} \times (1 - \frac{mk(mk-1)}{N}).$$

We have a similar result like Corollary 1. For each $op \notin C^*$, there are exactly $\frac{N!}{(N-|\mathbb{P}|)!}$ many injective f's which induces op and $f^+[k](\mathbf{x}) = \mathbf{y}$ (see the constructions of all f in Equation 2 in the proof of Lemma 5). Thus,

$$|\{f \in \mathrm{Func}^{\mathrm{inj}}(G,G) : f^+(\mathbf{x}) = \mathbf{y}\}| \geq N^{|\mathbb{P}_1|} \times (1 - \epsilon_1) \times \frac{N!}{(N - |\mathbb{P}_1| - k)!}$$

where $\epsilon_1 = \frac{mk(mk-1)}{N}$. Hence, $\Pr[v^+[k](\mathbf{x}) = \mathbf{y}] \geq N^{-k} \times (1 - \epsilon_1)$ for all $\mathbf{y} \in T := G[k] = \{\mathbf{y} \in G^k : y_1, \cdots, y_k \text{ are distinct}\}$ and $\mathbf{x} \in G^m[k]$. Now we have, $\Pr[u^{(m)}[k](\mathbf{x}) \notin T] \leq \frac{k(k-1)}{2N}$. Thus by Lemma 3 we have,

$$d_{\mathrm{stat}}(v_m^+[k]_\mathbf{x}, u^{(m)}[k]_\mathbf{x}) \leq \frac{k(k-1)}{N} + \frac{2mk(mk-1)}{N}$$

for any k-query functions $\mathbf{x}$ and hence

$$d_{v_m^+,u^{(m)}}(k) \leq \frac{k(k-1)}{N} + \frac{2mk(mk-1)}{N}.$$

Theorem 3. $d_{\mathrm{stat}}(v_m^+[k]_\mathbf{x}, u^{(m)}[k]_\mathbf{x}) \leq \frac{k(k-1)}{N} + \frac{2mk(mk-1)}{N}$ *for any k-query function* $\mathbf{x} = (x_1, \cdots, x_k)$ *and hence* $d_{v_m^+,u^{(m)}}(k) \leq \frac{k(k-1)}{N} + \frac{2mk(mk-1)}{N}$.

4 DAG (Directed Acyclic Graph) Based PRF [8]

In this section, we state a class of PRF based on DAG proposed by Jutla [8]. We modify the class slightly so that it contains many known constructions like PMAC, OMAC, TMAC, XCBC etc. The security analysis would be immediate from that of the general class. We first give some terminologies related to DAG.

Terminologies on DAG: Let $D = (V, E)$ be a directed acyclic graph with finite vertex set V and edges E. We say that $u \prec v$ if there is a directed path from u to v. Note that it is a partial order on V. Let D have exactly one sink node v_f (the maximum element with respect to $\prec$) and at most two source nodes (the minimum element with respect to $\prec$). If there are two such we call them as v_s and v_{iv}. In the original paper, Jutla considered only one source node. Here we extend it to two so that it can contain one more source node for initial value.

- For each node $v \in V$, define D_v by the subgraph induced by the vertex set $V_v = \{u : u \prec v\}$. We define, $N(v) = \{u \in V : (u, v) \in E\}$, the *neighborhood* of v.

- Any map $c : E \to \mathcal{M}$ is said to be color map on D where $\mathcal{M}$ is a field. A colored DAG is pair (D, c) where c is a color map on D.
- Two colored DAG (D_1, c_1) and (D_2, c_2) are said to be *isomorphic* if there is a graph isomorphism between D_1 and D_2 which preserves the color map. More precisely, a graph isomorphism $\rho : D_1 \to D_2$ satisfies $c_2(\rho(e)) = c_1(e) \ \forall \ e \in E_1$. In this case we write $(D_1, c_1) \cong (D_2, c_2)$.

Definition 1. *We say a colored graph* $G = (D, c)$ *is non-singular if for all* $u, v \in V$, $G_u := (D_u, c[u]) \cong (D_v, c[v]) := D_v$ *implies either* $u = v$ *or* $\{u, v\} = \{v_s, v_{iv}\}$ *with* $c(v_s, w) \neq c(v_{iv}, w)$ *whenever* (v_s, w) *and* $(v_{iv}, w) \in E$. *Here the color map* $c[u]$ *is the restriction of* c *on* D_u.

Definition 2. *We say a sequence of colored graph* $\mathcal{S} = \langle G^l = (D^l, c^l) = ((V^l, E^l) , c^l) \rangle_{l \geq 1}$ *is PRF-preserving if each* D^l *is non-singular and* $G^l \not\cong G_u^{l'} = (D_u^{l'}, c^{l'}[u])$ *for* $u \in V^{l'}$ *and* $l' \neq l$.

Functional Representation of Message: Given a sequence of colored graph $\langle G^l = (D^l, c^l) \rangle_{l \geq 1}$, let $U^l = V^l \backslash \{v_{iv}^l\}$. We fix a sequence of initial values $iv_l \in \mathcal{M}$, $l \geq 1$. Let $X : U^l \to \mathcal{M}$ be a function, called as a message function. We define its corresponding message-initial value function $\overline{X}$ on G^l as follows :

$$\left. \begin{aligned} \overline{X}(v) &= X(v) \text{ if } v \in U^l \\ &= iv_l \quad \text{ if } v = v_{iv}^l \end{aligned} \right\} \tag{4}$$

In the definition of $\overline{X}$ we include the graph G^l as a domain even if it is defined only on the set of vertices. Here, we look message in $\mathcal{M}^l$ as a message function on G^l. For any well order $<$ on U^l we can correspond $\mathcal{M}^l$ with a message function on U^l where $|U^l| = l$. Namely, $X(u_1) \| \cdots \| X(u_l) \in \mathcal{M}^l$ where $u_1 < \cdots < u_l$ and $U^l = \{u_1, \cdots, u_l\}$. Later we will see that each node of the DAG has the underlying function f. The input for the invocation of f at any node is the sum of previous output (outputs of neighborhood nodes) and the value of message-initial value function $\overline{X}$ at that node.

PRF (Pseudo Random Function) Domain Extension Algorithm: Let $f : \mathcal{M} \to \mathcal{M}$ be a function, $(\mathcal{M}, +, \cdot)$ be a field with $|\mathcal{M}| = N$. Let $\mathcal{S} = \langle G^l \rangle_{l \geq 1}$ be a PRF-preserving sequence of DAG. Given any $X : U^l \to \mathcal{M}$ we have message-initial value function, $\overline{X} : V^l \to \mathcal{M}$. We define two functions, $a_f, b_f : V^l \to \mathcal{M}$ recursively as follows :

$$a_f(v) = \overline{X}(v) + \sum_{w \in N(v)} c^l((w, v)) \cdot b_f(w) \quad \text{and} \quad b_f(v) = f(a_f(v)), v \in V^l.$$
$$\tag{5}$$

The output of $f^{\mathcal{S}}(X)$ is $b_f(v_f^l)$ where v_f^l is the unique sink node. When v is a source node, $N(v) = \emptyset$ and hence $a_f(v) = X(v)$.

Security Analysis: Two message-input functions on colored DAG, $X_1 : G_1 \to \mathcal{M}$ and $X_2 : G_2 \to \mathcal{M}$ are said to be *identical* if $G_1 \cong G_2$ and $X_1(u) = X_2(v)$ where v is the image of u under a graph isomorphism. If not then we say that they are *non-identical*. We identify all identical message-functions. Given $v \in V$ and a message-initial value function on $G = (D, c)$ we define $X[v]$ by the function X restricted on G_v.

Let $X_1, \cdots, X_k$ be k distinct functions, $X_i : U^{l_i} \to \mathcal{M}$ and let $\overline{X_i}$ be its corresponding message-initial value function. Let $\mathbb{P} := \mathbb{P}(\mathbf{X}) = \{X : X = X_i[v], v \in V^{l_i}\}$ where $\mathbf{X} = (X_1, \cdots, X_k)$. We call this also prefix set for $\mathbf{X}$. This is a generalized notion for prefixes of messages in CBC case (see Section 3). Here we similarly have $|\mathbb{P}| \leq Q$, where Q is the total number of message blocks from $\mathcal{M}$. Now we make similar analysis like CBC.

We fix any $\mathbf{X}$. Given any f, define the *intermediate induced output function* (or simply induced output function) $op_f : \mathbb{P}_1(\mathbf{X}) \to \mathcal{M}$ as $op_f(p) = b_f(v)$ where $p = X_i[v]$ and b_f is given as in Equation 5 while we compute $f^{\mathcal{S}}(\overline{X})$ using the colored graph G^{l_i}. Any function from $\mathbb{P}_1(\mathbf{X})$ to $\mathcal{M}$ is called as output function. Let $p = X_i[v] \in \mathbb{P}$, define $\mathrm{last}(p) = X_i(v)$ and $\mathrm{chop}(p) = \{X_i[u] : u \in N(v)\}$. It is an empty set for source node v. Let $X_i[u] = q \in \mathrm{chop}(p)$, then we denote the edge (u, v) by $e_{q,p}$. Given op, define a corresponding *input function* $ip : \mathbb{P} \to \mathcal{M}$ as

$$ip(p) = \mathrm{last}(p) + \sum_{q \in \mathrm{chop}(p)} c^{l_i}(e_{q,p}) \cdot op(q).$$

Now we state a analogous statement of Lemma 6 and Corollary 1

Lemma 7. *Let op be an induced output function such that $|ip(\mathbb{P}_1)| = q$ where ip is the corresponding input function and $ip(\mathbb{P}_1)$ is the range of it. Then for any $\mathbf{y} = (y_1, \cdots, y_k) \in G^k$ there are exactly N^{N-q} many f such that $op = op_f$ and $f^{\mathcal{S}}[k](\mathbf{x}) = \mathbf{y}$.*

Corollary 2. *If op is an induced output function such that corresponding input function ip is injective then there are $N^{N-|\mathbb{P}_1|}$ many f's such that $op_f = op$ and there are $N^{N-|\mathbb{P}_1|-k}$ many f's such that $op_f = op$ and $f^+[k](\mathbf{x}) = \mathbf{y}$.*

Now we give a lower bound of the number of output functions such that corresponding input function is injective. For each $p_1 \neq p_2 \in \mathbb{P}$, let C_{p_1,p_2} be the set of all output functions such that the induced input function has same value on p_1 and p_2. Let C be the set of all output functions such that the induced input function is not injective. Thus, $C = \bigcup_{p_1 \neq p_2 \in \mathbb{P}} C_{p_1,p_2}$. Let $X_{i_1}[v_1] = p_1 \neq p_2 = X_{i_2}[v_2]$, $\mathrm{chop}(p_1) = \{q_i = X_{i_1}[u_i] : 1 \leq i \leq l\}$ and $\mathrm{chop}(p_2) = \{q_i' = X_{i_2}[w_i] : 1 \leq i \leq l'\}$. Now we have three possible cases as given below :

Case-1: $\mathrm{chop}(p_1) = \mathrm{chop}(p_2) = \{q_1, \cdots, q_l\}$ and $c_1(e_{q_i,p_1}) = c_2(e_{q_i,p_2})$, $\forall\, i$ where c_i is the color function corresponding to p_i. Then the underlying graphs for p_1 and p_2 are identical. Since $p_1 \neq p_2$, $X(v_1) \neq X(v_2)$ and hence $C_{p_1,p_2} = \emptyset$.

Case-2: Let $\text{chop}(p_1) = \text{chop}(p_2) = Q$ but there exists $q \in Q$ such that $c_1(e_{q,p_1}) \neq c_2(e_{q_i,p_2})$. Then $ip(p_1) = ip(p_2)$ implies $X_{i_1}(v_1) + \sum_{q \in \text{chop}(p_1)} e_{q,p} \cdot op(q) = X_{i_2}(v_2) + \sum_{q \in \text{chop}(p_2)} e_{q,p} \cdot op(q)$. Hence, $\sum_{q \in Q} a_q \cdot op(q) = a$ for some constants a_q and a where all a_q's are not zero (since color functions are different on Q). Thus, $|C_{p_1,p_2}| = N^{|\mathbb{P}_1|-1}$.

Case-3: Let $\text{chop}(p_1) \neq \text{chop}(p_2)$. In this case $ip(p_1) = ip(p_2)$ implies $\sum_{q \in Q} a_q \cdot op(q) = a$ where Q is not empty and all a_q's are not zero. Thus, $|C_{p_1,p_2}| = N^{|\mathbb{P}_1|-1}$.

So we obtain that $|C_{p_1,p_2}| \leq N^{|\mathbb{P}_1|-1}$ and hence $|\neg C| \geq N^{|\mathbb{P}_1|}(1 - \frac{|\mathbb{P}|(|\mathbb{P}|-1)}{2N})$. By Corollary 1

$$|E| \geq |\neg C| \times N^{N-|\mathbb{P}_1|-k} \geq N^{N-k}(1 - \frac{|\mathbb{P}|(|\mathbb{P}|-1)}{2N}),$$

where $E = \{f \in \text{Func}(G, G) ; \; f^{\mathcal{S}}[k](\mathbf{X}) = \mathbf{y}\}$. Thus,

$$\Pr[u^{\mathcal{S}}[k](\mathbf{X}) = \mathbf{y}] \geq \frac{(1-\epsilon)}{N^k},$$

where $\epsilon = \frac{Q(Q-1)}{2N}$ and u is a uniform random function. By Theorem 1 we have the following main Theorem of this section.

Theorem 4. *For any* $\mathbf{X} = (X_1, \cdots, X_k)$ *and* $\mathbf{y} = (y_1, \cdots, y_k) \in G^k$ *where* X_i's *are distinct message function, we have* $\Pr[u^{\mathcal{S}}[k](\mathbf{X}) = \mathbf{y}] \geq \frac{(1-\epsilon)}{N^k}$, *where* $\epsilon = \frac{Q(Q-1)}{2N}$ *and* Q *is the total number of message blocks in queries. We also have,* $d_{\text{stat}}(u^{\mathcal{S}}[k]_{\mathbf{X}}, U[k]_{\mathbf{X}}) \leq \frac{mk(mk-1)}{2N}$ *and hence* $d_{u^{\mathcal{S}},U}(k) \leq \frac{mk(mk-1)}{2N}$ *where* U *is the uniform random function from the set of all message functions to* $\mathcal{M}$.

Remark 1. The same security analysis can be made for the PRF based on a uniform random injective function like CBC case. We leave the details to the reader as it is very much similar to the CBC case.

Remark 2. Let $\mathcal{M} = \{0,1\}^n := \text{GF}(2^n)$. To define a pseudo random function on $\{0,1\}^*$ one can pad 10^i (for minimum $i \geq 0$) so that the length is the multiple of n and then can apply the PRF algorithm as above. So for any distinct messages, the padded messages are also distinct and hence it would be a pseudo random function on the input set $\{0,1\}^*$. There is another way to pad it. We pad 10^i to a message X if it is not a multiple of n, otherwise we would not pad anything (this is the case for OMAC,TMAC, XCBC etc.). In this case we have two sequences of colored graph G_1^l and G_2^l (for all messages with size multiple of n and all messages with size not multiple of n respectively). Here, we require the combined sequence $\langle G_1^l, G_2^l \rangle_{l \geq 1}$ is *PRF-preserving* (thus, even if after padding the messages are equal the corresponding message functions are not *identical*). The similar analysis also can be made in this scenario.

Remark 3. In Appendix B we show that XCBC, TMAC, OMAC, PMAC are defined based on a PRF-preserving sequence of DAG. Thus, the pseudo-randomness of these functions are immediate from ourmain theorem.

5   A Simple Proof for On-line Cipher Hash-CBC [2]

In this section we define what is meant by on-line cipher and what is the ideal candidate for that. Then we give a simpler security proof of Hash-CBC [2] and note that in the original proof there is a flaw which could not not be easily taken care unless we make further assumptions.

An online cipher, Hash-CBC construction is given by Bellare *et. al.* [2]. In Crypto 2001 [2], the notion of On-Line cipher has been introduced and a secure Hash-CBC construction has been proposed. First we define what is meant by On-Line cipher and the definition of Hash-CBC construction.

1. Let G be a group and $G^{[1,m]} = \cup_{1 \le i \le n} G^i$ and $|G| = N$. A function $f : G^{[1,m]} \to G^{[1,m]}$ is called a *length preserving injective function* if f restricted to G^i is an injective map from G^i to G^i.

2. Let f be a length-preserving injective function and $M = M_1 \parallel \cdots \parallel M_m$, then we write $f(M) = (f^{(1)}(M), \cdots, f^{(m)}(M))$, where $f^{(i)}(M) \in G$. f is said to be *on-line* if there exists a function $X : G^{[1,m]} \to G$ such that for every $M = M[1] \parallel \cdots \parallel M[m]$, $f^{(i)}(M) = X(M[1] \parallel \cdots \parallel M[i])$. It says that first i blocks of cipher only depends on the first i blocks of message. Note that for each $i \ge 1$, and $(M[1] \parallel \cdots \parallel M[i-1]) \in G^{i-1}$, $X(M[1] \parallel \cdots \parallel M[i-1] \parallel x)$ is an injective function from G to G as a function of x since f is length-preserving injective function. We also say that X is an on-line function.

3. X^U is said to be uniform random on-line function if X is chosen uniformly from the set of all on-line functions from $G^{[1,m]}$ to G.

Hash-CBC

Let H be a random function from G to G which satisfies the following property. $\Pr[H(x_1) - H(x_2) = y] \le \epsilon$ for all $x_1 \ne x_2 \in G$ and $y \in G$. We say this random function by ϵ-almost universal random function. Thus for any (x_i, y_i), $1 \le i \le k$, with distinct x_i's we have,

$$\Pr[H(x_i) + y_i = H(x_i) + y_j \text{ for some } i \ne j] \le \frac{k(k-1)\epsilon}{2}. \tag{6}$$

Given an ϵ-almost universal random function and a uniform random injective function v on G we define a random on-line function F, known as **HCBC** (or **Hash-CBC**), as follows: $X(M[1] \cdots M[j]) = C[j]$, where $C[i] = v(H(C[i-1]) + M[i])$, $1 \le i \le j$ and $C[0] = 0$.

Note that X is a random on-line function. Let $x_1, \cdots, x_k \in G^{[1,m]}$ and $\mathbb{P}$ be the set of all non-empty prefixes of these messages. Let $y_p \in G$, where y_p's are distinct and not equal to 0 and $|\mathbb{P}| = q$. Now we want to compute $\Pr[X(p) = y_p, \forall p \in \mathbb{P}]$ where the probability is based on uniform random injective function v and ϵ-almost universal random function H. Let D be the event that for all p, $(H(y_{\text{chop}(p)}) + \text{last}(p))$'s are distinct where $y_\lambda := 0$ and λ is the empty string. Since y_p's are distinct and not equal to 0, $\Pr[D] \ge 1 - \frac{q(q-1)\epsilon}{2}$. Condition on D

all inputs of v are distinct. Thus, $\Pr[X(p) = y_p, \forall p \in \mathbb{P} \mid D] = \frac{1}{N(N-1)\cdots(N-q+1)}$ and hence $\Pr[X(p) = y_p, \forall p \in \mathbb{P}] \geq \frac{(1 - \frac{q(q-1)\epsilon}{2})}{N(N-1)\cdots(N-q+1)}$ (by Equation 6) $\geq (1 - \frac{q(q-1)\epsilon}{2}) \times \Pr[X^U(p) = y_p, \forall p \in \mathbb{P}]$ since $\Pr[X^U(p) = y_p, \forall p \in \mathbb{P}] \leq 1/N(N-1)\cdots(N-q+1)$. Given any query functions, let $X^U[q]$ and $X[q]$ denote the joint distribution of X^U and X on $\mathbb{P}$ respectively. Let $T = \{(y_p)_{p \in \mathbb{P}} : y_{\text{chop}(p)} \neq 0 \,\forall p,$ and y_p's are distinct $\}$ It is easy to check that $\Pr[X^U \notin T] \leq \frac{q(q-1)}{2N}$. Now by Lemma 3 we obtain the following main Theorem of this section.

Theorem 5. *For any query function, the statistical distance* $d_{\text{stat}}(X^U[q], X[q]) \leq q(q-1)\epsilon + \frac{q(q-1)}{N}$ *and hence* $\text{Adv}_{X^U, X}(q) \leq q(q-1)\epsilon + \frac{q(q-1)}{N}$.

Remark 4. In Appendix C, we note that in the original proof of the security of Hash-CBC has flaws. Thus a correction is must for this construction. Here we provide not only a correct proof but a simple proof for security.

Remark 5. In [2], authors also consider chosen-cipher text security for a variant of the above construction. In this scenario, there are two different types of queries. Let $\mathbb{P}$ denotes the set of all prefixes of the queries of on-line function X and $\mathbb{P}^*$ denotes the set of all prefixes of queries of corresponding inverse on-line function Y (say). Now one can similarly prove that

$$\Pr[X(p) = y_p, \forall p \in \mathbb{P} \text{ and } Y(p) = w_p \forall p \in \mathbb{P}^*]$$

$$\leq (1 - \epsilon) \times \Pr[X^U(p) = y_p, \forall p \in \mathbb{P} \text{ and } Y^U(p) = w_p \forall p \in \mathbb{P}^*],$$

where X^U and Y^U denote the uniform random on-line function and it's corresponding inverse function respectively. So we have same security analysis. We leave reader to verify all the details of the chosen cipher text security.

6   Conclusion and Future Work

In this paper we make a unifying approach to prove the indistinguishability of many existing constructions. This paper attempts to clean up several results regarding indistinguishability so that the researchers can feel and understand the subject in a better and simpler way. As a concluding remark we would like to say that one can view the security analysis in the way we have observed in this paper and can have better and simpler proof for it. Some cases people have wrong proofs due to length and complicity of it. Thus, a more concrete as well as simple proof is always welcome.

In future, this unifying idea may help us to make good constructions. It seems that one may find constructions where the security bound is more than the birth day attack bound. Till now, there is no known construction based on ideal function (having output n-bit) which has security close to 2^n. One may obtain a better bound for CBC as we have used only those output functions which induces an injective input functions. One can try to estimate the other output functions also.

Acknowledgement. I would like to thank Donghoon Chang, Dr. Kishan Chand Gupta and Professor Douglas R. Stinson for their helpful comments, suggestions and proofreading.

References

1. M. Bellare, K. Pietrzak and P. Rogaway. Improved Security Analysis for CBC MACs. Advances in Cryptology - CRYPTO 2005. Lecture Notes in Computer Science, Volume **3621**, pp 527-545.
2. M. Bellare, A. Boldyreva, L. Knudsen and C. Namprempre. On-Line Ciphers and the Hash-CBC constructions. Advances in Cryptology - CRYPTO 2000. Lecture Notes in Computer Science, Volume **2139**, pp 292-309.
3. M. Bellare, J. Killan and P. Rogaway. The security of the cipher block chanining Message Authentication Code. Advances in Cryptology - CRYPTO 1994. Lecture Notes in Computer Science, Volume **839**, pp 341-358.
4. Daniel J. Bernstein. A short proof of the unpredictability of cipher block chaining (2005). URL: http://cr.yp.to/papers.html#easycbc. ID 24120a1f8b92722b5e1 5fbb6a86521a0.
5. J. Black and P. Rogaway. CBC MACs for arbitrary length messages. Advances in Cryptology - CRYPTO 2000. Lecture Notes in Computer Science, Volume **1880**, pp 197-215.
6. J. Black and P. Rogaway. A Block-Cipher Mode of Operations for Parallelizable Message Authentication. Advances in Cryptology - Eurocrypt 2002. Lecture Notes in Computer Science, Volume **2332**, pp 384-397.
7. T. Iwata and K. Kurosawa. OMAC : One-Key CBC MAC. Fast Software Encryption, 10th International Workshop, FSE 2003. Lecture Notes in Computer Science, Volume **2887**, pp 129-153.
8. C. S. Jutla. PRF Domain Extension using DAG. Theory of Cryptography: Third Theory of Cryptography Conference, TCC 2006. Lecture Notes in Computer Science, Volume **3876** pp 561-580.
9. K. Kurosawa and T. Iwata. TMAC : Two-Key CBC MAC. Topics in Cryptology - CT-RSA 2003: The Cryptographers' Track at the RSA Conference 2003. Lecture Notes in Computer Science, Volume **2612**, pp 33-49.

Appendix A: Proofs of Lemmas in Section 2

Lemma 1. $d_{stat}(X, Y) = \Pr[X \in T_0] - \Pr[Y \in T_0] = \frac{1}{2} \times \sum_{a \in S} |\Pr[X = a] - \Pr[Y = a]|$, where $T_0 = \{a \in S : \Pr[X = a] \geq \Pr[Y = a]\}$.

Proof. For T_0 as given in the Lemma 1, it is easy to see that

$$\sum_{a \in S} |\Pr[X = a] - \Pr[Y = a]| = 2 \times (\Pr[X \in T_0] - \Pr[Y \in T_0]).$$

For any $T \subset S$, $2 \times (\Pr[X \in T] - \Pr[Y \in T])$

$$= \sum_{a \in T} (\Pr[X = a] - \Pr[Y = a]) - \sum_{a \notin T} (\Pr[X = a] - \Pr[Y = a])$$

$$\leq \sum_{a \in S} |\Pr[X = a] - \Pr[Y = a]|. \qquad \blacksquare$$

Lemma 2. Let X and Y be two random variable taking values on S and $\epsilon > 0$. If $\Pr[X = a] \geq (1 - \epsilon) \times \Pr[Y = a]$, $\forall a \in S$ or $\Pr[X = a] \leq (1 + \epsilon) \times \Pr[Y = a]$, $\forall a \in S$ then $d_{\mathrm{stat}}(X, Y) \leq \epsilon$.

Proof. For any subset $T \subset S$, $\Pr[X \in T] \geq (1 - \epsilon) \times \Pr[Y \in T]$ since $\Pr[X = a] \geq (1 - \epsilon) \times \Pr[Y = a] \ \forall \ a$. So, $\Pr[Y \in T] - \Pr[X \in T] \leq \epsilon \times \Pr[Y \in T] \leq \epsilon$. Thus, $d_{\mathrm{stat}}(X, Y) \leq \epsilon$. Similarly one can prove for the other case. ∎

Lemma 3. Let X and Y be two random variables taking values on S. Let for a subset $T \subset S$, $\Pr[X = a] \geq (1 - \epsilon_1) \times \Pr[Y = a]$, $\forall a \in T$ and $\Pr[Y \notin T] \leq \epsilon_2$ then $d_{\mathrm{stat}}(X, Y) \leq 2\epsilon_1 + 2\epsilon_2$.

Proof. For any subset $T_1 \subset T$, $\Pr[Y \in T_1] - \Pr[X \in T_1] \leq \epsilon_1 \times \Pr[Y \in T_1] \leq \epsilon_1$. From the given relation we also note that $\Pr[X \in T] \geq (1 - \epsilon_1) \times \Pr[Y \in T]$. Thus, $\Pr[X \notin T] \leq \epsilon_1 + \epsilon_2 - \epsilon_1 \epsilon_2$. Thus, $d_{\mathrm{stat}}(X, Y) \leq \epsilon_1 + \Pr[X \in \neg T] + \Pr[Y \in \neg T] \leq 2(\epsilon_1 + \epsilon_2)$. ∎

Lemma 4. For any $\mathcal{A}$, $d^{\mathcal{A}}(X, Y) \leq d_{\mathrm{stat}}(X, Y)$. Conversely, there exists an algorithm $\mathcal{A}_0$ (may not be efficient) such that $d^{\mathcal{A}_0}(X, Y) = d_{\mathrm{stat}}(X, Y)$.

Proof. Output of $\mathcal{A}$ is completely determined by a pair (r, z), where r is the random string chosen from $\mathcal{R}$ and z is the input. Let $S_{r_0} = \{a \in S : \mathcal{A}(r_0, a) = 1\}$. Thus, $d^{\mathcal{A}}(X, Y)$

$$= \left| \Pr[\mathcal{A}(r, X) = 1] - \Pr[\mathcal{A}(r, Y) = 1] \right|$$

$$= \left| \sum_{r_0 \in \mathcal{R}} \Pr[r = r_0] \left(\Pr[\mathcal{A}(r_0, X) = 1 \mid r = r_0] - \Pr[\mathcal{A}(r_0, Y) = 1 \mid r = r_0] \right) \right|$$

$$= \left| \sum_{r_0 \in \mathcal{R}} \Pr[r = r_0] \left(\Pr[\mathcal{A}(r_0, X) = 1] - \Pr[\mathcal{A}(r_0, Y) = 1] \right) \right|$$

$$= \left| \sum_{r_0 \in \mathcal{R}} \Pr[r = r_0] \left(\Pr[X \in S_{r_0}] - \Pr[Y \in S_{r_0}] \right) \right| \leq d_{\mathrm{stat}}(X, Y).$$

The equality holds if $S_{r_0} = T_0$ as in Lemma 1. Thus, on input z, $\mathcal{A}_0$ computes the probability $\Pr[X = z]$, $\Pr[Y = z]$ and outputs 1 if $\Pr[X = z] \geq \Pr[Y = z]$, otherwise 0. Hence $d^{\mathcal{A}_0}(X, Y) = d_{\mathrm{stat}}(X, Y)$. ∎

Theorem 1. If $\Pr[f[k](\mathbf{a}) = \mathbf{y}] \geq (1 - \epsilon) \times \Pr[g[k](\mathbf{a}) = \mathbf{y}]$ for each $\mathbf{a} \in H[k]$ and $\mathbf{y} \in G^k$, then for any k-query function $\mathbf{x} = (x_1, \cdots, x_k)$, $d_{\mathrm{stat}}(f[k]_{\mathbf{x}}, g[k]_{\mathbf{x}}) \leq \epsilon$ and hence $d_{f,g}(k) \leq \epsilon$.

Proof. $\Pr[f[k]_{x_1, \cdots, x_k} = (y_1, \cdots, y_k)]$

$= \Pr[f[k](a_1, \cdots, a_k) = (y_1, \cdots, y_k)]$ $((a_1, \cdots, a_k)$ is uniquely determined by $(y_1, \cdots, y_k))$

$\geq (1 - \epsilon) \times \Pr[g[k](a_1, \cdots, a_k) = (y_1, \cdots, y_k)]$

$= (1 - \epsilon) \times \Pr[g[k]_{x_1, \cdots, x_k} = (y_1, \cdots, y_k)]$, $\forall \ (y_1, \cdots, y_k) \in G^k$. The Theorem follows from Lemma 2. ∎

Appendix B: Some Known PRFs for Variable Length Input

There are three popularly known constructions which deals with variable size input and uses CBC mode. These are XCBC [5], TMAC [9], OMAC [7] and PMAC [6]. Let K_1 and K_2 be two secret constants from $\{0,1\}^n$. Given $M = M_1 \| \cdots M_{l-1} \| M_l$ with $|M_1| = \cdots = |M_{l-1}| = n$, $|M_l| = n_1$, $1 \leq n_1 \leq n$ and a random function f on $\{0,1\}^n$, define f^* as follows :

$$\left. \begin{array}{ll} f^*(M) = f_l^+(M_1 \| \cdots \| M_{l-1} \| (M_l \oplus K_1)) & \text{if } n_1 = n \\ = f_l^+(M_1 \| \cdots \| M_{l-1} \| (M_l 10^i \oplus K_2)) & \text{if } n_1 < n, i = n - n_1 - 1 \end{array} \right\} (7)$$

XCBC, **TMAC** and **OMAC** are defined on the basis of choices of K_1 and K_2.

– If K_1 and K_2 are chosen independently from f then it is known as XCBC.
– If $K_2 = c \cdot K_1$ and K_1 is chosen independently from f then it is known as TMAC where $c \in \{0,1\}^n$ is some fixed known constant not equal to 1 and 0, and $\cdot$ is a field multiplication on $\{0,1\}^n = GF(2^n)$.
– If $K_1 = c \cdot L$ and $K_2 = c^2 \cdot L$ where $L = f(0)$ then it is known as OMAC.

Security of OMAC

Here we only consider security for OMAC. For the other constructions, one can make a similar treatment as in CBC. For OMAC as in the previous Remark 2 we have two sequences of colored DAGs G_1^l and G_2^l. Each graph is a sequential graph with one more edge at the end. More precisely, $V^l = \{v_s = 1, \cdots, l = v_f, v_{iv}\}$ and $E^l = \{(1,2), \cdots, (l-1, l), (v_{iv}, l)\}$. The color function for G_1^l is as follows : $c^l((i, i+1)) = 1$ and $c^l((v_{iv}, l)) = c$, where $c \neq 0, 1$. Similarly, the color function for G_2^l is as follows : $c^l((i, i+1)) = 1$ and $c^l((v_{iv}, l)) = c^2$. We choose $iv_l = \mathbf{0} \in \{0,1\}^n$. It is easy to check that each colored DAG is non-singular. Any colored DAG can not be isomorphic to a colored subgraph as the sink node has inward degree 2 where as the other nodes have inward degree 1. Thus, the sequence is admissible. The pseudo randomness property follows from the Theorem 4.

Security of PMAC

One can similarly observe that PMAC also belong to this class. The underlying graph $D^l = (V^l, E^l)$, where $V^l = \{v_{iv}, 1, \cdots, l-1, v_f\}$ and $E^l = \{(v_{iv}, i), (i, v_f), 1 \leq i \leq l-1\} \cup \{(v_{iv}, v_f)\}$. There is two color functions depending on the message size. When message size is multiple of n, $c_1(v_{iv}, i) = c^i$ and $c_1(v_{iv}, v_f) = 0$, otherwise it takes constant 1. The other color function is same except that $c_2(v_{iv}, v_f) = a$. Here, c and a are some constants not equal to 0 and 1, and $iv_l = \mathbf{0} \in \{0,1\}^n$.

Appendix C: A Flaw in the Original Proof [2]

In the original paper due to Bellare *et. al.* [2], the security proof has some flaw. The Claim 6.5 of [2] says that if some bad event does not occur then the distribution of the view is identical for both classes of functions. More precisely, $X(p)$'s and $X^U(p)$'s are identically distributed condition on some bad event does not occur (i.e., the inputs of uniform random injective function v are distinct). In case of X^U, all conditional random variables $X^U(p)$'s are uniformly and identically distributed on the set T. But, conditional distribution of $X(p)$'s is not so as the condition is itself involved with $X(p)$ and an unknown distribution due to H. For example, when $p_1 = x_1$ and $p_2 = x_1 \parallel x_2$ then $X(p_1) = v(H(0) \oplus x_1)$ and $X(p_2) = v(H(X(p_1)) \oplus x_2)$. The conditional event E (the complement of bad event) is $H(0) \oplus x_1 \neq H(X(p_1)) \oplus x_2$ and $v(H(0) \oplus x_1) \neq 0$. According to their claim for any $0 \neq y_1 \neq y_2$, $p = \Pr[v(H(0) \oplus x_1) = y_1, v(H(X(p_1)) \oplus x_2) = y_2 \mid E] = \frac{1}{(N-1)(N-1)}$ (note that y_2 can be zero). Let $a := x_1 \oplus x_2$, $C = H(0) \oplus x_1$ and $\epsilon_{y,z,c} = \Pr[H(y) \oplus H(z) = c]$.

Now, $p_1 := \Pr[v(H(0) \oplus x_1) = y_1 \ \wedge \ v(H(y_1) \oplus x_2) = y_2 \ \wedge \ E]$

$$= \sum_{h_1,h_2 \ : \ h_1 \oplus h_2 \neq a} \Pr[v(h_1 \oplus x_1) = y_1, v(h_2 \oplus x_2)$$

$$= y_2, H(0) = h_1, H(y_1) = h_2 = \frac{\epsilon_{0,y_1,a}}{N(N-1)}$$

and $p_2 := \Pr[E] = \Pr[v(C) \neq 0, H(v(C) \oplus x_2) \neq C]$

$$= \sum_{z,h \ : \ z \neq 0} \Pr[v(h \oplus x_1) = z, H(0) = h, H(z) \neq h \oplus a]$$

$$= \frac{1}{N} \times \sum_{z \ : \ z \neq 0} \Pr[H(0) \oplus H(z) \neq a].$$

Thus, $p = \frac{\epsilon_{0,y_1,a}}{(N-1) \times \sum_{z \neq 0} \epsilon_{0,z,a}} \neq \frac{1}{(N-1)(N-1)}$ in general. This can occur if $\epsilon_{0,z,a} = \epsilon_{0,y_1,a}$ for all $z \neq 0$, but there is no such assumption for H in [2]. A similar flaw can be observed in the Claim 8.6 of [2] where the chosen cipher text security is considered.

Appendix D: Improved Security Bound of CBC [1]

In this section we will give a simple partial proof of improved security analysis given by Bellare *et. al.* [1]. We will follow same notation as in Section 3. We say an output function op is *induced* if there exists an u such that $op_u = op$. We define an event $D^*[k]$ where the corresponding input function of induced output function $ip : \mathbb{P} \to G$ satisfies the following property :

$$\forall \ p_1 \in \{x_1, \cdots, x_k\}, p_2 \in \mathbb{P} \text{ and } p_1 \neq p_2, ip(p_1) \neq ip(p_2). \tag{8}$$

In [1] for $k = 2$, it has been proved that $\Pr[\neg D^*[2]] \leq (8m/N + 64m^4/N^2)$. For $k \geq 2$ it is easy to check that $\Pr[\neg D^*[k]] \leq k(k-1)/2 \times (8m/N + 64m^4/N^2)$. Here we will assume this result as we have not found any simple proof of this. Secondly, one can translate this into a purely combinatorial problem which was solved rigorously by Bellare $et.$ $al.$ (Lemma 2 of [1]). Now $\Pr[u^+_{x_1,\cdots,x_k} = (y_1, \cdots, y_k) \mid D^*[k]] = 1/N^k$. This is true that for any $induced$ output function op with above property there exists N^{N-q_1} many u's which induces op and there are N^{N-q_1-k} many u's which induces op and $u^+(x_i) = y_i \ \forall \ i$, where q_1 denotes the size of range of induced input function of op (see Corollary 1). Here, we do not need that the corresponding input function is injective. We can still have a similar statement like in Corollary 1 as the input function is taking completely different values on $\{x_1, \cdots, x_k\}$ from the values on $\mathbb{P}_1$ (see Equation 8). Thus, $\Pr[u^+_{x_1,\cdots,x_k} = (y_1, \cdots, y_k)] \geq \left(1 - \frac{k(k-1) \times (8m/N + 64m^4/N^2)}{2}\right) \times \frac{1}{N^k}$. Thus we have,

Theorem 6. $\mathrm{Adv}_{u^+_m, u^{(m)}}(k) \leq k(k-1)/2 \times (8m/N + 64m^4/N^2)$.

Extended Double-Base Number System with Applications to Elliptic Curve Cryptography

Christophe Doche[1] and Laurent Imbert[2]

[1] Department of Computing
Macquarie University, Australia
`doche@ics.mq.edu.au`
[2] LIRMM, CNRS, Université Montpellier 2, UMR 5506, France
& ATIPS, CISaC, University of Calgary, Canada
`Laurent.Imbert@lirmm.fr`

Abstract. We investigate the impact of larger digit sets on the length of Double-Base Number system (DBNS) expansions. We present a new representation system called *extended DBNS* whose expansions can be extremely sparse. When compared with double-base chains, the average length of extended DBNS expansions of integers of size in the range 200–500 bits is approximately reduced by 20% using one precomputed point, 30% using two, and 38% using four. We also discuss a new approach to approximate an integer n by $d2^a3^b$ where d belongs to a given digit set. This method, which requires some precomputations as well, leads to realistic DBNS implementations. Finally, a left-to-right scalar multiplication relying on extended DBNS is given. On an elliptic curve where operations are performed in Jacobian coordinates, improvements of up to 13% overall can be expected with this approach when compared to window NAF methods using the same number of precomputed points. In this context, it is therefore the fastest method known to date to compute a scalar multiplication on a generic elliptic curve.

Keywords: Double-base number system, Elliptic curve cryptography.

1 Introduction

Curve-based cryptography, especially elliptic curve cryptography, has attracted more and more attention since its introduction about twenty years ago [1,2,3], as reflected by the abundant literature on the subject [4,5,6,7]. In curve-based cryptosystems, the core operation that needs to be optimized as much as possible is a scalar multiplication. The standard method, based on ideas well known already more than two thousand years ago, to efficiently compute such a multiplication is the double-and-add method, whose complexity is linear in terms of the size of the input. Several ideas have been introduced to improve this method; see [8] for an overview. In the remainder, we will mainly focus on two approaches:

- Use a representation such that the expansion of the scalar multiple is sparse. For instance, the non-adjacent form (NAF) [9] has a nonzero digit density of

R. Barua and T. Lange (Eds.): INDOCRYPT 2006, LNCS 4329, pp. 335–348, 2006.
© Springer-Verlag Berlin Heidelberg 2006

1/3 whereas the average density of a binary expansion is 1/2. This improvement is mainly obtained by adding -1 to the set $\{0,1\}$ of possible coefficients used in binary notation. Another example is the double-base number system (DBNS) [10], in which an integer is represented as a sum of products of powers of 2 and 3. Such expansions can be extremely sparse, *cf.* Section 2.

- Introduce precomputations to enlarge the set of possible coefficients in the expansion and reduce its density. The k-ary and sliding window methods as well as window NAF methods [11,12] fall under this category.

In the present work, we mix these two ideas. Namely, we investigate how precomputations can be used with the DBNS and we evaluate their impact on the overall complexity of a scalar multiplication.

Also, computing a sparse DBNS expansion can be very time-consuming although it is often neglected when compared with other representations. We introduce several improvements that considerably speed up the computation of a DBNS expansion, *cf.* Section 4.

The plan of the paper is as follows. In Section 2, we recall the definition and basic properties of the DBNS. In Section 3, we describe how precomputations can be efficiently used with the DBNS. Section 4 is devoted to implementation aspects and explains how to quickly compute DBNS expansions. In Section 5, we present a series of tests and comparisons with existing methods before concluding in Section 6.

2 Overview of the DBNS

In the *Double-Base Number System*, first considered by Dimitrov *et al.* in a cryptographic context in [13], any positive integer n is represented as

$$n = \sum_{i=1}^{\ell} d_i 2^{a_i} 3^{b_i}, \text{ with } d_i \in \{-1, 1\}. \tag{1}$$

This representation is obviously not unique and is in fact highly redundant. Given an integer n, it is straightforward to find a DBNS expansion using a greedy-type approach. Indeed, starting with $t = n$, the main task at each step is to find the $\{2,3\}$-integer z that is the closest to t (i.e. the integer z of the form $2^a 3^b$ such that $|t - z|$ is minimal) and then set $t = t - z$. This is repeated until t becomes 0. See Example 2 for an illustration.

Remark 1. *Finding the best $\{2,3\}$-approximation of an integer t in the most efficient way is an interesting problem on its own. One option is to scan all the points with integer coordinates near the line $y = -x \log_3 2 + \log_3 t$ and keep only the best approximation. A much more sophisticated method involves continued fractions and Ostrowski's number system, cf. [14]. It is to be noted that these methods are quite time-consuming. See Section 4 for a more efficient approach.*

Example 2. *Take the integer* $n = 841232$. *We have the sequence of approximations*

$$841232 = 2^7 3^8 + 1424,$$
$$1424 = 2^1 3^6 - 34,$$
$$34 = 2^2 3^2 - 2.$$

As a consequence, $841232 = 2^7 3^8 + 2^1 3^6 - 2^2 3^2 + 2^1$.

It has been shown that every positive integer n can be represented as the sum of at most $O\left(\frac{\log n}{\log \log n}\right)$ signed $\{2, 3\}$-integers. For instance, see [13] for a proof. Note that the greedy approach above-mentioned is suitable to find such short expansions.

This initial class of DBNS is therefore very sparse. When one endomorphism is virtually free, like for instance triplings on supersingular curves defined over $\mathbb{F}_3$, the DBNS can be used to efficiently compute $[n]P$ with max a_i doublings, a very low number of additions, and the necessary number of triplings [15]. Note that this idea has recently been extended to Koblitz curves [16]. Nevertheless, it is not really suitable to compute scalar multiplications in general. For generic curves where both doublings and triplings are expensive, it is essential to minimize the number of applications of these two endomorphisms. Now, one needs at least max a_i doublings and max b_i triplings to compute $[n]P$ using (1). However, given the DBNS expansion of n returned by the greedy approach, it seems to be highly non-trivial, if not impossible, to attain these two lower bounds simultaneously.

So, for generic curves the DBNS needs to be adapted to compete with other methods. The concept of *double-base chain*, introduced in [17], is a special type of DBNS. The idea is still to represent n as in (1) but with the extra requirements $a_1 \geqslant a_2 \geqslant \cdots \geqslant a_\ell$ and $b_1 \geqslant b_2 \geqslant \cdots \geqslant b_\ell$. These properties allow to compute $[n]P$ from right-to-left very easily. It is also possible to use a Horner-like scheme that operates from left-to-right. These two methods are illustrated after Example 3.

Note that, it is easy to accommodate these requirements by restraining the search of the best exponents (a_{j+1}, b_{j+1}) to the interval $[0, a_j] \times [0, b_j]$.

Example 3. *A double-base chain of* n *can be derived from the following sequence of equalities:*

$$841232 = 2^7 3^8 + 1424,$$
$$1424 = 2^1 3^6 - 34,$$
$$34 = 3^3 + 7,$$
$$7 = 3^2 - 2,$$
$$2 = 3^1 - 1.$$

As a consequence, $841232 = 2^7 3^8 + 2^1 3^6 - 3^3 - 3^2 + 3^1 - 1$.

In that particular case, the length of this double-base chain is strictly bigger than the one of the DBNS expansion in Example 2. This is true in general as well and the difference can be quite large. It is not known whether the bound $O\left(\frac{\log n}{\log\log n}\right)$ on the number of terms is still valid for double-base chains.

However, computing $[841232]P$ is now a trivial task. From right-to-left, we need two variables. The first one, T being initialized with P and the other one, S set to point at infinity. The successive values of T are then P, $[3]P$, $[3^2]P$, $[3^3]P$, $[2^13^6]P$, and $[2^73^8]P$, and at each step T is added to S. Doing that, we obtain $[n]P$ with 7 doublings, 8 triplings, and 5 additions. To proceed from left-to-right, we notice that the expansion that we found can be rewritten as

$$841232 = 3\big(3\big(3\big(2^13^3(2^63^2 + 1) - 1\big) - 1\big) + 1\big) - 1,$$

which implies that

$$[841232]P = [3]\big([3]\big([3]\big([2^13^3]([2^63^2]P + P) - P\big) - P\big) + P\big) - P.$$

Again, 7 doublings, 8 triplings, and 5 additions are necessary to obtain $[n]P$.

More generally, one needs exactly a_1 doublings and b_1 triplings to compute $[n]P$ using double-base chains. The value of these two parameters can be optimized depending on the size of n and the respective complexities of a doubling and a tripling (see Figure 2).

To further reduce the complexity of a scalar multiplication, one option is to reduce the number of additions, that is to minimize the density of DBNS expansions. A standard approach to achieve this goal is to enlarge the set of possible coefficients, which ultimately means using precomputations.

3   Precomputations for DBNS Scalar Multiplication

We suggest to use precomputations in two ways. The first idea, which applies only to double-base chains, can be viewed as a two-dimensional window method.

3.1   Window Method

Given integers w_1 and w_2, we represent n as in (1) but with coefficients d_i in the set $\{\pm 1, \pm 2^1, \pm 2^2, \ldots, \pm 2^{w_1}, \pm 3^1, \pm 3^2, \ldots, \pm 3^{w_2}\}$. This is an indirect way to relax the conditions $a_1 \geqslant a_2 \geqslant \cdots \geqslant a_\ell$ and $b_1 \geqslant b_2 \geqslant \cdots \geqslant b_\ell$ in order to find better approximations and hopefully sparser expansions. This method, called (w_1, w_2)-*double-base chain*, lies somewhere between normal DBNS and double-base chain methods.

Example 4. *The DBNS expansion of* $841232 = 2^73^8 + 2^13^6 - 2^23^2 + 2^1$, *can be rewritten as* $841232 = 2^73^8 + 2^13^6 - \mathbf{2} \times 2^13^2 + 2^1$, *which is a* $(1,0)$-*window-base chain. The exponent* a_3 *that was bigger than* a_2 *in Example 2 has been replaced by* a_2 *and the coefficient* d_3 *has been multiplied by 2 accordingly. As a result, we now have two decreasing sequences of exponents and the expansion is only four terms long.*

It remains to see how to compute $[841232]P$ from this expansion. The right-to-left scalar multiplication does not provide any improvement, but this is not the case for the left-to-right approach. Writing $841232 = 2\big(3^2\big(3^4(2^63^2+1)-2\big)+1\big)$, we see that

$$[841232]P = [2]\big([3^2]\big([3^4]([2^63^2]P + P) - [2]P\big) + P\big).$$

If $[2]P$ is stored along the computation of $[2^63^2]P$ then 7 doublings, 8 triplings and only 3 additions are necessary to obtain $[841232]P$.

It is straightforward to design an algorithm to produce (w_1, w_2)-double-base chains. We present a more general version in the following, *cf.* Algorithm 1. See Remark 6 (v) for specific improvements to (w_1, w_2)-double-base chains.

Also a left-to-right scalar multiplication algorithm can easily be derived from this method, *cf.* Algorithm 2.

The second idea to obtain sparser DBNS expansions is to generalize the window method such that any set of coefficients is allowed.

3.2   Extended DBNS

In a (w_1, w_2)-double-base chain expansion, the coefficients are signed powers of 2 or 3. Considering other sets $\mathcal{S}$ of coefficients, for instance odd integers coprime with 3, should further reduce the average length of DBNS expansions. We call this approach *extended DBNS* and denote it by $\mathcal{S}$-DBNS.

Example 5. *We have* $841232 = 2^73^8 + 5 \times 2^53^2 - 2^4$. *The exponents form two decreasing sequences, but the expansion has only three terms. Assuming that* $[5]P$ *is precomputed, it is possible to obtain* $[841232]P$ *as*

$$[2^4]\big([2^13^2]([2^23^6]P + [5]P) - P\big)$$

with 7 doublings, 8 triplings, and only 2 additions

This strategy applies to any kind of DBNS expansion. In the following, we present a greedy-type algorithm to compute extended double-base chains.

Algorithm 1.   Extended double-base chain greedy algorithm

INPUT: A positive integer n, a parameter a_0 such that $a_0 \leqslant \lceil \log_2 n \rceil$, and a set $\mathcal{S}$ containing 1.

OUTPUT: Three sequences $(d_i, a_i, b_i)_{1 \leqslant i \leqslant \ell}$ such that $n = \sum_{i=1}^{\ell} d_i 2^{a_i} 3^{b_i}$ with $|d_i| \in \mathcal{S}$, $a_1 \geqslant a_2 \geqslant \cdots \geqslant a_\ell$, and $b_1 \geqslant b_2 \geqslant \cdots \geqslant b_\ell$.

1.	$b_0 \leftarrow \lceil (\log_2 n - a_0) \log_2 3 \rceil$	[See Remark 6 (ii)]
2.	$i \leftarrow 1$ and $t \leftarrow n$	
3.	$s \leftarrow 1$	[to keep track of the sign]
4.	**while** $t > 0$ **do**	
5.	find the best approximation $z = d_i 2^{a_i} 3^{b_i}$ of t with $d_i \in \mathcal{S}$, $0 \leqslant a_i \leqslant a_{i-1}$, and $0 \leqslant b_i \leqslant b_{i-1}$	
6.	$d_i \leftarrow s \times d_i$	
7.	**if** $t < z$ **then** $s \leftarrow -s$	

8. $t \leftarrow |t - z|$
9. $i \leftarrow i + 1$
10.   **return** (d_i, a_i, b_i)

Remarks 6

(i) Algorithm 1 processes the bits of n from left-to-right. It terminates since the successive values of t form a strictly decreasing sequence.

(ii) The parameters a_0 and b_0 are respectively the biggest powers of 2 and 3 allowed in the expansion. Their values have a great influence on the density of the expansion, *cf.* Section 5 for details.

(iii) To compute normal DBNS sequences instead of double-base chains, replace the two conditions $0 \leqslant a_i \leqslant a_{i-1}$, $0 \leqslant b_i \leqslant b_{i-1}$ in Step 5 by $0 \leqslant a_i \leqslant a_0$ and $0 \leqslant b_i \leqslant b_0$.

(iv) In the following, we explain how to find the best approximation $d_i 2^{a_i} 3^{b_i}$ of t in a very efficient way. In addition, the proposed method has a time-complexity that is mainly independent of the size of $\mathcal{S}$ and not directly proportional to it as with a naïve search. See Section 4 for details.

(v) To obtain (w_1, w_2)-double-base chains, simply ensure that $\mathcal{S}$ contains only powers 2 and 3. However, there is a more efficient way. First, introduce two extra variables $a_{\max}$ and $b_{\max}$, initially set to a_0 and b_0 respectively. Then in Step 5, search for the best approximation z of t of the form $2^{a_i} 3^{b_i}$ with $(a_i, b_i) \in [0, a_{\max} + w_1] \times [0, b_{\max} + w_2] \setminus [a_{\max} + 1, a_{\max} + w_1] \times [b_{\max} + 1, b_{\max} + w_2]$. In other words, we allow one exponent to be slightly bigger than its current maximal bound, but the (exceptional) situation where $a_i > a_{\max}$ and $b_i > b_{\max}$ simultaneously is forbidden. Otherwise, we should be obliged to include in $\mathcal{S}$ products of powers of 2 and 3 and increase dramatically the number of precomputations. Once the best approximation has been found, if a_i is bigger than $a_{\max}$, then a_i is changed to $a_{\max}$ while d_i is set to $2^{a_i - a_{\max}}$. If b_i is bigger than $b_{\max}$, then b_i is changed to $b_{\max}$ while d_i is set to $3^{b_i - b_{\max}}$. Finally, do $a_{\max} \leftarrow \min(a_i, a_{\max})$ and $b_{\max} \leftarrow \min(b_i, b_{\max})$ and the rest of the Algorithm remains unchanged.

(vi) In the remainder, we discuss some results obtained with Algorithm 1 using different sets of coefficients. More precisely, each set $\mathcal{S}_m$ that we consider contains the first $m + 1$ elements of $\{1, 5, 7, 11, 13, 17, 19, 23, 25\}$.

We now give an algorithm to compute a scalar multiplication from the expansion returned by Algorithm 1.

Algorithm 2. Extended double-base chain scalar multiplication

---

INPUT: A point P on an elliptic curve E, a positive integer n represented by $(d_i, a_i, b_i)_{1 \leqslant i \leqslant \ell}$ as returned by Algorithm 1, and the points $[k]P$ for each $k \in \mathcal{S}$.

OUTPUT: The point $[n]P$ on E.

---

1. $T \leftarrow O_E$ $[O_E$ is the point at infinity on $E]$
2. set $a_{\ell+1} \leftarrow 0$ and $b_{\ell+1} \leftarrow 0$
3. **for** $i = 1$ **to** ℓ **do**
4. $T \leftarrow T \oplus [d_i]P$
5. $T \leftarrow [2^{a_i - a_{i+1}} 3^{b_i - b_{i+1}}]T$
6. **return** T

Example 7. *For* $n = 841232$, *the sequence returned by Algorithm 2 with* $a_0 = 8$, $b_0 = 8$, *and* $\mathcal{S} = \{1,5\}$ *is* $(1,7,8),(5,5,2),(-1,4,0)$. *In the next Table, we show the intermediate values taken by* T *in Algorithm 2 when applied to the above-mentioned sequence. The computation is the same as in Example 5.*

i	d_i	$a_i - a_{i+1}$	$b_i - b_{i+1}$	T
1	1	2	6	$[2^2 3^6]P$
2	5	1	2	$[2^1 3^2]([2^2 3^6]P + [5]P)$
3	-1	4	0	$[2^4]([2^1 3^2]([2^2 3^6]P + [5]P) - P)$

Remark 8. *The length of the chain returned by Algorithm 1 greatly determines the performance of Algorithm 2. However, no precise bound is known so far, even in the case of simple double-base chains. So, at this stage our knowledge is only empirical, cf. Figure 2. More work is therefore necessary to establish the complexity of Algorithm 2.*

4   Implementation Aspects

This part describes how to efficiently compute the best approximation of any integer n in terms of $d_1 2^{a_1} 3^{b_1}$ for some $d_1 \in \mathcal{S}$, $a_1 \leqslant a_0$, and $b_1 \leqslant b_0$. The method works on the binary representation of n denoted by $(n)_2$. It operates on the most significant bits of n and uses the fact that a multiplication by 2 is a simple shift.

To make things clear, let us explain the algorithm when $\mathcal{S} = \{1\}$. First, take a suitable bound B and form a two-dimensional array of size $(B + 1) \times 2$. For each $b \in [0, B]$, the corresponding row vector contains $[(3^b)_2, b]$. Then sort this array with respect to the first component using lexicographic order denoted by $\preccurlyeq$ and store the result.

To compute an approximation of n in terms of $2^{a_1} 3^{b_1}$ with $a_1 \leqslant a_0$ and $b_1 \leqslant b_0$, find the two vectors v_1 and v_2 such that $v_1[1] \preccurlyeq (n)_2 \preccurlyeq v_2[1]$. This can be done with a binary search in $O(\log B)$ operations.

The next step is to find the first vector v_1' that smaller than v_1 in the sorted array and that is suitable for the approximation. More precisely, we require that:

- the difference δ_1 between the binary length of n and the length of $v'_1[1]$ satisfies $0 \leqslant \delta_1 \leqslant a_0$,
- the corresponding power of 3, *i.e.* $v'_1[2]$, is less than b_0.

This operation is repeated to find the first vector v'_2 that is greater than v_2 and fulfills the same conditions as above. The last step is to decide which approximation, $2^{\delta_1} 3^{v'_1[2]}$ or $2^{\delta_2} 3^{v'_2[2]}$, is closer to n.

In case $|\mathcal{S}| > 1$, the only difference is that the array is of size $\left(|\mathcal{S}|(B+1)\right) \times 3$. Each row vector is of the form $[(d3^b)_2, b, d]$ for $d \in \mathcal{S}$ and $b \in [0, B]$. Again the array is sorted with respect to the first component using lexicographic order. Note that multiplying the size of the table by $|\mathcal{S}|$ has only a negligible impact on the time complexity of the binary search. See [18, Appendix A] for a concrete example and some improvements to this approach.

This approximation method ultimately relies on the facts that lexicographic and natural orders are the same for binary sequences of the same length and also that it is easy to adjust the length of a sequence by multiplying it by some power of 2. The efficiency comes from the sorting operation (done once at the beginning) that allows to retrieve which precomputed binary expansions are close to n, by looking only at the most significant bits.

For environments with constrained memory, it may be difficult or even impossible to store the full table. In this case, we suggest to precompute only the first byte or the first two bytes of the binary expansions of $d3^b$ together with their binary length. This information is sufficient to find two approximations A_1, A_2 in the table such that $A_1 \leqslant n \leqslant A_2$, since the algorithm operates only on the most significant bits. However, this technique is more time-consuming since it is necessary to actually *compute* at least one approximation and sometimes more, if the first bits are not enough to decide which approximation is the closest to n.

In Table 1, we give the precise amount of memory (in bytes) that is required to store the vectors used for the approximation for different values of B. Three situations are investigated, *i.e.* when the first byte, the first two bytes, and the full binary expansions $d3^b$, for $d \in \mathcal{S}$ and $b \leqslant B$ are precomputed and stored. See [19] for a PARI/GP implementation of Algorithm 1 using the techniques described in this section.

5   Tests and Results

In this section, we present some tests to help evaluating the relevance of extended double-base chains for scalar multiplications on generic elliptic curves defined over $\mathbb{F}_p$, for p of size between 200 and 500 bits. Comparisons with the best systems known so far, including ℓ-NAF$_w$ and normal double-base chains are given.

In the following, we assume that we have three basic operations on a curve E to perform scalar multiplications, namely addition/subtraction, doubling, and tripling. In turn, each one of these elliptic curve operations can be seen as a sequence of inversions I, multiplications M, and squarings S in the underlying field $\mathbb{F}_p$.

There exist different systems of coordinates with different complexities. For many platforms, projective-like coordinates are quite efficient since they do not

Table 1. Precomputations size (in bytes) for various bounds B and sets S

Bound B	25	50	75	100	125	150	175	200
				$S = \{1\}$				
First byte	33	65	96	127	158	190	221	251
First two bytes	54	111	167	223	279	336	392	446
Full expansion	85	293	626	1,084	1,663	2,367	3,195	4,108
				$S = \{1, 5, 7\}$				
First byte	111	214	317	420	523	627	730	829
First two bytes	178	356	534	712	890	1,069	1,247	1,418
Full expansion	286	939	1,962	3,357	5,122	7,261	9,769	12,527
				$S = \{1, 5, 7, 11, 13\}$				
First byte	185	357	529	701	873	1,045	1,216	1,381
First two bytes	300	597	894	1,191	1,488	1,785	2,081	2,366
Full expansion	491	1,589	3,305	5,642	8,598	12,173	16,364	20,972
				$S = \{1, 5, 7, 11, 13, 17, 19, 23, 25\}$				
First byte	334	643	952	1,262	1,571	1,881	2,190	2,487
First two bytes	545	1,079	1,613	2,148	2,682	3,217	3,751	4,264
Full expansion	906	2,909	6,026	10,255	15,596	22,056	29,630	37,947

require any field inversion for addition and doubling, *cf.* [20] for a comparison. Thus, our tests will not involve any inversion. Also, to ease comparisons between different scalar multiplication methods, we will make the standard assumption that S is equivalent to 0.8M. Thus, the complexity of a scalar multiplication will be expressed in terms of a number of field multiplications only and will be denoted by N_M.

Given any curve $E/\mathbb{F}_p$ in Weierstraß form, it is possible to directly obtain $[3]P$ more efficiently than computing a doubling followed by an addition. Until now, all these direct formulas involved at least one inversion, *cf.* [21], but recently, an inversion-free tripling formula has been devised for Jacobian projective coordinates [17]. Our comparisons will be made using this system. In Jacobian coordinates, a point represented by $(X_1 : Y_1 : Z_1)$ corresponds to the affine point $(X_1/Z_1^2, Y_1/Z_1^3)$, if $Z_1 \neq 0$, and to the point at infinity O_E otherwise. A doubling can be done with $4M + 6S$, a tripling with $10M + 6S$ and a mixed addition, *i.e.* an addition between a point in Jacobian coordinates and an affine point, using $8M + 3S$.

With these settings, we display in Figure 1, the number of multiplications N_M required to compute a scalar multiplication on a 200-bit curve with Algorithm 2, for different choices of a_0 and various DBNS methods. Namely, we investigate double-base chains as in [17], window double-base chains with 2 and 8 precomputations, and extended double-base chains with $S_2 = \{1, 5, 7\}$ and $S_8 = \{1, 5, 7, 11, 13, 17, 19, 23, 25\}$, as explained in Section 3.2. Comparisons are done on 1,000 random 200-bit scalar multiples. Note that the costs of the precomputations are not included in the results.

Figure 1 indicates that $a_0 = 120$ is close to the optimal choice for every method. This implies that the value of b_0 should be set to 51. Similar computations have been done for sizes between 250 and 500. It appears that a simple and good heuristic to minimize N_M is to set $a_0 = \lceil 120 \times \text{size}/200 \rceil$ and b_0 accordingly. These values of a_0 and b_0 are used in the remainder for sizes in $[200, 500]$.

In Figure 2, we display the average length of different extended DBNS expansions in function of the size of the scalar multiple n. Results show that the length of a classic double-base chain is reduced by more than 25% with only 2 precomputations and by 43% with 8 precomputations.

In Table 2, we give the average expansion length ℓ, as well as the maximal power a_1 (resp. b_1) of 2 (resp. 3) in the expansion for different methods and different sizes. The symbol $\#\mathcal{P}$ is equal to the number of precomputed points for a given method and the set $\mathcal{S}_m$ contains the first $m + 1$ elements of $\{1, 5, 7, 11, 13, 17, 19, 23, 25\}$. Again, $1,000$ random integers have been considered in each case.

In Table 3, we give the corresponding complexities in terms of the number of multiplications and the gain that we can expect with respect to a window NAF method involving the same number of precomputations.

See [18] for a full version including a similar study for some special curves.

6   Conclusion

In this work, we have introduced a new family of DBNS, called extended DBNS, where the coefficients in the expansion belong to a given digit set $\mathcal{S}$. A scalar multiplication algorithm relying on this representation and involving precomputations was presented. Also, we have described a new method to quickly find the best approximation of an integer by a number of the form $d2^a3^b$ with $d \in \mathcal{S}$. This approach greatly improves the practicality of the DBNS. Extended DBNS sequences give rise to the fastest scalar multiplications known to date for generic elliptic curves. In particular, given a fixed number of precomputations, the extended DBNS is more efficient than any corresponding window NAF method. Gains are especially important for a small number of precomputations, typically up to three points. Improvements larger than 10% over already extremely optimized methods can be expected. Also, this system is more flexible, since it can be used with any given set of coefficients, unlike window NAF methods.

Further research will include an extension of these ideas to Koblitz curves, for which DBNS-based scalar multiplication techniques without precomputations exist already, see [16,22,23]. This will most likely lead to appreciable performance improvements.

Acknowledgements

This work was partly supported by the Canadian NSERC strategic grant #73-2048, *Novel Implementation of Cryptographic Algorithms on Custom Hardware*

Platforms and by a French-Australian collaborative research program from the *Direction des Relations Européennes et Internationales du CNRS*, France.

References

1. Miller, V.S.: Use of elliptic curves in cryptography. In: Advances in Cryptology – Crypto 1985. Volume 218 of Lecture Notes in Comput. Sci. Springer-Verlag, Berlin (1986) 417–426
2. Koblitz, N.: Elliptic curve cryptosystems. Math. Comp. **48** (1987) 203–209
3. Koblitz, N.: Hyperelliptic cryptosystems. J. Cryptology **1** (1989) 139–150
4. Blake, I.F., Seroussi, G., Smart, N.P.: Elliptic curves in cryptography. Volume 265 of London Mathematical Society Lecture Note Series. Cambridge University Press, Cambridge (1999)
5. Hankerson, D., Menezes, A.J., Vanstone, S.A.: Guide to elliptic curve cryptography. Springer-Verlag, Berlin (2003)
6. Avanzi, R.M., Cohen, H., Doche, C., Frey, G., Nguyen, K., Lange, T., Vercauteren, F.: Handbook of Elliptic and Hyperelliptic Curve Cryptography. Discrete Mathematics and its Applications (Boca Raton). CRC Press, Inc. (2005)
7. Blake, I.F., Seroussi, G., Smart, N.P.: Advances in Elliptic Curve Cryptography. Volume 317 of London Mathematical Society Lecture Note Series. Cambridge University Press, Cambridge (2005)
8. Doche, C.: Exponentiation. In: [6] 145–168
9. Morain, F., Olivos, J.: Speeding up the Computations on an Elliptic Curve using Addition-Subtraction Chains. Inform. Theor. Appl. **24** (1990) 531–543
10. Dimitrov, V.S., Jullien, G.A., Miller, W.C.: Theory and applications of the double-base number system. IEEE Trans. on Computers **48** (1999) 1098–1106
11. Miyaji, A., Ono, T., Cohen, H.: Efficient Elliptic Curve Exponentiation. In: Information and Communication – ICICS'97. Volume 1334 of Lecture Notes in Comput. Sci., Springer (1997) 282–291
12. Takagi, T., Yen, S.M., Wu, B.C.: Radix-*r* non-adjacent form. In: Information Security Conference – ISC 2004. Volume 3225 of Lecture Notes in Comput. Sci. Springer-Verlag, Berlin (2004) 99–110
13. Dimitrov, V.S., Jullien, G.A., Miller, W.C.: An algorithm for modular exponentiation. Information Processing Letters **66** (1998) 155–159
14. Berthé, V., Imbert, L.: On converting numbers to the double-base number system. In: In F. T. Luk, editor, Advanced Signal Processing Algorithms, Architecture and Implementations XIV, volume 5559 of Proceedings of SPIE. (2004) 70–78
15. Ciet, M., Sica, F.: An Analysis of Double Base Number Systems and a Sublinear Scalar Multiplication Algorithm. In: Progress in Cryptology – Mycrypt 2005. Volume 3715 of Lecture Notes in Comput. Sci., Springer (2005) 171–182
16. Avanzi, R.M., Sica, F.: Scalar Multiplication on Koblitz Curves using Double Bases. In: proceedings of Vietcrypt 2006. Lecture Notes in Comput. Sci. (2006) See also Cryptology ePrint Archive, Report 2006/067, http://eprint.iacr.org/
17. Dimitrov, V.S., Imbert, L., Mishra, P.K.: Efficient and secure elliptic curve point multiplication using double-base chains. In: Advances in Cryptology – Asiacrypt 2005. Volume 3788 of Lecture Notes in Comput. Sci., Springer (2005) 59–78
18. Doche, C., Imbert, L.: Extended Double-Base Number System with Applications to Elliptic Curve Cryptography (2006) full version of the present paper, see Cryptology ePrint Archive, http://eprint.iacr.org/

19. Doche, C.: A set of PARI/GP functions to compute DBNS expansions `http://www.ics.mq.edu.au/~doche/dbns_basis.gp`
20. Doche, C., Lange, T.: Arithmetic of Elliptic Curves. In: [6] 267–302
21. Ciet, M., Joye, M., Lauter, K., Montgomery, P.L.: Trading inversions for multiplications in elliptic curve cryptography. Des. Codes Cryptogr. **39** (2006) 189–206
22. Dimitrov, V.S., Jarvine, K., Jr, M.J.J., Chan, W.F., Huang, Z.: FPGA Implementation of Point Multiplication on Koblitz Curves Using Kleinian Integers. In: proceedings of CHES 2006. Lecture Notes in Comput. Sci. (2006)
23. Avanzi, R.M., Dimitrov, V.S., Doche, C., Sica, F.: Extending Scalar Multiplication using Double Bases. In: proceedings of Asiacrypt 2006. Lecture Notes in Comput. Sci., Springer (2006)

Appendix: Graphs and Tables

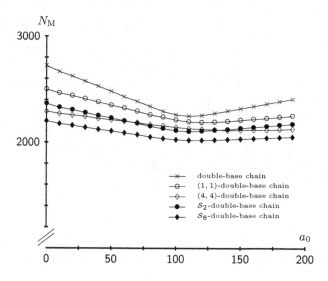

Fig. 1. Average number of multiplications to perform a random scalar multiplication on a generic 200-bit curve with various DBNS methods parameterized by a_0

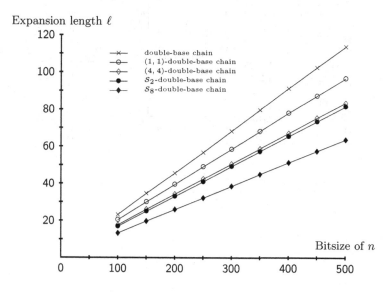

Fig. 2. Average expansion length ℓ of random scalar integers n with various DBNS

Table 2. Parameters for various scalar multiplication methods on generic curves

Size	#$\mathcal{P}$	200 bits ℓ	a_1	b_1	300 bits ℓ	a_1	b_1	400 bits ℓ	a_1	b_1	500 bits ℓ	a_1	b_1
2NAF$_2$	0	66.7	200	0	100	300	0	133.3	400	0	166.7	500	0
Binary/ternary	0	46.1	90.7	68.1	69.2	136.4	102.2	91.9	182.6	136.3	114.4	228.0	170.7
DB-chain	0	45.6	118.7	50.4	68.2	178.7	75.5	91.3	239.0	100.6	113.7	298.6	126.2
3NAF$_2$	1	50	200	0	75	300	0	100	400	0	125	500	0
$(1,0)$-DB-chain	1	46.8	118.9	50.2	70.5	179.1	75.1	94.5	239.3	100.3	117.7	298.8	125.9
$(0,1)$-DB-chain	1	42.9	118.7	50.4	63.8	178.7	75.5	85.4	239.0	100.6	106.4	298.6	126.2
S_1-DB chain	1	36.8	118.1	49.9	55.0	178.0	75.0	72.9	238.2	100.1	91.0	297.8	125.7
2NAF$_3$	2	50.4	0	126	75.6	0	189	100.8	0	252	126	0	315
$(1,1)$-DB-chain	2	39.4	118.9	50.2	58.5	179.1	75.1	77.9	239.3	100.3	96.6	298.8	125.9
S_2-DB chain	2	32.9	117.8	49.8	49.2	177.8	74.9	65.3	238	100.0	81.5	297.7	125.6
4NAF$_2$	3	40	200	0	60	300	0	80	400	0	100	500	0
S_3-DB chain	3	30.7	117.5	49.7	45.7	177.5	74.8	60.6	237.8	99.8	75.6	297.3	125.4
$(2,2)$-DB-chain	4	36.8	119.2	49.8	54.7	179.3	74.8	72.6	239.4	100.1	90.5	299.0	125.7
S_4-DB chain	4	28.9	117.3	49.6	43.2	177.3	74.7	57.6	237.6	99.8	71.5	297.1	125.4
$(3,3)$-DB-chain	6	35.3	119.3	49.5	52.2	179.4	74.6	69.2	239.5	99.6	86.1	299.2	125.2
S_6-DB chain	6	27.3	117.4	49.4	40.6	177.3	74.5	54.0	237.6	99.6	67.1	297	125.3
3NAF$_3$	8	36	0	126	54	0	189	72	0	252	90	0	315
$(4,4)$-DB-chain	8	34.2	119.3	49.3	50.5	179.5	74.2	67.0	239.6	99.3	83.5	299.3	125
S_8-DB chain	8	25.9	117.2	49.3	38.5	177.1	74.4	51.2	237.4	99.5	63.6	296.9	125.2

Table 3. Complexity of various extended DBNS methods for generic curves and gain with respect to window NAF methods having the same number of precomputations

Size	#$\mathcal{P}$	200 bits N_M	Gain	300 bits N_M	Gain	400 bits N_M	Gain	500 bits N_M	Gain
2NAF$_2$	0	2442.9	—	3669.6	—	4896.3	—	6122.9	—
Binary/ternary	0	2275.0	6.87%	3422.4	6.74%	4569.0	6.68%	5712.5	6.70%
DB-chain	0	2253.8	7.74%	3388.5	7.66%	4531.8	7.44%	5666.5	7.45%
3NAF$_2$	1	2269.6	—	3409.6	—	4549.6	—	5689.6	—
$(1,0)$-DB-chain	1	2265.8	0.17%	3410.3	−1.98%	4562.3	−1.72%	5707.4	−1.69%
$(0,1)$-D B-chain	1	2226.5	1.90%	3343.2	1.95%	4471.0	1.73%	5590.4	1.74%
S_1-DB chain	1	2150.4	5.25%	3238.1	5.03%	4326.3	4.91%	5418.1	4.77%
2NAF$_3$	2	2384.8	—	3579.3	—	4773.8	—	5968.2	—
$(1,1)$-DB-chain	2	2188.6	8.23%	3285.5	8.21%	4390.0	8.04%	5487.7	8.05%
S_2-DB chain	2	2106.5	11.67%	3174.1	11.32%	4243.6	11.11%	5314.8	10.95%
4NAF$_2$	3	2165.6	—	3253.6	—	4341.6	—	5429.6	—
S_3- DB chain	3	2078.1	4.04%	3132.8	3.71%	4189.8	3.50%	5248.5	3.34%
$(2,2)$-DB-chain	4	2158.2	—	3242.6	—	4333.1	—	5421.6	—
S_4-DB chain	4	2056.7	—	3105.0	—	4156.1	—	5204.0	—
$(3,3)$-DB-chain	6	2139.4	—	3215.0	—	4291.7	—	5371.9	—
S_6-DB chain	6	2036.3	—	3074.3	—	4115.4	—	5155.1	—
3NAF$_3$	8	2236.2	—	3355.8	—	4475.4	—	5595.0	—
$(4,4)$-DB-chain	8	2125.4	4.95%	3192.2	4.88%	4264.1	4.72%	5340.5	4.55%
S_8-DB chain	8	2019.3	9.70%	3049.8	9.12%	4084.3	8.74%	5116.8	8.55%

CMSS – An Improved Merkle Signature Scheme

Johannes Buchmann[1], Luis Carlos Coronado García[2], Erik Dahmen[1],
Martin Döring[1,*], and Elena Klintsevich[1]

[1] Technische Universität Darmstadt
Department of Computer Science
Hochschulstraße 10, 64289 Darmstadt, Germany
{buchmann, dahmen, doering, klintsev}@cdc.informatik.tu-darmstadt.de
http://www.sicari.de
[2] Banco de México
Av. 5 de Mayo No. 6, 5to piso
Col. Centro C.P. 06059, México, D.F.
coronado@banxico.org.mx

Abstract. The Merkle signature scheme (MSS) is an interesting alternative for well established signature schemes such as RSA, DSA, and ECDSA. The security of MSS only relies on the existence of cryptographically secure hash functions. MSS has a good chance of being quantum computer resistant. In this paper, we propose CMSS, a variant of MSS, with reduced private key size, key pair generation time, and signature generation time. We demonstrate that CMSS is competitive in practice by presenting a highly efficient implementation within the Java Cryptographic Service Provider FlexiProvider. We present extensive experimental results and show that our implementation can for example be used to sign messages in Microsoft Outlook.

Keywords: Java Cryptography Architecture, Merkle Signatures, One-Time-Signatures, Post-Quantum Signatures, Tree Authentication.

1   Introduction

Digital signatures have become a key technology for making the Internet and other IT infrastructures secure. Digital signatures provide authenticity, integrity, and support for non-repudiation of data. Digital signatures are widely used in identification and authentication protocols, for example for software downloads. Therefore, secure digital signature algorithms are crucial for maintaining IT security.

Commonly used digital signature schemes are RSA [RSA78], DSA [Elg85], and ECDSA [JM99]. The security of those schemes relies on the difficulty of factoring large composite integers and computing discrete logarithms. However, it is unclear whether those computational problems remain intractable in the

---

* Author supported by SicAri, a project funded by the German Ministry for Education and Research (BMBF).

R. Barua and T. Lange (Eds.): INDOCRYPT 2006, LNCS 4329, pp. 349–363, 2006.

future. For example, Peter Shor [Sho94] proved that quantum computers can factor integers and can calculate discrete logarithms in the relevant groups in polynomial time. Also, in the past thirty years there has been significant progress in solving the integer factorization and discrete logarithm problem using classical computers (Lenstra and Verheul). It is therefore necessary to come up with new signature schemes which do not rely on the difficulty of factoring and computing discrete logarithms and which are even secure against quantum computer attacks. Such signature schemes are called post-quantum signature schemes.

A very interesting post-quantum signature candidate is the Merkle signature scheme (MSS) [Mer89]. Its security is based on the existence of cryptographic hash functions. In contrast to other popular signature schemes, MSS can only verify a bounded number of signatures using one public key. Also, MSS has efficiency problems (key pair generation, large secret keys and signatures) and was not used much in practice.

Our contribution. In this paper, we present CMSS, a variant of MSS, with reduced private key size, key pair generation time, and signature generation time. We show that CMSS is competitive in practice by presenting a highly efficient CMSS Java implementation in the Java Cryptographic Service Provider Flexi-Provider [Flexi]. This implementation permits easy integration into applications that use the Java Cryptography Architecture [JCA02]. We present experiments that show: As long as no more that 2^{40} documents are signed, the CMSS key pair generation time is reasonable, and signature generation and verification times in CMSS are competitive or even superior compared to RSA and ECDSA. We also show that the CMSS implementation can be used to sign messages in Microsoft Outlook using our `FlexiS/MIME` plug-in [FOP03]. The paper specifies CMSS keys using Abstract Syntax Notation One (ASN.1) [Int02] which guarantees interoperability and permits efficient generation of X.509 certificates and PKCS#12 personal information exchange files. CMSS is based on the Thesis of Coronado [Cor05b] and incorporates the improvements of MSS from [Szy04, DSS05].

Related Work. Szydlo presents a method for the construction of authentication paths requiring logarithmic space and time in [Szy04]. Dods, Smart and Stam give the first complete treatment of practical implementations of hash based digital signature schemes in [DSS05]. In [NSW05], Naor et. al. propose a C implementation of MSS and give timings for up to 2^{20} signatures. A preliminary version of CMSS including security proofs appeared in the PhD thesis of Coronado [Cor05b] and in [Cor05a].

Organization. The rest of this paper is organized as follows: In Section 2, we describe the Winternitz one-time signature scheme and the Merkle signature scheme. In Section 3, we describe CMSS. Section 4 describes details of the CMSS implementation in the FlexiProvider and the ASN.1 specification of the keys. Section 5 presents experimental data including a comparison with standard signature schemes. Section 6 describes the integration of the CMSS implementation into Microsoft Outlook. Section 7 states our conclusions.

2 Preliminaries

Before we describe CMSS in Section 3, we first describe the Winterzitz one-time signature scheme used in CMSS and the Merkle signature scheme (MSS) which CMSS is based on.

2.1 The Winternitz One-Time Signature Scheme

In this section, we describe the Winternitz one-time signature scheme (OTSS) that was first mentioned in [Mer89] and explicitly described in [DSS05]. It is a generalization of the Merkle OTSS [Mer89], which in turn is based on the Lamport-Diffie OTSS [DH76]. The security of the Winternitz OTSS is based on the existence of a cryptographic hash function $H : \{0,1\}^* \rightarrow \{0,1\}^s$ [MOV96]. It uses a block size parameter w that denotes the number of bits that are processed simultaneously. Algorithms 1, 2, and 3 describe the Winternitz OTSS key pair generation, signature generation, and signature verification, respectively.

Algorithm 1. Winternitz OTSS Key Pair Generation

System Parameters: hash function $H : \{0,1\}^* \rightarrow \{0,1\}^s$, parameters $w \in \mathbb{N}$ and
$t = \lceil s/w \rceil + \lceil (\lfloor \log_2 \lceil s/w \rceil \rfloor + 1 + w)/w \rceil$
Output: signature key X, verification key Y
1: choose $x_1, \ldots, x_t \in_R \{0,1\}^s$ uniformly at random.
2: set $X = (x_1, \ldots, x_t)$.
3: compute $y_i = H^{2^w - 1}(x_i)$ for $i = 1, \ldots, t$.
4: compute $Y = H(y_1 || \ldots || y_t)$, where $||$ denotes concatenation.
5: **return** (X, Y).

Algorithm 2. Winternitz OTSS Signature Generation

System Parameters: hash function $H : \{0,1\}^* \rightarrow \{0,1\}^s$, parameters $w \in \mathbb{N}$
and $t = \lceil s/w \rceil + \lceil (\lfloor \log_2 \lceil s/w \rceil \rfloor + 1 + w)/w \rceil$
Input: document d, signature key X
Output: one-time signature σ of d
1: compute the s bit hash value $H(d)$ of document d.
2: split the binary representation of $H(d)$ into $\lceil s/w \rceil$ blocks $b_1, \ldots, b_{\lceil s/w \rceil}$ of length w, padding $H(d)$ with zeros from the left if required.
3: treat b_i as the integer encoded by the respective block and compute the checksum
$$C = \sum_{i=1}^{\lceil s/w \rceil} 2^w - b_i.$$
4: split the binary representation of C into $\lceil (\lfloor \log_2 \lceil s/w \rceil \rfloor + 1 + w)/w \rceil$ blocks $b_{\lceil s/w \rceil + 1}, \ldots, b_t$ of length w, padding C with zeros from the left if required.
5: treat b_i as the integer encoded by the respective block and compute $\sigma_i = H^{b_i}(x_i)$, $i = 1, \ldots, t$, where $H^0(x) := x$.
6: **return** $\sigma = (\sigma_1, \ldots, \sigma_t)$.

Algorithm 3. Winternitz OTSS Signature Verification

System Parameters: hash function $H : \{0,1\}^* \to \{0,1\}^s$, parameters $w \in \mathbb{N}$
  and $t = \lceil s/w \rceil + \lceil (\lfloor \log_2 \lceil s/w \rceil \rfloor + 1 + w)/w \rceil$
Input: document d, signature $\sigma = (\sigma_1, \ldots, \sigma_t)$, verification key Y
Output: TRUE if the signature is valid, FALSE otherwise
  1: compute $b_1, \ldots, b_t$ as in Algorithm 2.
  2: compute $\phi_i = H^{2^w - 1 - b_i}(\sigma_i)$ for $i = 1, \ldots, t$.
  3: compute $\Phi = H(\phi_1 || \ldots || \phi_t)$.
  4: **if** $\Phi = Y$ **then return** TRUE **else return** FALSE

The parameter w makes the Winternitz OTSS very flexible. It allows a trade-off between the size of a signature and the signature and key pair generation times. If w is increased, more bits of $H(d)$ are processed simultaneously and the signature size decreases. But more hash function evaluations are required during key and signature generation. Decreasing w has the opposite effect. In [DSS05], the authors show that using $w = 2$ requires the least number of hash function evaluations per bit.

Example 1. Let $w = 2$ and $H(d) = 110001110$. Hence $s = 9$ and $t = 8$. Therefore we have $(b_1, \ldots, b_5) = (01, 10, 00, 11, 10)$, $C = 12$ and $(b_6, b_7, b_8) = (00, 11, 00)$. The signature of d is $\sigma = \big(H(x_1), H^2(x_2), x_3, H^3(x_4), H^2(x_5), x_6, H^3(x_7), x_8\big)$.

2.2 The Merkle Signature Scheme

The basic Merkle signature scheme (MSS) [Mer89] works as follows. Let $H : \{0,1\}^* \to \{0,1\}^s$ be a cryptographic hash function and assume that a one-time signature scheme (OTSS) is given. Let $h \in \mathbb{N}$ and suppose that 2^h signatures are to be generated that are verifiable with one MSS public key.

MSS Key Pair Generation. At first, generate 2^h OTSS key pairs (X_i, Y_i), $i = 1, \ldots, 2^h$. The X_i are the signature keys. The Y_i are the verification keys. The MSS private key is the sequence of OTSS signature keys. To determine the MSS public key, construct a binary authentication tree as follows. Consider each verification key Y_i as a bit string. The leafs of the authentication tree are the hash values $H(Y_i)$ of the verification keys. Each inner node (including the root) of the tree is the hash value of the concatenation of its two children. The MSS public key is the root of the authentication tree.

MSS Signature Generation. The OTSS key pairs are used sequentially. We explain the calculation of the MSS signature of some document d using the ith key pair (X_i, Y_i). That signature consists of the index i, the ith verification key Y_i, the OTSS signature σ computed with the ith signature key X_i, and the authentication path A for the verification key Y_i. The authentication path A is a sequence of nodes $(a_h, \ldots, a_1)$ in the authentication tree of length h that is constructed as follows. The first node in that sequence is the leaf different from the ith leaf that has the same parent as the ith leaf. Also, if a node N in the sequence is not the last node, then its successor is the node different from N with the same parent as N. Figure 1 shows an example of an authentication path for $h = 2$. Here, the authentication path for Y_2 is the sequence $A_2 = (a_2, a_1)$.

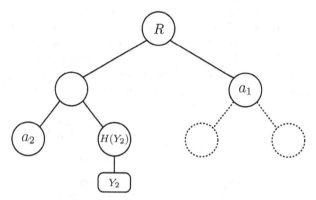

Fig. 1. Merkle's Tree Authentication

MSS Signature Verification. To verify a MSS signature (i, Y, σ, A), the verifier first verifies the one-time signature σ with the verification key Y. If this verification fails, the verifier rejects the MSS signature as invalid. Otherwise, the verifier checks the validity of the verification key Y by using the authentication path A. For this purpose, the verifier constructs a sequence of nodes of the tree of length $h+1$. The first node in the sequence is the ith leaf of the authentication tree. It is computed as the hash $H(Y)$ of the verification key Y. For each node N in the sequence which is not the last node, its successor is the parent P of N in the authentication tree. The verifier can calculate P since the authentication path A included in the signature contains the second child of P. The verifier accepts the signature, if the last node in the sequence is the MSS public key.

3   CMSS

In this section, we describe CMSS. It is an improvement of the Merkle signature scheme (MSS) [Mer89]. A preliminary version of CMSS including security proofs appeared in the PhD thesis of Coronado [Cor05b] and in [Cor05a].

For any $h \in \mathbb{N}$, MSS signs $N = 2^h$ documents using N key pairs of a one-time signature scheme. Unfortunately, for $N > 2^{25}$, MSS becomes impractical because the private keys are very large and key pair generation takes very long.

CMSS can sign $N = 2^{2h}$ documents for any $h \in \mathbb{N}$. For this purpose, two MSS authentication trees, a main tree and a subtree, each with 2^h leafs, are used. The public CMSS key is the root of the main tree. Data is signed using MSS with the subtree. But the root of the subtree is not the public key. That root is authenticated by an MSS signature that uses the main tree. After the first 2^h signatures have been generated, a new subtree is constructed and used to generate the next 2^h signatures. In order to make the private key smaller, the OTSS signature keys are generated using a pseudo random number generator (PRNG) [MOV96]. Only the seed for the PRNG is stored in the CMSS private key.

CMSS key pair generation is much faster than that of MSS, since key generation is dynamic. At any given time, only two trees, each with only 2^h leafs, have to be constructed. CMSS can efficiently be used to sign up to $N = 2^{40}$ documents. Also, CMSS private keys are much smaller than MSS private keys, since only a seed for the PRNG is stored in the CMSS private key, in contrast to a sequence of N OTSS signature keys in the case of MSS. So, CMSS can be used in any practical application. CMSS is illustrated in Figure 2 for $h = 2$.

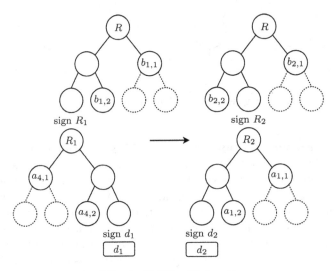

Fig. 2. CMSS with $h = 2$

In the following, CMSS is described in detail. First, we describe CMSS key pair generation. Then, we explain the CMSS signature generation process. In contrast to other signature schemes, the CMSS private key is updated after every signature generation. This is necessary in order to keep the private key small and to make CMSS forward secure [Cor05a]. Such signature schemes are called *key-evolving signature schemes* and were first defined in [BM99].

CMSS Key Pair Generation. Algorithm 6 describes CMSS key pair generation. The algorithm uses two subroutines described in Algorithms 4 and 5. CMSS uses the Winternitz OTSS described in Section 2.1. For the OTSS key pair generation, we use a pseudo random number generator (PRNG) $f : \{0,1\}^s \rightarrow \{0,1\}^s \times \{0,1\}^s$ [MOV96]. In our experiments, we use a PRNG based on SHA1 which is part of the SUN JCE provider [JCA02]. The modified Winternitz OTSS key pair generation process is described in Algorithm 4.

Algorithm 4. Winternitz OTSS key pair generation using a PRNG

System Parameters: PRNG $f : \{0,1\}^s \rightarrow \{0,1\}^s \times \{0,1\}^s$, hash function
   $H : \{0,1\}^* \rightarrow \{0,1\}^s$, parameters $w \in \mathbb{N}$ and $t = \lceil s/w \rceil + \lceil (\lfloor \log_2 \lceil s/w \rceil \rfloor + 1 + w)/w \rceil$
Input: a seed $seed_{in} \in_R \{0,1\}^s$ chosen uniformly at random
Output: a Winternitz OTSS key pair (X, Y) and a seed $seed_{out} \in \{0,1\}^s$
   1: compute $(seed_{out}, s_0) = f(seed_{in})$
   2: **for** $i = 1, \ldots, t$ **do**
   3:     compute $(s_i, x_i) = f(s_{i-1})$
   4: set $X = (x_1, \ldots, x_t)$
   5: compute the verification key Y as in steps 3 and 4 of Algorithm 1
   6: **return** (X, Y) and $seed_{out}$

Algorithm 5 is used to construct a binary authentication tree and its first authentication path. This is done leaf-by-leaf, using a stack for storing intermediate results. Algorithm 5 carries out the computation for one leaf. It is assumed that in addition to the node value, the height of a node is stored. The algorithm is inspired by [Mer89] and [Szy04].

Algorithm 5. Partial construction of an authentication tree

System Parameters: hash function $H : \{0,1\}^* \rightarrow \{0,1\}^s$
Input: a leaf value $H(Y)$, algorithm stack $stack$, sequence of nodes A
Output: updated stack $stack$ and updated sequence A
   1: set $in = H(Y)$
   2: **while** in has same height as top node from $stack$ **do**
   3:     **if** in has greater height than last node in A or A is empty **then**
   4:         append in to A
   5:     pop top node top from $stack$
   6:     compute $in = H(top \| in)$
   7: push in onto $stack$
   8: **return** $stack, A$

CMSS key pair generation is carried out in two parts. First, the first subtree and its first authentication path are generated using Algorithms 4 and 5. Then, the main tree and its first authentication path are computed. The CMSS public key is the root of the main tree. The CMSS private key consists of two indices i and j, three seeds for the PRNG, three authentication paths (of which one is constructed during signature generation), the root of the current subtree and three algorithm stacks for subroutines. The details are described in Algorithm 6.

CMSS Signature Generation. CMSS signature generation is carried out in four parts. First, the MSS signature of document d is computed using the subtree.

Algorithm 6. CMSS key pair generation

System Parameters: hash function $H : \{0,1\}^* \to \{0,1\}^s$, PRNG $f : \{0,1\}^s \to$
$\{0,1\}^s \times \{0,1\}^s$, Winternitz parameter w
Input: parameter $h \in \mathbb{N}$, two seeds $seed_{main}$ and $seed_{sub}$ chosen uniformly at random
in $\{0,1\}^s$
Output: a CMSS key pair $(priv, R)$
1: set $N = 2^h$ and $seed_0 = seed_{sub}$
2: initialize empty stack $stack_{sub}$ and empty sequence of nodes A_1
3: **for** $i = 1, \ldots, N$ **do**
4:      compute $((X_i, Y_i), seed_i) \leftarrow$ Algorithm 4$(seed_{i-1})$
5:      compute $(stack_{sub}, A_1) \leftarrow$ Algorithm 5$(H(Y_i), stack_{sub}, A_1)$
6: let R_1 be the single node in $stack_{sub}$; R_1 is the root of the first subtree
7: set $seed_{next} = seed_N$ and $seed_0 = seed_{main}$
8: initialize empty stack $stack_{main}$ and empty sequence of nodes B_1
9: **for** $j = 1, \ldots, N$ **do**
10:      compute $((X_j, Y_j), seed_j) \leftarrow$ Algorithm 4$(seed_{j-1})$
11:      compute $(stack_{main}, B_1) \leftarrow$ Algorithm 5$(H(Y_j), stack_{main}, B_1)$
12: let R be the single node in $stack_{main}$; R is the root of the main tree
13: initialize empty stacks $stack_{main}, stack_{sub}$, and $stack_{next}$ and empty sequence of
nodes C_1
14: set $priv = (1, 1, seed_{\{main,sub,next\}}, A_1, B_1, C_1, R_1, stack_{\{main,sub,next\}})$
15: **return** $(priv, R)$

Then, the MSS signature of the root of the subtree is computed using the main
tree. Then, the next subtree is partially constructed. Finally, the CMSS private
key is updated.

The CMSS signature generation algorithm uses an algorithm of Szydlo for the
efficient computation of authentication paths. We do not explain this algorithm
here but we refer to [Szy04] for details. We call the algorithm Szydlo.auth.
Input to Szydlo.auth are the authentication path of the current leaf, the seed
for the current tree and an algorithm stack. Output are the next authentication
path and the updated stack. Szydlo.auth needs to compute leaf values of leafs
with higher index than the current leaf. For this purpose, Algorithm 7 is used.
The details of CMSS signature generation are described in Algorithm 8.

Algorithm 7. leafcalc

System Parameters: hash function $H : \{0,1\}^* \to \{0,1\}^s$, PRNG $f : \{0,1\}^s \to$
$\{0,1\}^s \times \{0,1\}^s$
Input: current leaf index i, current seed $seed$, leaf index $j > i$
Output: leaf value $H(Y_j)$ of jth leaf
1: set $seed_0 = seed$
2: **for** $k = 1, \ldots, j - i$ **do** compute $(seed_k, s_0) = f(seed_{k-1})$
3: compute $((X_j, Y_j), seed_{out}) \leftarrow$ Algorithm 4$(seed_{j-i})$
4: **return** $H(Y_j)$

Algorithm 8. CMSS signature generation

System Parameters: hash function $H : \{0,1\}^* \to \{0,1\}^s$

Input: document d, CMSS private key $priv = (i, j, seed_{main}, seed_{sub}, seed_{next},$ $A_i, B_j, C_1, R_j, stack_{main}, stack_{sub}, stack_{next})$

Output: signature sig of d, updated private key $priv$, or STOP if no more signatures can be generated

1: **if** $j = 2^h + 1$ **then** STOP
2: obtain an OTSS key pair: $((X_i, Y_i), seed_{sub}) \leftarrow$ Algorithm $4(seed_{sub})$
3: compute the one-time signature of d: $\sigma_i \leftarrow$ Algorithm $2(d, X_i)$
4: obtain second OTSS key pair: $((X_j, Y_j), seed_{temp}) \leftarrow$ Algorithm $4(seed_{main})$
5: compute the one-time signature of R_j: $\tau_j \leftarrow$ Algorithm $2(R_j, X_j)$
6: set $sig = (i, j, \sigma_i, \tau_j, A_i, B_j)$

7: compute the next authentication path for the subtree:
   $(A_{i+1}, stack_{sub}) \leftarrow$ Szydlo.auth$(A_i, seed_{sub}, stack_{sub})$
   and replace A_i in $priv$ by A_{i+1}
8: partially construct the next subtree:
   $((X_i, Y_i), seed_{next}) \leftarrow$ Algorithm $4(seed_{next})$
   $(stack_{next}, C_1) \leftarrow$ Algorithm $5(H(Y_i), stack_{next}, C_1)$

9: **if** $i < 2^h$ **then** set $i = i + 1$
10: **else**
11:     let R_{j+1} be the single node in $stack_{next}$; R_{j+1} is the root of the $(j+1)$th subtree.
12:     compute the next authentication path for the main tree:
        $(B_{j+1}, stack_{main}) \leftarrow$ Szydlo.auth$(B_j, seed_{main}, stack_{main})$
        and replace B_j in $priv$ by B_{j+1}
13:     replace R_j in $priv$ by R_{j+1}, $seed_{main}$ by $seed_{temp}$, and A_i by C_1
14:     set $i = 1$ and $j = j + 1$
15: **return** the CMSS signature sig of d and the updated private key $priv$

CMSS Signature Verification. CMSS signature verification proceeds in two steps. First, the two authentication paths are validated, then the validity of the two one-time signatures is verified. The details are described in Algorithm 9.

Algorithm 9. CMSS signature verification

System Parameters: hash function $H : \{0,1\}^* \to \{0,1\}^s$

Input: document d, CMSS signature $sig = (i, j, \sigma_i, \tau_j, A_i, B_j)$, CMSS public key R

Output: TRUE if the signature is valid, FALSE otherwise.

1: repeat steps 1 to 3 of Algorithm 3 with input d and σ_i to obtain an alleged verification key Φ_i
2: using Φ_i and A_i, compute the root R_j of the current subtree as in the case of MSS signature verification (see Section 2.2).
3: repeat steps 1 to 3 of Algorithm 3 with input R_j and τ_j to obtain an alleged verification key Ψ_j
4: using Ψ_j and B_j, compute the root Q of the main tree as in the case of MSS.
5: **if** Q is not equal to the CMSS public key R **then return** FALSE
6: verify the one-time signature σ_i of d using Algorithm 3 and verification key Φ_i
7: verify the one-time signature τ_j of R_j using Algorithm 3 and verification key Ψ_j
8: **if** both verifications succeed **return** TRUE **else return** FALSE

4   Specification and Implementation

This section describes parameter choices and details of our CMSS implementation. CMSS is implemented as part of the Java Cryptographic Service Provider (CSP) FlexiProvider [Flexi]. It is therefore possible to integrate the implementation into any application that uses the Java Cryptographic Architecture [JCA02] and Java Cryptography Extension [JCE02]. Our CMSS implementation is available at [Flexi] as open source software.

Scheme Parameters. The hash function H used in the OTSS and the authentication trees can be chosen among SHA1, SHA256, SHA384, and SHA512. The Winternitz parameter w can be chosen among $1, 2, 3$, and 4. As PRNG f, we use a PRNG based on SHA1 which is part of the SUN JCE provider [JCA02]. For each choice, there exists a distinct object identifier (OID) that can be found in Appendix B.

As described earlier, CMSS makes use of the Winternitz OTSS. However, it is possible to replace the Winternitz OTSS by any other one-time signature scheme. If unlike in the case of Winternitz OTSS the verification keys can not be computed from the signature keys, they have to be part of the CMSS signature. Also, the PRNG based on SHA1 can be replaced by any other PRNG.

Key Generation. The CMSS private and public keys are stored using Abstract Syntax Notation One (ASN.1) [Int02]. ASN.1 ensures interoperability between different applications and also allows efficient generation of X.509 certificates and PKCS#12 personal information exchange files. The ASN.1 encoding of the keys can be found in Appendix A. In addition to what was described in Section 3, both the CMSS public and private key contain the OID of the algorithm they can be used with.

Signature Generation and Verification. For the computation of authentication paths, we use the preprint version of the algorithm `Szydlo.auth` which is more efficient than the conference version. See [Szy04] for details.

Each time a new CMSS signature is computed, the signature of the root of the current subtree is recomputed. This reduces the size of the CMSS private key. The time required to recompute this MSS signature is tolerable.

5   Experimental Results

This section compares the CMSS implementation with RSA, DSA, and ECDSA. We compare the times required for key pair generation, signature generation, and signature verification as well as the sizes of the private key, public key, and signatures. For RSA, DSA, and ECDSA, the implementations provided by the Java CSP FlexiProvider are used, which is available at [Flexi] as open source software.

The results are summarized in Table 1. In case of CMSS, the first column denotes the logarithm to the base 2 of the number of possible signatures N. For RSA, DSA, and ECDSA, the column *mod* denotes the size of the modulus. The size of the keys is the size of their DER encoded ASN.1 structure.

The experiments were made using a computer equipped with a Pentium M 1.73GHz CPU, 1GB of RAM and running Microsoft Windows XP.

Table 1. Timings for CMSS, RSA, DSA, and ECDSA

$\log N$	$s_{public\ key}$	$s_{private\ key}$	$s_{signature}$	t_{keygen}	t_{sign}	t_{verify}
CMSS with SHA1, $w = 1$						
20	46 bytes	1900 bytes	7168 bytes	2.9 s	10.2 ms	1.2 ms
30	46 bytes	2788 bytes	7368 bytes	1.5 min	13.6 ms	1.2 ms
40	46 bytes	3668 bytes	7568 bytes	48.8 min	17.5 ms	1.2 ms
CMSS with SHA1, $w = 2$						
20	46 bytes	1900 bytes	3808 bytes	2.6 s	9.2 ms	1.3 ms
30	46 bytes	2788 bytes	4008 bytes	1.4 min	12.4 ms	1.4 ms
40	46 bytes	3668 bytes	4208 bytes	43.8 min	14.9 ms	1.3 ms
CMSS with SHA1, $w = 3$						
20	46 bytes	1900 bytes	2688 bytes	3.1 s	9.7 ms	1.5 ms
30	46 bytes	2788 bytes	2888 bytes	1.5 min	13.2 ms	1.5 ms
40	46 bytes	3668 bytes	3088 bytes	47.8 min	16.9 ms	1.6 ms
CMSS with SHA1, $w = 4$						
20	46 bytes	1900 bytes	2128 bytes	4.1 s	12.5 ms	2.0 ms
30	46 bytes	2788 bytes	2328 bytes	2.0 min	17.0 ms	2.0 ms
40	46 bytes	3668 bytes	2528 bytes	62.3 min	21.7 ms	2.0 ms

mod	$s_{public\ key}$	$s_{private\ key}$	$s_{signature}$	t_{keygen}	t_{sign}	t_{verify}
RSA with SHA1						
1024	162 bytes	634 bytes	128 bytes	0.4 s	13.8 ms	0.8 ms
2048	294 bytes	1216 bytes	256 bytes	3.4 s	96.8 ms	3.0 ms
DSA with SHA1						
1024	440 bytes	332 bytes	46 bytes	18.2 s	8.2 ms	16.2 ms
ECDSA with SHA1						
192	246 bytes	231 bytes	55 bytes	5.1 ms	5.1 ms	12.9 ms
256	311 bytes	287 bytes	71 bytes	9.6 ms	9.8 ms	24.3 ms
384	441 bytes	402 bytes	102 bytes	27.3 ms	27.3 ms	66.9 ms

The table shows that the CMSS implementation offers competitive signing and verifying times compared to RSA, DSA, and ECDSA. The table also shows that a CMSS public key is significantly smaller than a RSA or a DSA public key.

In the case of $N = 2^{40}$, key pair generation takes quite long. However, this does not affect the usability of the implementation, since key pair generation has to be performed only once. Also, the size of the signature and the private key is larger compared to RSA and DSA. While this might lead to concerns regarding memory constrained devices, those sizes are still reasonable in an end-user scenario.

To summarize, CMSS offers a very good trade-off concerning signature generation and verification times compared to RSA and DSA while preserving a reasonable signature and private key size. Appendix C contains a table showing timings for CMSS with SHA256.

6   Signing Messages in Microsoft Outlook with CMSS

Section 5 showed that the space and time requirements of our CMSS implementation are sufficiently small for practical usage. Also, the number of signatures that can be generated is large enough for practical purposes.

The implementation can be easily integrated in applications that use the JCA. An example for such an application is the FlexiS/MIME Outlook plug-in [FOP03], which enables users to sign and encrypt emails using any Java Cryptographic Service Provider in a fast and easy way. The plugin is available at [FOP03] as a free download and is compatible with Microsoft Outlook 98, 2000, 2002, XP and 2003.

In addition to the basic functions like key pair generation, signature generation and verification, the plug-in also supports the generation of self-signed X.509 certificates and PKCS#10 conform certification requests for a certification authority. Furthermore, it is possible to import and export X.509 certificates and PKCS#12 personal information exchange files.

Using the FlexiS/MIME Outlook plug-in in conjunction with the FlexiProvider implementation, we are able to sign emails with CMSS. Furthermore, CMSS can be easily integrated into existing public-key infrastructures.

7   Conclusion

In this paper, we present CMSS, an improved Merkle signature scheme with significantly reduced private key size, key pair generation, and signature generation times. We describe an efficient CMSS FlexiProvider implementation. The implementation provides competitive or even superior timings compared to the commonly used signature schemes RSA, DSA, and ECDSA. This demonstrates that it is already possible today to use quantum computer resistant signature schemes without any loss of efficiency concerning signature generation and verification times and with reasonable signature and key lengths. Using CMSS, it is possible to sign up to 2^{40} messages, while preserving moderate key pair generation times. Because CMSS is implemented as part of a Java Cryptographic Service Provider, it can be used with any application that uses the JCA, e.g. the FlexiS/MIME plug-in, which can be used to sign emails with Microsoft Outlook.

References

[BM99]   M. Bellare and S. Miner. A Forward-Secure Digital Signature Scheme. In *Advances in Cryptology – CRYPTO '99*, number 1666 in LNCS, pages 431–448. Springer, 1999.

[Cor05a]   L. C. Coronado García. On the security and the efficiency of the Merkle signature scheme. Technical Report 2005/192, Cryptology ePrint Archive, 2005. Available at `http://eprint.iacr.org/2005/192/`.

[Cor05b]   L. C. Coronado García. *Provably Secure and Practical Signature Schemes*. PhD thesis, Computer Science Departement, Technical University of Darmstadt, 2005. Available at `http://elib.tu-darmstadt.de/diss/000642/`.

[DH76]   W. Diffie and M. Hellman. New directions in cryptography. *IEEE Transactions on Information Theory*, IT-22(6):644–654, 1976.

[DSS05]   C. Dods, N. P. Smart, and M. Stam. Hash Based Digital Signature Schemes. In *Cryptography and Coding*, number 3796 in LNCS, pages 96–115. Springer, 2005.

[Elg85]   T. Elgamal. A Public Key Cryptosystem and a Signature Scheme Based on Discrete Logarithms. In *Advances in Cryptology – CRYPTO '84*, number 196 in LNCS, pages 10–18. Springer, 1985.

[Flexi]   The FlexiProvider group at Technische Universität Darmstadt. *FlexiProvider, an open source Java Cryptographic Service Provider*, 2001–2006. Available at `http://www.flexiprovider.de/`.

[FOP03]   The FlexiPKI research group at Technische Universität Darmstadt. *The FlexiS/MIME Outlook Plugin*, 2003. Available at `http://www.informatik.tu-darmstadt.de/TI/FlexiPKI/FlexiSMIME/html`.

[Int02]   International Telecommunication Union. *X.680: Information technology - Abstract Syntax Notation One (ASN.1): Specification of basic notation*, 2002. Available at `http://www.itu.int/rec/T-REC-X.680/`.

[JCA02]   Sun Microsystems. *The Java Cryptography Architecture API Specification & Reference*, 2002. Available at `http://java.sun.com/j2se/1.4.2/docs/guide/security/CryptoSpec.html`.

[JCE02]   Sun Microsystems. *The Java Cryptography Extension (JCE) Reference Guide*, 2002. Available at `http://java.sun.com/j2se/1.4.2/docs/guide/security/jce/JCERefGuide.html`.

[JM99]   D. Johnson and A. Menezes. The Elliptic Curve Digital Signature Algorithm (ECDSA). Technical Report CORR 99-34, University of Waterloo, 1999. Available at `http://www.cacr.math.uwaterloo.ca`.

[Mer89]   R. Merkle. A certified digital signature. In *Advances in Cryptology – CRYPTO '89*, number 1462 in LNCS, pages 218–238. Springer, 1989.

[MOV96]   A. J. Menezes, P. C. van Oorschot, and S. A. Vanstone. *Handbook of applied cryptography*. CRC Press, Boca Raton, Florida, 1996. Available at `http://cacr.math.uwaterloo.ca/hac/`.

[NSW05]   D. Naor, A. Shenhav, and A. Wool. One-Time Signatures Revisited: Have They Become Practical? Technical Report 2005/442, Cryptology ePrint Archive, 2005. Available at `http://eprint.iacr.org/2005/442/`.

[RSA78]   R. L. Rivest, A. Shamir, and L. Adleman. A Method for Obtaining Digital Signatures and Public-Key Cryptosystems. *Communications of the ACM*, 21(2):120–126, 1978.

[Sho94]   P. W. Shor. Algorithms for Quantum Computation: Discrete Logarithms and Factoring. In *Proceedings of the 35th Annual IEEE Symposium on Foundations of Computer Science (FOCS 1994)*, pages 124–134. IEEE Computer Society Press, 1994.

[Szy04]   M. Szydlo. Merkle Tree Traversal in Log Space and Time. In *Advances in Cryptology – EUROCRYPT 2004*, number 3027 in LNCS, pages 541–554. Springer, 2004. Preprint version (2003) available at http://szydlo.com/.

A   ASN.1 Encoding

This section describes the specification of the CMSS public and private keys using Abstract Syntax Notation number One (ASN.1) [Int02].

```
CMSSPublicKey  ::= SEQUENCE {
    algorithm       OBJECT IDENTIFIER
    height          INTEGER
    root            OCTET STRING
}


CMSSPrivateKey ::= SEQUENCE {
    algorithm       OBJECT IDENTIFIER
    counterSub      INTEGER
    counterMain     INTEGER
    seedMain        OCTET STRING
    seedSub         OCTET STRING
    seedNext        OCTET STRING
    authMain        AuthPath
    authSub         AuthPath
    authNext        AuthPath
    stackMain       Stack
    stackSub        Stack
    stackNext       Stack
}


AuthPath        ::= SEQUENCE OF OCTET STRING
Stack           ::= SEQUENCE OF OCTET STRING
```

B   Object Identifiers

This section lists the object identifiers (OIDs) assigned to our CMSS implementation. The main OID for CMSS as well as the OID for the CMSSKeyFactory is

$$1.3.6.1.4.1.8301.3.1.3.2$$

The OIDs for CMSS are summarized in the following table, where the column "Hash function" denotes the hash function used in the OTSS and the authentication trees, and the column "w" denotes the Winternitz parameter w.

Table 2. OIDs assigned to CMSS

Hash function	w	Object Identifier (OID)
SHA1	1	1.3.6.1.4.1.8301.3.1.3.2.1
SHA1	2	1.3.6.1.4.1.8301.3.1.3.2.2
SHA1	3	1.3.6.1.4.1.8301.3.1.3.2.3
SHA1	4	1.3.6.1.4.1.8301.3.1.3.2.4
SHA256	1	1.3.6.1.4.1.8301.3.1.3.2.5
SHA256	2	1.3.6.1.4.1.8301.3.1.3.2.6
SHA256	3	1.3.6.1.4.1.8301.3.1.3.2.7
SHA256	4	1.3.6.1.4.1.8301.3.1.3.2.8
SHA384	1	1.3.6.1.4.1.8301.3.1.3.2.9
SHA384	2	1.3.6.1.4.1.8301.3.1.3.2.10
SHA384	3	1.3.6.1.4.1.8301.3.1.3.2.11
SHA384	4	1.3.6.1.4.1.8301.3.1.3.2.12
SHA512	1	1.3.6.1.4.1.8301.3.1.3.2.13
SHA512	2	1.3.6.1.4.1.8301.3.1.3.2.14
SHA512	3	1.3.6.1.4.1.8301.3.1.3.2.15
SHA512	4	1.3.6.1.4.1.8301.3.1.3.2.16

C   CMSS Timings Using SHA256

Table 3. Timings for CMSS with SHA256

$\log N$	$s_{public\ key}$	$s_{private\ key}$	$s_{signature}$	t_{keygen}	t_{sign}	t_{verify}
CMSS with SHA256, $w = 1$						
20	58 bytes	2884 bytes	17672 bytes	7.0 s	23.4 ms	2.9 ms
30	58 bytes	4244 bytes	17992 bytes	3.8 min	32.3 ms	3.3 ms
40	58 bytes	5604 bytes	18312 bytes	120.9 min	41.3 ms	3.3 ms
CMSS with SHA256, $w = 2$						
20	58 bytes	2884 bytes	9160 bytes	6.3 s	19.6 ms	2.8 ms
30	58 bytes	4244 bytes	9480 bytes	3.2 min	27.3 ms	2.8 ms
40	58 bytes	5604 bytes	9800 bytes	101.3 min	34.9 ms	2.9 ms
CMSS with SHA256, $w = 3$						
20	58 bytes	2884 bytes	6408 bytes	7.5 s	23.3 ms	3.7 ms
30	58 bytes	4244 bytes	6728 bytes	3.8 min	31.9 ms	3.7 ms
40	58 bytes	5604 bytes	7048 bytes	120.7 min	40.9 ms	3.7 ms
CMSS with SHA256, $w = 4$						
20	58 bytes	2884 bytes	4936 bytes	10.2 s	31.6 ms	5.1 ms
30	58 bytes	4244 bytes	5256 bytes	5.2 min	43.4 ms	5.1 ms
40	58 bytes	5604 bytes	5576 bytes	165.5 min	55.8 ms	5.1 ms

Constant-Size ID-Based Linkable and Revocable-iff-Linked Ring Signature

Man Ho Au[1], Joseph K. Liu[2], Willy Susilo[1], and Tsz Hon Yuen[3]

[1] Centre for Information Security Research
School of Information Technology and Computer Science
University of Wollongong
Wollongong 2522, Australia
{mhaa456, wsusilo}@uow.edu.au
[2] Department of Computer Science
University of Bristol
Bristol, BS8 1UB, UK
liu@cs.bris.ac.uk
[3] Department of Information Engineering
The Chinese University of Hong Kong
Shatin, N.T., Hong Kong
thyuen4@ie.cuhk.edu.hk

Abstract. In this paper, we propose a new notion called *Revocable-iff-Linked Ring Signature* (R-iff-L Ring Signature). In R-iff-L ring signatures, a signer can sign on behalf of the whole group, just like ordinary ring signatures. However, if he signs twice or more, he can be linked and his identity can be revoked by everyone. We formally define a new security model for the new notion in identity-based (ID-based) setting and propose a constant-size ID-based construction, that is, the size of the signature is *independent* of the size of the group. In addition, we enhance the security model of ID-based linkable ring signature scheme and provide an implementation with constant size setting. Both schemes are provably secure in our new model.

Keywords: Anonymity, Linkable, Revocable, Ring Signature.

1 Introduction

Group-oriented cryptography refers to cryptographic systems in which a group of users are involved. In schemes where participation of one or a proper subset of members is required to complete a process, anonymity refers to whether participants are distinguishable from non-participants. According to [2], anonymity for group-oriented cryptography can be divided into 7 different levels, namely, *Full Anonymity, Linkable Anonymity, Revocable-iff-Linked Anonymity, Revocable Anonymity, Linkable and Revocable Anonymity, Revocable-iff-Linked and Revocable Anonymity* and *No Anonymity*. Examples of group-oriented cryptographic schemes with different levels of anonymity are shown in the following table while interested readers can refer to [2] for a more detailed discussion.

R. Barua and T. Lange (Eds.): INDOCRYPT 2006, LNCS 4329, pp. 364–378, 2006.

Anonymity Level	Examples	Size	Event-Oriented	Ad-hoc
Full	Ring Sign[18]	$O(n)$	N/A	✓
	Anon Ident[11,16]	$O(1)$	N/A	✓
Linkable	Linkable Ring[13]	$O(n)$	×	✓
	Eo-Linkable Ring[24]	$O(n)$	✓	✓
Revocable-iff-Linked				
2-times	E-Cash[6,1],TbL[25]	$O(1)$	×	×
	this paper	$O(1)$	✓	✓
k-times	Compact E-Cash[7]	$O(1)$	×	×
	k-TAA[20]	$O(k)$	✓	×
	dynamic k-TAA[17]	$O(k)$	✓	✓
	constant-size K-TAA[21]	$O(1)$	✓	×
	k-Times Group Signature [2]	$O(1)$	✓	×
Full+OA	Group Signatures	$O(1)$	×	×
Link+OA	Fair E-Cash[8,22]	$O(1)$	×	×

Fig. 1. Examples of group-oriented cryptographic schemes with different levels of anonymity

Ring Signature. Ring signature allows a user to sign on behalf of the whole group, yet no one knows who the actual signer is. The idea was first proposed in by Cramer et al [10] and the notion was formalized by Rivest et al [18]. Variants include threshold setting [26,12,15] and enhanced security [4,9] have been proposed later.

Identity-based Cryptography. In 1984, Shamir [19] introduced the notion of Identity-based (ID-based) cryptography to simplify certificate management. The unique feature of ID-based cryptography is that a user's public key can be any arbitrary string. Since then, many other ID-based signature schemes have been proposed.

In the case of ID-based ring signature, we have to take extra care for the design of schemes. While some of the existing schemes provide anonymity *unconditionally*, others are computational only. The Private Key Generator (PKG) itself may have extra advantage in breaking the anonymity since it is in possession of all the private keys. This problem does not sound serious in normal ID-based ring signature scheme because almost all existing schemes is unconditionally anonymous. However, in the case of linkable ring signatures [13,24,14,23,3] where the verifier is able to determine whether two signatures are signed by the same signer, it is still an open problem to construct one with unconditional anonymity. Within the constraint of computational anonymity, it is a great challenge of providing privacy in an ID-based setting (to the PKG). We require special attention in the design of the scheme.

Contribution. In this paper, we propose a new notion called *Revocable-iff-Linked Ring Signature* which belongs to the Revocable-iff-Linked Anonymity category. In addition, we have the following contributions:

- We formally define a new security model for this notion, in an ID-based setting.
- We provide a constant size concrete implementation. When compared with the scheme in [2], we do not require any setup or group manager. The formation is spontaneous and is suitable for ad-hoc environment, which is a nice inherited property of ring signature.
- We propose a constant size ID-based ring signature scheme which is secure in the enhanced security model.

Organization. The rest of the paper is organized as follow. The enhanced security models of ID-based Linkable Ring Signature scheme and ID-based Revocable-iff-Linked Ring Signature scheme are given in Section 3. Our concrete implementations are presented in Section 4. We conclude the paper in Section 5.

2   Preliminaries

2.1   Notations

Let N be a product of two primes. N is a *safe prime product* if $N = pq = (2p' + 1)(2q' + 1)$ for some primes p, q, p', q' such that p' and q' are of the same length. Denote by $QR(N)$ the group of quadratic residues modulo a safe prime product N.

Let $\hat{e}$ be a bilinear map such that $\hat{e} : \mathbb{G}_1 \times \mathbb{G}_2 \to \mathbb{G}_T$.

- $\mathbb{G}_1$ and $\mathbb{G}_2$ are cyclic multiplicative groups of prime order p.
- each element of $\mathbb{G}_1$, $\mathbb{G}_2$ and $\mathbb{G}_T$ has unique binary representation.
- g_0, h_0 are generators of $\mathbb{G}_1$ and $\mathbb{G}_2$ respectively.
- $\psi : \mathbb{G}_2 \to \mathbb{G}_1$ is a computable isomorphism from $\mathbb{G}_2$ to $\mathbb{G}_1$, with $\psi(h_0) = g_0$.
- (Bilinear) $\forall x \in \mathbb{G}_1$, $y \in \mathbb{G}_2$ and $a, b \in \mathbb{Z}_p$, $\hat{e}(x^a, y^b) = \hat{e}(x, y)^{ab}$.
- (Non-degenerate) $\hat{e}(g_0, h_0) \neq 1$.

$\mathbb{G}_1$ and $\mathbb{G}_2$ can be same or different groups. We say that two groups $(\mathbb{G}_1, \mathbb{G}_2)$ are a bilinear group pair if the group action in $\mathbb{G}_1$, $\mathbb{G}_2$, the isomorphism ψ and the bilinear mapping $\hat{e}$ are all efficiently computable.

2.2   Mathematical Assumptions

Definition 1 (Decisional Diffie-Hellman). *The Decisional Diffie-Hellman (DDH) problem in $\mathbb{G}$ is defined as follows: On input a quadruple $(g, g^a, g^b, g^c) \in \mathbb{G}^4$, output 1 if $c = ab$ and 0 otherwise. We say that the (t, ϵ)-DDH assumption holds in $\mathbb{G}$ if no t-time algorithm has advantage at least ϵ over random guessing in solving the DDH problem in $\mathbb{G}$.*

Definition 2 (q-Strong Diffie-Hellman). *The q-Strong Diffie-Hellman (q-SDH) problem in $(\mathbb{G}_1, \mathbb{G}_2)$ is defined as follow: On input a $(q + 2)$-tuple $(g_0, h_0, h_0^x, h_0^{x^2}, \cdots, h_0^{x^q}) \in \mathbb{G}_1 \times \mathbb{G}_2^{q+1}$, output a pair (A, c) such that $A^{(x+c)} = g_0$ where $c \in \mathbb{Z}_p^*$. We say that the (q, t, ϵ)-SDH assumption holds in $(\mathbb{G}_1, \mathbb{G}_2)$ if no t-time algorithm has advantage at least ϵ in solving the q-SDH problem in $(\mathbb{G}_1, \mathbb{G}_2)$.*

The q-SDH assumption is shown to be true in the generic group model [5].

3 Security Model

3.1 Definition

The security definition of ID-Based Linkable Ring Signature and ID-Based Revocable-iff-Linked Ring Signature are very similar. Therefore we describe the security notions of them together, and their differences are specified at appropriate places.

An ID-Based Linkable (or Revocable-iff-Linked) Ring Signature scheme is a tuple of probabilistic polynomial-time (PPT) algorithms below:

- **Setup.** On input an unary string 1^λ where λ is a security parameter, the algorithm outputs a master secret key s and a list of system parameters param that includes λ and the descriptions of a user secret key space $\mathcal{D}$, a message space $\mathcal{M}$ as well as a signature space Ψ.
- **Extract.** On input a list param of system parameters, an identity $ID_i \in \{0,1\}^*$ for a user and the master secret key s, the algorithm outputs the user's secret key $d_i \in \mathcal{D}$. When we say identity ID_i corresponds to user secret key d_i or vice versa, we mean the pair (ID_i, d_i) is an input-output pair of **Extract** with respect to param and s. Usually this algorithm is executed by a trusted party called Private Key Generator (PKG).
- **Sign.** On input a list param of system parameters, a group size n of length polynomial in λ, a set $\{ID_i \in \{0,1\}^* | i \in [1,n]\}$ of n user identities, a message $m \in \mathcal{M}$, and a secret key $\{d_j \in \mathcal{D} | j \in [1,n]\}$, the algorithm outputs an ID-based linkable (or revocable-iff-linked) ring signature $\sigma \in \Psi$.
- **Verify.** On input a list param of system parameters, a group size n of length polynomial in λ, a set $\{ID_i \in \{0,1\}^* | i \in [1,n]\}$ of n user identities, a message $m \in \mathcal{M}$, a signature $\sigma \in \Psi$, it outputs either valid or invalid.
- **Link.** On input two signatures $\sigma_1, \sigma_2 \in \Psi$, it outputs either link or unlink.
- **Revoke.** (For ID-based revocable-iff-linked ring signature only.) On input two signatures $\sigma_1, \sigma_2 \in \Psi$ such that link $\leftarrow$ **Link**(σ_1, σ_2), it outputs ID.

Correctness. An ID-Based Linkable Ring Signature scheme should satisfy:

- *Verification Correctness* – Signatures signed by honest signers are verified to be invalid with negligible probability.
- *Linking Correctness* – If two signatures are linked, they must be generated from the same secret key of the same signer.

For ID-Based Revocable-iff-Linked Ring Signature, the *Revoking Correctness* require that the output of **Revoke** of two linked signatures must be the actual signer.

3.2 Security Requirement of ID-Based Linkable Ring Signature

A secure ID-Based Linkable Ring Signature scheme should possess *unforgeability*, *anonymity*, *linkability* and *non-slanderability* which will be defined below.

Unforgeability. An adversary should not be able to forge any signature just from the identities of the group members. We specify a security model which mainly captures the following two attacks:

1. Adaptive chosen message attack
2. Adaptive chosen identity attack

Adaptive chosen message attack allows an adversary to obtain message-signature pairs on demand during the forging attack. Adaptive chosen identity attack allows the adversary to forge a signature with respect to a group chosen by the adversary. To support adaptive chosen message attack, we provide the adversary the following oracle queries.

- **Extraction oracle ($\mathcal{EO}$):** On input ID_i, $d_i \leftarrow$ **Extract**(param, ID_i) is returned . The oracle is stateful, meaning that if $ID_i = ID_j$, then $d_i = d_j$.
- **Signing oracle ($\mathcal{SO}$):** $\mathcal{A}$ chooses a group of n identities $\{ID_i\}_{i \in [1,n]}$, a signer identity ID_j among them and a message m, the oracle outputs a valid ID-based linkable (or revocable-iff-linked) ring signature denoted by $\sigma \leftarrow$ **Sign**(param, n, $\{ID_i | i \in [1,n]\}$, m, d_j). The signing oracle may query the extraction oracle during its operation.
- **Hash oracle ($\mathcal{H}$):** $\mathcal{A}$ can ask for the values of the hash functions for any input.

We have the following unforgeability game:

1. A simulator $\mathcal{S}$ takes a sufficiently large security parameter λ and runs **Setup** to generate the public parameters param and master secret key s. The adversary $\mathcal{A}$ is given param.
2. $\mathcal{A}$ can make a polynomial number of oracle queries to $\mathcal{EO}$, $\mathcal{SO}$ and $\mathcal{H}$ adaptively.
3. $\mathcal{A}$ outputs a signature σ^* for message m^* and ring L^*.

$\mathcal{A}$ wins the above game if

1. **Verify**(param, $|L^*|$, L^*, m^*, σ^*) = valid;
2. (L^*, m^*) and σ^* should not be in the set of oracle queries and replies between $\mathcal{A}$ and $\mathcal{SO}$; and
3. $\mathcal{A}$ did not query $\mathcal{EO}$ on any identity $ID \in L^*$.

The advantage of $\mathcal{A}$ is defined as the probability that $\mathcal{A}$ wins.

Definition 3 (Unforgeability). *A scheme is unforgeable if no PPT adversary has non-negligible advantage in winning the above game.*

L-Anonymity. An adversary should not be able to tell the identity of the signer with a probability larger than $1/n$, where n is the cardinality of the ring. A crucial difference between Anonymity for ring signatures and L-Anonymity for linkable ring signatures is that in the latter, the adversary cannot query signatures of a user who appears in the challenge phase. The rationale is that if the adversary obtains signature of user i, it can tell if the challenge signature is generated by this user due to the linking property.

Different from a non-ID-based linkable ring signature scheme, the PKG who knows the master secret key (thus it knows the secret key of every user), may gain advantage on the anonymity of a signature. In order to capture this potential attack, we enhance our model in a way that the adversary is also given the master secret key.

In order to capture the potential attack, we further define the following oracle:

- **Reversed Extraction oracle** ($\mathcal{REO}$): The only difference between $\mathcal{REO}$ and the traditional $\mathcal{EO}$ is that, it is simulated by the adversary instead of the simulator. The initial request can be made by the adversary if the extracted protocol is an interactive one. In this case, the simulator acts as an honest user to provide interactions and the oracle records the necessary transcript of the interaction. Note that this maybe different from the final output of the interaction protocol due to some secret information which is only known to the honest user.

We have the following anonymity game:

1. A simulator $\mathcal{S}$ takes a sufficiently large security parameter λ and runs **Setup** to generate the public parameters param and master secret key s. The adversary $\mathcal{A}$ is given param and s.
2. $\mathcal{A}$ can make a polynomial number of oracle queries to $\mathcal{REO}$, $\mathcal{SO}$ and $\mathcal{H}$ adaptively.
3. In the challenge phase, $\mathcal{A}$ picks two identities ID_1^*, ID_2^*, which are not queried to the $\mathcal{SO}$ as a signer. $\mathcal{A}$ also picks a message m^* and a set of n identities L^*. Then $\mathcal{A}$ receives a challenge signature $\sigma^* = \textbf{Sign}(\text{param}, n + 2, L^* \cup \{ID_1^*, ID_2^*\}, m^*, d_{ID_b^*})$, where $b \in \{0, 1\}$.
4. $\mathcal{A}$ can queries oracles $\mathcal{REO}$, $\mathcal{SO}$ and $\mathcal{H}$ adaptively, where ID_1^*, ID_2^* are not queried to the $\mathcal{SO}$ as a signer.
5. Finally $\mathcal{A}$ outputs a guess $b' \in \{0, 1\}$.

$\mathcal{A}$ wins the above game if $b = b'$. The advantage of $\mathcal{A}$ is defined as the probability that $\mathcal{A}$ wins minus $1/2$.

Definition 4 (Anonymity). *A scheme is anonymous if no PPT adversary has non-negligible advantage in winning the above game.*

Note 1: Although the adversary has the master secret key and it can generate an additional secret key for ID_1^* or ID_2^*, this secret key is different from the one owned by ID_1^* or ID_2^* (generated by $\mathcal{REO}$). According to our definition of *Linking Correctness*, those signatures generated by these two secret keys cannot be linked, although they are corresponding to the same identity.

Linkability. An adversary should not be able to form two signatures with the same secret key without being linked by the **Link** protocol.

We have the following linkability game:

1. A simulator S takes a sufficiently large security parameter λ and runs **Setup** to generate the public parameters param and master secret key s. The adversary A is given param.
2. A can make a polynomial number of oracle queries to $\mathcal{EO}$, $\mathcal{SO}$ and $\mathcal{H}$ adaptively. ·
3. A outputs signatures σ_i^* for messages m_i^* and rings L_i^* for $i \in \{0,1\}$.

Let C be the set of identities queried to $\mathcal{EO}$. A wins the above game if:

- σ_0 and σ_1 are not outputs from $\mathcal{SO}$.
- **Verify**$(\text{param}, |L_i^*|, L_i^*, m_i^*, \sigma_i^*) = $ valid for $i \in \{0,1\}$;
- **Link**$(\sigma_0^*, \sigma_1^*) = $ Unlink; and
- $|(L_0^* \cup L_1^*) \cap C| \le 1$.

The advantage of A is defined as the probability that A wins.

Definition 5 (Linkability). *A scheme is linkable if no PPT adversary has non-negligible advantage in winning the above game.*

Non-slanderability. Informally speaking, non-slanderability ensure that no adversary, can frame an honest user for signing a signature. That is, an adversary cannot produce a valid signature that is linked to a signature generated by a user. In addition to the above oracles, we define one more:

- **Challenged Signing oracle** $(\mathcal{CSO})$: The only difference between $\mathcal{CSO}$ and the traditional $\mathcal{SO}$ is that, it requires the simulator to use the secret key queried from the $\mathcal{REO}$ and execute **Sign** algorithm specified in the scheme to generate the signature. $\mathcal{REO}$ should be queried before if necessary.

Formally it is defined as follow:

1. A simulator S takes a sufficiently large security parameter λ and runs **Setup** to generate param and master secret key s. S sends param and s to the Adversary A.
2. A makes a polynomial number of oracle queries to $\mathcal{REO}$ and $\mathcal{H}$ in an adaptive manner.
3. A submits a polynomial number of oracle queries to $\mathcal{CSO}$ adaptively for generating challenged signatures.
4. A outputs a signature σ^* for message m^* and ring L^*.

A wins the game if

- **Verify**$(\text{param}, |L^*|, L^*, m^*, \sigma^*)$ returns valid.
- σ^* is not an output of any $\mathcal{CSO}$ query.
- **Link**$(\sigma^*, \hat{\sigma}) = $ Link where $\hat{\sigma}$ is any signature outputted from $\mathcal{CSO}$.

Definition 6 (Non-slanderability). *A scheme is non-slanderability if no PPT adversary has non-negligible advantage in winning the above game.*

Note 2: Although the adversary may initialize the query of $\mathcal{REO}$, it cannot get the user secret key since it does not know some secret information which is only known to the honest user (that is, the simulator in this game). Thus it cannot generate a signature by that particular secret key which is linked together with some signatures outputted by $\mathcal{CSO}$. In addition, the remark of Note 1 also applies here.

Theorem 1. *For an ID-based linkable ring signature scheme, if it is linkable and non-slanderable, it implies that it is unforgeable.*

Proof. (sketch) We assume that the scheme is linkable and non-slanderable. Suppose there exists an adversary $\mathcal{A}$ who can forge the signature with non-negligible probability. $\mathcal{A}$ plays the game in Linkability. It submits one query to $\mathcal{EO}$ and produces a signature using this secret key. It forges another signature with another identity as the actual signer. Obviously these two signatures are unlink. That is, it breaks linkability, contradicts our assumption. $\qquad\square$

3.3 Security Requirement of ID-Based Revocable-iff-Linkable Ring Signature

The definitions of unforgeability and anonymity are the same as ID-based Linkable Ring Signature defined in Section 3.2. We skip here.

Revoke-iff-Linkability. An adversary should not be able to form two signatures with the same secret key without being linked by the **Link** protocol or pointed to a user outside the rings.

We have the following linkability game:

1. A simulator $\mathcal{S}$ takes a sufficiently large security parameter λ and runs **Setup** to generate the public parameters param and master secret key s. The adversary $\mathcal{A}$ is given param.
2. $\mathcal{A}$ can make a polynomial number of oracle queries to $\mathcal{EO}$, $\mathcal{SO}$ and $\mathcal{H}$ adaptively.
3. $\mathcal{A}$ outputs signatures σ_i^* for messages m_i^* and rings L_i^* for $i \in \{0,1\}$.

Let C be the set of identities queried to $\mathcal{EO}$. $\mathcal{A}$ wins the above game if it fulfils either condition:

1.   – σ_0 and σ_1 are not outputs from $\mathcal{SO}$.
   - **Verify**(param, $|L_i^*|, L_i^*, m_i^*, \sigma_i^*$) = valid for $i \in \{0,1\}$;
   - **Link**(σ_0^*, σ_1^*) = Unlink; and
   - $|(L_0^* \cup L_1^*) \cap C| \leq 1$.

   OR
2.   – σ_0 and σ_1 are not outputs from $\mathcal{SO}$.
   - **Verify**(param, $|L_i^*|, L_i^*, m_i^*, \sigma_i^*$) = valid for $i \in \{0,1\}$;
   - **Link**(σ_0^*, σ_1^*) = Link; and
   - **Revoke**(σ_0^*, σ_1^*) = ID' where $ID' \notin \{L_0^* \cup L_1^*\}$ or ID' has not been inputted to $\mathcal{EO}$.

The advantage of $\mathcal{A}$ is defined as the probability that $\mathcal{A}$ wins.

Definition 7 (Revoke-iff-Linkability). *A scheme is revocable-iff-linked if no PPT adversary has non-negligible advantage in winning the above game.*

Non-slanderability. The non-slanderability includes the one defined above in Section 3.2 (Def. 6) and the definition of Revoke-iff-linkability (Def. 7).

Definition 8 (Non-slanderability). *A scheme is non-slanderable if no PPT adversary has non-negligible advantage in winning the games defined in Def. 6 and Def. 7.*

4   Our Proposed Schemes

4.1   Construction

System Setup

- **Init (Common parameter):** Let λ be the security parameter. Let $(\mathbb{G}_1, \mathbb{G}_2)$ be a bilinear group pair with computable isomorphism ψ such that $|\mathbb{G}_1| = |\mathbb{G}_2| = p$ for some prime p of λ bits. Let $H : \{0,1\}^* \rightarrow \mathbb{Z}_p$, be a cryptographic hash function. Also assume $\mathbb{G}_p$ be a group of order p where DDH is intractable. Let g_0, g_1, g_2 be generators of $\mathbb{G}_1$, h_0, h_1, h_2 be generators of group $\mathbb{G}_2$ such that $\psi(h_i) = g_i$ for $i = 0, 1, 2$ and u_0, u_1, u_2 be generators of $\mathbb{G}_p$ such that relative discrete logarithm of the generators are unknown. This can be done by setting the generators to be output of a hash function of some publicly known seed.
- **Init (Accumulator):** Choose a generator h of $\mathbb{G}_2$. Randomly select $q \in_R \mathbb{Z}_p^*$ and compute $q_i = h^{(q^i)}$ for $i = 1 \cdots t_{max}$, where t_{max} is the maximum number of accumulation.

PKG Setup: The PKG randomly selects $\gamma \in_R \mathbb{Z}_p^*$ and compute $w = h_0{}^\gamma$. The master secret is γ while the public parameters are $(H, \psi, \mathbb{G}_1, \mathbb{G}_2, \mathbb{G}_p, p, g_0, g_1, g_2, h_0, h_1, h_2, u_0, u_1, u_2, h, q_1, \ldots, q_{t_{max}}, w)$.

Extract: User with identity ID_u obtain the corresponding secret key from PKG through the following interactive protocol.

1. User with identity ID_u randomly selects $s', r_s \in_R \mathbb{Z}_p^*$ and sends $C' = g_1^{s'} g_2^{r_s}$, along with the proof $\Pi_0 = SPK\{(s', r_s) : C' = g_1^{s'} g_2^{r_s}\}$ to PKG.
2. PKG verifies that Π_0 is valid. If it is valid, it randomly selects $s'' \in_R \mathbb{Z}_p^*$ and computes

$$C = C' g_1^{s''} \qquad e = H(ID_u) \qquad A = (g_0 C)^{\frac{1}{e+\gamma}}$$

and sends (A, e, s'') to the user.
3. User computes $s = s' + s''$ and checks if $e(A, wh_0^e) = e(g_0 g_1^s g_2^{r_s}, h_0)$. It then stores (A, e, s, r_s).

Sign(Link Version): For signing a message M, compute

$$v = h^{\Pi_{k=1}^{k=|\{ID\}|}(q+H(ID_k))} \qquad v_w = h^{\Pi_{k=1,k\neq u}^{k=|\{ID\}|}(H(ID_k)+q)} \qquad S = u_0{}^s$$

$$SPK\{(A,e,s,r_s,v_w) : A^{e+\gamma} = g_0 g_1^s g_2^{r_s} \wedge v_w^{e+q} = v \wedge S = u_0{}^s\}(M)$$

Note that S is the linkability tag and $v_w^{(q+H(ID_u))} = v$. This can be turned into event-oriented version by replacing u_0 with $G(event)$ where G is some suitable hash function. The signature contains (v, v_w, S) and the transcript of the SPK.

Sign(Revocable-iff-Link Version): Same as above except adding the following two elements. Compute $T = u_0{}^{r_T}u_1{}^s$ and $Y = u_0^{r_Y}u_1^e$ for some randomly generated $r_T, r_Y \in_R \mathbb{Z}_p^*$ and modify the above SPK to the following:

$$SPK\{(A,e,s,r_s,v_w,r_T,r_Y) : A^{e+\gamma} = g_0 g_1^s g_2^{r_s} \wedge v_w^{e+q} = v \wedge S = u_0{}^s$$
$$\wedge T = u_0{}^{r_T}u_1^s \wedge Y = u_0^{r_Y}u_1^e\}(M)$$

This can be efficiently constructed as a discrete-log relation SPK, by randomly generating some variables $r_1, r_2 \in_R \mathbb{Z}_p^*$ and computing

$$A_1 = g_1{}^{r_1}, \quad A_2 = Ag_2{}^{r_1}, \quad A_3 = g_1{}^{r_2}, \quad A_4 = v_w g_2{}^{r_2}, \quad \alpha = r_1 e, \quad \beta = r_2 e$$

$$SPK\{(r_1, r_2, \alpha, \beta, e, s, r_s, r_T, r_Y) : A_1 = g_1^{r_1} \wedge A_1^e = g_1^\alpha \wedge A_3 = g_1^{r_2} \wedge A_3^e = g_2^\beta$$
$$\wedge \frac{e(g_0, h_0)}{e(A_2, w)} = e(g_1, h_0)^s e(g_2, h_0)^{r_s} e(g_2, w)^{r_1} e(g_2, h_0)^\alpha e(A_2, h_0)^{-e} \wedge S = u_0^s \wedge$$
$$T = u_0^{r_T}u_1^s \wedge Y = u_0^{r_Y}u_1^e \wedge \frac{e(A_4, q_1)}{e(v, h)} = e(g_2, q_1)^{r_2} e(g_2, h)^{-\beta} e(A_4, h)^e\}(M)$$

Finally compute $s_r = r_T - cr_Y$ and $s_e = s - ce$ where c is the challenge used in the above SPK. The signature contains $(v, v_w, S, T, Y, A_1, A_2, A_3, A_4, \alpha, \beta, s_r, s_e)$ and the transcript of the SPK.

Verify: Verify the SPK. For revocable-iff-link version, also check if $T = Y^c u_0^{s_r} u_1^{s_e}$.

Link: Two signatures are linked if the share the same link tag S.

Revoke: Revoke can be done by computing $e = \frac{s_e - s_e'}{c' - c}$. Note that $e = H(ID)$.

Security Analysis is given in the Appendix.

5   Conclusion

In this paper, we proposed a new notion called *Revocable-iff-Linked Ring Signature*. We defined a new model in an ID-based setting and provide a constant-size concrete implementation. We also proposed an ID-based linkable ring signature scheme with constant size space complexity.

References

1. M. Au, S. S. Chow, and W. Susilo. Short e-cash. In *INDOCRYPT '05*, volume 3797 of *LNCS*, pages 332–346. Springer, 2005.
2. M. Au, W. Susilo, and S. Yiu. Event-oriented k-times revocable-iff-linked group signatures. In *ACISP '06*, volume 4058 of *LNCS*, pages 223–234. Springer, 2006.
3. M. H. Au, S. S. M. Chow, W. Susilo, and P. P. Tsang. Short linkable ring signatures revisited. In *EuroPKI '06*, volume 4043 of *LNCS*, pages 101–115. Springer, 2006.
4. A. Bender, J. Katz, and R. Morselli. Ring signatures: Stronger definitions, and constructions without random oracles. In *TCC '06*, volume 3876 of *LNCS*, pages 60–79. Springer, 2006.
5. D. Boneh and X. Boyen. Short Signatures Without Random Oracles. In *EURO-CRYPT '04*, volume 3027 of *LNCS*, pages 56–73. Springer, 2004.
6. S. Brands. Untraceable off-line cash in wallets with observers (extended abstract). In *CRYPTO '93*, volume 773 of *LNCS*, pages 302–318. Springer, 1993.
7. J. Camenisch, S. Hohenberger, and A. Lysyanskaya. Compact E-Cash. In *EURO-CRYPT '05*, volume 3494 of *LNCS*, pages 302–321. Springer, 2005.
8. S. Canard and J. Traoré. On Fair E-cash Systems Based on Group Signature Schemes. In *ACISP '03*, volume 2727 of *LNCS*, pages 237–248. Springer, 2003.
9. S. S. M. Chow, J. K. Liu, V. K. Wei, and T. H. Yuen. Ring signatures without random oracles. In *AsiaCCS '06*, pages 297–302. ACM, 2006.
10. R. Cramer, I. Damgård, and B. Schoenmakers. Proofs of Partial Knowledge and Simplified Design of Witness Hiding Protocols. In *CRYPTO '94*, volume 839 of *LNCS*, pages 174–187. Springer, 1994.
11. Y. Dodis, A. Kiayias, A. Nicolosi, and V. Shoup. Anonymous identification in ad hoc groups. In *EUROCRYPT '04*, volume 3027 of *LNCS*, pages 609–626. Springer, 2004.
12. J. K. Liu, V. K. Wei, and D. S. Wong. A Separable Threshold Ring Signature Scheme. In *ICISC '03*, volume 2971 of *LNCS*, pages 12–26. Springer, 2003.
13. J. K. Liu, V. K. Wei, and D. S. Wong. Linkable Spontaneous Anonymous Group Signature for Ad Hoc Groups (Extended Abstract). In *ACISP '04*, volume 3108 of *LNCS*, pages 325–335. Springer, 2004.
14. J. K. Liu and D. S. Wong. Linkable ring signatures: Security models and new schemes. In *ICCSA '05*, volume 3481 of *LNCS*, pages 614–623. Springer, 2005.
15. J. K. Liu and D. S. Wong. On The Security Models of (Threshold) Ring Signature Schemes. In *ICISC '04*, volume 3506 of *LNCS*, pages 204–217. Springer-Verlag, 2005.
16. L. Nguyen. Accumulators from Bilinear Pairings and Applications. In *CT-RSA '05*, volume 3376 of *LNCS*, pages 275–292. Springer, 2005.
17. L. Nguyen and R. Safavi-Naini. Dynamic k-Times Anonymous Authentication. Cryptology ePrint Archive, Report 2005/168, 2005. http://eprint.iacr.org/.
18. R. L. Rivest, A. Shamir, and Y. Tauman. How to leak a secret. In *ASIACRYPT '01*, volume 2248 of *LNCS*, pages 552–565. Springer, 2001.
19. A. Shamir. Identity-Based Cryptosystems and Signature Schemes. In *CRYPTO '84*, volume 196 of *LNCS*, pages 47–53. Springer, 1984.
20. I. Teranishi, J. Furukawa, and K. Sako. k-Times Anonymous Authentication (Extended Abstract). In *ASIACRYPT '04*, volume 3329 of *LNCS*, pages 308–322. Springer, 2004.
21. I. Teranishi and K. Sake. *k*-times Anonymous Authentication with a Constant Proving Cost. In *PKC '06*, volume 3958 of *LNCS*, pages 525–542. Springer, 2006.
22. M. Trolin. A universally composable scheme for electronic cash. Cryptology ePrint Archive, Report 2005/341, 2005. http://eprint.iacr.org/.

23. P. P. Tsang and V. K. Wei. Short Linkable Ring Signatures for E-Voting, E-Cash and Attestation. In *ISPEC '05*, volume 3439 of *LNCS*, pages 48–60. Springer, 2005.
24. P. P. Tsang, V. K. Wei, T. K. Chan, M. H. Au, J. K. Liu, and D. S. Wong. Separable Linkable Threshold Ring Signatures. In *INDOCRYPT '04*, volume 3348 of *LNCS*, pages 384–398. Springer, 2004.
25. V. K. Wei. Tracing-by-linking group signatures. In *ISC '05*, volume 3650 of *LNCS*, pages 149–163. Springer, 2005.
26. D. S. Wong, K. Fung, J. K. Liu, and V. K. Wei. On the RS-Code Construction of Ring Signature Schemes and a Threshold Setting of RST. In *ICICS '03*, volume 2836 of *LNCS*, pages 34–46. Springer, 2003.

A   Security Analysis

The Revocable-iff-Link Version can be regarded as a generalization of the Link Version. Thus we only show the security analysis of the Revocable-iff-Link Version and the security analysis of the Link Version is straightforward followed directly from the Revocable-iff-Link Version. In rest of this section, we refer "our scheme" as the proposed ID-Based Revocable-iff-Link Ring Signature scheme.

Theorem 2. *Our scheme is anonymous if the DDH assumption in $\mathbb{G}_1$ holds in the random oracle model.*

Proof. (sketch.) By the zero-knowledge property of the SPK in **Sign**, the parameters computed inside the SPK protocol reveal no information about the signer identity. Therefore only the parameters $(A_1, A_2, A_3, A_4, S, T, Y, s_r, s_e)$ may reveal such information.

For the case of (A_1, A_2, A_3, A_4) leaking information, suppose we are given a DDH tuple $(g, g^x, g^y, R) \in \mathbb{G}_1^4$ to determine if $R = g^{xy}$. $\mathcal{S}$ picks the master secret key and sets $g_1 = g, g_2 = g^y$. He simulates all oracles correctly with the master secret key. Then at the challenge phase, $\mathcal{S}$ picks $b \in_R \{0,1\}$ and $z \in_R \mathbb{Z}_p^*$. He sets:

$$A_1^* = g^x, \qquad A_2^* = A_b R, \qquad A_3^* = g^{xz}, \qquad A_4^* = e_{w_b} R^z$$

and simulates the rest of the signature. If $\mathcal{A}$ finally outputs $b' = b$, then $\mathcal{S}$ outputs 1 for the DDH problem. Otherwise, $\mathcal{S}$ outputs 0.

For the case of S leaking information, we can prove similarly by setting $g_1 = g, u_0 = g^y$. For all reversed extraction oracle queries, $\mathcal{S}$ picks $z_i \in_R \mathbb{Z}_p^*$ and sets $C' = g^{xz_i} g_2^{r_s}$. Then at the challenge phase, $\mathcal{S}$ picks $b \in_R \{0,1\}$. Suppose ID_b^* used $C' = g^{xz_b} g_2^{r_{s,b}}$ in $\mathcal{REO}$. Then $\mathcal{S}$ sets $S = R^{z_b} g^{ys_b''}$. and simulates the rest of the signature. If $\mathcal{A}$ finally outputs $b' = b$, then $\mathcal{S}$ outputs 1 for the DDH problem. Otherwise, $\mathcal{S}$ outputs 0.

For the case of (T, Y, s_r, s_e), T is redundant as it can be computed by the other three values and c. We have $s_r = r_T + cr_Y$. As the random number r_T only appears in s_r, s_r leaks no information about r_Y. Then we have $Y = u_0^{r_Y} u_1^e$. As r_Y only appears in s_r and Y, Y leaks no information about e. We also have $C' = g_1^{s'} g_2^{r_s}$. By the zero-knowledge property of the SPK in **Extract**, the SPK leaks no information about r_S. Hence C' leaks no information about s' as r_S

376   M.H. Au et al.

only appears in C'. Finally we have $s_e = s' + s'' - ce$. As s' only appears in C' and s_e, s_e leaks no information about e and s''.

As the parameters $(A_1, A_2, A_3, A_4, S, T, Y, s_r, s_e)$ do not reveal the information about the signer identity, anonymity is achieved if the DDH assumption holds in $\mathbb{G}_1$.

Theorem 3. *Our scheme is non-slanderable if the q-SDH assumption in holds in the random oracle model.*

According to definition 8, our scheme is non-slanderable if there is no PPT adversary can win the game in definition 6 and definition 7.

Lemma 1. *There is no PPT adversary has non-negligible advantage in winning the game defined in definition 6 if the DL assumption holds in the random oracle model.*

Proof. (Sketch.) We first simulate the game in definition 6. Assume there is an adversary $\mathcal{A}$ exists. We are going to construct another PPT $\mathcal{S}$ that makes use of $\mathcal{A}$ to solve the DL problem. $\mathcal{S}$ is given the DL tuple (y, g).

$\mathcal{A}$ setups by randomly picking param and s with $g_1 = g$, $u_0 = g^\mu$ for some random μ. $\mathcal{S}$ sends param and s to $\mathcal{A}$. $\mathcal{A}$ mostly simulates $\mathcal{REO}$, $\mathcal{CSO}$ and $\mathcal{H}$ honestly as a nornal signer. Except for one $\mathcal{REO}$ query ID_i, $\mathcal{S}$ sets $C_1' = yg_2^{r_s}$ for $r_s \in_R \mathbb{Z}_p^*$ and simulates the SPK in **Extract** using the random oracle $\mathcal{H}$. $\mathcal{S}$ obtains (A, e, s'', r_s)

Finally, $\mathcal{A}$ returns a valid signature σ^*, which is not the output from $\mathcal{CSO}$, but is linked to one of them. If it is linked to ID_i, $\mathcal{S}$ rewinds and extracts the SPK to obtain s. Then $\mathcal{S}$ returns $s - s''$ as the solution to the DL problem.

Lemma 2. *There is no PPT adversary has non-negligible advantage in winning the game defined in definition 7 if the q-SDH assumption in holds in the random oracle model.*

Proof. We then simulate the game in definition 7. Assume there is an adversary $\mathcal{A}$ exists. We are going to construct another PPT $\mathcal{S}$ that makes use of $\mathcal{A}$ to solve the q-SDH problem.

Setup. $\mathcal{S}$ receives a q-SDH tuple $(g_1', g_2', g_2'^x, \ldots, g_2'^{x^q})$. $\mathcal{S}$ randomly picks $e_1, \ldots$ $e_{q-1} \in \mathbb{Z}_p^*$ and computes $f(x) = \prod_{i=1}^{q-1}(x + e_i)$. If $x = -e_i$ for some i, $\mathcal{S}$ solves the q-SDH problem directly.

$\mathcal{S}$ uses the q-SDH tuple to compute:

$$h_0 = g_2'^{f(x)}, \qquad w = g_2'^{xf(x)}, \qquad g_0 = \psi(h_0).$$

$\mathcal{S}$ picks $e^*, a^*, k^* \in \mathbb{Z}_p^*$ and computes:

$$h_1 = [(wh_0^{e^*})^{k^*} h_0^{-1}]^{1/a^*} = h_0^{\frac{(e^*+x)k^*-1}{a^*}}, \qquad g_1 = \psi(h_1).$$

$\mathcal{S}$ randomly picks $\mu \in \mathbb{Z}_p^*, h \in \mathbb{G}_2$, sets $g_2 = g_0^\mu$ and sets q_i accordingly. $\mathcal{S}$ computes:

$$A_i = g_0^{1/x+e_i} = \psi(g_2'^{f(x)/x+e_i})$$

for $1 \le i \le q$. $\mathcal{A}$ is given param $= (g_0, g_1, g_2, h_0, w, h, q_1, \ldots, q_{t_{max}})$. For simplicity, denote $e^* = e_q$.

Oracle Simulation. $\mathcal{B}$ simulates the extraction and signing oracles as follow:

(*Hash oracle.*) With probability q/q_H, a new hash oracle query $H(ID)$ will return a new e_i that has never been returned by the hash oracle. Otherwise, $\mathcal{S}$ will return a random number in $\mathbb{Z}_p^*$.

(*Extraction oracle.*) For extraction oracle with input ID_i, if $H(ID_i) \ne e_i$, $\mathcal{S}$ declares failure and exits. Otherwise, $\mathcal{S}$ runs the **Extract** protocol with $\mathcal{A}$, rewinds and extracts (s', r_s) from the PoK. For $i = 1, \ldots, q-1$, $\mathcal{S}$ randomly picks $s'' \in \mathbb{Z}_p^*$ and computes:

$$A = (g_0 C g_2^{r_s})^{1/x+e_i}$$
$$= (g_0^{1+r_s\mu+\frac{(s'+s'')[(e^*+x)k^*-1]}{a^*}})^{1/x+e_i}$$
$$= A_i^{1+r_s\mu-\frac{(s'+s'')}{a^*}} g_0^{\frac{(s'+s'')k^*(e^*+x)}{a^*(e_i+x)}}$$
$$= A_i^{(1+r_s\mu-\frac{(s'+s'')}{a^*})}\left(g_0^{\frac{(s'+s'')k^*}{a^*}}\right)^{(1-\frac{e_i-e^*}{e_i+x})}$$
$$= A_i^{(1+r_s\mu-\frac{(s'+s'')}{a^*})-\frac{(s'+s'')k^*(e_i-e^*)}{a^*}}\left(g_0^{\frac{(s'+s'')k^*}{a^*}}\right)$$

$\mathcal{S}$ returns (A, e_i, s'') to $\mathcal{A}$.

For $i = n$, $\mathcal{S}$ returns $(A_n = g_0^{k^*}, e_n, s'' = a^* - s')$ to $\mathcal{A}$.

(*Signing oracle.*) By controlling the hash function used in the SPK in **Sign**, $\mathcal{S}$ can always generate a correct signature by the soundness property of the SPK.

Output Calculation. If $\mathcal{A}$ wins in the game in definition 7, $\mathcal{A}$ returns a signature σ_i^* for message m_i^* and ring L_i^* for $i = 0, 1$. Assume $\mathcal{A}$ wins by condition 1 of definition 7, then $\mathcal{A}$ must not query $\mathcal{KEO}$ for one ID^* before. WLOG, assume $\mathcal{A}$ didn't query for ID_0. Denote the secret keys for ID_0 used in σ^* as (A, e, s, v_w, r_T, r_Y). Then he must conducted a false proof in part of the SPK such that at least one of the following is fake:

1. $A^{e+x} = g_0 g_1^s g_2^{r_s}$
2. $v_w^{e+q} = v$
3. $S = u_0^s$
4. $T = u_0^{r_T} u_1^s$
5. $Y = u_0^{r_Y} u_1^e$

Item 2 happens with negligible probability under the assumption that the accumulator is secure [16] (which is also reduced to the q-SDH assumption). Item 3,4,5 happens with negligible probability under the DL assumption. Therefore we need to consider item 1.

$\mathcal{S}$ rewinds and extracts (A, e, s, r_s) from the SPK. We have the following possibilities:

- Case 1: $e \notin \{e_1, \ldots, e_n\}$. Then $\mathcal{S}$ computes:

$$A^{e+x} = g_0 g_1^s g_2^{r_s} = g_0^{1 + r_s \mu + \frac{s[(e^* + x)k^* - 1]}{a^*}}$$

$$A = \left(g_0^{\frac{a^* + r_s \mu a^* - s}{a^*(e+x)}}\right)\left[\left(g_0^{\frac{sk^*}{a^*}}\right)^{(1 - \frac{e - e^*}{e + x})}\right]$$

$$B = \left(A g_0^{-\frac{sk^*}{a^*}}\right)^{\frac{a^*}{a^* + r_s \mu a^* - s - sk^*(e - e^*)}}$$

$\mathcal{S}$ returns (B, e) as a new SDH pair.
- Case 2: $e = e_i$ and $A \neq A_i$ for some i. With probability $1/q$, $e = e^*$, $\mathcal{S}$ computes as in case 1:

$$A = \left(g_0^{\frac{a^* + r_s \mu a^* - s}{a^*(e+x)}}\right)\left(g_0^{\frac{sk^*}{a^*}}\right)$$

$$B = \left(A g_0^{-\frac{sk^*}{a^*}}\right)^{\frac{a^*}{a^* + r_s \mu a^* - s}}$$

$\mathcal{S}$ returns (B, e) as a new SDH pair.
- Case 3: $e = e_i$ and $A_0 = A_i$ for some i. We must have $A_i^{e_i + x} g_1^{-s} g_2^{r_s} = A_i^{e_i + x} g_1^{-s_i} g_2^{r_{s_i}}$, implies that $s + \mu r_s = s_i + \mu r_{s_i}$. If $\mathcal{S}$ simulates the game with $\mu = x$ and all other keys and parameters randomly chosen by $\mathcal{S}$, then $\mathcal{S}$ solves the discrete logarithm problem with respect to x. Hence $\mathcal{S}$ can solve the q-SDH problem.

From the new SDH pair, we can solve the q-SDH problem. We have:

$$B = g_1'^{f(x)/(x+e)} = g_1'^{\sum_{i=0}^{q-1} c_i x^i + c_{-1}/(x+e)}$$

where $c_{-1}, c_0, \ldots, c_{q-1}$ can be computed by $\mathcal{S}$ with $c_{-1} \neq 0$. Then $\mathcal{S}$ get:

$$g_1'^{1/(x+e)} = \left(B \prod_{i=0}^{q-1} \psi(g_2'^{x^i})^{-c_i}\right)^{1/c_{-1}}$$

which is the solution to the q-SDH problem.

Now assume $\mathcal{A}$ wins by condition 2 of definition 7. If $ID' \notin \{L_0^* \cup L_1^*\}$, then it contradicts the soundness property of the SPK. If ID' has not been input to the $\mathcal{EO}$, then $\mathcal{S}$ simulates as in the above case.

To conclude the proof of Theorem 3, the scheme is non-slanderable if DL assumption and q-SDH assumption holds in the random oracle model. However if one can solve the DL problem, he can obviously solve the q-SDH problem. Therefore we have the non-slanderability reduced to the q-SDH problem.

Theorem 4. *Our scheme is Revoke-iff-Link if the q-SDH assumption holds in the random oracle model.*

The proof of this theorem overlaps with the proof of Lemma 2.

Secure Cryptographic Workflow
in the Standard Model

M. Barbosa[1] and P. Farshim[2]

[1] Departamento de Informática, Universidade do Minho,
Campus de Gualtar, 4710-057 Braga, Portugal
mbb@di.uminho.pt

[2] Department of Computer Science, University of Bristol,
Merchant Venturers Building, Woodland Road,
Bristol BS8 1UB, United Kingdom
farshim@cs.bris.ac.uk

Abstract. Following the work of Al-Riyami *et al.* we define the notion of key encapsulation mechanism supporting cryptographic workflow (WF-KEM) and prove a KEM-DEM composition theorem which extends the notion of hybrid encryption to cryptographic workflow. We then generically construct a WF-KEM from an identity-based encryption (IBE) scheme and a secret sharing scheme. Chosen ciphertext security is achieved using one-time signatures. Adding a public-key encryption scheme we are able to modify the construction to obtain escrow-freeness. We prove all our constructions secure in the standard model.

Keywords: Cryptographic Workflow, Key Encapsulation, Secret Sharing, Identity-Based Encryption.

1 Introduction

The term *workflow* is used to describe a system in which actions must be performed in a particular order. In *cryptographic workflow* [23] this is achieved by making decryption a privileged action which can only be executed by users which possess an appropriate set of *authorisation credentials*, or simply *credentials*. Credentials are issued by a set of *authorisation authorities*, which can ensure that some action has been performed, or that some event has occurred, before granting them to users. Restricting access to encrypted messages in this way, workflow mechanisms can be implemented with cryptographic security guarantees.

An encryption scheme supporting cryptographic workflow should provide the following functionality [1]. Alice specifies the credentials that Bob should have in a *policy* that she decides before encrypting. Alice should be able to perform this encryption without knowing what credentials Bob actually has. A particular authorisation authority will validate that Bob is entitled to a given credential before awarding it. Each credential acts as a (partial) decryption key. Alice may also want to be sure that no colluding set of these authorisation authorities is able to decrypt and recover the message that she intended for Bob. If this is the case, the system should be *escrow-free*.

R. Barua and T. Lange (Eds.): INDOCRYPT 2006, LNCS 4329, pp. 379–393, 2006.
© Springer-Verlag Berlin Heidelberg 2006

In this paper we introduce the notion of KEMs supporting cryptographic workflow (WF-KEM) and their escrow-free counterparts (EFWF-KEM). We adapt the security models proposed in [1] for encryption schemes accordingly. We argue that the KEM-DEM paradigm introduced by Cramer and Shoup [14] for public-key encryption schemes also applies when one moves to encryption schemes supporting cryptographic workflow. In fact, we show that combining a secure WF-KEM (EFWF-KEM) with a secure DEM, one obtains a secure (escrow-free) encryption scheme supporting cryptographic workflow.

We present a generic construction that permits building WF-KEMs out of simpler cryptographic primitives. This is a generalisation of the construction presented in [1] based on the identity-based encryption (IBE) scheme of Boneh and Franklin. We show how one can construct analogous schemes by replacing its building blocks with other components providing the same functionality. More specifically, we prove that our transformation permits constructing a secure WF-KEM using secure IBE and Secret Sharing (SS) schemes. Finally, we extend our generic construction to obtain an EFWF-KEM using a secure public-key encryption scheme. Chosen ciphertext security is achieved via a one-time signature scheme. Our constructions are all secure in the standard model.

The paper is structured as follows. We first review related work in Section 2 and present the cryptographic primitives we use as building blocks in Section 3. Then in Section 4 we define precisely what we mean by secure WF-KEMs and EFWF-KEMs. In Section 5 we propose generic constructions of these primitives and prove them secure. Finally, in Section 6, we analyse the implications and efficiency of our results for cryptographic workflow and related problems.

2   Related Work

Identity-based cryptography was initially proposed by Shamir [26], who also introduced the first identity-based signature scheme. The first practical identity-based encryption (IBE) scheme is that proposed by Boneh and Franklin in [7], whose operation relies on the use of bilinear maps over groups of points on an elliptic curve. Sakai and Kasahara [24] later proposed another IBE scheme, also based on bilinear maps, but adopting a different key construction. The security of this scheme was established by Chen et al. in [11]. The latter scheme allows for more efficient encryption operation. Both these schemes are secure in the random oracle model (ROM). Recently, Waters [28], Kiltz [20] and Gentry [17] have proposed practical IBE schemes which are secure in the standard model.

The KEM-DEM construction was formalised by Cramer and Shoup in [14]. It captures the concept of hybrid encryption whereby one constructs a public-key encryption scheme by combining a symmetric Data Encapsulation Mechanism (DEM) with an asymmetric Key Encapsulation Mechanism (KEM). The security of the hybrid construction depends, of course, on the security of the KEM and DEM. In [14] it is shown that if the KEM and DEM constructions are individually secure, the resulting public-key encryption scheme will be also secure. The relations between the security notions for KEMs and the conditions for the security of KEM/DEM constructions are further discussed in [22,18] respectively. Dent [15] describes several constructions for secure KEMs. The KEM-DEM paradigm has been extended to the identity-based setting in [6].

Cryptographic workflow follows from the original ideas by Chen *et al.* in [12,13]. There the authors explored the possibilities of using the Boneh and Franklin IBE scheme in a setting where a user can extract different identity-based private keys from multiple TAs. They proposed using *credential descriptors* as public keys, in place of the usual identity strings, and showed that combining the master public keys of the TAs in different ways, one may securely send a message to a recipient and restrict her ability to decrypt it with a high degree of flexibility. Smart [27] applied the same principle to access control. Patterson [23] first employed the term *workflow* to describe this type of scheme.

Key escrow is an inherent property of identity-based cryptography, since it is the TA that computes private keys. This may be a problem in some applications. To solve this (and the issue of certificate management), Al-Riyami and Paterson [2] propose *certificateless public-key cryptography* (CL-PKC). CL-PKC is a modification of identity-based techniques which requires each user to have a (possibly unauthenticated) public key. Messages are encrypted using a combination of a user's public key and its identity.

Al-Riyami *et al.* [1] formalised the definitions of primitives and security models associated with cryptographic workflow and proposed an efficient escrow-free encryption scheme supporting cryptographic workflow. The scheme is based on the Boneh and Franklin IBE and it is proved secure under two security notions. The first one, called *receiver security*, ensures that only users with an appropriate set of credentials can decrypt the message. The second, called *external security*, captures the escrow-freeness notion: it must be unfeasible for any colluding set of TAs to decrypt the message. Unlike CL-PKC, however, escrow-freeness is achieved using a classical public-key encryption layer which relies on public key certification to achieve security.

Encryption schemes supporting cryptographic workflow are very close to those associated with *hidden credential* systems [9,19]. Both types of schemes typically employ a secret sharing layer and an identity-based encryption layer, although the goals in each case are different. In hidden credential systems one seeks to keep the access control policy secret, whereas in workflow schemes this is not the case. Secret sharing schemes are covered in [5,21,25].

A common feature of many schemes proposed for CL-PKC, cryptographic workflow and hidden credentials is that they are based on the concept of multiple encryption (or re-encryption). In multiple encryption, a ciphertext is created by combining the results of several instances of an encryption algorithm with different encryption keys. In the simplest case, where only two decryption keys are involved, the objective is that even if the adversary is in possession of one of those keys, she obtains no advantage. Recently, Dodis and Katz [16] have addressed the chosen ciphertext security of multiple encryptions in the general case, and have proposed generic constructions which are semantically secure. Our constructions build on these results.

3 Building Blocks

Due to space limitations we omit the public key encryption (PKE), data encapsulation mechanism (DEM) and identity-based encryption (IBE) primitive and security model definitions. We refer the reader to the full version of the paper [3], or alternatively to [14,7], for the details.

3.1 Secret Sharing

We follow the approach in [5] for secret sharing over general access structures.

Definition 1. *A collection $\mathcal{P}$ of subsets of a set $P = \{X_1, \ldots, X_n\}$ is called a monotone access structure on P if:*

$$\forall A \in \mathcal{P} \text{ and } \forall B \subseteq P, A \subseteq B \Rightarrow B \in \mathcal{P}.$$

A set $Q \subseteq P$ is called a qualifying subset of P if $Q \in \mathcal{P}$.

The access structures considered in this paper are all monotone and non-trivial *i.e.* $\mathcal{P} \neq \emptyset$. Note that non-triviality implies $P \in \mathcal{P}$.

A secret sharing scheme is defined as a pair of algorithms as follows:

- $\mathbb{S}(1^\kappa, \mathbf{s}, \mathcal{P})$: This is the probabilistic secret sharing algorithm which on input of the security parameter 1^κ, a string $\mathbf{s}$ and a (monotone) access structure $\mathcal{P}$, outputs a list of shares $\mathbf{shr} = (\mathbf{shr}_1, \ldots, \mathbf{shr}_n)$ one for each element in $P = \{X_1, \ldots, X_n\}$ as well as some auxiliary information $\mathbf{aux}$.
- $\mathbb{S}^{-1}(\mathbf{shr}, \mathbf{aux})$: This is the deterministic secret reconstruction algorithm. On input of a list of shares $\mathbf{shr}$ and some auxiliary information $\mathbf{aux}$, outputs a secret $\mathbf{s}$ or a failure symbol $\perp$.

A secret sharing scheme is *sound* if for all access structures $\mathcal{P}$ and strings $\mathbf{s} \in \{0,1\}^*$ of polynomial length in κ, we have:

$$\Pr\left(\mathbf{s} = \mathbb{S}^{-1}(\mathbf{shr}', \mathbf{aux}) \middle| \begin{array}{l} (\mathbf{shr}, \mathbf{aux}) \leftarrow \mathbb{S}(1^\kappa, \mathbf{s}, \mathcal{P}) \\ Q \leftarrow \mathcal{P} \\ \text{Parse } (X_{i_1}, \ldots, X_{i_k}) \leftarrow Q \\ [\mathbf{shr}']_j \leftarrow [\mathbf{shr}]_{i_j}, 1 \leq j \leq k \end{array} \right) = 1.$$

The level of security provided by the secret sharing scheme will influence the overall security of our constructions. We consider both perfect and computational (non-perfect) secret sharing schemes [21].

For perfect secret sharing we will not require a game-based security definition. When necessary, we use an information theoretical argument based on the following definition of security.

Definition 2. *(Perfect Secret Sharing) A secret sharing scheme provides perfect secrecy if every non-qualifying subset of shares does not contain any information about the secret (in the information-theoretic sense). Formally, for any non-empty and non-qualifying set $\{i_1, \ldots, i_n\}$ of an access structure $\mathcal{P}$, and for every* *two secrets* $\sec_0$ *and* $\sec_1$, *let* $(\mathbf{aux}_b, \mathbf{shr}_b) \leftarrow \mathbb{S}(\sec_b, \mathcal{P})$, *for* $b \in \{0,1\}$. *Then, for every possible share value* $\mathbf{shr}_{i_j}$, $1 \leq j \leq n$ *and for every possible* $\mathbf{aux}$ *value*

$$\Pr[\mathbf{shr}_{i_j} = [\mathbf{shr}_0]_{i_j}] = \Pr[\mathbf{shr}_{i_j} = [\mathbf{shr}_1]_{i_j}] \ and \ \Pr[\mathbf{aux} = \mathbf{aux}_0] = \Pr[\mathbf{aux} = \mathbf{aux}_1].$$

Note that, for perfect secret sharing schemes we do not have an asymptotic definition of security, and therefore we drop the security parameter in the primitive definition.

In perfect secret sharing schemes, the secret size constitutes a lower bound on the individual size of shares. To reduce this lower bound, one must relax the security definition and settle for polynomial-time indistinguishability. For computational secret sharing, we shall use the following definitions of semantic security: secret indistinguishability against selective share attacks (IND-SSA), and against adaptive share attacks (IND-CSA).

IND-SSA
1. $(s, \mathbf{s}_0, \mathbf{s}_1, \mathcal{P}^*, i_1, \ldots, i_k) \leftarrow A_1(1^\kappa)$
2. $b \leftarrow \{0,1\}$
3. $(\mathbf{shr}^*, \mathbf{aux}^*) \leftarrow \mathbb{S}(1^\kappa, \mathbf{s}_b, \mathcal{P}^*)$
4. $b' \leftarrow A_2(\mathbf{aux}^*, ([\mathbf{shr}^*]_{i_j})_{j=1}^k, s)$

IND-CSA
1. $(s, \mathbf{s}_0, \mathbf{s}_1, \mathcal{P}^*) \leftarrow A_1(1^\kappa)$
2. $b \leftarrow \{0,1\}$
3. $(\mathbf{shr}^*, \mathbf{aux}^*) \leftarrow \mathbb{S}(1^\kappa, \mathbf{s}_b, \mathcal{P}^*)$
4. $b' \leftarrow A_2^{\mathcal{O}}(\mathbf{aux}^*, s)$

$$\mathrm{Adv}_{\mathsf{SS}}^{\mathsf{IND-atk}}(A) := |\Pr[b' = b] - 1/2|.$$

Here $\mathsf{atk} \in \{\mathsf{SSA}, \mathsf{CSA}\}$. In the SSA model, $k \leq n$, and $\{i_1, \ldots, i_k\}$ must not include a qualifying set of shares in $\mathcal{P}^*$. In the CSA model, $\mathcal{O}$ is a share extraction oracle subject to the condition that the adversary cannot extract a set of shares corresponding to a qualifying set in $\mathcal{P}^*$.

3.2   One-Time Signature

In our constructions we achieve chosen ciphertext security using an adaptation of the technique by Canetti *et al.* [10] which is based on a one-time signature (OTS) scheme. An OTS is a weak form of signature in which the signing/verification key pair can only be used once. More specifically, an OTS is defined by a three-tuple of PPT algorithms:

- $\mathbb{G}_{\mathsf{OTS}}(1^\kappa)$: This is the key generation algorithm which, on input of the security parameter, outputs a key pair $(\mathbf{vk}, \mathbf{sk})$.
- $\mathsf{Sig}(\mathbf{m}, \mathbf{sk})$: This is the signature algorithm, which takes a message $\mathbf{m}$ and a secret key $\mathbf{sk}$, and returns a signature σ.
- $\mathsf{Ver}(\mathbf{m}, \sigma, \mathbf{vk})$: This is the deterministic verification algorithm which, given a message, a signature σ and a verification key returns either 0 (*reject*) or 1 (*accept*).

The *strong unforgeability* security of an OTS is defined through the following game in which any PPT adversary must have negligible advantage.

UF
1. $(\mathrm{vk}, \mathrm{sk}) \leftarrow \mathbb{G}_{\mathrm{OTS}}(1^\kappa)$
2. $(\mathrm{m}, s) \leftarrow A_1(\mathrm{vk})$
3. $\sigma \leftarrow \mathrm{Sig}(\mathrm{sk}, \mathrm{m})$
4. $(\mathrm{m}', \sigma') \leftarrow A_2(s, \sigma)$

$$\mathrm{Adv}_{\mathrm{OTS}}^{\mathrm{UF}}(A) := \Pr[(\sigma', \mathrm{m}') \neq (\mathrm{m}, \sigma) \wedge \mathrm{Ver}(\mathrm{m}', \sigma', \mathrm{vk}) = 1].$$

Note that this unforgeability definition implies that it must be unfeasible to create a new valid signature for a previously signed message. OTS schemes meeting this security definition can be constructed from any one-way function.

4   KEM Primitives for Cryptographic Workflow

4.1   Access Structures, Policies and Credentials

We first explain how we treat access structures in our constructions. We follow an approach similar to that in [1], but we briefly clarify this point stating our assumptions on their meaning in real life.

Suppose that we would like to encrypt a message such that only British nationals can read. To achieve this, we need a TA who issues *credentials* only to those who possess British nationality. For example, the Home Office would be the obvious TA to issue British Nationality certificates. However, it could be the case that two or more TAs are able to issue such a credential. For instance, the user's employer could, after checking the appropriate documentation, grant her a similar credential. We therefore need to specify precisely which authority we are trusting. The need for this is even more apparent when the policy is more complex. Consider the policy English $\wedge$ English $\wedge$ Adult, where the first two terms refer to nationality and language with credentials issued by the Home Office and the British Council, respectively. It could also be the case that the same authority issues credentials on age and nationality: it is up to the authority to interpret the semantics.

For this reason, we view a policy term as a pair $(\mathrm{ID}, \mathrm{Mpk})$ where $\mathrm{ID} \in \{0, 1\}^*$ is an identifier for the policy term and Mpk is the public key of the authority issuing the credential described in ID. We denote by m the number of distinct TAs present in the system, by n the number of distinct policy terms in an access structure and by k the number of distinct policy terms in a qualifying set.

4.2   KEMs Supporting Cryptographic Workflow

A *key encapsulation mechanism supporting cryptographic workflow* (WF-KEM) is defined as a four-tuple of polynomial time (PT) algorithms as follows:

- $\mathbb{G}_{\mathrm{WF-KEM}}(1^\kappa, m)$: This is the probabilistic authority key generation algorithm which on input of a security parameter 1^κ outputs m authority secret/public key pairs $((\mathrm{Msk}_i, \mathrm{Mpk}_i))_{i=1}^m$, as well as the descriptions of the key, randomness and ciphertext spaces. These are denoted by $\mathbb{K}_{\text{WF-KEM}}$, $\mathbb{R}_{\text{WF-KEM}}$ and $\mathbb{C}_{\text{WF-KEM}}$, respectively[1].

- $\mathbb{X}_{\text{WF-KEM}}(X, \text{Msk})$: This is the probabilistic credential extraction algorithm which on input of a policy term X, consisting of a policy identifier/authority public key pair (ID, Mpk), and the secret key Msk corresponding to Mpk, outputs a pair (crd, X) which we call a *credential*.

- $\mathbb{E}_{\text{WF-KEM}}(\mathcal{P})$: This is the probabilistic key encapsulation algorithm which on input of an access structure $\mathcal{P}$ on n policy terms $P = \{X_1, \ldots, X_n\}$ outputs a pair $(\mathbf{k}, \mathbf{c})$ where $\mathbf{k} \in \mathbb{K}_{\text{WF-KEM}}$ and $\mathbf{c}$ is an encapsulation of $\mathbf{k}$.

- $\mathbb{D}_{\text{WF-KEM}}(\mathbf{c}, \mathbf{crd})$: This is the deterministic decapsulation algorithm which on input of an encapsulation $\mathbf{c}$ and a list of k credentials $\mathbf{crd}$, outputs a key or a failure symbol $\perp$.

A WF-KEM scheme is called *sound* if for every policy $\mathcal{P}$ on n terms with $m, n \in \mathbb{N}$ we have:

$$
\Pr\left(\mathbf{k} = \mathbb{D}_{\text{WF-KEM}}(\mathbf{c}, \mathbf{crd}) \,\middle|\,
\begin{array}{l}
((\text{Msk}_i, \text{Mpk}_i))_{i=1}^{m} \leftarrow \mathbb{G}_{\text{WF-KEM}}(1^\kappa, m) \\
(\mathbf{k}, \mathbf{c}) \leftarrow \mathbb{E}_{\text{WF-KEM}}(\mathcal{P}) \\
Q \leftarrow \mathcal{P} \\
\text{Parse } (X_{i_1}, \ldots, X_{i_k}) \leftarrow Q \\
[\mathbf{crd}]_j \leftarrow \mathbb{X}_{\text{WF-KEM}}(X_{i_j}, \text{Msk}_{i_j}), 1 \le j \le k
\end{array}
\right) = 1.
$$

The security games against chosen credential and ciphertext attacks for a WF-KEM are defined as follows. As in [1] we call this notion *recipient security*.

(m, n)-IND-atk
1. $((\text{Msk}_i, \text{Mpk}_i))_{i=1}^{m} \leftarrow \mathbb{G}_{\text{WF-KEM}}(1^\kappa, m)$
2. $(s, \mathcal{P}^*) \leftarrow A_1^{\mathcal{O}_1}(\text{Mpk}_1, \ldots, \text{Mpk}_m)$
3. $\mathbf{k}_0 \leftarrow \mathbb{K}_{\text{WF-KEM}}$
4. $(\mathbf{k}_1, \mathbf{c}^*) \leftarrow \mathbb{E}_{\text{WF-KEM}}(\mathcal{P}^*)$
5. $b \leftarrow \{0, 1\}$
6. $b' \leftarrow A_2^{\mathcal{O}_2}(\mathbf{k}_b, \mathbf{c}^*, s)$

$\text{Adv}_{\text{WF-KEM}}^{(m,n)-\text{IND-atk}}(A) := |\Pr[b' = b] - 1/2|.$

Here $\mathcal{P}^*$ must be on n terms; $\mathcal{O}_1$ and $\mathcal{O}_2$ contain credential extraction and decapsulation oracles subject to the following restrictions: the set of queries that the adversary makes to the credential extraction oracle must not form a qualifying set of $\mathcal{P}^*$; the adversary cannot query the decapsulation oracle on $\mathbf{c}^*$.

We distinguish adaptive ($\text{atk} = \text{CCCA}$) and non-adaptive ($\text{atk} = \text{CCCA}^-$) attacks. The difference is that in non-adaptive attacks, the adversary is not allowed to query the extraction oracle on any $X = (\text{ID}, \text{Mpk})$ with $X \in P^*$ in the second stage of the game. A WF-KEM scheme is called IND-CCCA (IND-CCCA$^-$)

---

[1] From this point on we assume that the public keys of various primitives in this paper include these as well as the security parameter.

secure if all PPT attackers have negligible advantage in the above game as a function of the security parameter.

Note that WF-KEMs are intrinsically multi-user, as anyone who is able to obtain a qualifying set of credentials will be capable of decapsulating. However, in most practical cases this probably will not be the case, as the credential policy term semantics will include the intended recipient's identity. This is related to another important characteristic of WF-KEMs. Any colluding set of TAs who can produce a qualifying set of credentials are also able to invert the encapsulation, and this means that a WF-KEM is not escrow-free.

4.3   KEMs Supporting Escrow-Free Cryptographic Workflow

The notion of a *KEM supporting escrow-free cryptographic workflow* (EFWF-KEM) implies modifying the previous primitive to remove recipient ambiguity. We follow an approach similar to [1] and [2] whereby the primitive is extended to include a recipient public and private key pair.

EFWF-KEMs are defined through five PT algorithms. Four of these algorithms are analogous to those defined for WF-KEMs. In addition to these we add an extra user key generation algorithm:

- $\mathbb{G}_{\text{EFWF-KEM}}(1^\kappa, m)$: This is the probabilistic authority key generation algorithm which on input of a security parameter 1^κ outputs m authority secret/public key pairs $((\text{Msk}_i, \text{Mpk}_i))_{i=1}^m$, as well as the descriptions of the key, randomness and ciphertext spaces. These are denoted by $\mathbb{K}_{\text{EFWF-KEM}}$, $\mathbb{R}_{\text{EFWF-KEM}}$ and $\mathbb{C}_{\text{EFWF-KEM}}$, respectively.
- $\mathbb{G}_{\text{EFWF-KEM}}^{\text{U}}(1^\kappa)$: This is the probabilistic user key generation algorithm which on input of the security parameter 1^κ outputs a private/public key pair (SK, PK).
- $\mathbb{X}_{\text{EFWF-KEM}}(X, \text{Msk})$: This is the probabilistic credential extraction algorithm which on input of a policy term X, consisting of a policy identifier/authority public key pair (ID, Mpk), and the secret key Msk corresponding to Mpk, outputs a pair (crd, X) which we call a *credential*.
- $\mathbb{E}_{\text{EFWF-KEM}}(\mathcal{P}, \text{PK})$: This is the probabilistic key encapsulation algorithm which on input of an access structure $\mathcal{P}$ on n policy terms $P = \{X_1, \ldots, X_n\}$ and a public key PK, outputs a pair (k, c) where $\text{k} \in \mathbb{K}_{\text{EFWF-KEM}}$ and c is an encapsulation of k.
- $\mathbb{D}_{\text{EFWF-KEM}}(\text{c}, \text{crd}, \text{SK})$: The deterministic decapsulation algorithm which on input of an encapsulation c, a list of k credentials crd, and a secret key SK outputs a key or a failure symbol $\perp$.

An EFWF-KEM scheme is called *sound* if for every policy $\mathcal{P}$ on n terms with $m, n \in \mathbb{N}$ we have:

$$\Pr\left(\text{k} = \mathbb{D}_{\text{EFWF-KEM}}(\text{c}, \text{crd}, \text{SK}) \middle| \begin{array}{l} ((\text{Msk}_i, \text{Mpk}_i))_{i=1}^m \leftarrow \mathbb{G}_{\text{EFWF-KEM}}(1^\kappa, m) \\ (\text{SK}, \text{PK}) \leftarrow \mathbb{G}_{\text{EFWF-KEM}}^{\text{U}}(1^\kappa) \\ (\text{k}, \text{c}) \leftarrow \mathbb{E}_{\text{EFWF-KEM}}(\mathcal{P}, \text{PK}) \\ Q \leftarrow \mathcal{P}; \text{Parse } (X_{i_1}, \ldots, X_{i_k}) \leftarrow Q \\ [\text{crd}]_j \leftarrow \mathbb{X}_{\text{EFWF-KEM}}(X_{i_j}, \text{Msk}_{i_j}), 1 \le j \le k \end{array}\right) = 1.$$

Recipient security for an EFWF-KEM is defined through a game very similar to that presented for a WF-KEM. The only difference is that here the adversary is provided with a user key pair which is generated at the beginning of the game. This captures the notion that even the user who knows the private key must possess a qualifying set of credentials to decapsulate. The game is specified below on the left. Again, $\mathcal{P}^*$ must be on at most n terms; the $\mathcal{O}_1$ and $\mathcal{O}_2$ oracles are exactly as in the previous game for adaptive ($\mathtt{atk} = \mathtt{CCCA}$) and non-adaptive chosen credential attacks ($\mathtt{atk} = \mathtt{CCCA}^-$).

(m,n)-IND-atk

1. $((\mathrm{Msk}_i, \mathrm{Mpk}_i))_{i=1}^{m} \leftarrow \mathbb{G}_{\mathrm{EFWF-KEM}}(1^\kappa)$
2. $(\mathrm{SK}, \mathrm{PK}) \leftarrow \mathbb{G}_{\mathrm{EFWF-KEM}}^{\mathrm{U}}(1^\kappa)$
3. $(s, \mathcal{P}^*) \leftarrow A_1^{\mathcal{O}_1}((\mathrm{Mpk}_1)_{i=1}^{m}, \mathrm{SK}, \mathrm{PK})$
4. $\mathrm{k}_0 \leftarrow \mathbb{K}_{\mathrm{EFWF-KEM}}$
5. $(\mathrm{k}_1, \mathbf{c}^*) \leftarrow \mathbb{E}_{\mathrm{EFWF-KEM}}(\mathcal{P}^*, \mathrm{PK})$
6. $b \leftarrow \{0, 1\}$
7. $b' \leftarrow A_2^{\mathcal{O}_2}(\mathrm{k}_b, \mathbf{c}^*, s)$

(m,n)-IND-CCA2

1. $((\mathrm{Msk}_i, \mathrm{Mpk}_i))_{i=1}^{m} \leftarrow \mathbb{G}_{\mathrm{EFWF-KEM}}(1^\kappa)$
2. $(\mathrm{SK}, \mathrm{PK}) \leftarrow \mathbb{G}_{\mathrm{EFWF-KEM}}^{\mathrm{U}}(1^\kappa)$
3. $(s, \mathcal{P}^*) \leftarrow A_1^{\mathcal{O}_1}((\mathrm{Msk}_i, \mathrm{Mpk}_i)_{i=1}^{m}, \mathrm{PK})$
4. $\mathrm{k}_0 \leftarrow \mathbb{K}_{\mathrm{EFWF-KEM}}$
5. $(\mathrm{k}_1, \mathbf{c}^*) \leftarrow \mathbb{E}_{\mathrm{EFWF-KEM}}(\mathcal{P}^*, \mathrm{PK})$
6. $b \leftarrow \{0, 1\}$
7. $b' \leftarrow A_2^{\mathcal{O}_2}(\mathrm{k}_b, \mathbf{c}^*, s)$

To capture escrow-freeness, we follow the approach in [1] and define *external security* through the indistinguishability game shown above on the right. Note that the adversary controls everything except the user secret key. Here $\mathcal{P}^*$ must be on n terms; $\mathcal{O}_1$ and $\mathcal{O}_2$ denote a decapsulation oracle subject to the restriction that the adversary cannot query it on $\mathbf{c}^*$. An EFWF-KEM scheme is called IND-CCCA (IND-CCCA$^-$) and IND-CCA2 secure if all PPT attackers have negligible advantage in the above games as a function of the security parameter, where advantages are defined as

$$\mathrm{Adv}_{\mathrm{EFWF-KEM}}^{(m,n)-\mathrm{IND-atk}}(A) := |\Pr[b' = b] - 1/2|,$$

$$\mathrm{Adv}_{\mathrm{EFWF-KEM}}^{(m,n)-\mathrm{IND-CCA2}}(A) := |\Pr[b' = b] - 1/2|.$$

4.4 Hybrid Encryption Supporting Cryptographic Workflow

The concept and security model of an encryption scheme supporting escrow-free cryptographic workflow (EFWF-ENC), as proposed in [1], are defined in a very similar manner to an EFWF-KEM. We refer the reader to [3] for the details. Using an EFWF-KEM and a standard DEM with compatible key spaces, one can construct a hybrid encryption scheme supporting escrow-free cryptographic workflow in the usual way:

$\mathbb{E}_{\mathrm{EFWF-ENC}}(\mathbf{m}, \mathcal{P}, \mathrm{PK})$
 $-$ $(\mathbf{k}, \bar{\mathbf{c}}) \leftarrow \mathbb{E}_{\mathrm{EFWF-KEM}}(\mathcal{P}, \mathrm{PK})$
 $-$ $\mathbf{c} \leftarrow \mathbb{E}_{\mathrm{DEM}}(\mathbf{m}, \mathbf{k})$
 $-$ $\mathbf{c} \leftarrow (\bar{\mathbf{c}}, \mathbf{c})$
 $-$ Return $\mathbf{c}$

$\mathbb{D}_{\mathrm{EFWF-ENC}}(\mathbf{c}, \mathbf{crd}, \mathrm{SK})$
 $-$ $(\bar{\mathbf{c}}, \mathbf{c}) \leftarrow \mathbf{c}$
 $-$ $\mathbf{k} \leftarrow \mathbb{D}_{\mathrm{EFWF-KEM}}(\bar{\mathbf{c}}, \mathbf{crd}, \mathrm{SK})$
 $-$ If $\mathbf{k} = \perp$ then return $\perp$
 $-$ $\mathbf{m} \leftarrow \mathbb{D}_{\mathrm{DEM}}(\mathbf{c}, \mathbf{k})$
 $-$ Return $\mathbf{m}$

In the full version of paper [3] we prove the following theorem relating the security of this hybrid encryption scheme to that of its EFWF-KEM and DEM components. We use a technique similar to that in [14]. A similar result holds for non-escrow free primitives.

Theorem 1. *The hybrid EFWF-ENC scheme as constructed above is secure in the recipient and external security models if the underlying EFWF-KEM and DEM are secure. More precisely, for* atk $\in$ {CCCA, CCCA$^-$} *we have:*

$$\text{Adv}_{\text{EFWF-ENC}}^{(m,n)-\text{IND}-\text{atk}}(A) \leq 2 \cdot \text{Adv}_{\text{EFWF-KEM}}^{(m,n)-\text{IND}-\text{atk}}(B_1) + \text{Adv}_{\text{DEM}}^{\text{FG}-\text{CCA}}(B_2),$$

$$\text{Adv}_{\text{EFWF-ENC}}^{(m,n)-\text{IND}-\text{CCA2}}(A) \leq 2 \cdot \text{Adv}_{\text{EFWF-KEM}}^{(m,n)-\text{IND}-\text{CCA2}}(B_1) + \text{Adv}_{\text{DEM}}^{\text{FG}-\text{CCA}}(B_2).$$

5   Generic Constructions

5.1   A WF-KEM Construction

We first present a construction of a WF-KEM using an IBE, a secret sharing scheme and a one-time signature scheme.

The authority key generation and credential extraction algorithms of the resulting WF-KEM are direct adaptations of the master key generation and secret key extraction algorithms of the underlying IBE:

- $\mathbb{G}_{\text{WF-KEM}}(1^\kappa, m)$: Runs the $\mathbb{G}_{\text{IBE}}(1^\kappa)$ algorithm m times obtaining (Mpk, Msk). The key space is $\mathbb{K}_{\text{WF-KEM}} = \{0,1\}^\kappa$.
- $\mathbb{X}_{\text{WF-KEM}}(X, \text{Msk})$: Parses X to get (ID, Mpk), extracts $\text{crd} = \mathbb{X}_{\text{IBE}}(\text{ID}, \text{Msk})$ and returns (crd, X).

The encapsulation and decapsulation algorithms are as follows.

$\mathbb{E}_{\text{WF-KEM}}(\mathcal{P})$
- (vk, sk) $\leftarrow \mathbb{G}_{\text{OTS}}(1^\kappa)$
- k $\leftarrow \mathbb{K}_{\text{WF-KEM}}$
- (shr, aux) $\leftarrow \mathbb{S}(1^\kappa, \text{k}, \mathcal{P})$
- For $j = 1, \ldots, n$ do
  (ID, Mpk) $\leftarrow X_j$
  $c_j \leftarrow \mathbb{E}_{\text{IBE}}([\text{shr}]_j || \text{vk}, \text{ID}, \text{Mpk})$
- c $\leftarrow (c_1, \ldots, c_n, \text{vk}, \text{aux}, \mathcal{P})$
- $\sigma \leftarrow \text{Sig}(\text{c}, \text{sk})$
- Return (k, c$||\sigma$)

$\mathbb{D}_{\text{WF-KEM}}(\text{c}||\sigma, \text{crd})$
- $(c_1, \ldots, c_n, \text{vk}, \text{aux}, \mathcal{P}) \leftarrow$ c
- If $\text{Ver}(\text{c}, \sigma, \text{vk}) \neq 1$ return $\perp$
- For $j = 1, \ldots, k$ do
  (crd, X) $\leftarrow [\text{crd}]_j$
  Find c_i corresponding to X
  $([\text{shr}]_j || \text{vk}_j) \leftarrow \mathbb{D}_{\text{IBE}}(c_i, \text{crd})$
  If $([\text{shr}]_j || \text{vk}_j) = \perp$ return $\perp$
  If $\text{vk}_j \neq \text{vk}$ return $\perp$
- k $\leftarrow \mathbb{S}^{-1}(\text{shr}, \text{aux})$
- If k $= \perp$ return $\perp$
- Return k

Note that, similarly to what is done in [16] for multiple encryption in the public-key setting, one could use an IBE primitive modified to include non-malleable public labels to bind vk to each individual c_j. We chose not to do this so that we could base our construction on the more standard IBE primitive and security model. The security of the above construction is captured via the following theorem which is proved in [3].

Theorem 2. *The above construction is (m, n)-IND-CCCA secure if the underlying IBE is IND-CCA2 secure, the OTS is UF secure, and the secret sharing scheme is information theoretically secure. More precisely we have:*

$$\text{Adv}_{\text{WF-KEM}}^{\text{IND-CCCA}}(A) \leq \text{Adv}_{\text{OTS}}^{\text{UF}}(B_1) + 2mn^2 \cdot \text{Adv}_{\text{IBE}}^{\text{IND-CCA2}}(B_2).$$

The best result we obtain in the standard model for computational secret sharing schemes is the following. In Section 6 we explain why this is the case. A sketch proof is also provided in [3].

Theorem 3. *The above construction is (m, n)-IND-CCCA$^-$ secure if the underlying IBE is IND-CCA2 secure, the OTS is UF secure, and the secret sharing scheme is IND-SSA secure. More precisely we have:*

$$\text{Adv}_{\text{WF-KEM}}^{\text{IND-CCCA}^-}(A) \leq \text{Adv}_{\text{OTS}}^{\text{UF}}(B_1) + 2mn^2 \cdot \text{Adv}_{\text{IBE}}^{\text{IND-CCA2}}(B_2) + \text{Adv}_{\text{SS}}^{\text{IND-SSA}}(B_3).$$

5.2   An EFWF-KEM Construction

We now extend the previous generic construction to achieve escrow-freeness. We build an EFWF-KEM using an additional component: a PKE scheme. The authority key generation and credential extraction algorithms are as in the WF-KEM construction. The user key generation algorithm is that of the underlying PKE. Finally, the encapsulation and decapsulation algorithms are:

$\mathbb{E}_{\text{EFWF-KEM}}(\mathcal{P}, \text{PK})$
- $(\text{vk}, \text{sk}) \leftarrow \mathbb{G}_{\text{OTS}}(1^\kappa)$
- $k_1, k_2 \leftarrow \mathbb{K}_{\text{EFWF-KEM}}$
- $(\text{shr}, \text{aux}) \leftarrow \mathbb{S}(1^\kappa, k_1, \mathcal{P})$
- $\bar{c} \leftarrow \mathbb{E}_{\text{PKE}}(k_2 || \text{vk}, \text{PK})$
- For $j = 1, \ldots, n$ do
  $(\text{ID}, \text{Mpk}) \leftarrow X_j$
  $c_j \leftarrow \mathbb{E}_{\text{IBE}}([\text{shr}]_j || \text{vk}, \text{ID}, \text{Mpk})$
- $c \leftarrow (\bar{c}, c_1, \ldots, c_n, \text{vk}, \text{aux}, \mathcal{P})$
- $\sigma \leftarrow \text{Sig}(c, \text{sk})$
- Return $(k_1 \oplus k_2, c || \sigma)$

$\mathbb{D}_{\text{EFWF-KEM}}(c || \sigma, \text{crd}, \text{SK})$
- $(\bar{c}, c_1, \ldots, c_n, \text{vk}, \text{aux}, \mathcal{P}) \leftarrow c$
- If $\text{Ver}(c, \sigma, \text{vk}) \neq 1$ return $\perp$
- For $j = 1, \ldots, k$ do
  $(\text{crd}, X) \leftarrow [\text{crd}]_j$
  Find c_i corresponding to X
  $([\text{shr}]_j || \text{vk}_j) \leftarrow \mathbb{D}_{\text{IBE}}(c_i, \text{crd})$
  If $([\text{shr}]_j || \text{vk}_j) = \perp$ return $\perp$
  If $\text{vk}_j \neq \text{vk}$ return $\perp$
- $k_1 \leftarrow \mathbb{S}^{-1}(\text{shr}, \text{aux})$
- $k_2 \leftarrow \mathbb{D}_{\text{PKE}}(\bar{c}, \text{SK})$
- If $k_1 = \perp$ or $k_2 = \perp$ return $\perp$
- Return $k_1 \oplus k_2$

Again we have two security results which depend on the security provided by the underlying secret sharing scheme. The following theorems are proved in [3].

Theorem 4. *The above EFWF-KEM construction is (m, n)-IND-CCCA and (m, n)-IND-CCA2 secure if the underlying PKE and IBE are IND-CCA2 secure, the OTS is UF secure, and the secret sharing scheme is information-theoretically secure. More precisely we have:*

$$\text{Adv}_{\text{EFWF-KEM}}^{\text{IND-CCCA}}(A) \leq \text{Adv}_{\text{OTS}}^{\text{UF}}(B_1) + 2mn^2 \cdot \text{Adv}_{\text{IBE}}^{\text{IND-CCA2}}(B_2),$$

$$\text{Adv}_{\text{EFWF-KEM}}^{\text{IND-CCA2}}(A) \leq \text{Adv}_{\text{OTS}}^{\text{UF}}(B_1) + 2\text{Adv}_{\text{PKE}}^{\text{IND-CCA2}}(B_2).$$

Theorem 5. *The above EFWF-KEM construction is (m,n)-IND-CCCA$^-$ and (m,n)-IND-CCA2 secure if the underlying PKE is IND-CCA2 secure, the underlying IBE is IND-CCA2 secure, the OTS is UF secure, and the secret sharing scheme is IND-SSA secure. More precisely we have:*

$$\mathrm{Adv}_{\mathrm{EFWF-KEM}}^{\mathrm{IND-CCCA}^-}(A) \leq \mathrm{Adv}_{\mathrm{OTS}}^{\mathrm{UF}}(B_1) + 2mn^2 \cdot \mathrm{Adv}_{\mathrm{IBE}}^{\mathrm{IND-CCA2}}(B_2) + \mathrm{Adv}_{\mathrm{SS}}^{\mathrm{IND-SSA}}(B_3),$$

$$\mathrm{Adv}_{\mathrm{EFWF-KEM}}^{\mathrm{IND-CCA2}}(A) \leq \mathrm{Adv}_{\mathrm{OTS}}^{\mathrm{UF}}(B_1) + 2\mathrm{Adv}_{\mathrm{PKE}}^{\mathrm{IND-CCA2}}(B_2).$$

6   Discussion

The main contribution in this work is the fact that, through the generic constructions that we propose, and using underlying components which achieve the required levels of security in the standard model, we obtain the first WF-KEM and EFWF-KEM schemes provably secure in the standard model.

There are, however, other interesting aspects to the results presented in the previous sections, which we now discuss.

Relation with the original construction in [1]: The concrete EFWF-KEM scheme in [1] is originally defined as a full encryption scheme, although internally it is structured as a KEM-DEM construction. The basic building block in the KEM part is a weak version of the IBE scheme by Boneh and Franklin [7]. Chosen ciphertext security is achieved globally through a transformation akin to that used in the KEM constructions in [15], which is valid in the random oracle model. We require fully chosen ciphertext secure individual components, and the way we achieve global CCA2 security in the standard model comes from the IBE to PKE transformation in [10], adapted to multiple encryption in [16].

Our constructions do inherit the combination of a secret sharing scheme, an IBE scheme and a PKE scheme. However, if we allow for computational secret sharing, then we can only achieve IND-CCCA$^-$ security. This is true even if the underlying secret sharing scheme tolerates adaptive chosen share attacks. This is the main difference between the security of our construction and that in [1]. Intuitively this can be explained as follows. Using the RO heuristic one can perform a *late binding* between challenge share values and the challenge ciphertext. This makes it possible to construct the challenge without explicitly knowing the shares, and directly map the adversary's credential extraction queries to external calls to a share extraction oracle.

The standard model does not allow the same proof strategy, so we cannot prove the security of our constructions against adaptive credential extracting attackers unless we adopt perfect secret sharing. This will only be an issue in terms of the overall efficiency of the constructions, which we discuss below.

Finally, it is interesting to note the very effective application of the randomness reuse paradigm [4] in [1] to achieve impressive computational and ciphertext length savings.

Relation with multiple encryption: This work builds on the general results by Dodis and Katz [16] for chosen ciphertext security of multiple encryption. However our constructions require that we extend these results in three different aspects: (1) to consider adaptive user corruption attacks, (2) to consider generalised access structures and (3) we require a mix of identity-based and public-key encryption techniques. Our results imply that equivalent extensions can be derived in the context of generic multiple encryption.

Relation with certificateless encryption: We will not explore this connection in detail due to space constraints. However, we do note the similarity between the security models of a CL-KEM scheme [6] and the EFWF-KEM security models introduced. This similarity implies that a simplified version of our construction considering only one credential and a single authority can be seen as a CL-KEM scheme which can be proved IND-CCA2 secure against Type I- and Type II adversaries [6].

Efficiency considerations: We analyse the efficiency of our constructions by looking at the computational load and ciphertext length that they produce. A high level analysis shows that the computational weight associated with encapsulation and decapsulation is that of sharing the secret key, encrypting the n shares using the IBE scheme, possibly encrypting another secret key with the PKE scheme, and generating a one-time signature. The corresponding ciphertexts include the public sharing information, n IBE ciphertexts, possibly one PKE ciphertext, the OTS verification key and a signature string.

An obvious way to optimise the end-result is to choose underlying components which are themselves efficient. For example, adopting the IBE scheme of Sakai and Kasahara [11] one obtains a solution which is computationally more efficient than the original construction in [1]. However, there are three techniques which can further improve the efficiency of our constructions.

The enhanced IBE to PKE transformation proposed in [8], which replaces the OTS component by a MAC and a weak form of commitment has also been adapted to achieve chosen ciphertext security in [16] for a weak form of multiple encryption. It turns out that this weak form of multiple encryption is sufficient to allow an extension to WF-KEMs similar to what we achieved with the OTS-based technique. We chose not to include these results in this paper as they lead to more involved proofs and they are less intuitive.

The randomness reuse paradigm [4] can also be applied in this context, although to the best of our knowledge there is currently no IBE scheme which is IND-CCA2 secure in the standard model, and which allows reuse of randomness. This, in itself, is an interesting open problem. However, if we settle for the fully secure version of the Boneh and Franklin IBE scheme, then we can obtain bandwidth and computational (point multiplication) savings by re-using the first component in all IBE ciphertexts. Further improvements may be attainable by re-using the same randomness in the PKE component as in [1].

Our constructions can be easily adapted to work with IBE and PKE schemes extended to take labels as additional parameters, and bind them non-malleably to the ciphertext. This adaptation reduces to using the OTS verification key as the label parameter. Potential benefits of this would arise from labelled IBE or PKE schemes which achieve this functionality more efficiently than the direct non-malleable labelling that we adopted in our constructions.

As a final note on efficiency, we look at the potential benefits of using a computational secret sharing scheme rather than a perfect secret sharing scheme. The main advantage in this is to obtain share sizes which are smaller than the shared secret, which is important when the secret is large. For example, the scheme in [21] uses a perfect secret sharing scheme as an underlying component to split an auxiliary secret key. This key is then used to encrypt the results of partitioning the (large) secret using an information dispersal algorithm. This provides share sizes which asymptotically approach the optimal $|S|/n$ by detaching the size of the (large) secret from the input to the perfect secret sharing scheme. In our case this is an invalid argument, as the secrets we share are themselves secret keys.

Acknowledgements

The authors would like to thank Nigel Smart for his helpful comments. Part of this work was carried out when the second author visited University of Minho in Portugal, funded by the European Commission through the IST Programme under Contract IST-2002-507932 ECRYPT.

References

1. S.S. Al-Riyami, J. Malone-Lee and N.P. Smart. Escrow-Free Encryption Supporting Cryptographic Workflow. *Cryptology ePrint Archive*, Report 2004/258. 2004.
2. S.S. Al-Riyami and K.G. Paterson. Certificateless Public-Key Cryptography. *Advances in Cryptology - ASIACRYPT 2003*, LNCS 2894:452–473. Springer-Verlag, 2003.
3. M. Barbosa and P. Farshim. Secure Cryptographic Workflow in the Standard Model. Full Version. *Cryptology ePrint Archive*, Report 2006/???. 2006.
4. M. Bellare, A. Boldyreva and J. Staddon. Randomness Re-Use in Multi-Recipient Encryption Schemes. *Public Key Cryptography - PKC 2003*, LNCS 2567:85–99. Springer-Verlag, 2003.
5. J. Benaloh and J. Leichter. Generalized Secret Sharing and Monotone Functions. *Advances in Cryptology - CRYPTO '88*, LNCS 403:27–35. Springer-Verlag, 1990.
6. K. Bentahar, P. Farshim, J. Malone-Lee and N.P. Smart. Generic Constructions of Identity-Based and Certificateless KEMs. *Cryptology ePrint Archive*, Report 2005/058, 2005.
7. D. Boneh and M. Franklin. Identity-Based Encryption from the Weil Pairing. *SIAM Journal on Computing*, 32:586–615. 2003.
8. D. Boneh and J. Katz. Improved Efficiency for CCA-Secure Cryptosystems Built Using Identity-Based Encryption. *Cryptology ePrint Archive*, Report 2004/261, 2004.
9. R.W. Bradshaw, J.E. Holt and K.E. Seamons. Concealing Complex Policies with Hidden Credentials. *11th ACM Conference on Computer and Communications Security*, 2004.

10. R. Canetti, S. Halevi and J. Katz. Chosen-Ciphertext Security from Identity-Based Encryption. *Cryptology ePrint Archive*, Report 2003/182, 2003.
11. L. Chen and Z. Cheng. Security Proof of Sakai-Kasahara's Identity-Based Encryption Scheme. *Cryptography and Coding*, LNCS 3796:442–459. Springer-Verlag, 2005.
12. L. Chen and K. Harrison. Multiple Trusted Authorities in Identifier Based Cryptography from Pairings on Elliptic Curves. *Technical Report, HPL-2003-48*, HP Laboratories, 2003.
13. L. Chen, K. Harrison, D. Soldera and N.P. Smart. Applications of Multiple Trusted Authorities in Pairing Based Cryptosystems. *Proceedings InfraSec 2002*, LNCS 2437:260–275. Springer-Verlag, 2002.
14. R. Cramer and V. Shoup. A Practical Public-Key Cryptosystem Provably Secure against Adaptive Chosen Ciphertext Attack. *Advances in Cryptology - CRYPTO '98*, LNCS 1462:13–25. Springer-Verlag, 1998.
15. A.W. Dent. A Designer's Guide to KEMs. *Coding and Cryptography*, LNCS 2898:133–151. Springer-Verlag, 2003.
16. Y. Dodis and J. Katz. Chosen-Ciphertext Security of Multiple Encryption. *TCC 2005*, LNCS 3378:188–209. Springer-Verlag, 2005.
17. C. Gentry. Practical identity-based encryption without random oracles. *Advances in Cryptology - EUROCRYPT 2006*, LNCS 4004:445–464. Springer-Verlag, 2006.
18. J. Herranz and D. Hofheinz and E. Kiltz. KEM/DEM: Necessary and Sufficient Conditions for Secure Hybrid Encryption. *Cryptology ePrint Archive*, Report 2006/265. 2006.
19. J.E. Holt, R.W. Bradshaw, K.E. Seamons and H. Orman. Hidden Credentials. *2nd ACM Workshop on Privacy in the Electronic Society*, pp. 1–8, 2003.
20. E. Kiltz. Chosen-Ciphertext Secure Identity-Based Encryption in the Standard Model with short Ciphertexts. *Cryptology ePrint Archive*, Report 2006/122, 2006.
21. H. Krawczyk. Secret Sharing Made Short. *Proceedings of Crypto'93 - Advances in Cryptology*, LNCS. Springer-Verlag, 1993.
22. W. Nagao, Y. Manabe and T. Okamoto. On the Equivalence of Several Security Notions of Key Encapsulation Mechanism. *Cryptology ePrint Archive*, Report 2006/268. 2006.
23. K.G. Paterson. Cryptography from Pairings: A Snapshot of Current Research. *Information Security Technical Report*, 7:41–54, 2002.
24. R. Sakai and M. Kasahara. ID-Based Cryptosystems with Pairing on Elliptic Curve. *Cryptology ePrint Archive*, Report 2003/054, 2003.
25. A. Shamir. How to Share a Secret. *Communications of the ACM*, 22:612–613, 1979.
26. A. Shamir. Identity-Based Cryptosystems and Signature Schemes. *Proceedings of CRYPTO '84 on Advances in Cryptology*, LNCS 196:47–53. Springer-Verlag, 1985.
27. N.P. Smart. Access Control Using Pairing Based Cryptography. *Topics in Cryptology - CT-RSA 2003*, LNCS 2612:111–121. Springer-Verlag, 2003.
28. B.R. Waters. Efficient Identity-Based Encryption Without Random Oracles. *Cryptology ePrint Archive*, Report 2004/180, 2004.

Multi-receiver Identity-Based Key Encapsulation with Shortened Ciphertext

Sanjit Chatterjee[1] and Palash Sarkar[2]

[1] Indian Institute of Science Education and Research
HC VII, Sector III, Salt Lake City
(IIT Kharagpur Kolkata Campus)
India 700106
[2] Applied Statistics Unit
Indian Statistical Institute
203, B.T. Road, Kolkata
India 700108
palash@isical.ac.in

Abstract. This paper describes two identity based encryption (IBE) protocols in the multi-receiver setting. The first protocol is secure in the selective-ID model while the second protocol is secure in the full model. The proofs do not depend on the random oracle heuristic. The main interesting feature of both protocols is that the ciphertext size is $|S|/N$, where S is the intended set of receivers and N is a parameter of the protocol. To the best of our knowledge, in the multi-receiver IBE setting, these are the first protocols to achieve sub-linear ciphertext sizes. There are three previous protocols for this problem – two using the random oracle heuristic and one without. We make a detailed comparison to these protocols and highlight the advantages of the new constructions.

Keywords: Multi-receiver encryption, identity based encryption, bilinear pairing.

1 Introduction

In a multi-recipient public key encryption scheme [4,21,5] all users use a common public key encryption system. Suppose there are n users indexed by $1, \ldots, n$; user i having public and private key pair (pk_i, sk_i). A sender who wants to send messages $M_1, \ldots, M_n$ to users $1, \ldots, n$, where M_i is intended for the user i, encrypts M_i using pk_i and sends the resulting ciphertexts $C_1, \ldots, C_n$. This general setting is referred to as multi-plaintext, multi-recipient public key encryption scheme in the literature [21]. If a single message is encrypted, i.e., $M_1 = \cdots = M_n = M$, then we get a single-plaintext, multi-recipient public key encryption scheme. In terms of functionality the later is same as a public key broadcast encryption [17,18].

Alternatively, one can send an encapsulated session key K to multiple parties, whereas the original message M is encrypted through a symmetric encryption scheme using K. In this case, the ciphertext consists of the encapsulation of K,

R. Barua and T. Lange (Eds.): INDOCRYPT 2006, LNCS 4329, pp. 394–408, 2006.

together with an encryption of M using K. Smart [24] considered this notion of mKEM, i.e., an efficient key encapsulation mechanism for multiple parties in the KEM-DEM philosophy.

In the identity-based setting [22,9], the public key corresponding to each user is her/his identity. Given an identity v, a trusted private key generator (PKG) creates the secret key corresponding to v using its own master secret. Now consider the problem of encrypting the same message M for a large set of identities, for example in a group mail. One can either directly encrypt M or use a key-encapsulation mechanism. A trivial solution would be to encrypt (resp. encapsulate) M (resp. K) separately for each individual identities and then transmit them separately. Let the set of identities be S. Then one has to perform $|S|$ many independent encryptions/encapsulation, where $|S|$ denotes the cardinality of S. This solution is clearly too expensive in terms of bandwidth requirement as well as pairing computation.

Baek, Safavi-Naini and Susilo considered this problem in [1]. Along with a formal definition and security model for MR-IBE, they proposed a construction based on the Boneh-Franklin IBE using bilinear pairing. This protocol was proved secure in the selective-ID model *using* the random oracle heuristic. Independent of this work, Barbosa and Farshim [2] proposed an identity-based key encapsulation scheme for multiple parties. This is an extension of the concept of mKEM of Smart [24] to the identity-based setting. Their construction was inspired by the "OR" construction of Smart for access control [23] using bilinear pairing. Security of this scheme also uses the random oracle heuristic, though in the full model. A construction without using the random oracle heuristic has been described in [14]. The construction is based on the Boneh-Boyen (H)IBE [6].

OUR CONTRIBUTION: One common limitation of all the above protocols – be it encryption or key encapsulation and whether they use the random oracle heuristic or not – is that the ciphertext size becomes large as the set S of intended recipients increase. For all three protocols, the ciphertext consists of approximately $|S|$ many elements of an elliptic curve group of suitable order.

The context of the current work is based on the following scenario.

- The sender uses a broadcast channel for transmission. Each receiver picks out the part relevant to him/her from the entire broadcast.
- Each recipient gets to know the entire set of receivers. In other words, each receiver knows who are the other persons receiving the same message.

In a broadcast transmission, it is of interest to lower the amount of data to be transmitted. Secondly, since each receiver gets to know the entire set of receivers, this set has to be broadcast along with the message. Thus, the only way of reducing the amount of transmission is by reducing the size of the ciphertext (or the encapsulation of the secret key).

In this work, we concentrate on the problem of reducing the ciphertext size in multi-receiver identity based key encapsulation (mID-KEM). We give constructions where the expected ciphertext size is a fraction of $|S|$. This, comes at a cost of increasing the private key size. In other words, what we achieve is a controllable trade-off between the ciphertext size and the private key size.

Our first construction is proved secure in the selective-ID model while the second construction is secure in the full model. Both the protocols are proved to be secure *without* using the random oracle heuristic. Also, for both the protocols, we first prove security against chosen plaintext attacks and then adapt the techniques of Boyen-Mei-Waters [11] to attain CCA-security.

Our technique for constructing the mID-KEM is based on the constant size ciphertext hierarchical identity based encryption (HIBE) protocol (BBG-HIBE) in [8]. The algebraic ideas behind the construction are drawn from this work though there are a few differences to be taken care of. Some of these are discussed below. First, an mID-KEM does not require the key delegation property of a HIBE. Though this is not directly required, the simulation of key-extraction queries in the security proof has to use the techniques from [8]. Second, in a HIBE an encryption is to an identity whose maximum length is equal to the depth of the HIBE. In an mID-KEM, the set of users to which a key has to be encapsulated can be large and has no relation to the public parameters of the system. The third difference is that while the security of our protocol depends on the hardness of the DBDHE problem introduced in [8], the security of the BBG-HIBE itself can be based on the weaker problem wDBDHI* (see the full version of [8] at the eprint server).

Due to lack of space, the proofs are omitted. These are available in the full version of the paper at the eprint sever maintained by IACR.

2   Definitions

2.1   Cryptographic Bilinear Map

Let G_1 and G_2 be cyclic groups of same prime order p and $G_1 = \langle P \rangle$, where we write G_1 additively and G_2 multiplicatively. A mapping $e : G_1 \times G_1 \rightarrow G_2$ is called a cryptographic bilinear map if it satisfies the following properties:

- Bilinearity: $e(aP, bQ) = e(P, Q)^{ab}$ for all $P, Q \in G_1$ and $a, b \in \mathbb{Z}_p$.
- Non-degeneracy: If $G_1 = \langle P \rangle$, then $G_2 = \langle e(P, P) \rangle$.
- Computability: There exists an efficient algorithm to compute $e(P, Q)$ for all $P, Q \in G_1$.

Since $e(aP, bP) = e(P, P)^{ab} = e(bP, aP)$, $e()$ also satisfies the symmetry property. Modified Weil pairing [9] and Tate pairing [3,19] are examples of cryptographic bilinear maps where G_1 is an elliptic curve group and G_2 is a subgroup of a finite field. This motivates our choice of the additive notation for G_1 and the multiplicative notation for G_2. In papers on pairing implementation [3,19], it is customary to write G_1 additively and G_2 multiplicatively. Initial "pure" protocol papers such as [9,20] followed this convention. Later works such as [6,7,25] write both G_1 and G_2 multiplicatively. Here we follow the first convention as it is closer to the known examples of bilinear maps.

2.2   mID-KEM Protocol

Following [1,2], we define a multi-receiver Identity-Based Key Encapsulation (mID-KEM) scheme as a set of four algorithms: Setup, Key Generation, Encapsulation and Decapsulation.

Setup: It takes input 1^κ, where κ is a security parameter and returns the system public parameters together with the master key. The system parameters include a description of the groups G_1, G_2 and $e()$. These are publicly known while the master key is known only to the private key generator (PKG).

Key Generation: It takes as input an identity v, the system public parameters and returns a private key d_v, using the master key. The identity v is used as a public key while d_v is the corresponding private key.

Encapsulation: It takes as input the public parameters and a set of identities S and produces a pair (K, Hdr), where K is a key for a symmetric encryption algorithm and Hdr is a header which encapsulates K. An actual message M is encrypted by a symmetric encryption algorithm under K to obtain C_M. The actual broadcast consists of (S, Hdr, C_M), where (S, Hdr) is called the full header and C_M is called the broadcast body.

Decapsulation: It takes as input a pair (S, Hdr); an identity v and a private key d_v of v. If v $\in S$, then it returns the symmetric key K which was used to encrypt the message. This K can be used to decrypt the broadcast body C_M to obtain the actual message M.

2.3   Hardness Assumption

Security of our mID-KEM scheme is based on the *decisional bilinear Diffie-Hellman exponent* (DBDHE) problem introduced by Boneh-Boyen-Goh in [8]. The l-DBDHE problem is stated as follows.

> Given $P, Q, aP, \ldots, a^{l-1}P, a^{l+1}P, \ldots, a^{2l}P$ for random $a \in \mathbb{Z}_p$ and $Z \in G_2$, decide whether $Z = e(P, Q)^{a^l}$ or whether Z is random.

Let $\mathcal{B}$ be a probabilistic algorithm which takes this instance as input and produces a bit as output. The advantage of $\mathcal{B}$ in solving this decision problem is defined to be

$$\mathsf{Adv}_{\mathcal{B}}^{\mathsf{DBDHE}} = |\Pr[\mathcal{B}(P, Q, \overrightarrow{R}, e(P,Q)^{a^l}) = 1] - \Pr[\mathcal{B}(P, Q, \overrightarrow{R}, Z) = 1]|$$

where $\overrightarrow{R} = (aP, a^2P, \ldots a^{l-1}P, a^{l+1}P, \ldots, a^{2l}P)$ and Z is a random element of G_2. The probability is calculated over the random choices of $a \in \mathbb{Z}_p$, $Z \in G_2$ and also the random bits used by $\mathcal{B}$.

The (t, ϵ, l)-DBDHE assumption holds if $\mathsf{Adv}_{\mathcal{B}}^{\mathsf{DBDHE}} \leq \epsilon$ for any algorithm $\mathcal{B}$ for the l-DBDHE problem, where the runtime of $\mathcal{B}$ is at most t.

2.4   Security Model of mID-KEM

We define indistinguishability under chosen ciphertext attack for multi-receiver identity-based key encapsulation scheme. The adversarial behaviour is defined by the following game between an adversary $\mathcal{A}$ and a simulator $\mathcal{B}$.

$\mathcal{A}$ is allowed to query two oracles – a decryption oracle and a key-extraction oracle. At the initiation, it is provided with the system public parameters.

Phase 1: $\mathcal{A}$ makes a finite number of queries where each query is addressed either to the decryption oracle or to the key-extraction oracle. In a query to the decryption oracle, it provides the full broadcast header as well as the identity under which it wants the decryption. In return, the simulator $\mathcal{B}$ provides $\mathcal{A}$ with either the corresponding symmetric key or bad. In a query to the key-extraction oracle, it provides an identity and the corresponding private key is given to it by the simulator $\mathcal{B}$. $\mathcal{A}$ is allowed to make these queries adaptively, i.e., any query may depend on the previous queries as well as their answers.

Challenge: At this stage, $\mathcal{A}$ fixes a set of identities S^*, under the (obvious) constraint that it has not asked for the private key of any identity in S^*. The simulator $\mathcal{B}$ generates a proper pair (K^*, Hdr^*) corresponding to the set of identities S^* as defined by the encryption algorithm. It then chooses a random bit γ and sets $K_0 = K^*$ and sets K_1 to be a random symmetric key of equal length. $\mathcal{B}$ returns $(K_\gamma, \mathsf{Hdr}^*)$ to $\mathcal{A}$.

Phase 2: $\mathcal{A}$ now issues additional queries just like Phase 1, with the (obvious) restriction that it cannot ask the decryption oracle for the decryption of Hdr^* under any identity in S^* nor the key-extraction oracle for the private key of any identity in S^*.

Guess: $\mathcal{A}$ outputs a guess γ' of γ.

Adversarial Advantage: The advantage of the adversary $\mathcal{A}$ in attacking the mID-KEM scheme is defined as:

$$\mathsf{Adv}_{\mathcal{A}}^{\mathsf{mID\text{-}KEM}} = |\Pr[(\gamma = \gamma')] - 1/2|.$$

The mID-KEM protocol is said to be $(\epsilon, t, q_{\mathsf{v}}, q_C, \sigma)$-secure against chosen ciphertext attacks (IND-mID-CCA secure), if for any adversary running in time t; making q_{v} key-extraction queries; making q_C decryption queries; and with $|S^*| \le \sigma$, we have $\mathsf{Adv}_{\mathcal{A}}^{\mathsf{mID\text{-}KEM}} \le \epsilon$.

We include the upper bound on the size of set of target identities S^* as part of the adversary's resources. The set S^* is the set of identities under which the adversary wants the encryption in the challenge stage. Intuitively, increasing the size of this set allows the possibility of the adversary receiving more information.

2.5 Selective-ID Model

We can have a weaker version of the above model by restricting the adversary. In this model, the adversary has to commit to the set of target identities S^* even before the protocol is set-up. During the actual game, it cannot ask the key-extraction oracle for the private key of any identity in S^* and in the challenge stage the set S^* is used to generate K^*.

Henceforth, we will call this restricted model the selective-ID (sID) model and the unrestricted model to be the full model.

2.6 CPA Security

We may impose another restriction on the adversary, namely, we do not allow the adversary access to the decryption oracle. This restriction can be made on both the full and the sID models. One can define the advantage of such an adversary in a manner similar to above.

As above, a mID-KEM protocol is said to be (ϵ, t, q, σ) IND-mID-CPA secure if for any adversary running in time t; making q queries to the key-extraction oracle; and with $|S^*| \leq \sigma$, we have $\mathsf{Adv}_{\mathcal{A}}^{\mathsf{mID-KEM}} \leq \epsilon$.

In the identity based setting, there are generic methods [13,10] for transforming a CPA-secure protocol into a CCA-secure protocol. Thus, one can simply prove the CPA-security of a protocol and then apply known transformations to obtain CCA-security. Recently [11], a non-generic method has been obtained for transforming the CPA-secure protocols in [25,6] into CCA-secure protocols.

3 Construction of CPA-Secure mID-KEM in the sID Model

Let, $e(), G_1, G_2$ be as defined in Section 2.1 and identities are assumed to be elements of $\mathbb{Z}_p^*$.

Setup: Let P be a generator of G_1. Choose a random secret $x \in \mathbb{Z}_p$ and set $P_1 = xP$. Also choose random elements P_2, P_3 in G_1 and a random vector $\vec{U} = (U_1, \ldots, U_N)$ with entries in G_1. The significance of the parameter N is discussed later. The public parameters are

$$\langle P, P_1, P_2, P_3, \vec{U}, H() \rangle$$

where $H()$ is a publicly computable surjective function $H : \mathbb{Z}_p^* \to \{1, \ldots, N\}$.
The master secret key is xP_2.

Key Generation: Given any identity $\mathsf{v} \in \mathbb{Z}_p^*$, this algorithm generates the private key d_{v} of v as follows.

Compute $k = H(\mathsf{v}) \in \{1, \ldots, N\}$. Choose a random element $r \in \mathbb{Z}_p$ and output

$$d_{\mathsf{v}} = (d_0, d_1, b_1, \ldots, b_{k-1}, b_{k+1}, \ldots, b_N)$$

where $d_0 = (xP_2 + r(P_3 + \mathsf{v}U_k))$; $d_1 = rP$; and for $1 \leq i \leq N$, $b_i = rU_i$. Note that b_k is not part of the private key for any identity v, such that $k = H(\mathsf{v})$. The private key for any identity consists of $(N + 1)$ elements of G_1.

A Notation: For a set S of identities, we introduce the notation

$$V(S) = P_3 + \sum_{\mathsf{v} \in S} \mathsf{v}U_{H(\mathsf{v})}. \tag{1}$$

The expression $\mathsf{v}U_{H(\mathsf{v})}$ denotes the scalar multiplication of U_k by v, where $k = H(\mathsf{v})$. The use of this notation will simplify the description of the protocol.

Encapsulation: Let S be a set of identities for which we want to encapsulate a session key K. We partition S into several subsets in the following manner.

Let $H(S) = \{j_1, \ldots, j_k\}$ be the set of distinct indices obtained by applying $H()$ to the elements of the set S. For $1 \leq i \leq k$, let $\{s_{i,1}, \ldots, s_{i,\tau_i}\}$ be the subset of all elements in S which map to j_i. Let $\tau = \max_{1 \leq i \leq k}(\tau_i)$. We view S as a (possibly incomplete) $k \times \tau$ matrix having entries $s_{i,j}$ where $1 \leq i \leq k$ and $1 \leq j \leq \tau_i$. For $1 \leq j \leq \tau$, define the set S_j to be the jth column of this matrix. Then S is a disjoint union of $S_1, \ldots, S_\tau$ and for all j, $|S_j| = |H(S_j)|$, i.e., H is injective on S_j.

Choose a random $s \in \mathbb{Z}_p$, compute $K = e(P_1, P_2)^s$ and then set the header as

$$\mathsf{Hdr} = (sP, sV(S_1), \ldots, sV(S_\tau)).$$

The header consists of $\tau + 1$ elements of G_1. The full header is the tuple $(S_1, \ldots, S_\tau, \mathsf{Hdr})$, where K is used to obtain the broadcast body C_M by encrypting the message M using symmetric encryption. The entire broadcast consists of $(S_1, \ldots, S_\tau, \mathsf{Hdr}, C_M)$.

Discussion

1. There is no security assumption on $H()$. We need $H()$ to be surjective to ensure that the entire set of public parameters are used. While there is no security requirement on $H()$, we still expect the output of $H()$ to be uniformly distributed. The expected size of the header is equal to $|S|/N$ under this assumption. We would like to emphasize that $H()$ is *not* assumed to be a random oracle. In particular, this assumption is *not* used in the security proof. Rather, this should be seen as the usual assumption on a hash function used in data/file structure.

2. The parameter N in the protocol controls the trade-off between header size and the size of the public parameters as well as the size of the private key. For each i in the range $\{1, \ldots, N\}$, there is a component of the public key corresponding to this index i. Any identity is mapped to an index in the range $\{1, \ldots, N\}$ using $H()$ and the corresponding component of the public key is used to obtain the component d_0 of the private key of the identity. If S is a random set of identities, then under the assumption that the output of $H()$ is uniformly distributed, the expected size of the header is $1 + \lceil |S|/N \rceil$. N still has another role – in the security reduction this is the maximum number of identities that the adversary is allowed to target.

Decapsulation: An individual user does not need the full header

$$(S_1, \ldots, S_\tau, \mathsf{Hdr} = (C_0, C_1, \ldots, C_\tau))$$

for decapsulation. For a user with identity v, it is sufficient for him to obtain (S_j, C_0, C_j), such that $\mathsf{v} \in S_j$. This can be easily picked out by the user from the general broadcast of the full header. Note that by construction, for $\mathsf{v}, \widehat{\mathsf{v}} \in S_j$ with $\mathsf{v} \neq \widehat{\mathsf{v}}$, we have $H(\mathsf{v}) \neq H(\widehat{\mathsf{v}})$, i.e., in other words $H()$ is injective on S_j.

Thus, the input to the decapsulation algorithm is a tuple $(\widehat{\mathsf{v}}, S, C, D)$, where $\widehat{\mathsf{v}} \in S$; $H()$ is injective on S; $C = sP$ and $D = sV(S)$. The private key $d_{\widehat{\mathsf{v}}}$ for $\widehat{\mathsf{v}}$ is

$$d_{\widehat{\mathsf{v}}} = (d_0, d_1, b_1, \ldots, b_{k-1}, b_{k+1}, \ldots, b_N)$$

where $k = H(\widehat{\mathsf{v}})$. Suppose that r was used during the generation of the private key $d_{\widehat{\mathsf{v}}}$ for $\widehat{\mathsf{v}}$. Then $d_0 = xP_2 + r(P_3 + \widehat{\mathsf{v}}U_{H(\widehat{\mathsf{v}})})$, $d_1 = rP$ and $b_i = rU_i$. Compute

$$\begin{aligned}
\mathsf{key}_{\widehat{\mathsf{v}}} = d_0 &+ \sum_{\mathsf{v} \in S \wedge \mathsf{v} \neq \widehat{\mathsf{v}}} \left(\mathsf{v} b_{H(\mathsf{v})} \right) \\
&= xP_2 + r(P_3 + \widehat{\mathsf{v}}U_{H(\widehat{\mathsf{v}})}) + r \sum_{\mathsf{v} \in S \wedge \mathsf{v} \neq \widehat{\mathsf{v}}} \left(\mathsf{v}U_{H(\mathsf{v})} \right) \\
&= xP_2 + r \left(P_3 + \sum_{\mathsf{v} \in S} \mathsf{v}U_{H(\mathsf{v})} \right).
\end{aligned}$$

It then obtains the session key K from the components C_0, C_j of Hdr, $\mathsf{key}_{\widehat{\mathsf{v}}}$ and d_1 as

$$\frac{e(\mathsf{key}_{\widehat{\mathsf{v}}}, C_0)}{e(C_j, d_1)} = \frac{e\left(xP_2 + r(P_3 + \sum_{\mathsf{v} \in S} \mathsf{v}U_{H(\mathsf{v})}), sP\right)}{e\left(s(P_3 + \sum_{\mathsf{v} \in S} \mathsf{v}U_{H(\mathsf{v})}), rP\right)} = e(P_1, P_2)^s.$$

In the above protocol, each identity is mapped to an index in the range $\{1, \ldots, N\}$ using the function $H()$. Each index has an associated public parameter. While generating the private key for an identity, the identity itself as well as the public parameter part corresponding to the index of this identity is "mixed" to the master secret xP_2, while the parts corresponding to all other public parameters are provided individually to the users. In the case, where two identities have the same index, the corresponding private keys will be different. One reason for this is that the randomizers r will be different for the two identities. More importantly, the first component of the private key depends on the actual value of the identity and hence will be different for the two distinct identities.

Now suppose that during encryption, an identity v in the broadcast set S has an index i. Then as shown above, the entity possessing a private key corresponding to v can decrypt the message. On the other hand, suppose there is an identity $\mathsf{v}' \notin S$ such that $H(\mathsf{v}) = i = H(\mathsf{v}')$. Then we must be assured that possessing the private key of v' does not allow decryption. Suppose that during encryption v is assigned to set S_j. Then the $(j+1)$th entry of the header is of the form $s \sum_{\mathsf{v} \in S_j} V_{\mathsf{v}}$. In particular, this sum includes the value V_{v}. During decryption, v constructs $\mathsf{key}_{\mathsf{v}} = xP_2 + r \sum_{\mathsf{v} \in S_j} V_{\mathsf{v}}$, where r was used to generate the private key for v. Let $S_j' = (S_j \setminus \{\mathsf{v}\}) \cup \{\mathsf{v}'\}$. Then v' can compute the value $\mathsf{key}_{\mathsf{v}'} = xP_2 + r' \sum_{\mathsf{v} \in S_j'} V_{\mathsf{v}}$, where r' was used to generate the private key for v'. In general, there is no way to compute $\mathsf{key}_{\mathsf{v}}$ from $\mathsf{key}_{\mathsf{v}'}$.

Theorem 1. *The mID-KEM protocol is (t, q, ϵ, N)-secure against chosen plaintext attacks in the sID model under the assumption that $(t+t', \epsilon', N+1)$-DBDHE assumption holds for $\langle G_1, G_2, e() \rangle$, where $\epsilon \le \epsilon'$ and t' is the time required for $O(q)$ scalar multiplications in G_1 and $O(N)$ multiplications in $\mathbb{Z}_p$.*

4   Construction of CPA-Secure mID-KEM in the Full Model

We first discuss the modifications required in the construction of Section 3 to make it secure against adaptive adversary. Here, identities are assumed to be n-bit strings. Let ℓ be a size parameter, $1 < \ell \le n$, chosen a-priori.

Set-up: The public parameter consists of N vectors, each of length ℓ. Let, $\overrightarrow{P_3} = (P_1, \ldots, P_N)$ and $\mathcal{U} = (\overrightarrow{U_1}, \ldots, \overrightarrow{U_N})$, where $\overrightarrow{U_i} = (U_{i,1}, \ldots, U_{i,\ell})$ with each P_i and each $U_{i,j}$ in G_1. So the public parameters are

$$\langle P, P_1, P_2, \overrightarrow{P_3}, \mathcal{U}, H() \rangle.$$

The function $H : \{0,1\}^n \to \{1, \ldots, N\}$ is a surjective map whose role is as in the protocol of Section 3.

A Notation: For any identity $\mathsf{v} = (\mathsf{v}_1, \ldots, \mathsf{v}_\ell)$, where each v_i is a bit string of length n/ℓ, we define

$$\left. \begin{array}{ll} V(\mathsf{v}) = P_{3,k} + \sum_{j=1}^{\ell} \mathsf{v}_j U_{k,j} & \text{where } k = H(\mathsf{v}); \\ V(S) = \sum_{\mathsf{v} \in S} V(\mathsf{v}) & \text{for any set } S \text{ of identities.} \end{array} \right\} \quad (2)$$

When the context is clear we use V_{v} in place of $V(\mathsf{v})$.

The parameters (N, n, ℓ) control the configuration of the mID-KEM. Hence, we will refer to the construction as (N, n, ℓ) mID-KEM construction.

Key Generation: Given any identity v this algorithm generates the private key d_{v} of v as follows. Compute $k = H(\mathsf{v}) \in \{1, \ldots, N\}$. Choose a random element $r \in \mathbb{Z}_p$ and output

$$d_{\mathsf{v}} = (d_0, d_1, a_1, \ldots, a_{k-1}, a_{k+1}, \ldots, a_N, \overrightarrow{b_1}, \ldots, \overrightarrow{b_{k-1}}, \overrightarrow{b_{k+1}}, \ldots, \overrightarrow{b_N})$$

where $d_0 = (xP_2 + rV_{\mathsf{v}})$; $d_1 = rP$; and for $1 \le i \le N$, $a_i = rP_{3,i}$; $\overrightarrow{b_i} = r\overrightarrow{U_i}$. Note that a_k and $\overrightarrow{b_k}$ are not part of the private key for any identity v, such that $k = H(\mathsf{v})$. The private key for any identity consists of $N(\ell+1) - (\ell-1)$ elements of G_1.

Encapsulation: Let S be the set of identities under which we want to perform the encryption. Form the sets $S_1, \ldots, S_\tau$ based on the function $H()$ as in the encapsulation algorithm of the protocol in Section 3. Choose a random $s \in \mathbb{Z}_p$. The header consists of

$$\mathsf{Hdr} = (sP, sV(S_1), \ldots, sV(S_\tau)).$$

The full header is formed from Hdr as in Section 3.

Decapsulation: As in Section 3, a user with identity $\widehat{\mathsf{v}} = (\widehat{\mathsf{v}}_1, \ldots, \widehat{\mathsf{v}}_\ell)$ does not require the full header for decapsulation. From the full header, the user forms the tuple $(\widehat{\mathsf{v}}, S, sP, sV(S))$. The user also has the private key $d_{\widehat{\mathsf{v}}}$ corresponding to $\widehat{\mathsf{v}}$. Note that $\widehat{\mathsf{v}} = (\widehat{\mathsf{v}}_1, \ldots, \widehat{\mathsf{v}}_\ell)$ and suppose r was used to generate $d_{\widehat{\mathsf{v}}}$. Recall that $d_0 = xP_2 + r\left(P_{3,H(\widehat{\mathsf{v}})} + \sum_{j=1}^{\ell} \widehat{\mathsf{v}}_j U_{H(\widehat{\mathsf{v}}),j}\right)$. Compute

$$\mathsf{key}_{\widehat{\mathsf{v}}} = d_0 + \sum_{\mathsf{v} \in S, \mathsf{v} \neq \widehat{\mathsf{v}}} \left(a_{H(\mathsf{v})} + \sum_{j=1}^{\ell} \mathsf{v}_j U_{H(\mathsf{v}),j}\right).$$

It can be shown that $\mathsf{key}_{\widehat{\mathsf{v}}} = xP_2 + rV(S)$. Using $\mathsf{key}_{\widehat{\mathsf{v}}}$, it is possible to obtain $e(P_1, P_2)^t$ as in Section 3.

The following theorem shows that the above scheme is secure against an adaptive adversary.

Theorem 2. *The (N, n, ℓ) mID-KEM protocol is (ϵ, t, q, h)-secure against chosen plaintext attacks in the full model under the assumption that $(t', \epsilon', (N+1))$-DBDHE assumption holds, where*

$$\epsilon \leq 2(2\sigma(\mu_l + 1))^h \epsilon'; \text{ and } t' = t + \chi + O(\epsilon^{-2} \ln(\epsilon^{-1})\lambda^{-1} \ln(\lambda^{-1}))$$

with $\mu_l = l((2^n)^{1/l} - 1)$, $\lambda = 1/(2(2\sigma(\mu_l + 1))^h)$, $\sigma = max(2q, 2^{n/\ell})$ and χ is the time required for $O(q)$ scalar multiplications in G_1 and $O(N)$ multiplications in $\mathbb{Z}_p$.

5   Security Against Chosen Ciphertext Attack

We adapt the technique of Boyen-Mei-Waters [11] for attaining CCA-security. This technique applies to both the sID-secure and the full model secure protocols giving rise to the following two protocols.

1. **M_1:** A CCA-secure mID-KEM in the sID model.
2. **M_2:** A CCA-secure mID-KEM in the full model.

Below we describe how to obtain M_1 by modifying the construction in Section 3. Essentially the same thing also holds for the full model secure protocol.

Set-Up: In addition to the set-up for the CPA-secure scheme, we choose a random element W from G_1. This W is part of the public parameters. There is also an injective encoding $H_1 : G_1 \to \mathbb{Z}_p$.

Key Generation: Remains same as before.

Encapsulation: In addition to what was provided earlier, one more element B is provided as part of the header. This element is computed in the following manner. Let the full header for the CPA-secure protocol be $(S_1, \ldots, S_\tau, sP, sA_1, \ldots, sA_\tau)$. Let $\nu = H_1(sP)$ and set $B = s(\nu P_2 + W)$. The full header for the new protocol is

$$(S_1, \ldots, S_\tau, sP, sA_1, \ldots, sA_\tau, B).$$

Note that one B is used for all the users, i.e., B is independent of the identity.

Decapsulation: The input to the decapsulation algorithm is a tuple $(\mathsf{v}, S, C = sP, D = sA, B)$. The portion (v, S, sP, sA) was used as input to the decapsulation algorithm for the CPA-secure protocol. The new entry is B. Compute $A = V(S) = P_3 + \sum_{\mathsf{v} \in S} \mathsf{v} U_{H(\mathsf{v})}$, $\nu = H_1(C)$ and $E = \nu P_2 + W$. The following public verification checks are performed:

1. Does $\mathsf{v} \in S$?
2. Is H injective on S?
3. Is $e(P, D) = e(C, A)$?
4. Is $e(P, B) = e(C, E)$?

If the answer to any of the above questions is no, then return bad. Otherwise, obtain the encapsulated key as in the case of the CPA-secure algorithm.

6   Comparison

Here we make a comparison of $\mathbf{M}_1$ and $\mathbf{M}_2$ with the previous three protocols proposed so far, i,e., BSS [1], BaFa [2] and CS [14] protocols.

We would like to note that the various protocols are based on different hardness assumptions. Strictly speaking, a comparison between them is not possible. The comparison we make assumes that all the hard problems are "equally hard". Under this assumption, one can perhaps make a meaningful comparison.

For the sake of uniformity we assume that the BSS protocol is used for key encapsulation and the header (Hdr) consists of only elements of G_1. During decryption the symmetric key (K) is decapsulated from this header as in our construction of Section 3. The BSS protocol was proved secure in the sID model using the random oracle assumption. This protocol is in a sense an extension of the Boneh-Franklin IBE [9] to the multi-receiver setting. Here the public parameter consists of three elements of G_1, including the generator of the group. The private key corresponding to a given identity is generated using a map-to-point function modeled as a random oracle and consists of a single element of G_1.

The protocol of [14] is based on the Boneh-Boyen HIBE [6]. The protocol is proved to be CPA-secure but can be made CCA-secure using a technique similar to the one described in this paper (based on the Boyen-Mei-Waters [11] construction). The protocol does not use the random oracle heuristic. Here the public parameter consist of $4 + N$ elements of G_1, where N is the number of target identities committed by the adversary [14]. The private key is similar to that in the Boneh-Boyen scheme [6] and consists of two elements of G_1.

In Table 1, we make a comparison between the public parameter size, private key size and the header size for the three protocols. The public parameter size and the private key size are smaller for the BSS protocol – that is because they *use* a random oracle to generate the private key. For both CS as well as $\mathbf{M}_1$ the public parameter is linear in N, where N is a parameter of the model and once chosen is constant. Also, the private key size for the new construction is N, while that for CS is only 2. The real advantage of the new protocol is in shortening the header size. The header size for both BSS and CS protocols are $|S| + 1$. In contrast, the expected header size for the new protocol is $\lceil |S|/N \rceil + 1$.

A typical value of N would be 16. For this value of N, the public parameter of $\mathbf{M}_1$ consists of 19 elements (elliptic curve points) of G_1 and the private key consists of 16 elements of G_1. Both these sizes are within acceptable limits. On the other hand, a broadcast to a group of around 1000 users will consist of around 60 elements of G_1, which is also within reasonable limits. On the other hand, for both the BSS and the CS protocols, the broadcast will consist of 1000 elements of G_1, which can be prohibitively costly. Thus, in certain situations, the reduction in the size of the broadcast can be more significant than the increases in the sizes of the public parameters and that of the private key.

Efficiency: Let us consider the efficiency of different operations (key generation, encapsulation and decapsulation) for the various protocols. The efficiency of key generation is proportional to the size of the private key for all the protocols.

Similarly, the decapsulation efficiency is proportional to the part of the full header required by an individual user for decapsulation. For M_1, this is N scalar multiplications plus two pairing computations, while for the other protocols this is only two pairing computations. The reason is that in attempting to reduce the header size below $|S|$, we are grouping users during encapsulations. Thus, during decapsulation, each user has to use the information about which of the other groups have been grouped with it. Since each group has at most N users, decapsulation requires at most N scalar multiplications.

The encapsulation efficiency for the BSS protocol is $1 + |S|$ scalar multiplications. On the other hand, the CS protocol requires $N \times |S|$ scalar multiplications. Thus, the cost of removing the random oracle heuristic is a decrease in the efficiency of encapsulation. This cost is even more than the trivial protocol of encrypting separately to all the identities in S using (say) the BB-IBE. The cost in this case will be $2|S|$ scalar multiplications. However, the header size in this trivial protocol is going to be $2|S|$. If we want to reduce the header size to $|S|$ (and avoid the random oracle heuristic), then the CS protocol increases the cost of encapsulation.

The encapsulation efficiency for $\mathbf{M}_1$ is approximately $|S|(N + 1)/N$. For $N > 1$, this value is less than $2|S|$ and hence our protocol is more efficient than the trivial protocol. On the other hand, it is less efficient than the BSS protocol. Thus, compared to the BSS protocol, our contribution is to decrease the header size and avoid the random oracle heuristic.

Table 1. Comparison of mID-KEM protocols secure in the selective-ID model. Here S is the set of identities to which the encapsulation is formed.

Protocol	Hardness Assumption	Random Oracle	Pub. Para. size (elts. of G_1)	Pvt. Key size (elts. of G_1)	Header size (elts. of G_1)
BSS [1]	DBDH	Yes	3	1	$|S| + 1$
CS [14]	DBDH	No	$4 + N$	2	$|S| + 1$
$\mathbf{M_1}$	DBDHE	No	$3 + N$	N	$\lceil |S|/N \rceil + 1$

Table 2. Comparison of mID-KEM protocols secure in the full model. The security degradation of both protocols is exponential in N, which is the number of attacked users. Here S is the set of identities to which the encapsulation is formed.

Protocol	Hardness Assumption	Random Oracle	Pub. Para. size (elts. of G_1)	Pvt. Key size (elts. of G_1)	Header size (elts. of G_1)
BaFa [2]	GBDH	Yes	2	1	$|S| + 2$
$\mathbf{M_2}$	DBDHE	No	$3 + N + N\ell$	$N(\ell+1) - (\ell-1)$	$\lceil |S|/N \rceil + 1$

Full Model Secure Protocols: The comparison of the full model secure protocols is given in Table 2. The security of the BaFa protocol was proved in the full model using the random oracle heuristic assuming the hardness of gap bilinear Diffie-Hellman problem. The system suffers from an exponential security degradation which is around q^N, where q is the number of random oracle queries and N is the number of target identities. The construction of Section 4 is also secure in the full model but *without* using the random oracle heuristic. The security degradation is again q^N. The large security degradation of both these protocols imply that these protocols are not really useful when N is around 10 or so. Thus, our contribution in the full model is really to show that it is possible to obtain security without using random oracle and simultaneously obtain sublinear header size.

7   Conclusion

We present two protocols for multi-receiver identity-based key encapsulation system. The first protocol is secure in the selective-ID model while the second protocol is secure in the full model. The security proofs do not use the random oracle heuristic and are based on the hardness of decisional bilinear Diffie-Hellman exponentiation problem. The main advantage of the new protocols over the previous ones is that for encryption to a set S of users, the header size in the new protocols is sub-linear in $|S|$, whereas for the previous protocols this is approximately $|S|$.

Acknowledgement. We would like to thank the reviewers for their comments and for pointing out several typos.

References

1. Joonsang Baek, Reihaneh Safavi-Naini, and Willy Susilo. Efficient Multi-receiver Identity-Based Encryption and Its Application to Broadcast Encryption. In Serge Vaudenay, editor, *Public Key Cryptography*, volume 3386 of *Lecture Notes in Computer Science*, pages 380–397. Springer, 2005.
2. M. Barbosa and P. Farshim. Efficient identity-based key encapsulation to multiple parties. In Nigel P. Smart, editor, *IMA Int. Conf.*, volume 3796 of *Lecture Notes in Computer Science*, pages 428–441. Springer, 2005.
3. Paulo S. L. M. Barreto, Hae Yong Kim, Ben Lynn, and Michael Scott. Efficient Algorithms for Pairing-Based Cryptosystems. In Moti Yung, editor, *CRYPTO*, volume 2442 of *Lecture Notes in Computer Science*, pages 354–368. Springer, 2002.
4. Mihir Bellare, Alexandra Boldyreva, and Silvio Micali. Public-key encryption in a multi-user setting: Security proofs and improvements. In *EUROCRYPT*, pages 259–274, 2000.
5. Mihir Bellare, Alexandra Boldyreva, and Jessica Staddon. Randomness re-use in multi-recipient encryption schemeas. In Desmedt [16], pages 85–99.
6. Dan Boneh and Xavier Boyen. Efficient Selective-ID Secure Identity-Based Encryption Without Random Oracles. In Cachin and Camenisch [12], pages 223–238.
7. Dan Boneh and Xavier Boyen. Secure Identity Based Encryption Without Random Oracles. In Matthew K. Franklin, editor, *CRYPTO*, volume 3152 of *Lecture Notes in Computer Science*, pages 443–459. Springer, 2004.
8. Dan Boneh, Xavier Boyen, and Eu-Jin Goh. Hierarchical Identity Based Encryption with Constant Size Ciphertext. In Cramer [15], pages 440–456. Full version available at Cryptology ePrint Archive; Report 2005/015.
9. Dan Boneh and Matthew K. Franklin. Identity-Based Encryption from the Weil Pairing. *SIAM J. Comput.*, 32(3):586–615, 2003. Earlier version appeared in the proceedings of CRYPTO 2001.
10. Dan Boneh and Jonathan Katz. Improved Efficiency for CCA-Secure Cryptosystems Built Using Identity-Based Encryption. In Alfred Menezes, editor, *CT-RSA*, volume 3376 of *Lecture Notes in Computer Science*, pages 87–103. Springer, 2005.
11. Xavier Boyen, Qixiang Mei, and Brent Waters. Direct Chosen Ciphertext Security from Identity-Based Techniques. In Vijay Atluri, Catherine Meadows, and Ari Juels, editors, *ACM Conference on Computer and Communications Security*, pages 320–329. ACM, 2005.
12. Christian Cachin and Jan Camenisch, editors. *Advances in Cryptology - EUROCRYPT 2004, International Conference on the Theory and Applications of Cryptographic Techniques, Interlaken, Switzerland, May 2-6, 2004, Proceedings*, volume 3027 of *Lecture Notes in Computer Science*. Springer, 2004.
13. Ran Canetti, Shai Halevi, and Jonathan Katz. Chosen-Ciphertext Security from Identity-Based Encryption. In Cachin and Camenisch [12], pages 207–222.
14. Sanjit Chatterjee and Palash Sarkar. Generalization of the Selective-ID Security Model for HIBE Protocols. In Moti Yung, Yevgeniy Dodis, Aggelos Kiayias, and Tal Malkin, editors, *Public Key Cryptography*, volume 3958 of *Lecture Notes in Computer Science*, pages 241–256. Springer, 2006. Revised version available at Cryptology ePrint Archive, Report 2006/203.
15. Ronald Cramer, editor. *Advances in Cryptology - EUROCRYPT 2005, 24th Annual International Conference on the Theory and Applications of Cryptographic Techniques, Aarhus, Denmark, May 22-26, 2005, Proceedings*, volume 3494 of *Lecture Notes in Computer Science*. Springer, 2005.

16. Yvo Desmedt, editor. *Public Key Cryptography - PKC 2003, 6th International Workshop on Theory and Practice in Public Key Cryptography, Miami, FL, USA, January 6-8, 2003, Proceedings*, volume 2567 of *Lecture Notes in Computer Science*. Springer, 2002.

17. Yevgeniy Dodis and Nelly Fazio. Public Key Broadcast Encryption for Stateless Receivers. In Joan Feigenbaum, editor, *Digital Rights Management Workshop*, volume 2696 of *Lecture Notes in Computer Science*, pages 61–80. Springer, 2002.

18. Yevgeniy Dodis and Nelly Fazio. Public key trace and revoke scheme secure against adaptive chosen ciphertext attack. In Desmedt [16], pages 100–115.

19. Steven D. Galbraith, Keith Harrison, and David Soldera. Implementing the Tate Pairing. In Claus Fieker and David R. Kohel, editors, *ANTS*, volume 2369 of *Lecture Notes in Computer Science*, pages 324–337. Springer, 2002.

20. Craig Gentry and Alice Silverberg. Hierarchical ID-Based Cryptography. In Yuliang Zheng, editor, *ASIACRYPT*, volume 2501 of *Lecture Notes in Computer Science*, pages 548–566. Springer, 2002.

21. Kaoru Kurosawa. Multi-recipient public-key encryption with shortened ciphertext. In David Naccache and Pascal Paillier, editors, *Public Key Cryptography*, volume 2274 of *Lecture Notes in Computer Science*, pages 48–63. Springer, 2002.

22. Adi Shamir. Identity-Based Cryptosystems and Signature Schemes. In G. R. Blakley and David Chaum, editors, *CRYPTO*, volume 196 of *Lecture Notes in Computer Science*, pages 47–53. Springer, 1984.

23. Nigel P. Smart. Access control using pairing based cryptography. In Marc Joye, editor, *CT-RSA*, volume 2612 of *Lecture Notes in Computer Science*, pages 111–121. Springer, 2003.

24. Nigel P. Smart. Efficient Key Encapsulation to Multiple Parties. In Carlo Blundo and Stelvio Cimato, editors, *SCN*, volume 3352 of *Lecture Notes in Computer Science*, pages 208–219. Springer, 2004.

25. Brent Waters. Efficient Identity-Based Encryption Without Random Oracles. In Cramer [15], pages 114–127.

Identity-Based Parallel Key-Insulated Encryption Without Random Oracles: Security Notions and Construction*

Jian Weng[1], Shengli Liu[1,2], Kefei Chen[1], and Changshe Ma[3]

[1] Dept. of Computer Science and Engineering
Shanghai Jiao Tong University, Shanghai 200240, P.R. China
[2] Key Laboratory of CNIS
Xidian University, Xian 710071, P.R. China
[3] School of Computer
South China Normal University Shipai, Guangzhou, 510631, P.R. China
{jianweng, slliu, kfchen}@sjtu.edu.cn, JuanJuansmcs@gmail.com

Abstract. In this paper, we apply the parallel key-insulation mechanism to identity-based encryption (IBE) scenarios, and minimize the damage caused by key-exposure in IBE systems. We first formalize the definition and security notions for ID-based parallel key-insulated encryption (IBPKIE) systems, and then propose an IBPKIE scheme based on Water's IBE scheme. To the best of our knowledge, this is the first IBPKIE scheme up to now. Our scheme enjoys two attractive features: (i) it is provably secure without random oracles; (ii) it not only allows frequent key updating, but also does not increase the risk of helper key-exposure.

Keywords: Parallel Key-Insulation, Identity-Based Encryption, Key-Exposure, Bilinear Pairings.

1 Introduction

1.1 Background and Previous Work

The traditional public key infrastructure involves complex construction of certification authority(CA), and requires expensive communication and computation cost for certification verification. To relieve this burden, Shamir [27] introduced an innovative concept called identity-based cryptography, where user's public-key is determined as his identity such as e-mail address and telephone number. The identity is a natural link to a user, hence it simplifies the certification management in public key infrastructures. The first usable IBE schemes are independently proposed by Boneh and Franklin [3] and Cocks [12], followed by many other elegant IBE schemes (see [2] for some of these). These classical IBE schemes rely on the assumption that secret keys are kept perfectly secure. In

---

* Supported by the National Science Foundation of China under Grant Nos.60303026, 60403007, 60573030 and 60673077.

R. Barua and T. Lange (Eds.): INDOCRYPT 2006, LNCS 4329, pp. 409–423, 2006.

practice, however, it is easier for an adversary to obtain the secret key from a naive user than to break the computational assumption on which the system is based. With more and more cryptographic primitives are deployed on insecure environments (e.g. mobile devices), the key-exposure problem becomes an ever-greater threat. No matter how strong these IBE systems are, once the secret keys are exposed, their security is entirely lost.

In conventional public key encryption scenarios, certificate revocation list (CRL) can be utilized to revoke the public key in case of key-exposure. Users can become aware of other users' revoked keys by referring to the CRL. However, straightforward implementation of CRL will not be the best solution to IBE schemes. Remember that utilizing the CRL, the public key will also be renewed. However, the public key in IBE system represents a user's identity and is not desired to be changed. One exemplification as shown in [22] is the application of IBE systems in a mobile phone scenario, where the phone number represents a user's identity, and it will be simple and convenient for the mobile phone users to identify and communicate with each other only by their phone numbers. Hence renewing the phone number is not a practical solution.

To mitigate the damage caused by key-exposure, several key-evolving protocols have been studied. This mechanism includes forward security [1,5,8], intrusion-resilience [24,13] and key-insulation [15,14]. The latter was introduced by Dodis, Katz, Xu and Yung [15], followed by several elegant key-insulated systems [7,14,11,26,19,22,25,20,29,21,16]. In this model, a physically-secure but computationally-limited device, named base or helper, is involved. The full-fledged secret key is divide into two parts: a helper key and an initial temporary secret key. The former is stored in the helper while the latter is kept by the user. The lifetime of the system is divided into discrete periods. The public key is fixed for all the lifetime, while temporary secret key is updated periodically: at the beginning of each period, the user obtains from the helper an update key for the current period. By combining this update key with the temporary secret key for the previous period, the user can derive the temporary secret key for the current period. Cryptographic operations (such as signing and decryption) in a given period only involve the corresponding temporary secret key in this period. Exposure of the temporary secret keys at some periods will not enable an adversary to derive temporary secret keys for the remaining periods. Thus there is no need to change the public key, which is a desirable property for IBE systems. Hanaoka, Hanaoka, Shikata and Imai [22] applied the key-insulation mechanism to IBE system and proposed an ID-based hierarchical strongly key-insulated encryption scheme which is secure in the random oracle model.

Hanaoka et al. [20] first noticed the following situations in public key-insulated encryption (PKIE) schemes: when key-exposure occurs in key-insulated cryptosystems, to alleviate the damage, temporary secret key has to be updated at very short intervals; however, this will in turn increase the frequency of helper's connection to insecure environments and increase the risk of helper key-exposure. Keep in mind that once the helper key is exposed, the PKIE scheme will be broken if one of the temporary secret key is also exposed. Is it possible to increase the security of both temporary secret keys and helper key simultaneously? Hanaoka, Hanaoka and Imai [20] provided a very clever method named parallel key-insulation to deal with this problem for PKIE sysems: based on Boneh-Franklin's IBE scheme [3], they proposed a parallel key-insulated public key encryption scheme. Their scheme differs from the original key-insulated model in that two distinct helpers are introduced and alternately used to update the temporary secret keys. The two helper keys are independent of each other, and they can successfully increase the security of users and helpers by allowing for frequent key updates without increasing the risk of helper key-exposure. Since it's a worthwhile task to deal with the key-exposure problem in IBE scenarios, a natural question is to construct an ID-based parallel key-insulated encryption (IBPKIE) scheme. We notice that Hanaoka et al's [20] PKIE scheme is secure in the random oracle model. As pointed in [9], the security of a cryptographic scheme in the random oracle model does not always imply its security of implementation. Thus another natural question to construct an IBPKIE whose security does not rely on the random oracle model.

1.2 Our Contributions

In this paper, we try to answers the aforementioned questions. We first formalize the definition and security notions for IBPKIE systems, and then propose an IBPKIE scheme which is provably secure without random oracles. To the best of our knowledge, this is the firs IBPKIE scheme up to now. The proposed scheme supports frequent key updates without increasing the risk of helper key-exposure. This is an attractive advantage which the standard ID-based key-insulated encryption scheme can not possess.

1.3 Organization

The rest of this paper is organized as follows. Section 2 gives an introduction to bilinear pairings and related complexity assumptions. We formalize the definition and security notions for IBPKIE in Section 3. A concrete IBPKIE scheme is proposed in Section 4. Section 5 gives the security proof for our proposed scheme and Section 6 concludes this paper.

2   Preliminaries

2.1   Bilinear Pairings

We first briefly review bilinear pairings. Let $\mathbb{G}_1$ and $\mathbb{G}_2$ be two cyclic multiplicative groups with the same prime order q. A bilinear pairing is a map $\hat{e} : \mathbb{G}_1 \times \mathbb{G}_1 \to \mathbb{G}_2$ with the following properties:

- Bilinearity: $\forall u, v \in \mathbb{G}_1, \forall a, b \in \mathbb{Z}_q^*$, we have $\hat{e}(u^a, v^b) = \hat{e}(u, v)^{ab}$;
- Non-degeneracy: there exist $u, v \in \mathbb{G}_1$ such that $\hat{e}(u, v) \neq 1_{\mathbb{G}_2}$;
- Computability: there exists an efficient algorithm to compute $\hat{e}(u, v)$ for $\forall u, v \in \mathbb{G}_1$.

As shown in [3], such non-degenerate admissible maps over cyclic groups can be obtained from the Weil or Tate pairing over supersingular elliptic curves or abelian varieties.

2.2 Decisional Bilinear Diffie-Hellman Assumption

We proceed to recall the definition of decisional bilinear Diffie-Hellman (DBDH) assumption on which our scheme is based.

Definition 1. *The **DBDH problem** in $(\mathbb{G}_1, \mathbb{G}_2)$ is to distinguish the distributions $(g, g^a, g^b, g^c, \hat{e}(g,g)^{abc}) \in \mathbb{G}_1^4 \times \mathbb{G}_2$ and $(g, g^a, g^b, g^c, \hat{e}(g,g)^z) \in \mathbb{G}_1^4 \times \mathbb{G}_2$, where $a, b, c, z \in_R \mathbb{Z}_q^*$. For a probabilistic polynomial-time (PPT) adversary $\mathcal{B}$, we define his **advantage** against the DBDH problem in $(\mathbb{G}_1, \mathbb{G}_2)$ as*

$$Adv_{\mathcal{B}}^{DBDH} \triangleq \left| Pr\left[\mathcal{B}(g, g^a, g^b, g^c, \hat{e}(g,g)^{abc}) = 1\right] - Pr\left[\mathcal{B}(g, g^a, g^b, g^c, \hat{e}(g,g)^z) = 1\right] \right|,$$

where the probability is taken over the random choice of a, b, c, z and the random bits consumed by $\mathcal{B}$.

Definition 2. *We say the (t, ϵ)-**DBDH assumption** holds in $(\mathbb{G}_1, \mathbb{G}_2)$ if no t-time adversary has at least advantage ϵ in solving the DBDH problem.*

3 Framework of IBPKIE

3.1 Syntax

An **IBPKIE scheme** Π consists of a tuple of six polynomial-time algorithms:

Setup (k, N): a probabilistic setup algorithm takes as input the security parameter k and (possibly) the total number of periods N. It returns a system parameter $param$ and a master key msk;

Extract $(msk, param, ID)$: a probabilistic key extraction algorithm takes as input msk, $param$ and an identity ID. It returns this user's initial temporary secret key $TSK_{ID,0}$ and two helper keys $(HK_{ID,1}, HK_{ID,0})$;

UpdH $(t, ID, HK_{ID,i})$: a (possibly) probabilistic helper key update algorithm which takes as input a period index t, a user's identity ID and the i-th helper key $HK_{ID,i}$. This algorithm, run by the i-th (here $i = t \mod 2$) helper for user ID, returns an update key $UK_{ID,t}$;

UpdT $(t, ID, UK_{ID,t}, TSK_{ID,t-1})$: a deterministic temporary secret key update algorithm which takes as input the index t of the next period, a user's identity ID, the temporary secret key $TSK_{ID,t-1}$ and the update key $UK_{ID,t}$. It returns the temporary secret key $TSK_{ID,t}$;

Encrypt (t, ID, m): a probabilistic encryption algorithm which takes as input a period index t, an identity ID and a plaintext m. It returns a pair (t, C) composed of the period t and a ciphertext C;

Decrypt $((t, C), TSK_{ID,t})$: a deterministic decryption algorithm which takes as input a ciphertext (t, C) and the matching temporary secret key $TSK_{ID,t}$. It returns either a plaintext m or "$\perp$".

Consistency requires that for $\forall t \in \{1, \cdots, N\}, \forall m \in \mathcal{M}, \forall ID \in \{0,1\}^*$, $\mathsf{Decrypt}((t, C), TSK_{ID,t}) = m$ holds, where $(t, C) = \mathsf{Encrypt}(t, ID, m)$ and $\mathcal{M}$ denotes the plaintext space.

3.2   Security Notions for IBPKIE

In this subsection, we formalize the security notions for IBPKIE schemes. This is based on the security definitions in (parallel) key-insulated encryption [15,20] and IBE systems [3].

Chosen-Ciphertext Security. We first consider the basic (i.e., non-stong) key-insulation security for IBPKIE. On the one hand, as standard IBE systems, the extract queries and decryption queries should be considered. On the other hand, as traditional key-insulated encryption schemes, the temporary secret key exposure should be addressed. Moreover, we provide the helper key queries for the adversary: we allow him to compromise all the helper keys for the non-challenged identities, and even allow him to compromise one of the helper keys for the challenged identity. More precisely, the semantic security against an adaptive chosen-ciphertext attack in the sense of key-insulation (IND-ID&KI-CCA) is defined by the following game between an adversary $\mathcal{A}$ and a challenger $\mathcal{C}$:

- **Setup:** The challenger $\mathcal{C}$ runs the Setup algorithm and gives adversary $\mathcal{A}$ the resulting system parameter *param*, keeping the master key *mst* itself.
- **Phase 1:** $\mathcal{A}$ adaptively issues a set of queries as below:
  - Extract query $\langle ID \rangle$: $\mathcal{C}$ runs algorithm Extract to generate the initial temporary secret key $TSK_{ID,0}$ and the two helper keys $(HK_{ID,1}, HK_{ID,0})$. $\mathcal{C}$ then sends $(TSK_{ID,0}, (HK_{ID,1}, HK_{ID,0}))$ to $\mathcal{A}$.
  - Helper key query $\langle ID, j \rangle$: $\mathcal{C}$ runs algorithm Extract to generate $HK_{ID,j}$ and sends it to $\mathcal{A}$.
  - Temporary secret key query $\langle ID, t \rangle$: $\mathcal{C}$ runs algorithm UpdT to obtain $TSK_{ID,t}$ and sends it to $\mathcal{A}$.
  - Decryption query $\langle (t, C), ID \rangle$: $\mathcal{C}$ responds by running algorithm UpdT to generate the temporary secret key $TSK_{ID,t}$. It then runs algorithm Decrypt to decrypt the ciphertext (t, C) using $TSK_{ID,t}$ and sends the resulting plaintext to $\mathcal{A}$.
- **Challenge:** Once $\mathcal{A}$ decides that Phase 1 is over, it outputs an identity ID^*, a period index t^* and two equal length plaintext $m_0, m_1 \in \mathcal{M}$ on which it wishes to be challenged. $\mathcal{C}$ flips a random coin $\beta \in_R \{0, 1\}$ and sets the challenged ciphertext to $C^* = (t^*, \mathsf{Encrypt}(t^*, ID^*, m_\beta))$, which is sent to $\mathcal{A}$.
- **Phase 2:** $\mathcal{A}$ adaptively issues queries as Phase 1, and $\mathcal{C}$ answers these queries in the same way as Phase 1.
- **Guess:** Finally, $\mathcal{A}$ outputs a guess $\beta' \in \{0, 1\}$. $\mathcal{A}$ wins the game if $\beta' = \beta$ and the following conditions are satisfied:
  - $\langle ID^* \rangle$ does not appear in extract queries;
  - $\langle ID^*, t^* \rangle$ does not appear in temporary secret key queries;
  - $\mathcal{A}$ can not issue both temporary secret key query $\langle ID^*, t^* - 1 \rangle$ and helper key query $\langle ID^*, t^* \bmod 2 \rangle$;

- $\mathcal{A}$ can not issue both temporary secret key query $\langle ID^*, t^*+1 \rangle$ and helper key query $\langle ID^*, 1-t^* \mod 2 \rangle$;
- Both $\langle ID^*, 1 \rangle$ and $\langle ID^*, 0 \rangle$ do not simultaneously appear in helper key queries.
- $\mathcal{A}$ can not issue decryption query on $\langle (t^*, C^*), ID^* \rangle$.

We refer to such an adversary $\mathcal{A}$ as an IND-ID&-KI-CCA adversary. We define $\mathcal{A}$'s advantage in attacking scheme Π as $Adv_{\mathcal{A}}^{\Pi} \triangleq \left| \Pr[\beta' = \beta] - \frac{1}{2} \right|$.

Remark 1. *For those non-challenged identities, the temporary secret key query is of no help for adversary $\mathcal{A}$, since he can derive any temporary secret key for these identities by issuing extract query. Therefore, without loss of generality, we require that in the above game, adversary $\mathcal{A}$ only issue temporary secret key queries on the challenged identity.*

Definition 3. *We say that an IBPKIE scheme Π is $(t, q_e, q_h, q_t, q_d, \epsilon)$-secure if for any t-time IND-ID&KI-CCA adversary $\mathcal{A}$ who makes at most q_e extract queries, q_h helper key queries, q_t temporary secret key queries and q_d decryption queries, we have that $Adv_{\mathcal{A}}^{\Pi} < \epsilon$. As shorthand, we say that Π is $(t, q_e, q_h, q_t, q_d, \epsilon)$-IND-ID&KI-CCA secure.*

Chosen-Plaintext Security. As usual, we define chosen-plaintext security for IBPKIE systems as in the preceding game, except that adversary $\mathcal{A}$ is disallowed to issue any decryption query. The security notion is termed as IND-ID&KI-CPA.

Definition 4. *We say that an IBPKIE system Π is $(t, q_e, q_h, q_t, \epsilon)$-IND-ID&KI-CPA secure if Π is $(t, q_e, q_h, q_t, 0, \epsilon)$-IND-ID&KI-CCA secure.*

Strong Key-Insulation. In [15,14], Dodis et al. introduced a notion named *strong key-insulation* by addressing attacks that compromise the the helper (this includes attacks by the helper itself, in case it is untrustworthy). They model this attack by giving the helper key to the adversary, whereas the adversary is prohibited to issue any temporary secret key query. Here we can also deal with this attack in IBPKIE scenarios, and we name it by IND-ID&SKI-CCA. We allow him to query both the two helper keys for any identity, even including the challenged identity. However, as the strong key-insulated security in [15,14], the adversary is disallowed to issue temporary secret key query on the challenged identity for any period. Note that we allow the adversary to query temporary secret key on any other identities for any period. Since these queries is implied by the extract queries, we do not explicitly provide them for the adversary. Concretely, the IND-ID&SKI-CCA security is defined by the following game between an adversary $\mathcal{A}$ and a challenger $\mathcal{C}$:

- **Setup:** The same as the IND-ID&KI-CCA game.
- **Phase 1:** Adversary $\mathcal{A}$ adaptively issues extract queries, helper key queries and decryption queries. $\mathcal{C}$ responds to these queries in the same way as the IND-ID&KI-CCA game.
- **Challenge:** The same as the IND-ID&KI-CCA game.

- **Phase 2:** The same as in Phase 1.
- **Guess:** Eventually, $\mathcal{A}$ outputs a guess $\beta' \in \{0, 1\}$. $\mathcal{A}$ wins the game if $\beta' = \beta$ and the following two conditions are both satisfied:
  - $\langle ID^* \rangle$ does not appear in extract queries;
  - $\mathcal{A}$ can not issue decryption query on $\langle (t^*, C^*), ID^* \rangle$.

Similarly to Definition 3 and 4, we can define the notions of $(t, q_e, q_h, q_d, \epsilon)$-IND-ID&SKI-CCA security and (t, q_e, q_h, ϵ)-IND-ID&SKI-CPA security.

Finally, as in [14], we address an adversary who compromises the user's storage while a key is being updated from $TSK_{ID,t-1}$ to $TSK_{ID,t}$, and we call it a key-update exposure at (ID, t). When this occurs, the adversary gets $TSK_{ID,t-1}, UK_{ID,t}$ and $TSK_{ID,t}$ (actually, the latter can be computed from the formers). We say an IBKIS scheme has secure key-updates if a key-update exposure at (ID, t) is of no more help to the adversary than compromising $TSK_{ID,t-1}$ and $TSK_{ID,t}$.

Definition 5. *An IBKIS scheme has **secure key-updates** if the view of any adversary $\mathcal{A}$ making a key-update exposure at (ID, t) can be perfectly simulated by an adversary $\mathcal{A}'$ making temporary secret key queries on $\langle ID, t - 1 \rangle$ and $\langle ID, t \rangle$.*

4   Our Proposed Scheme

In this section, based on Water's IBE scheme [28], we present an IBPKIE scheme.

Let $\mathbb{G}_1$ and $\mathbb{G}_2$ be two groups with prime order q of size k, g be a random generator of $\mathbb{G}_1$, and $\hat{e}$ be a bilinear map such that $\hat{e} : \mathbb{G}_1 \times \mathbb{G}_1 \to \mathbb{G}_2$. Let H be a collision-resistant hash function such that $H: \{0, 1\}^* \to \{0, 1\}^{n_u}$. Inspired by the cryptographic applications of pseudo-random function (PRF) in [17], we also use a PRF F such that given a k-bit seed s and a k-bit argument x, it outputs a k-bit string $F_s(x)$. The proposed IBPKIE scheme consists of the following six algorithms:

Setup: Given a security parameter k, this algorithm works as follows:
1. Pick $\alpha \in_R \mathbb{Z}_q^*, g_2 \in_R \mathbb{G}_1$ and define $g_1 = g^\alpha$.
2. Choose $u' \in_R \mathbb{G}_1$ and a vector $U = (u_i)$ with $u_i \in_R \mathbb{G}_1$ for $i = 1, \cdots, n_u$.
3. For clarity, we define a function L such that for any set $\mathcal{S} \subseteq \{1, \cdots, n_u\}$,
   $$L(\mathcal{S}) = u' \prod_{i \in \mathcal{S}} u_i.$$
4. Return the master key $mst = g_2^\alpha$ and the public parameters
   $$param = (\mathbb{G}_1, \mathbb{G}_2, \hat{e}, q, g, g_1, g_2, u', U, H, F, L).$$

To make the notation easy to follow, hereafter, we use $\mathcal{U}_{ID,t}$ and $\mathcal{U}'_{ID}$ to denote the following sets for a given identity ID and a given period index t as below.

$$\mathcal{U}_{ID,t} = \{i \,|\, S_1[i] = 1, S_1 = H(ID, t)\} \subseteq \{1, \cdots, n_u\},$$
$$\mathcal{U}'_{ID} = \{j \,|\, S_2[j] = 1, S_2 = H(ID)\} \subseteq \{1, \cdots, n_u\}.$$

Extract: Given an identity ID, the PKG constructs the private key as below:

1. Randomly choose two helper keys $HK_{ID,1}, HK_{ID,0} \in_R \{0,1\}^k$. Compute $k_{ID,-1} = F_{HK_{ID,1}}(-1\|ID)$, $k_{ID,0} = F_{HK_{ID,0}}(0\|ID)$. Note that if the length of the input for F is less than k, we can add some "0"s as the prefix to meet the length requirement.
2. Choose $r \in_R \mathbb{Z}_q^*$, compute the initial temporary secret key $TSK_{ID,0}$ as

$$\left(g_2^\alpha L(\mathcal{U}'_{ID})^r L\left(\mathcal{U}_{ID,-1}\right)^{k_{ID,-1}} L\left(\mathcal{U}_{ID,0}\right)^{k_{ID,0}}, g^{k_{ID,-1}}, g^{k_{ID,0}}, g^r \right). \quad (1)$$

3. Return $(TSK_{ID,0}, HK_{ID,1}, HK_{ID,0})$.

UpdH: Given an identity ID and a period index t, the i-th (here $i = t \mod 2$) helper for ID acts as follows: compute $k_{ID,t-2} = F_{HK_{ID,i}}(t-2\|ID)$, $k_{ID,t} = F_{HK_{ID,i}}(t\|ID)$, define and return the update key as

$$UK_{ID,t} = \left(\frac{L(\mathcal{U}_{ID,t})^{k_{ID,t}}}{L(\mathcal{U}_{ID,t-2})^{k_{ID,t-2}}}, g^{k_{ID,t}} \right).$$

UpdT: Given a period index t, an update key $UK_{ID,t}$ and a temporary secret key $TSK_{ID,t-1}$, user ID acts as follows:

1. Parse $UK_{ID,t}$ as $(\hat{U}_{ID,t}, \hat{R}_{ID,t})$, and $TSK_{ID,t-1}$ as $(U_{ID,t-1}, R_{ID,t-2}, R_{ID,t-1}, R)$.
2. Define and return $TSK_{ID,t}$ as $\left(U_{ID,t-1} \cdot \hat{U}_{ID,t}, R_{ID,t-1}, \hat{R}_{ID,t}, R \right)$.

Note that if we let $i = t \mod 2$ and $j = 1 - i$, then the temporary secret key $TSK_{ID,t}$ is always set to

$$\left(g_2^\alpha L(\mathcal{U}'_{ID})^r L\left(\mathcal{U}_{ID,t-1}\right)^{k_{ID,t-1}} L\left(\mathcal{U}_{ID,t}\right)^{k_{ID,t}}, g^{k_{ID,t-1}}, g^{k_{ID,t}}, g^r \right), \quad (2)$$

where $k_{ID,t-1} = F_{HK_{ID,j}}(t-1\|ID), k_{ID,t} = F_{HK_{ID,i}}(t\|ID)$.

Encrypt: In period t, a message $m \in \mathbb{G}_1$ is encrypted for an identity ID as follows: choose $s \in_R \mathbb{Z}_q^*$ and then define the ciphertext as

$$C = \left(t, \hat{e}(g_1, g_2)^s \cdot m, g^s, L(\mathcal{U}'_{ID})^s, L(\mathcal{U}_{ID,t-1})^s, L(\mathcal{U}_{ID,t})^s \right). \quad (3)$$

Decrypt: Given a ciphertext $C = (t, C_1, C_2, C_3, C_4, C_5)$ for the identity ID, C can be decrypted by $TSK_{ID,t} = (U_{ID,t}, R_{ID,t-1}, R_{ID,t}, R)$ as

$$m = C_1 \frac{\hat{e}(R, C_3)\hat{e}(R_{ID,t-1}, C_4)\hat{e}(R_{ID,t}, C_5)}{\hat{e}(U_{ID,t}, C_2)}.$$

It is easy to see that the above scheme satisfies the correctness requirement.

5   Security Analysis

To support our scheme, in this section, we will show how to achieve the provable security for our scheme in the standard model.

5.1   Chosen Plaintext Security

Theorem 1. *The proposed IBPKIE scheme is $(T, q_e, q_h, q_t, \epsilon)$-IND-ID&KI-CPA secure in the standard model, assuming the (T', ϵ')-DBDH assumption holds in $(\mathbb{G}_1, \mathbb{G}_2)$ with*

$$T' \leq T + \mathcal{O}\big((q_e + q_t)(t_e + n_u t_m)\big), \quad \epsilon' \geq \frac{9\epsilon}{512(q_e + q_t)^3(n_u + 1)^3},$$

where t_e and t_m denote the running time of an exponentiation and a multiplication in group $\mathbb{G}_1$ respectively.

Proof. Suppose there exist a $(T, q_e, q_h, q_t, \epsilon)$-IND-ID&-KI-CPA adversary against our scheme, we can construct a (T', ϵ')-adversary $\mathcal{B}$ against the DBDH assumption in $(\mathbb{G}_1, \mathbb{G}_2)$. On input $(g, g^a, g^b, g^c, h) \in \mathbb{G}_1^4 \times \mathbb{G}_2$ for some unknown $a, b, c \in_R \mathbb{Z}_q^*$, $\mathcal{B}$'s goal is to decide whether $h = \hat{e}(g, g)^{abc}$. $\mathcal{B}$ flips a fair coin $\mathcal{COIN} \in \{1, 2\}$. $\mathcal{B}$ plays Game 1 with $\mathcal{A}$ if $\mathcal{COIN} = 1$ and else Game 2.

Game 1: In this game, $\mathcal{B}$ acts as a challenger expecting that $\mathcal{A}$ will never corrupt the helper key on the challenged identity. $\mathcal{B}$ interacts with $\mathcal{A}$ as follows:

Setup: $\mathcal{B}$ constructs the public parameters for $\mathcal{A}$ in the following way:
1. Set $l_u = \frac{4(q_e + q_t)}{3}$, randomly choose an integer k_u with $0 \leq k_u \leq n_u$. We assume that $l_u(n_u + 1) < q$.
2. Randomly choose the following integers:
$$x' \in_R \mathbb{Z}_{l_u}, y' \in_R \mathbb{Z}_q,$$
$$\hat{x}_i \in_R \mathbb{Z}_{l_u}, \text{ for } i = 1, \cdots, n_u. \text{ Let } \hat{X} = \{\hat{x}_i\}.$$
$$\hat{y}_i \in_R \mathbb{Z}_q, \text{ for } i = 1, \cdots, n_u. \text{ Let } \hat{Y} = \{\hat{y}_i\}.$$
3. Construct a set of public parameters as below:
$$g_1 = g^a, g_2 = g^b, u' = g_2^{x' - l_u k_u} g^{y'},$$
$$U = (\hat{u}_i) \text{ with } \hat{u}_i = g_2^{\hat{x}_i} g^{\hat{y}_i} \text{ for } i = 1, \cdots, n_u.$$

All these public parameters are passed to $\mathcal{A}$.

Observe that from the perspective of the adversary, the distribution of these public parameters are identical to the real construction. Note that the master key is implicitly set to be $g_2^\alpha = g_2^a = g^{ab}$.

To make the notation easy to follow, we also define two functions J and K such that for any set $\mathcal{S} \subseteq \{1, \cdots, n_u\}$, $K(\mathcal{S}) = x' - l_u k_u + \sum_{i \in \mathcal{S}} \hat{x}_i$, $J(\mathcal{S}) = y' + \sum_{i \in \mathcal{S}} \hat{y}_i$. Note that for any set $\mathcal{S} \subseteq \{1, \cdots, n_u\}$, the following equality always holds:

$$g_2^{K(\mathcal{S})} g^{J(\mathcal{S})} = L(\mathcal{S}).$$

Before giving the oracles simulation, we point out that some implicit relations exist in our scheme: (i) according to Equ. (1) and (2), as for a given identity, his initial temporary secret key and all the temporary secret keys share the same exponent r; (ii) all the temporary secret keys for a given user ID are mutually dependent on one another, i.e., $TSK_{ID,t-1}$ and $TSK_{ID,t}$ share the same exponent $k_{ID,t-1}$, while $TSK_{ID,t}$ and $TSK_{ID,t+1}$ share the same exponent $k_{ID,t+1}$. To embody these relations in the simulation, $\mathcal{B}$ forms a list named R^{list} as explained below. For easy explanation, an algorithm named RQuery is also defined.

Algorithm RQuery(ID, t):

Begin

> If there exists a tuple in R^{list} for this input
>> then output the predefined value.
>
> Else if $t =' -'$ then
>> choose $\hat{r} \in_R \mathbb{Z}_q^*$, add $(ID,' -', \hat{r})$ on R^{list}, return $\hat{r}$.
>
>> Else
>>> choose $\hat{k}_{ID,t} \in_R \mathbb{Z}_q^*$, add $(ID, t, \hat{k}_{ID,t})$ on R^{list}, return $\hat{k}_{ID,t}$.
>>
>> End if
>
> End if

End.

Phase 1: $\mathcal{B}$ answers a series of queries for $\mathcal{A}$ in the following way:

- *Helper key queries.* $\mathcal{B}$ maintains a list HK^{list} which is initially empty. Upon receiving a helper key query $\langle ID, i \rangle$ with $i \in \{1, 0\}$. $\mathcal{B}$ first checks whether HK^{list} contains a tuple $(ID, i, HK_{ID,i})$. If it does, $HK_{ID,i}$ is returned to $\mathcal{A}$. Otherwise, $\mathcal{B}$ chooses $HK_{ID,i} \in_R \mathbb{Z}_q^*$, adds $(ID, i, HK_{ID,i})$ on HK^{list} and returns $HK_{ID,i}$ to $\mathcal{A}$.

- *Extract queries.* Upon receiving an extract query for identity ID, $\mathcal{B}$ outputs "failure" and aborts if $K(\mathcal{U}'_{ID}) \equiv 0 \mod q$ (denote this event by **E1**). Otherwise, $\mathcal{B}$ acts as follows:

  1. Issue helper key queries on $\langle ID, 1 \rangle$ and $\langle ID, 0 \rangle$ to obtain $HK_{ID,1}$ and $HK_{ID,0}$.
  2. Compute $\hat{r} =$ RQuery$(ID,' -')$ and define the initial temporary secret key $TSK_{ID,0}$ as

$$\left(g_1^{-\frac{J(\mathcal{U}'_{ID})}{K(\mathcal{U}'_{ID})}} L(\mathcal{U}'_{ID})^{\hat{r}} L(\mathcal{U}_{ID,-1})^{k_{ID,-1}} L(\mathcal{U}_{ID,0})^{k_{ID,0}}, g^{k_{ID,-1}}, g^{k_{ID,0}}, g_1^{\frac{-1}{K(\mathcal{U}'_{ID})}} g^{\hat{r}} \right),$$

where $k_{ID,-1} = F_{HK_{ID,1}}(-1 \| ID)$ and $k_{ID,0} = F_{HK_{ID,0}}(0 \| ID)$.
  3. Return $(TSK_{ID,0}, HK_{ID,1}, HK_{ID,0})$ to $\mathcal{A}$.

  Observe that if let $r = \hat{r} - \frac{a}{K(\mathcal{U}'_{ID})}$, then it can be verified that $TSK_{ID,0}$ has the correct form as Equ. (1).

- *Temporary secret key queries.* As argued in Remark 1, we require that $\mathcal{A}$ just issue temporary secret key queries for the challenged identity. To maintain the mutually dependent relation among all the temporary secret keys for a given user, we do not make use of the case $L(\mathcal{U}_{ID,t'}) \not\equiv 0 \mod q$ for an odd t', where t' is equal to either t or $t - 1$ (note that not making use of an even t' can be handled similarly). Concretely, upon receiving a temporary secret key query $\langle ID, t \rangle$ (wlog, we assume t is even, since an odd t can be handled in a similar manner), $\mathcal{B}$ outputs "failure" and aborts if $L(\mathcal{U}'_{ID}) \equiv L(\mathcal{U}_{ID,t}) \equiv 0 \mod q$ holds (denote this event by **E2**). Otherwise,

$\mathcal{B}$ first computes $\hat{r} = \mathsf{RQuery}(ID,'-')$, $\hat{k}_{ID,t-1} = \mathsf{RQuery}(ID, t-1)$, $\hat{k}_{ID,t} = \mathsf{RQuery}(ID, t)$, and then constructs the temporary secret key $TSK_{ID,t}$ for $\mathcal{A}$ according to the following tow cases: if $L(\mathcal{U}'_{ID}) \neq 0 \mod q$, define $TSK_{ID,t}$ as

$$\left(g_1^{-\frac{J(\mathcal{U}'_{ID})}{K(\mathcal{U}'_{ID})}} L(\mathcal{U}'_{ID})^{\hat{r}} L(\mathcal{U}_{ID,t-1})^{\hat{k}_{ID,t-1}} L(\mathcal{U}_{ID,t})^{\hat{k}_{ID,t}}, g^{\hat{k}_{ID,t-1}}, g^{\hat{k}_{ID,t}}, g_1^{\frac{-1}{K(\mathcal{U}'_{ID})}} g^{\hat{r}} \right),$$

otherwise, i.e., $\left(L(\mathcal{U}'_{ID}) \equiv 0 \mod q \right) \wedge \left(L(\mathcal{U}_{ID,t}) \not\equiv 0 \mod q \right)$, define $TSK_{ID,t}$ as

$$\left(g_1^{-\frac{J(\mathcal{U}_{ID,t})}{K(\mathcal{U}_{ID,t})}} L(\mathcal{U}'_{ID})^{\hat{r}} L(\mathcal{U}_{ID,t-1})^{\hat{k}_{ID,t-1}} L(\mathcal{U}_{ID,t})^{\hat{k}_{ID,t}}, g^{\hat{k}_{ID,t-1}}, g_1^{\frac{-1}{K(\mathcal{U}_{ID,t})}} g^{\hat{k}_{ID,t}}, g^{\hat{r}} \right).$$

Observe that in both cases, $TSK_{ID,t}$ has the correct form as Equ. (2). Moreover, the aforementioned relations are also satisfied. Note that since F is a PRF and adversary $\mathcal{A}$ does not know the corresponding seeds $HK_{ID,1}$ and $HK_{ID,0}$, the exponents $k_{ID,t-1}$ and $k_{ID,t}$ are indistinguishable from the real construction in $\mathcal{A}$'s view.

Challenge: Once $\mathcal{A}$ decides that Phase 1 is over, he outputs an identity ID^*, a period index t^* and two equal-length messages $m_0, m_1 \in \mathbb{G}_1$ on which it wishes to be challenged. $\mathcal{B}$ outputs "failure" and aborts if $L(\mathcal{U}'_{ID^*}) \equiv L(\mathcal{U}_{ID^*,t^*-1}) \equiv L(\mathcal{U}_{ID^*,t^*}) \equiv 0 \mod q$ does not holds (denote this event by **E3**). Otherwise, $\mathcal{B}$ picks a random bit $\beta \in \{0, 1\}$ and responds with the challenged ciphertext

$$C^* = \left(t^*, h \cdot m_\beta, g^c, (g^c)^{J(\mathcal{U}'_{ID^*})}, (g^c)^{J(\mathcal{U}_{ID^*,t^*-1})}, (g^c)^{J(\mathcal{U}_{ID^*,t^*})} \right).$$

Note that if $h = \hat{e}(g, g)^{abc}$, then we can see that C^* is indeed a valid challenged ciphertext for $\mathcal{A}$ due to the following equalities:

$$h = \hat{e}(g_1, g_2)^c, \qquad\qquad (g^c)^{J(\mathcal{U}'_{ID^*})} = L(\mathcal{U}'_{ID^*})^c,$$
$$(g^c)^{J(\mathcal{U}_{ID^*,t^*-1})} = L(\mathcal{U}_{ID^*,t^*-1})^c, \qquad (g^c)^{J(\mathcal{U}_{ID^*,t^*})} = L(\mathcal{U}_{ID^*,t^*})^c.$$

On the other hand, if h is uniform and independent in $\mathbb{G}_1$, then C^* is independent of β in adversary $\mathcal{A}$'s view.

Phase 2: $\mathcal{A}$ issues the rest of queries as in Phase 1 with the restriction described in Section 3.2. $\mathcal{B}$ responds to these queries for $\mathcal{A}$ in the same way as Phase 1.

Guess: Eventually, $\mathcal{A}$ outputs a guess $\beta' \in \{0, 1\}$. $\mathcal{B}$ first checks whether $\mathcal{A}$ has corrupted one of ID^*'s helper keys during this game. If it does, $\mathcal{B}$ outputs "failure" and aborts (denote this event by **E4**). Otherwise, $\mathcal{B}$ concludes its own game by outputting a guess as follows: if $\beta' = \beta$ then $\mathcal{B}$ outputs 1 meaning $h = \hat{e}(g, g)^{abc}$; otherwise, it outputs 0 meaning that h is a random element in $\mathbb{G}_2$.

Game 2: In this game, $\mathcal{B}$ acts as a challenger expecting that $\mathcal{A}$ will corrupt exactly one of the helper keys on the challenged identity. $\mathcal{B}$ picks $\gamma \in_R \{1, 0\}$ and bets on that $\mathcal{A}$ queries on the γ-th helper. Wlog, we assume $\gamma = 1$ (the case of $\gamma = 0$ can be handled in a similar manner). Note that for the challenged identity ID and a period index t, $\mathcal{A}$ has no information on the exponent $k_{ID,t}$ since he does not know $HK_{ID,0}$. Algorithm RQuery is defined as in Game1. $\mathcal{B}$ conducts the phases of **Setup**, *helper key queries*, *extract queries* and **Challenge** for $\mathcal{A}$ in the same way as Game 1. Here we let **F1** and **F3** denote the abort events in *extract queries* and **Challenge** respectively. $\mathcal{B}$ answers the *temporary secret key queries* for $\mathcal{A}$ as follows:

- *Temporary secret key queries:* As in Game 1, we also assume that $\mathcal{A}$ just issues temporary secret key queries for the challenged identity. For a temporary secret key query $\langle ID, t \rangle$, we explain how to deal with the case of an odd t (the case of an even t can be handled in a similar manner). Note that $\mathcal{A}$ can not compute $k_{ID,t-1}$ since he does not know $HK_{ID,0}$. $\mathcal{B}$ outputs "failure" and aborts if $L(\mathcal{U}'_{ID}) \equiv L(\mathcal{U}_{ID,t-1}) \equiv 0$ mod q holds (denote this event by **F2**). Otherwise, $\mathcal{B}$ first computes $\hat{r} =$ RQuery$(ID,'-')$, $\hat{k}_{ID,t-1} =$ RQuery$(ID, t-1)$, $k_{ID,t} = F_{HK_{ID,1}}(t\|ID)$, and then constructs the temporary secret key $TSK_{ID,t}$ for $\mathcal{A}$ according to two cases: if $L(\mathcal{U}'_{ID}) \not\equiv 0$ mod q, define $TSK_{ID,t}$ as

$$
\left(g_1^{-\frac{J(\mathcal{U}'_{ID})}{K(\mathcal{U}'_{ID})}} L(\mathcal{U}'_{ID})^{\hat{r}} L(\mathcal{U}_{ID,t-1})^{\hat{k}_{ID,t-1}} L(\mathcal{U}_{ID,t})^{k_{ID,t}}, g^{\hat{k}_{ID,t-1}}, g^{k_{ID,t}}, g_1^{\frac{-1}{K(\mathcal{U}'_{ID})}} g^{\hat{r}} \right),
$$

otherwise, define and $TSK_{ID,t}$ as

$$
\left(g_1^{-\frac{J(\mathcal{U}_{ID,t-1})}{K(\mathcal{U}_{ID,t-1})}} L(\mathcal{U}'_{ID})^{\hat{r}} L(\mathcal{U}_{ID,t-1})^{\hat{k}_{ID,t-1}} L(\mathcal{U}_{ID,t})^{k_{ID,t}}, g_1^{\frac{-1}{K(\mathcal{U}_{ID,t-1})}} g^{\hat{k}_{ID,t-1}}, g^{k_{ID,t}}, g^{\hat{r}} \right).
$$

It can be verified that in both cases, $TSK_{ID,t}$ has the correct form as Equ. (2).

Guess: Eventually, $\mathcal{A}$ outputs a guess $\beta' \in \{0, 1\}$. $\mathcal{B}$ first checks whether $\mathcal{A}$ has corrupted $HK_{ID^*,1}$ during this game. If not, $\mathcal{B}$ outputs "failure" and aborts (denote this event by **F4**). Otherwise, $\mathcal{B}$ concludes its own game by outputting 0 or 1 in the same way as Game1.

This completes the simulation. From the description of $\mathcal{B}$, we know that the time complexity of $\mathcal{B}$ is dominated by the exponentiations and the multiplications in the extract queries and temporary secret key queries. Since there are $\mathcal{O}(1)$ exponentiations and $\mathcal{O}(n_u)$ multiplications in each stages, we known that the time complexity of $\mathcal{B}$ is bounded by $T' \leq T + \mathcal{O}\big((q_e + q_t)(t_e + n_u t_m)\big)$.

We can also bound $\mathcal{B}$'s advantage against the DBDH assumption in $(\mathbb{G}_1, \mathbb{G}_2)$ by $\epsilon' \geq \dfrac{9\epsilon}{512(q_e + q_t)^3(n_u + 1)^3}$. Due to the space limit, we do not provide the details here.

This completes the proof of the theorem.   □

Our IBPKIE scheme aslo provides strong key-insulated security. Concretely, we have the following theorem.

Theorem 2. *The proposed scheme is* (T, q_e, q_h, ϵ)*-IND-ID&SKI-CPA secure in the standard model, assuming the* (T', ϵ')*-DBDH assumption holds in* $(\mathbb{G}_1, \mathbb{G}_2)$ *with*

$$T' \leq T + \mathcal{O}\big(q_e(t_e + n_u t_m)\big), \quad \epsilon' \geq \frac{9}{512 q_e^3 (n_u + 1)^3},$$

where t_e *and* t_m *denote the same meanings as Theorem 1.*

Proof. (Sketch) On input $(g, g^a, g^b, g^c, h) \in \mathbb{G}_1^4 \times \mathbb{G}_2$ for some unknown $a, b, c \in_R \mathbb{Z}_q^*$, $\mathcal{B}$'s goal is to decide whether $h = \hat{e}(g, g)^{abc}$. $\mathcal{B}$ interacts with $\mathcal{A}$ as follows:

Setup: The same as Game 1 in Theorem 1 except that l_u is set to be $l_u = \frac{4q_e}{3}$.

Phase 1: $\mathcal{B}$ answers the helper key queries and extract queries for $\mathcal{A}$ in the same way as Game 1 in Theorem 1. Note that $\mathcal{A}$ is allowed to query all the challenged identity's helper keys, whereas the temporary secret key queries are no longer provided for $\mathcal{A}$.

Challenge: The same as Theorem 1.

Phase 2: $\mathcal{B}$ answers the helper key queries and extract queries for $\mathcal{A}$ as in Phase 1 with the restriction described in Section 3.2.

Guess: Eventually, $\mathcal{A}$ outputs a guess $\beta' \in \{0, 1\}$. $\mathcal{B}$ outputs 1 if $\beta' = \beta$ and 0 otherwise.

We can bound the time complexity of $\mathcal{B}$ by $T' \leq T + \mathcal{O}\big(q_e(t_e + n_u t_m)\big)$ and the advantage of $\mathcal{B}$ by $\epsilon' \geq \frac{9\epsilon}{512 q_e^3 (n_u + 1)^3}$. □

Theorem 3. *The proposed IBPKIE scheme has secure key-updates.*

This theorem follows from the fact that for any period index t and any identity ID, the update key $UK_{ID,t}$ can be derived from $TSK_{ID,t}$ and $TSK_{ID,t-1}$.

5.2   On Achieving Chosen-Ciphertext Security

Recent results of Canetti et al. [10], further improved by [4,6], show how to build a CCA-secure IBE scheme from a 2-level HIBE scheme[18,23]. Similarly to Water's IBE scheme, we can also transform the scheme in Section 4 into a hybrid 2-level HIBE, and then get an IND-ID&KI-CCA(IND-ID&SKI-CCA) secure IBKIE scheme.

6   Conclusion

Classical IBE schemes rely on the assumption that secret keys are kept perfectly secure. In practice, however, key-exposure seems inevitable. No matter how strong these IBE systems are, once the secret keys are exposed, their security is entirely lost. Thus it's a worthwhile task to deal with the key-exposure problem in IBE systems.

In this paper, we have applied the parallel key-insulation mechanism to IBE scenario and minimized the damage caused by key-exposure in IBE systems. We propose an IBPKIE scheme which is provably secure without resorting to the random oracle methodology. This is a desirable property since a proof in the random oracle model can only serve as a heuristic argument and can not imply the security in the implementation. The proposed scheme can allow frequent key updates without increasing the risk of helper key-exposure. This is an attractive advantage which the standard ID-based key-insulated encryption schemes can not possess.

Acknowledgements

The authors thank very much to the anonymous reviewers for their very helpful comments.

References

1. R. Anderson. Two Remarks on Public-Key Cryptology. Invited lecture, CCCS'97. Available at http://www.cl.cam.ac.uk/users/rja14/.
2. Paulo Barreto. The pairing-based crypto lounge. http://paginas.terra.com.br/ informatica/paulobarreto/pblounge.html.
3. D. Boneh and M. Franklin. Identity Based Encryption From the Weil Pairing. In Proc. of Crypto'2001, LNCS 2139, pp. 213-229. Springer, 2001.
4. D. Boneh and J. Katz. Improved Efficiency for CCA-Secure Cryptosystems Built Using Identity-Based Encryption. In Proc. of CT-RSA'2005, LNCS 3376, PP.87-103. Springer, 2005.
5. M. Bellare and S. Miner. A Forward-Secure Digital Signature Scheme. In Proc. of CRYPTO'1999, LNCS 1666, pp. 431-448, Springer, 1999.
6. X. Boyen, Q. Mei and B.Waters. Direct Chosen Ciphertext Security from Identity-Based Techniques. In Proc. of ACM CCS'05, ACM Press, pp. 320-329, 2005.
7. M. Bellare, and A. Palacio. Protecting against Key Exposure: Strongly Key-Insulated Encryption with Optimal Threshold. Available at http://eprint.iacr.org/2002/064.
8. R. Canetti, S. Halevi and J. Katz. A Forward-Secure Public-Key Encryption Scheme. In Proc. of Eurocrypt'2003, LNCS 2656, pp. 255-271, Springer, 2003.
9. R. Canetti, O. Goldreich and S. Halevi. The Random Oracle Methodology, Revisited. Journal of the ACM 51(4), ACM press, pp.557-594, 2004.
10. R. Canetti, S. Halevi and J. Katz. Chosen-Ciphertext security from Identity-Based Encryption. In Proc. of Eurocrypt'2004, LNCS 3027, pp. 207-222, Springere,2004.
11. J. H. Cheon, N. Hopper, Y. Kim, and I. Osipkov. Authenticated Key-Insulated Public Key Encryption and Timed-Release Cryptography. Available at http://eprint.iacr.org/2004/231.
12. C. Cocks: An Identity Based Encryption Scheme Based on Quadratic Residues. In Proc. of IMA'2001, LNCS 2260, pp. 360-363, Springer, 2001.
13. Y. Dodis, M. Franklin, J. Katz, A. Miyaji and M. Yung. Intrusion-Resilient Public-Key Encryption. In Proc. of CT-RSA'03, LNCS 2612, pp. 19-32. Springer, 2003.
14. Y. Dodis, J. Katz, S. Xu, and M. Yung. Strong Key-Insulated Signature Schemes. In Proc. of PKC'2003, LNCS 2567, pp. 130-144. Springer, 2003.

15. Y. Dodis, J. Katz, S. Xu and M. Yung. Key-Insulated Public-Key Cryptosystems. In Proc. of Eurocrypt'2002, LNCS 2332, pp.65-82, Springer, 2002.
16. Y. Dodis and M. Yung. Exposure-Resilience for Free: The Hierarchical ID-based Encryption Case. In Proc. of IEEE Security in Storage Workshop 2002, pp.45-52, 2002.
17. O. Goldreich, S. Goldwasser, S. Micali. On the cryptographic applications of random functions. In Proc. of Crypto'1984. LNCS 196, pp. 276-288, Springer, 1985.
18. C. Gentry and A. Silverberg, Hierarchical ID-Based Cryptography. In Proc. of Asiacrypt'02, LNCS 2501, pp.548-566, Springer, 2002.
19. N. González-Deleito, O. Markowitch, and E. Dall'Olio. A New Key-Insulated Signature Scheme. In Proc. of ICICS'2004, LNCS 3269, pp. 465-479, Springer, 2004.
20. G. Hanaoka, Y. Hanaoka and H. Imai. Parallel Key-Insulated Public Key Encryption. In Proc. of PKC'2006, LNCS 3958, pp. 105-122, Springer, 2006.
21. Y. Hanaoka, G. Hanaoka, J. Shikata, H. Imai. Unconditionally Secure Key Insulated Cryptosystems: Models, Bounds and Constructions. In Proc. of ICICS'2002, LNCS2513, pp. 85-96, Springer, 2002.
22. Y. Hanaoka, G. Hanaoka, J. Shikata, and H. Imai. Identity-Based Hierarchical Strongly Key-Insulated Encryption and Its Application. In Proc. of ASIACRYPT'2005, LNCS 3788, pp.495-514, Springer, 2005.
23. Jeremy Horwitz, Ben Lynn. Toward hierarchical identity-based encryption. In Proc. of EUROCRYPT 2002, LNCS 2332, pp. 466-481, Springer, 2002.
24. G. Itkis and L. Reyzin. SiBIR: Signer-Base Intrusion-Resilient Signatures. In Proc. of Crypto'2002, LNCS 2442, pp. 499-514, Springer, 2002.
25. J. Liu, and D. Wong. Solutions to Key Exposure Problem in Ring Signature. Available at http://eprint.iacr.org/2005/427.
26. Z. Le, Y. Ouyang, J. Ford and F. Makedon. A Hierarchical Key-Insulated Signature Scheme in the CA Trust Model. In Proc. of ISC'2004, LNCS 3225, pp. 280-291. Springer, 2004.
27. A. Shamir. Identity-Based Cryptosystems and Signature Schemes. In Proc. of Crypto'1984, LNCS 196, pp. 47-53, Springer, 1984.
28. B. Waters. Efficient Identity-Based Encryption Without Random Oracles. In Proc. of Cryptology-Eurocrypt'05, LNCS 3494, pp. 114-127, Springer, 2005.
29. D. H. Yum and P. J. Lee. Efficient Key Updating Signature Schemes Based on IBS. In Cryptography and Coding'2003, LNCS 2898, pp. 167-182, Springer, 2003.

AES Software Implementations on ARM7TDMI

Matthew Darnall[1,*] and Doug Kuhlman[2]

[1] Department of Mathematics
University of Wisconsin, Madison, 480 Lincoln Drive Madison, WI 53706
darnall@math.wisc.edu
[2] Motorola Labs, 1301 E Algonquin Rd Schaumburg, IL 60173
Doug.Kuhlman@motorola.com

Abstract. Information security on small, embedded devices has become a necessity for high-speed business. ARM processors are the most common for use in embedded devices. In this paper, we analyze speed and memory tradeoffs of AES, the leading symmetric cipher, on an ARM7TDMI processor. We give cycle counts as well as RAM and ROM footprints for many implementation techniques. By analyzing the techniques, we give the options we found which are the most useful for certain purposes. We also introduce a new implementation of AES that saves ROM by not explicitly storing all the SBOX data.

1 Introduction

Rijndael was selected as the Advanced Encryption Standard (AES) in 2001 by the National Institute of Standards and Technology (NIST) for use in government cryptographic purposes [1]. Before and after its adoption by the NIST, Rijndael received much study in the area of optimization. Many papers give software methods for optimizing certain portions of the algorithm. These papers usually focus on only one part of the algorithm. However, real world implementations are often a trade-off between speed, program size, and memory available. Rarely is one aspect fully optimized at the expense of the others.

ARM processors are the most common for embedded systems, with over 1 billion sold worldwide [6]. Their use in applications like digital radios and PDAs capable of electronic funds transfers has necessitated software implementations of AES for the ARM core.

The contribution of this paper is a detailed study of software optimization options of the Rijndael cipher on ARM processors. Previous work has focused on specific ideas for limited resource environments. We bring together the previous methods and some new ideas to give accurate timings and memory calculations for the AES cipher. Our work covers the key setup and encryption/decryption of AES with all three key sizes approved by the NIST. This paper focuses on the 128-bit key size, but the methods extend in a natural way.

---

* The first author was supported by the University of Wisconsin Mathematics Department's NSF VIGRE grant.

R. Barua and T. Lange (Eds.): INDOCRYPT 2006, LNCS 4329, pp. 424–435, 2006.

The rest of the paper has the following format. In section 2, we give a short description of AES and set some notation. In section 3, we describe the ARM7TDMI and the diagnostic tools we used to test our implementations. In section 4, we discuss options for the key expansion of AES. In section 5, we give code optimizations for the actual encryption and decryption. Section 6 details some results from combining various implementation tricks. In section 7, we summarize our results.

2   Description of AES

A full description of AES is included in [4], and we will maintain the same notation. AES is a 128-bit block cipher with 128-, 192-, or 256- bit keys. The data is operated on as a 16-byte state, which is thought of as four 32-bit column vectors. The number of rounds (Nr) is 10, 12, and 14 for the three respective key sizes (128, 192, and 256). The cipher starts with an XOR of the first 128 bits of the key with the plaintext. Then there are Nr - 1 rounds consisting of four parts: the S-BOX substitution, a byte permutation, a MixCol operation that operates on the state as four separate four-byte column vectors, and a key addition (XOR). The final round has no MixCol part. The MixCol operation is multiplication by a constant fourbyte vector, as described in [4]. The key for each round is determined by transformations of the key for the previous round, with the first key being the cipher key. Each part of a round is invertible. We shall denote the inverse of the MixCol operation as InvMixCol.

AES was designed to give excellent performance on many platforms. In particular, the cipher performs well on 32-bit platforms such as the ARM. The state and key can be implemented as an integer number of 32-bit variables. The key addition is a straightforward XOR with the state. The SBOX substitution has an algebraic description, but implementing it is more efficient in time and space when using a table with 256 one-byte entries. The byte permutation can be implemented when performing the S-BOX. As noted in [3], the best method for optimizing the algorithm lies in the implementation of the MixCol and InvMixCol operations.

3   ARM Information

The ARM processor family, created in 1985, has established itself as the leading embedded systems processor. The processors use RISC (Reduced Instruction Set Computer) architecture design philosophy. Due to the constrained environments the processors are used in, the processors have little power consumption and a high code density.

We tested all of our implementations for the ARM7TDMI. This processor is one of the most popular ARM designs, used mainly in mobile embedded systems such as cell phones, pagers, and mp3 players. The processor has a three stage pipeline. It uses a Von Neumann architecture design, so the data and the instructions use the same bus. This ARM core is particularly low power, using just .06 mW/Mhz. The lower power usage is partly due to the fact there is no cache, which can sacrifice performance. The use of encryption in cellular technology and Digital Rights Management (DRM) issues for mobile media players necessitates efficient encryption on the ARM7TDMI.

To give memory and cycles usage, we used a program provided by ARM Ltd. called ARMulator [7]. The software can provide ROM footprints for each portion of the program, even separating code and static variables. The software gives cycle counts to calculate time requirements. To simulate a real world environment, we started counting cycles before a call to the function performing the desired operation, such as encryption, decryption, or a key setup. We stopped counting cycles after the function was exited. For RAM measurements, we needed to include all data that was rewritable in the program, as well as the stack and heap. The ARMulator provides RAM measurements for the dynamic variables. The stack was measured by initializing the stack space to the constant value 0xcc. Then, after the program completed, the stack space was checked to find the position of the first byte not equal to 0xcc. The heap was treated similarly.

4   Key Scheduling

As with most block ciphers, the key for each round can be computed and stored before the cipher is performed on a specific plaintext. This section details the methods for computing the key schedule. Later, we shall give methods that compute the key for a particular round when it is needed to save on storage costs.

4.1   Large Decryption Tables

The key for each round is a function of the key from the previous round. When using large decryption tables, most of the key expansion will need to be operated on by InvMixCol, for ease in the actual decryption code. These operations can be calculated on one 32-bit word per iteration of the loop. The best way to perform the InvMixCol is a lookup to the SBOX table followed by a lookup to the full decryption table. This method is superior to the standard method given in [4] in both speed and code size, as shown in Table 1. We use this method for all code versions with the large decryption table.

Table 1. Cost of InvMixCol Methods for Dec. Key Expansion

Method	Cycles	ROM
Table Lookups	58	176
Standard	60	240

4.2 Loops in Key Expansion

When calculating the key expansion, we found the most compact option was to have one rounds key computed in each iteration of a loop. It is possible to unroll this loop to lower the cycle cost, but the code size is drastically larger. Given the large gains in ROM compared to the small cost in cycles, and the fact that the key expansion is usually performed once for many encryptions/decryptions, we suggest using a loop to perform the key expansion for most applications. All of our future comparisons that use a key expansion will assume that the key expansion was performed using a loop. The results for various key expansion options are given in Table 2.

Table 2. Effects of Loops on Key Expansion Stage

Type of Key Expansion	Loop	Cycles	ROM
Encryption	Yes	628	176
Encryption	No	442	900
Decryption	Yes	1903	144
Decryption	No	1627	3688

5 Cipher Code Optimizations

In this section, we examine a number of potential techniques to provide trade-offs between ROM, RAM, and/or cycle count. The general outline of each subsection is to describe the technique and then provide a table showing how implementing some of the techniques affect performance.

5.1 Using Pre-computed Tables

As noted in [4], the SBOX substitution has a nice algebraic description. For implementation purposes, the SBOX and Inverse SBOX substitutions are best performed by two 256-byte table lookups. In [4], the designers of Rijndael describe how each encryption round, except for the round key addition, can be tabularized as well. The SBOX substitution followed by MixCol can be implemented in four tables with 256 entries of 32 bits each, for a total of 4096 bytes of storage. The four tables are simply rotations of each other, so in fact one table can be used to save ROM, at the expense of extra rotations. The same method can be used to tabularize the decryption, but the round keys must be altered by the InvMixCol operation prior to use (see Equivalent Inverse Cipher in [1]).

As mentioned in section 3 and [6], the ARM processor allows registers to be rotated by the barrel shifter prior to an operation. This makes the use of four tables disadvantageous, as it uses more ROM and has no benefit in clock cycles. In fact, we found that the use of tables added cycles to each implementation we tested.

Table 3. Effects of Several Enc/Dec Tables vs. One Enc/Dec Table

Number of Tables	Enc Cycles	Dec Cycles	ROM for Tables (Kb)
1	1197	1199	2
4	1281	1283	8

This is easily seen by using a public version of AES code by Gladman [7]. The results of his code with one or four tables are shown in Table 3 for 128-bit keys.

The constant four-byte vector used in the MixCol operation has two 01 entries. This means that two of the bytes in each entry of the encryption table are precisely the same as the corresponding SBOX table entry. We propose using only the encryption table, and rotating it as necessary to recover the SBOX table. This will save 256 bytes for the SBOX table storage in ROM, with the disadvantage of slightly more code and cycles for the rotations. As shown in Table 4, the penalty in cycles for the proposed method is only 17 for encryption and 50 for the key setup, while a net gain of 140 bytes is achieved in ROM. This gain is identical for all three key sizes, since the SBOX is only used for the last round of the cipher.

Table 4. Effects of the Proposed No-SBOX Method vs. Standard Method

Method	Enc Cycles	Enc Key Setup Cycles	Total ROM
Standard	2094	628	3924
Proposed	2111	678	3784

5.2   Use of Loops

Since the cipher has (Nr -1) identical rounds, the use of a loop will save on ROM footprint size. The costs are the extra variable on the stack as a counter for the loop and the cycles to increment and check the counter. Partial unrolling is also possible, where each iteration of the loop performs more than one round of the cipher. We tried performing two and three rounds per iteration for the 128-bit key. Doing three rounds per iteration has the advantage that no check is necessary to ensure the correct number of rounds is performed, since three divides the number of total normal rounds (9).

5.3   Use of Functions

The use of functions has the potential to greatly reduce the code size of a program. The cost will be in the possible increase of the stack and the cycles to

Table 5. Effects of Different Loop Rolling on 128-bit AES

Loop Status	Total Cycles	Total ROM	Total RAM
Unrolled Full	1994	6900	280
Unrolled Thrice	2052	4488	284
Unrolled Twice	2085	4068	284

pass variables and enter the function. Many portions of the cipher can be implemented as functions. We tried implementing an entire round as a function as well as several byte manipulations that occur frequently in the cipher. When a function is used for a small task, such as selecting a particular byte out of a 32-bit word, the added code to enter and exit the function was too great to see a benefit in code size. The performance was slowed also, making the technique useless for improving the implementation. There was, however, a small gain in ROM size when the macros to turn four bytes into a word were made into a function (Table 6).

Table 6. Effects of Macros vs. Functions for Byte Manipulations

Method	Total Cycles	Total ROM	Total RAM
Macros	2094	3604	284
Function	2149	3572	284

When the round was implemented as a function, the code was still larger than when using a loop to run the rounds. Also, the performance was much worse than simply using a loop. For this reason, we recommend not using functions for the entire round.

5.4   Compact Versions

When ROM is the main constraint, using full encryption/decryption tables can be undesirable due to their large size. To save the 2 kilobytes used by the tables, an implementation can use tables for the SBOX and its inverse, while performing the Mix-Col and InvMixCol in full. There is a significant cost in cycles, but in constrained environments this can be necessary. Several techniques have been given for speeding up the original methods given in [5], such as the ideas in [2] and [3]. We have implemented the ideas from these papers and tested their performance. Given the necessity in this environment for a small ROM footprint, we used a loop for all the tests. Unrolling the loop would have resulted in a larger ROM footprint and worse performance than using tables with a loop.

An optimized version of the MixCol operation for ARM processors is given in [2]. Since the MixCol operation acts independently on each byte, it is possible to make each round calculation a loop that executes four times, once for each column of the state. Our results for the no encryption table option are given in Table 7. Looping within the round saves 116 bytes of ROM, at a loss of 1366 cycles. The decryption in the Table 7 options is done using the idea from [3].

Table 7. Statistics for Options That Dont Use Encryption Tables

Loop within each Round	Total Cycles	Total ROM	Total RAM
Yes	4981	2172	280
No	3615	2296	280

For the InvMixCol operation, [3] gives a method of transposing the state matrix to make parallel calculations easier. There is an extra cost in cycles and code for the key set up, as the key expansion must be transposed as well. Also, there is more code in the decryption function to transpose the state array. The benefit is in clock cycles for decryption. This method cannot be looped as the original method can; because once the state is transposed the InvMixCol operation no longer acts on columns independently.

Alternatively, the decryption round can be executed in a loop similar to the encryption round loop. Our results for different decryption options are given in Table 8. We designate as Loop 1 the method of within each round performing the InvMixCol and key addition inside a loop. We designate as Loop 2 the method of performing the InvSBOX substitution as well in a loop. The encryption in Table 8 calculations is done using the method described in [2]. Since the InvMixCol is so costly, we included the option of a full decryption table for comparison.

Table 8. Statistics for Options That Dont Use Decryption Tables

Method	Total Cycles	Total ROM	Total RAM
Transposed State	3615	2296	280
Loop 1	5119	2108	284
Loop 2	5542	1972	284
Full Dec. Table	2589	2800	284

5.5   No Key Expansion

To save RAM, the key for each round can be computed during the encryption and decryption, rather than before. This method is very costly in clock cycles if the same key is used for consecutive encryptions or decryptions, as it repeats operations for each encryption or decryption. The method also makes use of decryption tables, since the key for each round must be transformed by the InvMixCol operation before each use. Then the key must be transformed by the MixCol operation so that it can be used to derive the next rounds key. We give the statistics on the best options we found in Table 9.

Table 9. Statistics for Options That Dont Use Key Expansion

Full Enc Table	Total Cycles	Total ROM	Total RAM
Yes	4763	3760	120
No	5760	3040	120

5.6   Table Generation

The tables used in AES can be generated rather than stored in ROM. The code to generate these tables is surprisingly small, due to choice of SBOX and constant polynomial for the MixCol operation. The downside, as illustrated by Table 10, is a large cycle cost and RAM footprint. To store the complete tables takes just as much memory in RAM as it would require storing the tables in ROM. In addition, to generate the SBOX and Inverse SBOX, 512 bytes of temporary storage are needed. The SBOX and Inverse SBOX can be generated simultaneously. Due to the large overhead in cycles, generating the tables should only be done when a significant number of encryptions/decryptions are needed and ROM is the main limiting factor. The following table shows only the cycles, ROM, and RAM required for the table generation, so it would be an upfront one-time cost to the results given previously using various tables, minus the ROM cost of the tables themselves.

Table 10. Statistics for Generating the Tables

Table Type	Total Cycles	Total ROM	Total RAM
SBOX + Inv. SBOX	12528	216	1024
Full Enc. Table	2271	116	1024
Full Dec. Table	14091	232	1024

6   Combining Techniques

Many of the various techniques described in this paper can be combined to form a single version. Doing so gives various cost trade-offs between the three critical counts (cycles, ROM, and RAM). Actually, a fourth trade-off occurs, too, as some implementations will perform a single encryption and/or decryption faster than another but fall behind for bulk encryption/decryption.

Some versions are clearly better than others, too. We provide two of the more interesting examples. One possible implementation is to not have an SBOX (using the large encryption table for that), partially unroll the round loop (two rounds/loop iteration), and to use one large table each for encryption and decryption with a rolled key schedule (SRTxP in Table 11). This program uses 4228 bytes of ROM, takes 4863 cycles for key set-up, plus one encryption, and one decryption, and uses 284 bytes of RAM. However, including the SBOX but leaving the round tightly looped (SRTSN) only uses 3924 bytes of ROM, 4625 cycles, and 284 bytes of RAM, so it is clearly a superior option.

A second example shows a bit more subtlety. A program which unrolls both loops during key expansion but keeps only puts two rounds per loop in encryption/decryption (SBTSP) only requires 8192 bytes of ROM, 284 bytes of RAM, and 4163 cycles for a single set-up, plus encrypt and decrypt. A similar program which unrolls the decryption portion of the key scheduling and puts three rounds in each loop (SDTST) requires 8332 bytes of ROM, 284 bytes of RAM, and 4307 cycles for a single set-up, encrypt, decrypt trio. However, the second program is fasterfor repeated operations, so that 10 encryptions/decryptions take the first 23009 cycles but the second only 22775. Thus, no conclusion of pure superiority of one against the other can be made.

7   Conclusions and Future Work

This work identified many options available to implement AES on an ARM7TDMI processor. We have shown the actual cost of the various tradeoffs available to the programmer. It has been shown that for relatively small losses in the time for encryption, significant gains in ROM size can be achieved. Additionally, larger gains in ROM footprint can be made if the speed of the algorithm is significantly reduced. One of the methods for tradeoff, the use of no SBOX, has not yet been seen in the literature. The tradeoff for cutting the RAM by more than 50% is very costly in time, especially if multiple encryptions/decryptions shall be performed.

We have included Table 11 to compare many of the techniques mentioned before. The Total ROM includes all key setup and the encryption and decryption functions. The Total RAM includes rewritable data and stack/heap readings. The cycles are given for only one encryption and decryption, as well as for ten. As shown in the table, some methods offer good performance only when performing one or two encryptions.

Future work in the area should focus on several ideas. New embedded processors, such as Intels Xscale, will continue to be developed. While many of the ideas from work like this can be transferred to other processors, each new processor offers new methods of optimization. For maximum ease for the end user, each processor can be optimized individually. Software implementations of other cryptographic algorithms, such as Elliptic Curve Cryptography, on the ARM7TDMI should also be researched to complete the knowledge of this processor.

References

1. Announcing the ADVANCED ENCRYPTION STANDARD (AES): Federal Information Processing Standards FIPS 197, National Institute of Standards and Technology (NIST), Nov 2001
2. K. Atasu, L. Breveglieri, M. Macchetti: Efficient AES implementations for ARM based platforms, Proceedings of the 2004 ACM Symposium on Applied Computing, pp. 841-845 2004.
3. G. Bertoni, et al.: Efficient Software Implementation of AES on 32-bit platforms. In B. S. K. Jr., Cetin Kaya Koc, and C. Paar, editors, Cryptographic Hardware and Embedded Systems   CHES 2002, volume 2523 of Lecture Notes in Computer Science, pages 159-171. Springer, Berlin, Aug. 2002.
4. J. Daemen, V. Rijmen: The Design of Rijndael. Springer 2002.
5. B. Gladman: AES with support for the VIA Nehemiah Advanced Cryptography. Available at http://fp.gladman.plus.com, March 2004.
6. A. Sloss, D. Symes, C, Wright: ARM System Developers Guide. Elsevier Inc. 2004.
7. ARM Software Development Toolkit. Version 2.50 User Guide, ARM Ltd. 1998

A   Appendix

We include here a comprehensive table of interesting implementations. We intentionally left in some sub-optimal results for comparison purposes, but extremely bad options (like not using tables but completely unrolling the decryption key setup) are not included.

The names in this table require some explanation. The first letter or two determines the treatment of key expansion. If the first letter is S, then the key is set-up before any encryptions or decryptions. The format is then Sabcd, where a tells what was rolled in key setup (R means all rolled, E means encryption key unrolled only, D means decryption key unrolled only, and B means both unrolled). The third position, b, tells whether large tables were used, with T denoting Table and S denoting only the S-box and its inverse stored in ROM. The final two letters meanings depend on whether tables were used or not. When c is T, d tells whether the S-box is additionally present (S indicating presence and x indicating its absence) and e indicates the number of rounds looped (F is fully unrolled, P indicates two rounds in each loop, T means three rounds in each loop, and N means no unrolling was done, so that one round is done each pass through the loop). In the cases were c is S, d denotes whether encryption was rolled (E means encryptions was rolled, x means not rolled), and e tells how decryption was rolled (similar notation to Table 7, where x means not rolled, 1 and 2 are certain rollings, B is the entire process looped, and T meaning the large decryption table was used).

Table 11. Combined Data

Name	Cycles for 1	Cycles for 10	ROM	RAM
SRSE2	9646	73681	2164	304
SRSEB	9540	72621	2172	308
SRSx2	8073	57951	2292	304
SRSE1	9223	69451	2300	304
SRSB	7967	56891	2300	308
SRSx1	7650	53721	2428	304
SRSEx	7719	54411	2488	300
SRSxx	6146	38681	2616	280
SESE2	9460	73495	2888	304
SESEB	9354	72435	2896	308
SRSET	6693	44151	2992	300
SESx2	7887	57765	3016	304
SESE1	9037	69265	3024	304
SESxB	7781	56705	3024	308
FRxx	5760	57600	3040	120
SRSxT	5120	28421	3120	284
SESx1	7464	53535	3152	304
SESEx	7533	54225	3212	300
SESxx	5960	38495	3340	280
SESET	6507	43965	3716	300
FREx	4763	47630	3760	120
SRTxN	4866	23811	3784	284
SESxT	4934	28235	3844	284
SRTSN	4625	23471	3924	284
FRxD	6366	63660	4164	124
SRTxP	4863	23781	4228	284
SRTSP	4616	23381	4368	284
SRTxT	4830	23451	4568	284
SETSN	4439	23285	4648	284
SRTST	4583	23051	4788	284
FRED	5514	55140	4808	124
SETSP	4430	23195	5092	284
SETST	4397	22865	5512	284
SRTxF	4772	22871	7080	280
SRTSF	4525	22471	7220	280
SDTSN	4349	23195	7468	284
SDTSP	4340	23105	7912	284
SETSF	4339	22285	7944	280
SBTSP	4163	23009	8192	284
SDTST	4307	22775	8332	284
SBTSN	4154	22919	8636	284
SBTSN	4121	22589	9056	284
SDTSF	4249	22195	10764	280
SBTSF	4063	22009	11488	280
FxEx	4770	47700	14676	120
FxED	5345	53450	15468	120
Fxxx	5655	56550	16224	120
FxxD	6230	62300	17020	120

When the key is not set-up before encryption and decryption, the first letter in the name is F, for keying on the fly. The rest of the name indicates what other options were set. If the second letter is R, then the rounds were done in a large loop, else they were fully unrolled (x). The third letter indicates the presence of the large encryption table (E yes, x no), and the final letter the same for the decryption table.

Galois LFSR, Embedded Devices and Side Channel Weaknesses

Antoine Joux[1,2] and Pascal Delaunay[2,3]

[1] DGA
[2] Université de Versailles St-Quentin-en-Yvelines
PRISM
45, avenue des Etats-Unis
78035 Versailles Cedex, France
Antoine.Joux@m4x.org
[3] Thales
Pascal.DELAUNAY@fr.thalesgroup.com

Abstract. A new side channel attack against a simple LFSR is presented. The proposed attack targets a single Galois LFSR running on an embedded device where the only accessible information is the side channel leakage. Even if it is made only of simple XOR gates, such an object is vulnerable to side channel cryptanalysis depending on its implementation. Our attack combines simple side channel analysis and statistical analysis to guess output bits and fast correlation attack to recover the initial state. In practice, even if a LFSR is never used alone, this attack shows that simple XOR gates can reveal significant information in some circumstances.

1 Introduction

Since the introduction of Power Analysis Attacks by Kocher et al., Side Channel Analysis and Side Channel Resistance of cryptographic algorithms performed on embedded devices are being deeply studied. These attacks allow full recovery of secret data with relatively low complexity and small investment compared to mathematical cryptanalysis. Thus, the security of cryptographic algorithms on embedded devices is usually not only studied in a mathematical way but also in a side channel based approach.

While these attacks focus on key-dependent operations, values handled by the device and non-linear functions, elementary logic gates have rarely been studied in the literature. Moreover, stream ciphers, which are mainly composed of such gates, are not prime targets for side channel attacks. Still, there are a few publications on this topic such as [7] and [14]. We show in this paper that in the context of stream ciphers based on *Linear Feedback Shift Register* (LFSR) the leakage of XOR gates can be exploited in a Simple Side Channel Attack.

From the mathematical point of view, the security of LFSR based stream cipher as been deeply studied. In particular, correlation attacks are often considered. This class of attack studies how to recover the internal state of the

R. Barua and T. Lange (Eds.): INDOCRYPT 2006, LNCS 4329, pp. 436–451, 2006.

underlying LFSR by viewing the output of the stream cipher as a noisy version of the LFSR output (modeled by a binary symmetric channel). It is especially useful when a good linear approximation of the output function can be found. Ordinary correlation attacks can only be applied to small LFSR where the internal state can by exhaustively searched. Thus they are often used in a divide and conquer manner against stream ciphers which rely on several small LFSRs. With larger LFSRs, it is possible to use fast correlation attacks which overcome this length limitation at the cost of using longer output sequences. These *Fast Correlation Attacks*, were introduced by Meier and Staffelbach [9], and have two different phases. The first and most time consuming consists in finding as many *parity check equations* as possible (low-weight linear relations between the output bits and the initial state). Many papers (e.g. see [10], [3], [4]) describe efficient methods to compute such equations. The second phase decodes the sequence observed to reconstruct the initial state of the LFSR.

LFSRs are often implemented as *divisor registers* (also known as Galois LFSRs) (Fig. 2) because they offer minimal latency and higher speed for the same output sequence. We show in the following that such an implementation is vulnerable to *Simple Side Channel Attacks*. From now on, we consider a n-bit length divisor register running on an embedded device where the only available information is the side channel leakage. This roughly models a strong mathematical LFSR based stream cipher, where the output function is cryptographically strong. In that case, the stream output does not offer any usable information to an attacker wishing to recover the secret key or equivalently the LFSR state. In this model, we exhibit a significant bias between the side channel leakage of the Galois LFSR and the value of the bit which is shifted out of the LFSR.

If the bias is large enough to correctly predict n consecutive output bits then simple linear algebra allows direct recovery of the initial state. However, the side channel analysis is often noisy so this extremely favorable case is unlikely to happen.

As a consequence, we mix the basic side channel approach with cryptanalytic tools in order to go further. More precisely, the side channel analysis produces a biased prediction of the Galois LFSR output bits. In order to make a simplified analysis, we assume that the prediction of different bits is independent and thus behave as in the Binary Symmetric Channel model (BSC). Since correlation and fast correlation attacks are already described in this model, it is of course natural to use them as a tool to amplify the side channel predictions and recover the LFSR initial state. In other words, the basic idea in this paper is to replace the correlation between LFSR output and stream cipher output usually used in correlation attacks by a side channel correlation. Note that in some cases, it could also be possible to combine both the side channel information and the stream cipher output in order to get an even better bias. We do not consider this variant in depth.

The present paper is organized as follows : in section 2 we recall the basics on LFSRs and divisor registers. Section 3 is devoted to Side Channel Attacks. We analyze the leakage induced by the XOR gates of our Galois LFSR with respect to two often used models : the *Hamming weight model* and the *Hamming distance*

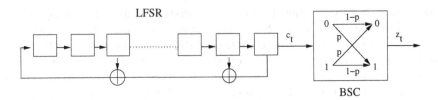

Fig. 1. LFSR in the Binary Symmetric Channel (BSC) model

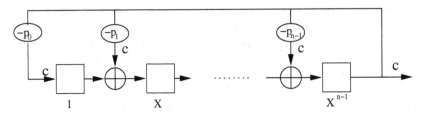

Fig. 2. Divisor register or Galois LFSR

model. In section 4, we correlate the leakage observed with prediction among output bits. We link the expected population of output bits that can be guessed with some non negligible advantage to the number of gates τ of the LFSR. Finally we explain how to recover the initial state using traditional cryptanalysis and illustrate the approach with some numerical evidence on various LFSRs.

2   LFSR and Divisor Register

LFSRs are among the simplest objects used in cryptography. They are the basis of many stream ciphers and Random Number Generators. Two points are important to provide sufficient security to a n-bit length LFSR. In order to avoid many mathematical attacks, the value n must be high enough and its feedback polynomial P should be irreducible in order to generate a maximum length sequence.

There is an interesting map between LFSRs (Fig. 1) and divisor registers (Fig. 2). Let $P(X)$ be the feedback polynomial of a LFSR over $\mathbb{F}_2$:

$$P(X) = \sum_{i=0}^{n} p_i X^i, p_i \in \mathbb{F}_2$$

Then there is an equivalent divisor register with feedback polynomial:

$$P^\star(X) = \sum_{i=0}^{n} p_{n-i} X^i, p_i \in \mathbb{F}_2$$

Fig. 2 and Fig. 1 respectively show a divisor register and its equivalent LFSR. Swapping from one implementation to the other is not a difficult task.

So if we are trying to attack a divisor register, we can consider its equivalent LFSR where mathematical cryptanalysis is concerned. All known attacks against LFSRs can be applied to divisor registers, this includes all forms of Fast Correlation Attacks.

In a hardware environment, LFSRs are often implemented in their Galois form. Each implementation has the same number of XOR cells but the latter form offers an important advantage. The XORs are computed in parallel in the Galois LFSR whereas there are at least $\log_\alpha(\tau)$ stages to compute the shifted in value in a traditional LFSR, where τ is the number of taps and α the number of inputs to a XOR gate (depending on the hardware). Thus the Galois form offers a shorter critical path, allowing higher speed.

When using LFSR in hardware, one often seeks to reduce the weight of the feedback polynomial in order to minimize the number of gates. On the other hand, this is usually not pushed to an extreme in LFSR based stream ciphers. This is mainly due to historical reasons. Indeed, older correlation attacks worked better on low weight feedback polynomials and avoiding these polynomials was seen as a safety measure. Ironically the side channel attack presented here works better when the feedback polynomial has a relatively high weight.

3   Side Channel Attacks

3.1   Background on Side Channel Attacks

The *Side Channel Attacks* appeared in preprints in 1998; the first publication appeared at Crypto 1999 [6] and revealed that a naive implementation of cryptographic algorithms on an embedded device (*e.g.* smart card) allows an easy recovery of the sensitive data (such as the private keys) by observing some side channels. There are several types of side channels considered in the literature, timing attacks, power leakage, electromagnetic leakage [1] and others less frequently encountered such as acoustic analysis.

The equipment required to implement Side Channel Attacks is reasonably simple and quite widely available. For example, power attacks require a single computer and a digital oscilloscope measuring the electrical current consumed by the chip during the computation. This measurement is typically performed by inserting a small resistor (*e.g.* 50Ω) on the power pin V_{CC} or on the ground pin V_{GND} and by measuring the voltage drop induced at the resistor. Electromagnetic attacks require almost the same equipment, swapping the resistor and the measurement setup for an electromagnetic probe measuring the electromagnetic emanations of the chip during the computation. Of course, finer measurements require more costly equipment, however, in all cases, it remains much more affordable than specific cryptanalytic hardware intended for fast exhaustive key searches or other mathematical cryptanalysis.

Power and electromagnetic attacks can further by divided in a few categories among which the most known are the Simple Attacks and the Differential Attacks. The first category gathers attacks for which the secret data is directly recovered from a single curve of observed leakage. On the contrary, the Differential Attacks require a large number of curves in order to exhibit biases correlated with the values handled by the device.

In this paper, we focus on a *Simple attack*, involving a single leakage curve, of either power or electromagnetic origin.

In order to explain the validity of the Side Channel Attacks, many papers model the leakage of an electronic device with the data handled ([2], [13], [12]). Different models coexist but two of them are often encountered: the *Hamming weight* model and the *Hamming distance* model, where the leakages are respectively described as functions of the Hamming weight of current data or of the Hamming distance between current and previous data. More precisely, these two models describe the side channel leakage as follows.

$$W = a \times \mathcal{H}(x) + b \tag{1}$$

for the Hamming weight model and

$$W = a \times \mathcal{H}(x \oplus p) + b \tag{2}$$

for the Hamming distance model, where a and b are reals, x is the data currently handled, p the previous one, and $\mathcal{H}(y)$ denotes the Hamming weight of the data y (*i.e.* the number of bits equal to 1 in y). In the next section, we analyze the behavior of LFSRs in these two models.

3.2   A Side Channel Attack Building Block Against Galois LFSR

Taking a closer look at the *divisor register* (Fig. 2), we see that at the beginning of each clock cycle, all the XORs are performed in parallel. Moreover, all these XORs share a common bit, the output bit c. Thanks to this simple remark it is possible to construct a bit oriented side channel attack that can predict the (internal) output bit c produced by the LFSR. This basic attack by itself is not sufficient to predict the complete initial state of the LFSR, however, we show in section 4 how fast correlation attacks can be used to recover the initial state from a stream of such predictions for many consecutive bits.

The key idea behind this side channel prediction of bit c is that depending on the value of c, two different cases occur:

1. If $c = 0$, only $0 \oplus 0$ and $0 \oplus 1$ are computed in the XOR gates.
2. If $c = 1$, $1 \oplus 0$ and $1 \oplus 1$ are computed.

Moreover, the number of XOR gates used in the implementation is equal to the Hamming weight τ of the feedback polynomial thus the common bit c is used in τ (essentially) independent XOR gates during each clock cycle. As a consequence, the total leakage observed while the XORs are computed is going to reflect these two cases. Moreover, since it is a sum of τ independent leakages (one for each XOR gate) all related to c, we expect a good Signal To Noise Ratio (SNR). Informally, even though a single XOR gate does not leak enough information to recover c, many XOR gates with one common operand and independent second operands may leak enough information to predict c with a good bias. More precisely, we expect the bias to grow as a function of the number of gates τ.

In the rest of this section, we study the theoretical probability of success of our basic approach in two widely used models of side channel leakage (described in section 3). We show that in both models the above intuitive result about the signal to noise ratio holds. More precisely, we show that the part of the leakage correlated to c grows linearly with τ, while the associated noise is proportional to $\sqrt{\tau}$. As a consequence, this implies that for large enough values of τ the LFSR output can be predicted with a good bias. In order to reach this conclusion, we first consider the leakage of a single XOR gate in each model, then add up the contributions of the individual gates to obtain the global side channel leakage.

Hamming Weight Model. Considering the Hamming weight model (equation 1) and assuming that retrieving the input values and computing the XOR are roughly simultaneous, for each XOR gate $i \in [\![1, n]\!]$, with inputs E_1 and E_2, we can write:

$$W^{(i)}(E_1, E_2) = \alpha \mathcal{H}(E_1) + \beta \mathcal{H}(E_2) + \gamma \mathcal{H}(E_1 \oplus E_2) + b^{(i)} \qquad (3)$$

where $b^{(i)}$ denotes some noise induced by the measurement and has a Gaussian distribution with mean μ_b and variance σ_b^2, while α, β and γ are reals and denote the importance of each value. Since we deal with XOR gates on one bit values, we can see that $\mathcal{H}(E_i) = E_i$.

Since XOR gates are symmetric at the mathematical level, it is at first tempting to consider that $0 \oplus 1$ and $1 \oplus 0$ behave in the same way and to assume that $\alpha = \beta$. However, since the common bit is reused in many gates, this reduces the symmetry, thus we keep these two coefficients in the equations. Equation 3 can then be rewritten as:

$$W^{(i)}(E_1, E_2) = \alpha E_1 + \beta E_2 + \gamma (E_1 \oplus E_2) + b^{(i)} \qquad (4)$$

In the sequel, we assume that the common input to the XOR gates is E_1. Thus the equation becomes:

$$W^{(i)}(0, E_2) = (\beta + \gamma)E_2 + b^{(i)}$$
$$W^{(i)}(1, E_2) = \alpha + \beta E_2 + \gamma(1 - E_2) + b^{(i)}$$

Now, considering the influence of the τ XOR gates of the divisor register, we can express the leakage according to the output bit c (the common operand):

$$W_{(c=0)} = (\beta + \gamma) \sum_{i=1}^{\tau} E_2^{(i)} + \sum_{i=1}^{\tau} b^{(i)}$$

$$W_{(c=1)} = \tau(\alpha + \gamma) + (\beta - \gamma) \sum_{i=1}^{\tau} E_2^{(i)} + \sum_{i=1}^{\tau} b^{(i)}$$

Looking more precisely at the source of the noises $b^{(i)}$, we can distinguish two main sources. A global noise, that is due to the overall circuit environment and a gate by gate noise which depends on the exact individual configuration of each gate. Assuming that the global noise contribution is G and that the gate by gate measurement noises $b^{(i)}_{\text{ind}}$ are independent, we find that $b = G + \sum_{i=1}^{\tau} b^{(i)}_{\text{ind}}$ follows a normal distribution of mean $\mu_G + \tau\mu_b$ and standard deviation $(\sigma_G^2 + \tau\sigma_b^2)^{1/2}$ (also noted $\mathcal{N}(\mu_G + \tau\mu_b, \sigma_G^2 + \tau\sigma_b^2)$).

Thus, in this model, we find the expected result about the signal to noise ratio induced by the τ XOR gates for the physical measurement noise. Indeed, the contribution of the global noise decreases very quickly (linearly with τ), while the contribution of the gate by gate noise decreases proportionally to $\sqrt{\tau}$. Moreover, viewing the other bits of the Galois LFSR entering the XOR gates as independent random bits, the Hamming weight of the resulting τ-bit word, follows the binomial distribution with parameter $\frac{1}{2}$ and order τ. This can be approximated by the normal distribution $\mathcal{N}(\frac{\tau}{2}, \frac{\tau}{4})$. Finally, if we consider this Hamming weight and the measurement noise to be independent, we find:

$$W_{(c=0)} = \mathcal{N}\left((\beta + \gamma)\frac{\tau}{2} + \tau\mu_b + \mu_G, \frac{(\beta + \gamma)\tau}{4} + \tau\sigma_b^2 + \sigma_G^2\right) \tag{5}$$

$$W_{(c=1)} = \mathcal{N}\left((2\alpha + \beta + \gamma)\frac{\tau}{2} + \tau\mu_b + \mu_G, \frac{(\beta - \gamma)\tau}{4} + \tau\sigma_b^2 + \sigma_G^2\right) \tag{6}$$

Examining these two formulas, we find that:

- They differ in terms of mean and variance thus the two Gaussian curves induced are distinct. This leads to two different profiles of observed leakage depending on the output value, as we expected.
- The bigger τ is, the wider the difference between the means. A large number of XOR cells in the divisor register leads to a non-negligible difference in leaked signal when two different bit values are shifted out.
- The total contribution of the noise has a fixed part and a part proportional to $\sqrt{\tau}$ whereas the distance between the means is proportional to τ. Thus the Signal to Noise Ratio increases with τ by a factor $\sqrt{\tau}$ or better.

Thus, for given values α, β and γ, determined by the gate technology, these two distributions highly differ for sufficiently large τ. We can expect noticeable differences in terms of power consumption or EM emanations depending on the shifted-out bit. Fig. 3 and Fig. 4 respectively show theoretical results with 10 and 40 XOR gates when the global noise G contribution is neglected.

In practice, the exact values of α, β and γ are not easy to determine. However, it seems that for CMOS gates, the contribution of power consumption (and thus the leakage) depends more on the output value than on the input values. This means that α and β is quite small compared to γ and thus that the two Gaussian curves overlap a lot. As a consequence, the prediction bias of a single XOR gate is expected to be low.

Hamming Distance Model. In this model, the data leakage is proportional to the Hamming distance between the last value and the current value. Thus we

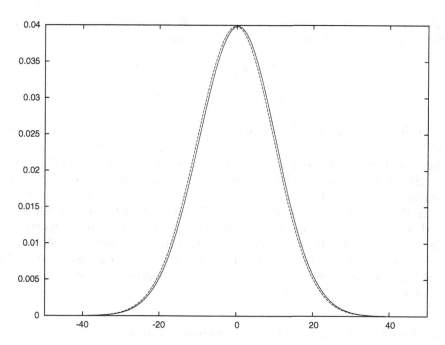

Fig. 3. Theoretical repartition of leakage observed, 10 *XOR* gates

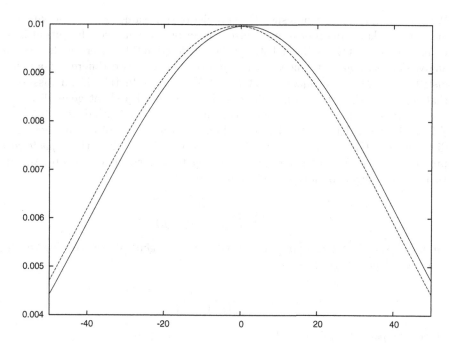

Fig. 4. Theoretical repartition of leakage observed, 40 *XOR* gates

introduce the time parameter T. We denote $E_j^{(i,T)}$ the entry E_j, $j \in \{1,2\}$ of XOR gate i at time T. With this notation, equation 2 becomes:

$$W^{(i,T)}(E_1, E_2) = \alpha \left(E_1^{(i,T)} \oplus E_1^{(i,T-1)} \right) + \beta \left(E_2^{(i,T)} \oplus E_2^{(i,T-1)} \right)$$
$$+ \gamma \left(E_1^{(i,T)} \oplus E_1^{(i,T-1)} \oplus E_2^{(i,T)} \oplus E_2^{(i,T-1)} \right) + b^{(i,T)}$$

In the above formula, we see that the common bit contributes in the form $E_1^{(i,T)} \oplus E_1^{(i,T-1)}$ instead of the simpler form $E_1^{(i)}$. Since this is the only significant difference between this equation and the corresponding equation in the Hamming weight model, we can skip the intermediate steps and directly conclude. After summing up the contributions of τ gates, we obtain a biased prediction of $c_T \oplus c_{T-1}$, the XOR of the output bit c at time T and $T-1$, instead of a prediction of c. However, it is well known that due to linearity predicting a linear function of the output of an LFSR is as good as predicting the LFSR output itself. Indeed, the linear output function can be transformed into a linear change of initial state and then removed. Thus, for the rest of our analysis, we can consider both theoretical leakage models as equivalent.

4   Distinguishers and Correlation Attacks

4.1   Optimal Distinguisher

We have to choose a distinguisher $\mathcal{D}$ to predict output values c_T by observing the measured leakage. In order to maximize correct predictions over the output bits, we use a variant of the optimal distinguisher depicted in [11]. After the theoretical analysis of section 3, we need to distinguish between two different Gaussian distributions $\mathcal{D}_0 = \mathcal{N}(m_0, \sigma_0^2)$ and $\mathcal{D}_1 = \mathcal{N}(m_1, \sigma_1^2)$. In [11], it was assumed that $\sigma_0 = \sigma_1$, since this is not the case here, we need a slight generalization. Assume that a value of v_T is chosen according either to $\mathcal{D}_0$ (with density function f_0) or $\mathcal{D}_1$ (with density f_1), for a random uniform choice of one of these two distributions. Given a measured v_T value, we want to analyze the *a posteriori* probability that v_T was generated according to $\mathcal{D}_0$, which we denote $p_0(v_T)$. Using Bayes' inversion formula, we find that:

$$p_0(v_T) = \frac{f_0(v_T)}{f_0(v_T) + f_1(v_T)}.$$

Similarly, the *a posteriori* probability that v_T was generated according to $\mathcal{D}_1$, which we denote $p_1(v_T)$, is:

$$p_1(v_T) = \frac{f_1(v_T)}{f_0(v_T) + f_1(v_T)}.$$

Of course, $p_0(v_T) + p_1(v_T) = 1$.

This analysis of *a posteriori* probabilities gives an optimal distinguisher. When v_T is measured, predict that $\mathcal{D}_0$ was used when $p_0(v_T) > 1/2$ and predict that

$\mathcal{D}_1$ was used when $p_1(v_T) > 1/2$. Since the boundary case $p_0(v_T) = p_1(v_T) = 1/2$ corresponds to the finite[1] set of intersection points of the two Gaussian curves, it is encountered with null probability and thus is not a problem.

In order to say more about the above optimal distinguisher, we need to analyze the intersection points of the two Gaussian curves. We recall that:

$$f_0(x) = \frac{1}{\sigma_0\sqrt{2\pi}} \, e^{-\frac{1}{2}\left(\frac{x-m_0}{\sigma_0}\right)^2} \quad \text{and} \quad f_1(x) = \frac{1}{\sigma_1\sqrt{2\pi}} \, e^{-\frac{1}{2}\left(\frac{x-m_1}{\sigma_1}\right)^2}.$$

When $\sigma_0 = \sigma_1$, the curves intersect in a single point $(m_0+m_1)/2$ and the analysis simplifies. When $\sigma_0 \neq \sigma_1$, the two Gaussians intersect in two distinct points. Between these two intersection points, the distinguisher predicts the distribution with the lowest σ value, outside, it predicts the one with the highest σ value.

4.2   Using the Distinguisher in Fast Correlation Attacks

Since the side channel measurements give a biased prediction of the LFSR output bits, they can be used to replace the usual correlations in the stream cipher output function in a fast correlation attack. However, with side channel correlation there is a essential difference, we do not only get a predicted value, we also know a precise measure. This can be very helpful since it allows us to learn additional information about the quality of the prediction. As a consequence, in order to use a set of measurements in a subsequent fast correlation attack, three approaches are now possible:

- The most basic approach is to directly use the optimal distinguisher output as a biased prediction for the fast correlation attack. In other words, we forget the measured value and only keep the corresponding prediction. In that case, given $\mathcal{D}_0$ and $\mathcal{D}_1$ it suffices to compute the probability of success of the optimal distinguisher and to directly apply a fast correlation attack on the output of the distinguisher. In this attack, the distinguisher prediction directly replaces the stream cipher bits. With this small change, all known correlation attacks can be used.
- Alternatively, it is also possible to predict only a good fraction of the LFSR output bits. More precisely, when p_0 is far enough from $1/2$ we give a prediction, when it is too near we refuse to predict. Using this approach, we predict a fraction of the output bits with a better bias. With a little care, we can still apply a fast correlation attack, keeping only parity checks corresponding to predicted bits. Using a good trade-off, we can improve the basic attack. This approach is very similar to the conditional correlation attacks which is described in [8].
- Finally, it is also possible to fully use the extra information present in the measured value v_T and to develop a specific fast correlation attack in order to fully use this information.

In the sequel, we address each approach in turn.

---

[1] Unless $\mathcal{D}_0 = \mathcal{D}_1$, in which case distinguishing is of course not possible.

Direct fast correlation attack approach. In this first approach, we need given the two distributions $\mathcal{D}_0 = \mathcal{N}(m_0, \sigma_0^2)$ and $\mathcal{D}_1 = \mathcal{N}(m_1, \sigma_1^2)$ to compute the bias of the optimal distinguisher. When $\sigma_0 = \sigma_1$, the probability of success is $p = \int_{-\infty}^{(m_0+m_1)/2} f_0(t)dt$ and assuming that $m_0 < m_1$ the bias is:

$$\epsilon = 2p - 1 = 2 \int_{m_0}^{(m_0+m_1)/2} f_0(t)dt = \frac{2}{\sigma_0\sqrt{2\pi}} \int_{m_0}^{(m_0+m_1)/2} e^{-\frac{1}{2}\left(\frac{t-m_0}{\sigma_0}\right)^2} dt.$$

The rightmost integral can easily be computed in many computer algebra systems, using either the *erf* or the *erfc* function.

When $\sigma_0 \neq \sigma_1$, let a and b, with $a < b$, denote the two roots of the equation $f_0(x) = f_1(x)$. Assuming that $\sigma_0 < \sigma_1$, a measured value between a and b corresponds to a prediction of $\mathcal{D}_0$. The probability of correctly predicting $\mathcal{D}_0$ and $\mathcal{D}_1$ are respectively:

$$p^{(0)} = \int_a^b f_0(t)dt \quad \text{and} \quad p^{(1)} = 1 - \int_a^b f_1(t)dt.$$

The total bias is obtained from the average probability and can be expressed as:

$$\epsilon = p^{(0)} + p^{(1)} - 1 = \int_a^b (f_0(t) - f_1(t))dt.$$

It can also be computed using the *erf* or the *erfc* function.

Once the prediction bias is know, we can directly apply any fast correlation attack, for example, the attack from [4]. In fact, for short enough LFSRs, it would also be possible to use a correlation attack instead. When feasible, this is the approach which requires the smallest amount of measurements. Given approximately $1/\epsilon^2$ measured bits, one can exhaustively search among the 2^n possible initial states. The correct initial state gives a sequence that matches the prediction corresponding to the measurement on more points than incorrect states. One possible application of this correlation attack approach could be to recover the state of a short LFSR based internal pseudo-random generator.

Conditional fast correlation attack approach. In this section, keeping the previous notations, we need an extra parameter $\zeta > 1$ that serves as a cutoff threshold for the distinguisher. When $f_0(v_T) > \zeta f_1(v_T)$, $\mathcal{D}_0$ is predicted, when $f_1(v_T) > \zeta f_0(v_T)$, $\mathcal{D}_1$ is predicted. The rest of the time, the distinguisher refuses to output a prediction. In this context, two parameters are essential, the fraction F_ζ of predicted bits and the expected bias ϵ_ζ for predicted bits. Both values can be computed in a similar way using various integrals. The integrations bounds involve the roots of $f_1 = \zeta f_0$ and of $f_0 = \zeta f_1$. When optimizing this approach, two main effects need to be accounted for, on the one hand the fraction of parity checks of weight[2] k being kept is F_ζ^{k-1}, on the other hand the number of parity

---

[2] Usually a parity check of weight k as $k - 1$ terms corresponding to measured bits and one term to the bit to predict.

checks needed for evaluation is ϵ_ζ^{2-2k}. As a consequence, we see that $F_\zeta \epsilon_\zeta^2$ should be maximized to reduce the number of parity checks and get the best possible efficiency. In that case, we see that the conditional correlation attack essentially behaves like a basic correlation attack with a different bias $\epsilon' = \epsilon_\zeta \sqrt{F_\zeta}$. Note that we gain memory in the final phase, since we perform the parity checks evaluation on a smaller set.

The refined approach: parity-check equations with overall probability.
As we previously mentioned, for each value v_T of measured leakage, we associate a probability $p_T = g(v_T)$ of false prediction. In this approach, we use this probability to obtain a bias ϵ_T for each output bit. Instead of using the traditional bias $\epsilon_T = 1 - 2p_T$, it is convenient to use a signed representation. More precisely, when the predicted output bit is 0 we keep ϵ_T as above, when the predicted output bit is 1 we replace ϵ_T by its opposite. This is convenient since multiplying signed biases allows to multiply the unsigned biases and XOR the predicted bits in a single operation. Following [4], where the output bits of the LFSR (our c_T) are denoted by z_i, we obtain parity-check equations of the form:

$$z_i = z_{m_1} \oplus \ldots \oplus z_{m_{k-1}} \oplus \sum_{j=0}^{B-1} c_{m,i}^j x_j \qquad (7)$$

As a consequence, if each output bit z_{m_j} has ϵ_{m_j} as its signed bias, the overall signed bias for this equation is, for a given (guessed) value X of length B:

$$\epsilon_i = \prod_{j \in [\![1,k-1]\!]} \epsilon_{m_j} \times (-1)^{c.X} \qquad (8)$$

After collecting many parity checks giving predictions of the same value z_i, we need to decide whether $z_i = 0$ or 1 depending on the signed biases of all predictions. We proceed as follows, we let $\mathcal{P}_0$ denote the product of the probabilities of having $z_i = 0$ for each individual parity check. Likewise, we denote by $\mathcal{P}_1$ the product of the probabilities of having $z_i = 1$. If $\mathcal{P}_0 > \mathcal{P}_1$ our overall prediction is $z_i = 0$, otherwise it is $z_i = 1$. Thus, we simply compute:

$$w = \frac{\mathcal{P}_0}{\mathcal{P}_1} = \prod_{\epsilon_i} \frac{1 + \epsilon_i}{1 - \epsilon_i}$$

Taking logarithms, we find:

$$\log(w) = \sum_{\epsilon_i} \log\left(\frac{1 + \epsilon_i}{1 - \epsilon_i}\right) \approx \sum_{\epsilon_i} 2\epsilon_i \text{ when } \epsilon_i \text{ is small}$$

Now, computing the overall prediction of one target bit is very simple, by summing the contributions of all parity check equations and taking the sign of the result. Moreover, the contributions of each parity checks behave nicely with respect to changes in the guessed part X of the LFSR. Indeed, when a bit X_j changes, two things can happen in a parity check. Either X_j is not present in its expression and the prediction and bias do not change. Or it is present, the prediction changes and this change is reflected by negating the signed bias. Furthermore, when ϵ_i takes the opposite sign, it is easy to see that $\log\left((1 + \epsilon_i)/(1 - \epsilon_i)\right)$ undergoes the same change. As a consequence, this additive way of dealing with the signed biases is fully compatible with the algorithms described in [4], we simply need to replace equation counts by weighted counts, where the weights are computed from the parity checks biases.

Thus, for each value of X, we can efficiently compute the corresponding $\log(w)$. If $\log(w)$ is far enough from 0 we decide that the value of X is correct and set z_i according to the sign of $\log(w)$. Compared to the previous approaches, we gain for two important reasons. First, equations associated to extreme leakage have a large weight and count much more than they do in the basic approach. Second, even the equations with bad weight still give some information instead of being thrown away. Moreover, we no longer need to optimize the threshold choice. As a consequence, this improved algorithm is both more efficient and simpler to use. Its drawback is that when replacing equation counters by weighted counters, we usually need to replace integers by floating point numbers, thus requiring about twice as much memory as the basic approach.

5   Simulated Results on 80 and 128-Bit LFSR

In this section, we study the practical application of our attack on LFSRs. One essential factor to the success of this attack is the amount of data leaked by individual XOR gates. Clearly, if XOR gates do not leak any signal, our approach fails. In the sequel, we assume that the leakage is low but still present. More precisely, we assume that for an individual gate, observing its leakage level allows to recover the considered input bit with a probability $p_0 > 1/2$. To illustrate the power of the technique we use, we show how the attack can work for a low value such as $p_0 = 0.51$. In order to simplify the analysis, we assume that the side channel leakage corresponding to 0 and 1 are two Gaussian distributions with the same deviation but with different mean values.

Note that in the other extreme case where the two distributions have different deviations but the same means, having access to many copies of the bit c in several XOR gates does not help. In order to improve the bias when adding Gaussian distributions, the mean values should be different.

Since the efficiency of the attack also depends on the LFSR length and the weight of the feedback polynomial, we propose to first consider a 80-bit LFSR with feedback weight 9. This illustrates our technique on a cryptographically significant size when the feedback polynomial has a relatively low weight. For example, one of the eStream phase 2 focus algorithm: Grain [5] uses a 80-bit LFSR with a feedback polynomial of weight 7. We chose 9 instead of 7 in order to work with a square. For this example, we computed theoretical complexity and also run the fast correlation attacks on simulated data. The simulated leakage measurement was generated using a simple program which adds together the $\tau = 9$ leakage levels corresponding to each output bit.

On this first example, we wanted to choose a value of p_0 near 0.51, in order to show that the attack already works with a small leakage level. We achived this goal by modeling the leakage of a single XOR gate with a 0 value on input c by a Gaussian of mean -0.025 and variance 1, while a 1 value on input c correspond to mean 0.025 and variance 1. Applying the computations of section 4, this choice yields $p_0 = 0.50997$. Having the same variance for both $c = 0$ and $c = 1$ is a simplifying hypothesis that makes the analysis easier, but does not significantly change the overall result. Similarly, shifting the leakage base level to get an overall 0 average also is a simplifying hypothesis.

From a theoretical point of view, putting together 9 gates with common gives input c, we obtain total leakage modeled by Gaussians with mean 0.225 or -0.225 and variance 9, i.e. normal deviation 3. From this data, we can compute the theoretical efficiency of the three attack variants. Using these data, we run a fast correlation attack on the 80-bit LFSR using parity checks of weight $k = 4$, where the target bit is correlated to 3 measured bits. As in [4], we have a certain number of bits which are guessed during the attack and removed from the LFSR size for the parity check computations. We omit the part of the fast correlation attack that aims at recovering the full register and simply focus on predicting a single bit z_i together with the correct guess for the removed bits. Let us compare our three variants on this numerical example:

- The basic approach is very similar to a numerical example given in [4] (with $1 - p = 0.469$, which corresponds to probability 0.531 and register size 89). With our proposed parameters, the value of c can be predicted with probability 0.52989, i.e. bias 0.05978. As a consequence, we roughly need 22 000 000 parity checks. With this number of parity checks we find a probability of correct around 84 percent if we keep the false alarm rate at $1/2$. From a complexity point of view, it is possible to get enough parity checks with k as above using 2^{27} bits of keystream, if we choose to guess $B = 26$ bits. We need a precomputation step of 2^{54} to build the parity checks (with memory requirements, around 2^{27}). The attack itself requires about $24 \cdot 2^B \approx 2^{30.6}$ operations to recover one bit z_i.

- With the conditional correlation approach, we want to minimize the ratio $\epsilon' = \epsilon_\zeta \sqrt{F_\zeta}$. Using numerical analysis, we find a maximum value $\epsilon' \approx 0.06738$ with $\zeta \approx 1.0958$. This corresponds to a predicted fraction $F_\zeta \approx 0.543$ and a bias for predicted bits of $\epsilon_\zeta \approx 0.09144$. Compared to the basic approach, we can either use half as many parity checks or get a better prediction probability around 92 percents. From a complexity point of view, we roughly gain a factor of 3, because the Walsh transform phase is perform on a much smaller set of equations. Moreover, we require about $1/8$ of the memory of the basic approach.

- For the refined approach, we do not know how to compute the theoretical success probability, however, practical experiments on these parameters show that with 22 000 000 parity checks, the probability of correct prediction is higher than 97 percents. However, this approach requires about twice as much memory as the basic approach.

In practice, it is always better not to use the basic approach and to prefer either the conditional or refined approach. On parameters where the attack is limited by the amount of measurement, the refined approach should be the preferred choice. When the limiting factor is the computer memory, the conditional approach should be used.

In the following table, we summarize our results and also give theoretical results for 128-bit LFSR with feedback polynomials of weight 16 and 64. We only ran simulations for the 80-bit LFSR.

LFSR	k	B	N	Pre	Att	ϵ	ϵ'	ζ	F_ζ	ϵ_ζ	Par. Checks
80–9	3	26	2^{27}	2^{54}	$2^{30.6}$	0.05978	0.06738	1.0958	0.543	0.09144	22 000 000
128–16	3	48	2^{35}	2^{70}	2^{51}	0.07966	0.08973	1.12943	0.545	0.1216	4 000 000
128–64	4	32	2^{31}	2^{71}	$2^{36.4}$	0.1585	0.1779	1.2706	0.557	0.2383	2 500 000

6   Conclusion

We have presented a new side channel attack against LFSRs and more precisely against the Galois implementation of LFSRs. Galois LFSRs (or divisor registers) present an important side channel leakage growing with the Hamming weight of their feedback polynomial. This implementation (Fig. 2) is more interesting than the traditional one (Fig. 1) in terms of speed and critical path but the leakage induced allows an attacker to recover the initial state with a Simple Side Channel Analysis paired with a Fast Correlation Attack.

As a consequence, simple Galois implementation of LFSRs in embedded devices should either be thoroughly tested or be avoided. One way to reduce the attack efficiency is to use very low weight feedback polynomials. Moreover, as of now, the traditional implementation (Fig. 1) seems immune to this attack since the XOR gates do not share a common operand. It is also possible to use a simple, but costly, masking technique by running several copies of the Galois implementation. Every copy but the last one can be initialized with a random value and the final copy should be chosen to ensure that the XOR of all initial values correspond to initial value in a single register implementation. By linearity, the XOR of all outputs yields the expected value.

More generally, the attack presented here shows that in any circuit reusing the same logical output many times may induce powerful side channel analysis by reducing the overall noise ratio of this output. Thus, use logical gates with large fan-out should better be limited where possible.

Acknowledgements

We would like to thank the anonymous referees for their comments and Kerstin Lemke-Rust for her help while writing the final version of this paper.

References

1. D. Agrawal, B. Archambeault, J. Rao, and P. Rohatgi. The EM Side-Channel(s). In *CHES 2002*, number 2523 in LNCS, pages 29–45, 2002.
2. E. Brier, C. Clavier, and F. Olivier. Optimal Statistical Power Analysis. Cryptology ePrint Archive: Report 2003/152.
3. A. Canteaut and M. Trabbia. Improved fast correlation attacks using parity-check equations of weight 4 and 5. In *Eurocrypt 2000*, number 1807 in Lecture Notes in Comput. Sci., pages 573–588, 2000.
4. P. Chose, A. Joux, and M. Mitton. Fast correlation attacks: An algorithmic point of view. In *Eurocrypt 2002*, number 2332 in Lecture Notes in Comput. Sci., pages 209–221, 2002.
5. M. Hell, T. Johansson, and W. Meier. Grain – a stream cipher for constrained environments. ECRYPT/eStream submission. Phase II, HW focus.
6. P. Kocher, J. Jaffe, and B. Jun. Differential Power Analysis. In *CRYPTO'99*, number 1666 in LNCS, pages 388–397, 1999.
7. J. Lano, N. Mentens, B. Preneel, and I. Verbauwhede. Power analysis of synchronous stream ciphers with resynchronization mechanism. In *The State of the Art of Stream Ciphers*, 2004.
8. L. Lu, W. Meier, and S. Vaudenay. The conditional correlation attack. In *CRYPTO 2005*, number 3621 in LNCS, pages 97–117, 2005.
9. W. Meier and O. Staffelbach. Fast correlation attacks on certain stream ciphers. *J. Cryptology*, 1(3):159–176, 1989.
10. M. Mihaljević, M. Fossorier, and H. Imai. Fast correlation attack algorithm with list decoding and an application. In *FSE 2001*, number 2355 in Lecture Notes in Comput. Sci., pages 196–210, 2002.
11. F. Muller. *Analyse d'algorithmes en cryptographie symétrique*. PhD thesis, Ecole Polytechnique, April 2005.
12. E. Peeters, F. Standaert, and J. Quisquater. Power and Electromagnetic Analysis: Improved Model, Consequences and Comparisons. *Special issue of Integration, The VLSI journal, Embedded Cryptographic Hardware*, September 2006.
13. E. Prouff. DPA Attacks and S-Boxes. In *FSE 2005*, number 3557 in LNCS, pages 424–442, 2005.
14. C. Rechberger and E. Oswald. Stream ciphers and side-channel analysis. In *The State of the Art of Stream Ciphers*, 2004.

Author Index

Lecture Notes in Computer Science

For information about Vols. 1–4231

please contact your bookseller or Springer